Lecture Notes in Networks and Systems

Volume 396

The series "Lecture Notes in Networks and Systems" publishes the latest developments in Networks and Systems—quickly, informally and with high quality. Original research reported in proceedings and post-proceedings represents the core of LNNS.

Volumes published in LNNS embrace all aspects and subfields of, as well as new challenges in, Networks and Systems.

The series contains proceedings and edited volumes in systems and networks, spanning the areas of Cyber-Physical Systems, Autonomous Systems, Sensor Networks, Control Systems, Energy Systems, Automotive Systems, Biological Systems, Vehicular Networking and Connected Vehicles, Aerospace Systems, Automation, Manufacturing, Smart Grids, Nonlinear Systems, Power Systems, Robotics, Social Systems, Economic Systems and other. Of particular value to both the contributors and the readership are the short publication timeframe and the world-wide distribution and exposure which enable both a wide and rapid dissemination of research output.

The series covers the theory, applications, and perspectives on the state of the art and future developments relevant to systems and networks, decision making, control, complex processes and related areas, as embedded in the fields of interdisciplinary and applied sciences, engineering, computer science, physics, economics, social, and life sciences, as well as the paradigms and methodologies behind them.

Indexed by SCOPUS, INSPEC, WTI Frankfurt eG, zbMATH, SCImago.

All books published in the series are submitted for consideration in Web of Science.

For proposals from Asia please contact Aninda Bose (aninda.bose@springer.com).

More information about this series at https://link.springer.com/bookseries/15179

Yu-Dong Zhang · Tomonobu Senjyu ·
Chakchai So-In · Amit Joshi
Editors

Smart Trends in Computing and Communications

Proceedings of SmartCom 2022

 Springer

Editors
Yu-Dong Zhang
University of Leicester
Leicester, UK

Chakchai So-In
Department of Computer Science
Khon Kaen University
Khon Kaen, Thailand

Tomonobu Senjyu
Faculty of Engineering
University of the Ryukyus
Nishihara, Okinawa, Japan

Amit Joshi
Global Knowledge Research Foundation
Ahmedabad, Gujarat, India

ISSN 2367-3370   ISSN 2367-3389 (electronic)
Lecture Notes in Networks and Systems
ISBN 978-981-16-9966-5   ISBN 978-981-16-9967-2 (eBook)
https://doi.org/10.1007/978-981-16-9967-2

This Springer imprint is published by the registered company Springer Nature Singapore Pte Ltd.
The registered company address is: 152 Beach Road, #21-01/04 Gateway East, Singapore 189721,
Singapore

Preface

The Sixth Edition of the SmartCom 2022—Smart Trends in Computing and Communications was held during 17 and 18 December 2021 physically at Four Points by Sheraton Jaipur, India, and digitally on Zoom which is organised by Global Knowledge Research Foundation. The associated partners were Springer, KCCI and InterYIT IFIP. The conference will provide a useful and wide platform both for display of the latest research and for exchange of research results and thoughts. The participants of the conference will be from almost every part of the world, with background of either academia or industry, allowing a real multinational multicultural exchange of experiences and ideas.

A great pool of more than 480 papers were received papers for this conference from across seven countries among which around 75 papers were accepted with this Springer Series and were presented through digital platform during the two days. Due to overwhelming response, we had to drop many papers in hierarchy of the quality. A total of 12 technical sessions were organised in parallel in 2 days along with few keynotes and panel discussions. The conference will be involved in deep discussion and issues which will be intended to solve at global levels. New technologies will be proposed, experiences will be shared, and future solutions for enhancement in systems and security will also be discussed. The final papers will be published in proceedings by Springer LNNS Series.

Over the years, this conference has been organised and conceptualised with collective efforts of a large number of individuals. I would like to thank each of the committee members and the reviewers for their excellent work in reviewing the papers. Grateful acknowledgements are extended to the team of Global Knowledge Research Foundation for their valuable efforts and support.

I look forward to welcome you on the 7th Edition of this SmartCom Conference in 2023.

Leicester, UK   Yu-Dong Zhang
Nishihara, Japan   Tomonobu Senjyu
Khon Kaen, Thailand   Chakchai So-In
Ahmedabad, India   Amit Joshi

Contents

Contents

Editors and Contributors

About the Editors

Prof. Yu-Dong Zhang worked as a postdoc from 2010 to 2012 with Columbia University, USA, and as an Assistant Research Scientist from 2012 to 2013 with Research Foundation of Mental Hygiene (RFMH), USA. He served as a Full Professor from 2013 to 2017 with Nanjing Normal University. Now, he serves as Professor with School of Computing and Mathematical Sciences, University of Leicester, UK. His research interests include deep learning and medical image analysis. He is Fellow of IET (FIET) and Senior Members of IEEE, IES, and ACM. He was the 2019 recipient of "Web of Science Highly Cited Researcher." He is the author of over 300 peer-reviewed articles, including more than 50 ESI Highly Cited Papers, and four ESI Hot Papers. His citation reached 16626 in Google Scholar (h-index 72) and 9878 in Web of Science (h-index 55). He has conducted many successful industrial projects and academic grants from NIH, Royal Society, GCRF, EPSRC, MRC, British Council, and NSFC.

Tomonobu Senjyu received his B.S. and M.S. degrees in electrical engineering from the University of the Ryukyus in 1986 and 1988, and Ph.D. degree in electrical engineering from Nagoya University in 1994. Since 1988, he has been with the Department of Electrical and Electronics Engineering, University of the Ryukyus, where he is currently a professor. His research interests are in the areas of power system optimization and operation, advanced control, renewable energy, IoT for energy management, ZEH/ZEB, smart city, and power electronics. He is a member of the Institute of Electrical Engineers of Japan and Senior member of IEEE.

Dr. Chakchai So-In [SM, IEEE (14); SM, ACM (15)] is Professor of Computer Science at the Department of Computer Science at Khon Kaen University. He received B.Eng. and M.Eng. from KU (TH), in 1999 and 2001, respectively; and M.S. and Ph.D. from WUSTL (MO, USA) in 2006 and 2010; all in computer engineering. In 2003, Dr. So-In was Internet Working Trainee in a CNAP program at NTU (SG) and obtained Cisco/Microsoft Certifications. In 2006 to 2010, Dr. So-In was an intern at mobile IP division, Cisco Systems, WiMAX Forum, and Bell Labs (USA). His research interests include mobile computing/sensor networks, Internet of things, computer/wireless/distributed networks, cybersecurity, and intelligent systems and future Internet. He is/was on Editor/Guest members in *IEEE Access, PLOS One, PeerJ (CS), Wireless and Mobile Computing*, and *ECTI-CIT*. Dr. So-In has authored/co-authored over 100 publications and 10 books including *IEEE JSAC, IEEE Communications/Wireless Communications Magazine, Computer Networks*; and *(Advanced) Android Application Development, Windows Phone Application Development, Computer Network Lab, Network Security Lab*. He has served as a committee member and reviewer for many prestigious conferences and journals such as ICNP, WCNC, Globecom, ICC, ICNC, PIMRC; and IEEE Transactions (Wireless Communications, Computers, Vehicular Technology, Mobile Computing, Industrial Informatics); IEEE Communications Magazine, Letters, Systems Journal, Access; and Computer Communications, Computer Networks, Mobile and Network Applications, Wireless Network.

Dr. Amit Joshi is currently Director of Global Knowledge Research Foundation and also Entrepreneur and Researcher who has completed his masters and research in the areas of cloud computing and cryptography in medical imaging. Dr. Joshi has an experience of around ten years in academic and industry in prestigious organizations. Dr. Joshi is an active member of ACM, IEEE, CSI, AMIE, IACSIT-Singapore, IDES, ACEEE, NPA, and many other professional societies. Currently, Dr. Joshi is International Chair of InterYIT at International Federation of Information Processing (IFIP, Austria). He has presented and published more than 50 papers in national and international journals/conferences of IEEE and ACM. Dr. Joshi has also edited more than 40 books which are published by Springer, ACM, and other reputed publishers. Dr. Joshi has also organized more than 50 national and international conferences and programs in association with ACM, Springer, IEEE to name a few across different countries including India, UK, Europe, USA, Canada, Thailand, Vietnam, Egypt, and many more.

Contributors

Adhikary Sriyanjana Department of Information Technology, Jadavpur University, Jadavpur, India

Adhvaryu Rachit C. U. Shah University, Surendranagar, Gujarat, India

Aeggegn Dessalegn Bitew Department of Electrical and Computer Engineering, Debre Markos University, Debre Markos, Ethiopia

Agajie Takele Ferede Department of Electrical and Computer Engineering, Debre Markos University, Debre Markos, Ethiopia

Agarwal Meenakshi BML Munjal University, Gurugram, Haryana, India

Agarwal Shalini Shri Ramswaroop Memorial University, Barabanki, Uttar Pradesh, India

Akindadelo Adedeji Tomide Department of Basic Sciences, Babcock University, Ilishan Remo, Nigeria

Ali Asfak ETCE Department, Jadavpur University, Kolkatta, West Bengal, India

Alsubari Saleh Nagi Department of Computer Science & IT, Dr. Babasaheb Ambedkar, Marathwada University, Aurangabad, M.S., India

Amin Meet Parul University, Vadodara, Gujarat, India

Arolkar Harshal GLS University, Ahmedabad, India

Arora Shweta Graphic Era Hill University, Bhimtal Campus, India

Bach Pham Son FPT University, Hanoi, Vietnam

Baig M. M. J.D. College of Engineering and Management, Nagpur, Maharashtra, India

Baig S. Q. Department of Computer Science and Engineering, ABIT, Cuttack, Odisha, India

Banerjee Bidisha Department of Computer Science and Engineering, Jadavpur University, Jadavpur, India

Baruah Dhrubajyoti Department of Computer Application, Jorhat Engineering College, Jorhat, Assam, India

Bhagat Esha Department of Computer Science and Engineering, Devang Patel Institute of Advance Technology and Research, Faculty of Technology Engineering (FTE), Charotar University of Science and Technology (CHARUSAT), Changa, India

Bhagat Komal Mehta Bhagwan Parshuram Institute of Technology, Delhi, India

Bhogal Rosepreet Kaur School of Electronics and Electrical Engineering, Lovely Professional University, Phagwara, India

Bin Musirin Ismail Faculty of Electrical Engineering, Centre of Electrical Power Engineering, Universiti Teknologi, Skudai, Malaysia

Biswal Bisakha Department of Computer Science and Engineering, ABIT, Cuttack, Odisha, India

Bodhankar Jahnavi Centre for Development of Advanced Computing, Pune, India

Chatiwala Aliasgar Parul University, Vadodara, Gujarat, India

Chatterjee Mou KIIT School of Management, KIIT University (Institution of Eminence), Bhubaneswar, India

Chattyopadhyay Samiran Department of Information Technology, Jadavpur University, Jadavpur, India

Chaudhari Nandini M. KSET, DRS Kiran and Pallavi Patel Global University, Vadodara, Gujarat, India

Chaudhary Prashant AAI, Centre for Development of Advanced Computing, Pune, India

Chavan Suchita A. KBC North Maharashtra University, Jalgaon, Maharashtra, India

Chitrao Pradnya Symbiosis Institute of Management Studies, A Constituent of Symbiosis International University, Pune, Maharashtra, India

Chitre Mandar Anil Pacific Academy of Higher Education and Research Society, Udaipur, Rajasthan, India

Choudhary Asha Department of Mathematics and Statistics, Manipal University Jaipur, Jaipur, India

Dadwani Krish Parul University, Vadodara, Gujarat, India

Das Ankita Birla Institute of Technology, Noida, India

Das Debesh Kumar CSE Department, Jadavpur University, Kolkatta, West Bengal, India

Das Subhraneil Department of Electronics and Communication, Manipal Institute of Technology, Manipal, India;
Manipal Academy of Higher Education, Manipal, India

David Valerie School of Electronics Engineering, Vellore Institute of Technology, Vellore, India

Deepa Thilak K. Department of Networking and Communications, SRM Institute of Science and Technology, Kattankulathur, Chennai, Tamil Nadu, India

Deora Mahipal Singh Bhupal Nobel's University, Udaipur, India

Desai Nishant Parul University, Vadodara, Gujarat, India

Deshmukh Sachin N. Department of Computer Science & IT, Dr. Babasaheb Ambedkar, Marathwada University, Aurangabad, M.S., India

Deshpande Santosh L. Visveswaraya Technological University, Belagavi, India

Devendran V. School of Computer Science and Engineering, Lovely Professional University, Phagwara, India

Devi Gayatri Department of Computer Science and Engineering, ABIT, Cuttack, Odisha, India

Diep Vu Thu Hanoi University of Science and Technology, Hanoi, Vietnam

Doe Jane Anonymous University, Atlanta, GA, USA

Doe John Anonymous University, Atlanta, GA, USA

Dubey Nilesh Department of Computer Science and Engineering, Devang Patel Institute of Advance Technology and Research, Faculty of Technology Engineering (FTE), Charotar University of Science and Technology (CHARUSAT), Changa, India

Dutta Suparna Birla Institute of Technology, Noida, India

Duy Ha Long FPT University, Hanoi, Vietnam

Faridi Arman Rasool Aligarh Muslim University, Aligarh, Uttar Pradesh, India

Garia Prakash Graphic Era Hill University, Bhimtal Campus, India

Gaurav A. Vallurupalli Nageswara Rao Vignana Jyothi Institute of Engineering and Technology, Bachupally, Nizampet, Hyderabad, Telangana, India

Gayan Kamal Department of Computer Application, Jorhat Engineering College, Jorhat, Assam, India

Gebru Yalew Werkie Department of Electrical and Computer Engineering, Debre Markos University, Debre Markos, Ethiopia

Geethu N. Department of Computer Science and Engineering, Amrita School of Engineering, Amrita Vishwa Vidyapeetham, Bengaluru, India

Giriprasad M. N. Department of ECE, JNTUA College of Engineering Anantapur, Anantapur, Andhra Pradesh, India

Gopalan Sundararaman Department of Electronics and Communication Engineering, Amrita Vishwa Vidyapeetham, Amritapuri, India

Gupta Anjani Dr. Akhilesh Das Gupta Institute of Technology & Management, GGSIP University, New Delhi, India

Gupta Atrayee Department of Computer Science and Engineering, Jadavpur University, Jadavpur, India

Gupta Bhoomi Maharaja Agrasen Institute of Technology, GGSIPU, Delhi, India

Gupta Sachin SoET, MVN University, Haryana, India

Hambarde Aparna Department of Computer Engineering, KJ College of Engineering and Management Research, Savitribai Phule Pune University, Pune, Maharashtra, India

Handur Vidya S. KLE Technological University, Hubballi, India

Heddallikar ArunKumar RADAR Division, SAMEER, IIT B CAMPUS, Mumbai, India

Hung Phan Duy FPT University, Hanoi, Vietnam

Iyer Nalini C. KLE Technological University, Hubballi, India

Jagannath Bhople Yogesh Department of Information Technology, Government Polytechnic Washim, Washim, India

Jain Dhyanendra Dr. Akhilesh Das Gupta Institute of Technology & Management, GGSIP University, New Delhi, India

Jawale Shila Datta Meghe College of Engineering, Navi Mumbai, Maharashtra, India

Jeyalakshmi J. Department of CSE, Rajalakshmi Engineering College, Chennai, Tamil Nadu, India

Jha Ravi Shankar KIIT School of Management, KIIT University (Institution of Eminence), Bhubaneswar, India

Joshi Kavita Ajay Graphic Era Hill University, Bhimtal Campus, India

Joshi Manuj Pacific Academy of Higher Education and Research Society, Udaipur, Rajasthan, India

Joshi Pulkit AAI, Centre for Development of Advanced Computing, Pune, India

Juneja Pradeep Kumar Graphic Era University, Dehradun, India

K Shamshuddin School of Electronics and Communication Engineering, KLE Technological University, Hubballi, Karnataka, India

Kalaiselvi K. Department of Networking and Communications, SRM Institute of Science and Technology, Kattankulathur, Chennai, Tamil Nadu, India

Kalele Shraddha Amit AAI, Centre for Development of Advanced Computing, Pune, India

Kamakshi Pille Kakatiya Institute of Technology and Science, Warangal, India

Kamdar Dipesh V. V. P. Engineering College, Rajkot, Gujarat, India

Karki Ashtha KIIT School of Management, KIIT University (Institution of Eminence), Bhubaneswar, India

Kaur Arshpreet Department of Computer Science and Engineering (ACED), Alliance University, Bangalore, Karnataka, India

Kausar Sayema JDCOEM, An Autonomous Institute, Nagpur, Maharashtra, India

Khanvilkar Manasi MIT Institute of Design, MIT-ADT University, Pune, MH, India

Khokhariya Uday Pandit Deendayal Energy University (PDEU), Gandhinagar, India

Khot Suhas Department of Computer Engineering, KJ College of Engineering and Management Research, Savitribai Phule Pune University, Pune, Maharashtra, India

Kirubakaran M. K. Department of Information Technology, St. Joseph's Institute of Technology, Chennai, Tamil Nadu, India

Krishna Sree V. Vallurupalli Nageswara Rao Vignana Jyothi Institute of Engineering and Technology, Bachupally, Nizampet, Hyderabad, Telangana, India

Kubasadgoudar Ashwin R. KLE Technological University Hubballi, Hubballi, India

Kumar Ajai AAI, Centre for Development of Advanced Computing, Pune, India

Kumar Ashish Department of Mathematics and Statistics, Manipal University Jaipur, Jaipur, India

Kumar Bandari Pranay Kakatiya Institute of Technology and Science, Warangal, India

Kumar Binod Department of MCA, Rajarshi Shahu College of Engineering, Pune, India

Kumar Shambhavi Pandit Deendayal Energy University (PDEU), Gandhinagar, India

Kumar Sharan Department of ECE, JNTUA College of Engineering Anantapur, Anantapur, Andhra Pradesh, India

Kumar Umesh Govt. Women Engineering College, Ajmer, India

Kumaresan K. Department of CSE, K.S.R. College of Engineering, Tiruchengode, Tamil Nadu, India

Kumbhar Rasika Department of Computer Engineering, KJ College of Engineering and Management Research, Savitribai Phule Pune University, Pune, Maharashtra, India

Kurariya Pavan Centre for Development of Advanced Computing, Pune, India

Kuthadi Venu Madhav Department of CS&IS, School of Sciences, BIUST, Palapye, Botswana

Lakshmikanthan C. Department of Electronics and Communication Engineering and Department of Mechanical Engineering, Amrita Vishwa Vidyapeetham, Coimbatore, India

Luqman Mohammad Aligarh Muslim University, Aligarh, Uttar Pradesh, India

Mageshwari M. Department of Computer Science and Engineering, SRM Institute of Science and Technology, Kattankulathur, Chengalpattu, Chennai, Tamil Nadu, India

Maheshwari Akshat BML Munjal University, Gurugram, Haryana, India

Malathi M. Department of ECE, Vivekananda College of Engineering for Women (Autonomous), Namakkal, Tamil Nadu, India

Malavika R. Department of Management, Amrita Vishwa Vidyapeetham, Amritapuri, Kollam, Kerala, India

Mali Rajkumar Department of Computer Engineering, KJ College of Engineering and Management Research, Savitribai Phule Pune University, Pune, Maharashtra, India

Mamatha Gannera Department of ECE, JNTUA College of Engineering Anantapur, Anantapur, Andhra Pradesh, India

Mandla Mandeepsingh Parul Institute of Engineering and Technology, Parul University, Vadodara, India

Mane Pranoti JDCOEM, An Autonomous Institute, Nagpur, Maharashtra, India

Mane Venkatesh KLE Technological University, Hubballi, India

Mankad Urja L.J. University, Ahmedabad, India

Maurya Sudhanshu Graphic Era Hill University, Bhimtal Campus, India

Meena Shyam Sundar Department of Computer Science and Engineering, Shri Vaishnav Institute of Information Technology, Shri Vaishnav Vidyapeeth Vishwavidyalaya, Indore, India

Menon Remya Vivek Department of Management, Amrita Vishwa Vidyapeetham, Amritapuri, Kollam, Kerala, India

Mishra Shreeya Govt. Women Engineering College, Ajmer, India

Mishra Wricha MIT Institute of Design, MIT-ADT University, Pune, MH, India

Mittal Amit Graphic Era Hill University, Bhimtal Campus, India

Modh Jatin C. Gujarat Technological University, Ahmedabad, India

Mukherjee Nandini Department of Computer Science and Engineering, Jadavpur University, Jadavpur, India

Nagaraja K. V. Department of Mathematics, Amrita School of Engineering, Amrita Vishwa Vidyapeetham, Bengaluru, India

Narendra Hardik Parul Institute of Engineering and Technology, Parul University, Vadodara, India

Naresh R. Department of Computer Science and Engineering, SRM Institute of Science and Technology, Kattankulathur, Chengalpattu, Chennai, Tamil Nadu, India

Navya K. T. Department of Electronics and Communication, Manipal Institute of Technology, Manipal, India;
Manipal Academy of Higher Education, Manipal, India

Neogy Sarmistha Department of Computer Science and Engineering, Jadavpur University, Jadavpur, India

Nghia Ha Minh FPT University, Hanoi, Vietnam

Nidhi Department of Computer Science and Applications, Kurukshetra University, Kurukshetra, India

Nikhil Vemu School of Electronics Engineering, Vellore Institute of Technology, Vellore, India

Nikita P. KLE Technological University, Hubballi, India

Nitnaware Pranali J.D. College of Engineering and Management, Nagpur, Maharashtra, India

Pabla Simranjitsingh Parul Institute of Engineering and Technology, Parul University, Vadodara, India

Padmavathi K. Computer Applications, Hindusthan College of Arts and Science, Coimbatore, India

Panchal D. S. Department of Computer Science & IT, Dr. Babasaheb Ambedkar, Marathwada University, Aurangabad, M.S., India

Pancholi Nidhay Pandit Deendayal Energy University (PDEU), Gandhinagar, India

Pandu Ranga Avinash Srikhakollu Vallurupalli Nageswara Rao Vignana Jyothi Institute of Engineering and Technology, Bachupally, Nizampet, Hyderabad, Telangana, India

Pareek Jyoti Department of Computer Science, Gujarat University, Ahmedabad, India

Parekh Viral C. U. Shah University, Surendranagar, Gujarat, India

Patel Amit Singh KLE Technological University, Hubballi, Karnataka, India

Patel Parul Department of ICT, VNSGU, Surat, India

Patel Swasti Parul Institute of Technology, Parul University, Vadodara, India

Pathak Kirtan Department of Computer Science and Engineering, Devang Patel Institute of Advance Technology and Research, Faculty of Technology Engineering (FTE), Charotar University of Science and Technology (CHARUSAT), Changa, India

Patil Prakashgoud KLE Technological University, Hubballi, Karnataka, India

Pattanashetty Vishal B. School of Electronics and Communication Engineering, KLE Technological University, Hubballi, Karnataka, India

Pawar Manjula K. KLE Technological University, Hubballi, Karnataka, India

Phukon Satya Ranjan Department of Computer Application, Jorhat Engineering College, Jorhat, Assam, India

Poongothai T. Department of Computer Science and Engineering, St. Martin's Engineering College, Secunderabad, Telangana, India

Poonkuzhali S. Department of CSE, Rajalakshmi Engineering College, Chennai, Tamil Nadu, India

Pothina Harshini Department of Electronics and Communication Engineering, Amrita School of Engineering, Amrita Vishwa Vidyapeetham, Bengaluru, India

Prasad Keerthana Manipal School of Information Sciences, Manipal, India; Manipal Academy of Higher Education, Manipal, India

Ragu Harini School of Electronics Engineering, Vellore Institute of Technology, Vellore, India

Rajan Annie Department of Computer Science, University of Mumbai, Mumbai, India;
DCT's Dhempe College, Panaji, India

Rajesh M. Department of Computer Science and Engineering, Amrita School of Engineering, Amrita Vishwa Vidyapeetham, Bengaluru, India

Rajeshkumar T. Department of CSE, Saveetha School of Engineering, Saveetha Institute of Medical and Technical Sciences, Chennai, Tamil Nadu, India

Rampalli Gautham Kakatiya Institute of Technology and Science, Warangal, India

Ramteke Rakesh J. KBC North Maharashtra University, Jalgaon, Maharashtra, India

Rana Atul SoET, MVN University, Haryana, India;
Research Scholar, MVN University, Haryana, India

Rana Swapnil Department of Computer Science and Engineering, Devang Patel Institute of Advance Technology and Research, Faculty of Technology Engineering (FTE), Charotar University of Science and Technology (CHARUSAT), Changa, India

Ranjan Department of Computer Science and Engineering (ACED), Alliance University, Bangalore, Karnataka, India

Rastradhipati Atreya MIT Institute of Design, MIT-ADT University, Pune, MH, India

Rath Arabinda Department of Computer Science and Engineering, ABIT, Cuttack, Odisha, India

Rathod D. P. Department EnTC, VJTI (Veermata Jijabai Institute of Technology), Mumbai, India

Ratnam Tatini Shanmuki Krishna Department of Electronics and Communication Engineering and Department of Mechanical Engineering, Amrita Vishwa Vidyapeetham, Coimbatore, India

Rejikumar G. Department of Management, Amrita Vishwa Vidyapeetham, Kochi, Kerala, India

Roy Himadri Shekhar Department of Computer Science and Engineering, Jadavpur University, Jadavpur, India

Saahith Reddy M. Vallurupalli Nageswara Rao Vignana Jyothi Institute of Engineering and Technology, Bachupally, Nizampet, Hyderabad, Telangana, India

Sahoo Priti Ranjan KIIT School of Management, KIIT University (Institution of Eminence), Bhubaneswar, India

Saini Jatinderkumar R. Symbiosis Institute of Computer Studies and Research, Symbiosis International (Deemed University), Pune, India

Saini Monika Department of Mathematics and Statistics, Manipal University Jaipur, Jaipur, India

Saini Pawankumar Symbiosis International University, Pune, Maharashtra, India

Sakthi U. Department of Computer Science and Engineering, Saveetha School of Engineering, SIMATS, Chennai, Tamil Nadu, India

Salau Ayodeji Olalekan Department of Electrical/Electronics and Computer Engineering, Afe Babalola University, Ado-Ekiti, Nigeria

Salgaonkar Ambuja Department of Computer Science, University of Mumbai, Mumbai, India

Sangpal Ravivanshikumar Department of Computer Engineering, KJ College of Engineering and Management Research, Savitribai Phule Pune University, Pune, Maharashtra, India

Saranya S. Department of ECE, SRM Easwari Engineering College, Chennai, India

Sarkar Ram CSE Department, Jadavpur University, Kolkatta, West Bengal, India

Sarmah Joydip Department of Computer Application, Jorhat Engineering College, Jorhat, Assam, India

Sasikumar P. School of Electronics Engineering, Vellore Institute of Technology, Vellore, India

Sawarkar S. D. Datta Meghe College of Engineering, Navi Mumbai, Maharashtra, India

Sawwashere Supriya J.D. College of Engineering and Management, Nagpur, Maharashtra, India

Senthil Murugan T. Kakatiya Institute of Technology and Science, Warangal, India

Senthilkumar A. V. Computer Applications, Hindusthan College of Arts and Science, Coimbatore, India

Shah Axita Department of Computer Science, Gujarat University, Ahmedabad, India

Shah Harshal Parul University, Vadodara, Gujarat, India

Shah Kaushal Pandit Deendayal Energy University (PDEU), Gandhinagar, India

Shaikh Arshad Learning and Personalization, BYJU'S, Bangalore, India

Sharma Anupam Kumar Dr. Akhilesh Das Gupta Institute of Technology & Management, GGSIP University, New Delhi, India

Sharma Somesh Graphic Era Hill University, Bhimtal Campus, India

Sharma Ved Prakash Dr. Akhilesh Das Gupta Institute of Technology & Management, GGSIP University, New Delhi, India

Shashidhar N. KLE Technological University, Hubballi, India

Shashvat Kumar Department of Computer Science and Engineering (ACED), Alliance University, Bangalore, Karnataka, India

Shelke Mahesh B. Department of Computer Science & IT, Dr. Babasaheb Ambedkar, Marathwada University, Aurangabad, M.S., India

Shell Michael Department of Electrical and Computer Engineering, Georgia Institute of Technology, Atlanta, GA, USA

Shet Raghavendra KLE Technological University Hubballi, Hubballi, India

Shinde Shraddha S. Department EnTC, VJTI (Veermata Jijabai Institute of Technology), Mumbai, India

Shraddha B. H. KLE Technological University, Hubballi, India

Shukla Madhu Department of Computer Engineering, Marwadi University, Rajkot, India

Singhal Vineet BML Munjal University, Gurugram, Haryana, India

Singh Lenali AAI, Centre for Development of Advanced Computing, Pune, India

Singh Prashant Dr. Akhilesh Das Gupta Institute of Technology & Management, GGSIP University, New Delhi, India

Singh Preeti MMM University of Technology, Gorakhpur, India

Singh Sarvpal MMM University of Technology, Gorakhpur, India

Singh Shashi Pal AAI, Centre for Development of Advanced Computing, Pune, India

Singh Shivoham Pacific Academy of Higher Education and Research Society, Udaipur, Rajasthan, India

Smrutirekha KIIT School of Management, KIIT University (Institution of Eminence), Bhubaneswar, India

Sonekar Shrikant JDCOEM, An Autonomous Institute, Nagpur, Maharashtra, India

Sonekar Shrikant V. J.D. College of Engineering and Management, Nagpur, Maharashtra, India

Sood Ajay Kumar BML Munjal University, Gurugram, Haryana, India

Sowmia K. R. Department of IT, Rajalakshmi Engineering College, Chennai, Tamil Nadu, India;
Department of CSE, Rajalakshmi Engineering College, Chennai, Tamil Nadu, India

Sravani Kumari K. Vallurupalli Nageswara Rao Vignana Jyothi Institute of Engineering and Technology, Bachupally, Nizampet, Hyderabad, Telangana, India

Sreenivasulu G. Computer Science and Engineering, ACE Engineering College, Hyderabad, India

Sridharan Aadityan Department of Physics, Amrita Vishwa Vidyapeetham, Amritapuri, India

Sunori Sandeep Kumar Graphic Era Hill University, Bhimtal Campus, India

Suresh A. Department of Management, Amrita Vishwa Vidyapeetham, Amritapuri, Kollam, Kerala, India

Swaminarayan Priya Parul Institute of Computer Applications, Parul University, Vadodara, India

Tamizharasan M. Electronics and Communication Engineering, Amrita Vishwa Vidyapeetham, Coimbatore, India

Tembhurne Snehal J.D. College of Engineering and Management, Nagpur, Maharashtra, India

Thangaraj K. Department of Information Technology, Sona College of Technology, Salem, Tamil Nadu, India

Thulasi Chitra N. Department of CSE, MLR Institute of Technology, Secunderabad, Telangana, India

Tien Nguyen Huu FPT University, Hanoi, Vietnam

Tokekar Vrinda Department of Information Technology, Institute of Engineering and Technology, Devi Ahilya Vishwavidyalaya, Indore, India

Trivedi Tanvi Bhupal Nobel's University, Udaipur, India

Upadhyaya Shuchita Department of Computer Science and Applications, Kurukshetra University, Kurukshetra, India

Vallapure Pratiksha Department of Computer Engineering, KJ College of Engineering and Management Research, Savitribai Phule Pune University, Pune, Maharashtra, India

Vartika Department of Computer Science and Engineering (ACED), Alliance University, Bangalore, Karnataka, India

Vats Prashant Dr. Akhilesh Das Gupta Institute of Technology & Management, GGSIP University, New Delhi, India

Vatsal Saumitra Shri Ramswaroop Memorial University, Barabanki, Uttar Pradesh, India

Vijay Anand M. Department of Computer Science and Engineering, Saveetha Engineering College, Chennai, Tamil Nadu, India

Vinh Bui Trong Hanoi Procuratorate University, Hanoi, Vietnam

Vishal P. KLE Technological University, Hubballi, India

Viswanadha Raju S. CSE, JNTUHCEJ, Jagityal, Karimnagar, Telangana, India

Yadav Babita SoET, MVN University, Aurangabad, India

Zala Kirtirajsinh Department of Computer Engineering, Marwadi University, Rajkot, India

Implementation of Tweet Stream Summarization for Malicious Tweet Detection

Pranoti Mane, Shrikant Sonekar, and Sayema Kausar

Abstract In contrast to traditional media, social media is populated by anonymous individuals who have the freedom to broadcast whatever they choose. This online social media culture is dynamic in nature, and the move from traditional media to digital media is growing increasingly popular among people. While conventional media will continue to be used less frequently in the future, the increasing use of online social networks (OSNs) will obfuscate the real information provided by traditional media. Genuine users provide material that is beneficial to the broader public; on the other hand, spammers transmit irrelevant or misleading content that turns social media into a front for spreading false information. Consequently, undesirable text or susceptible links might be delivered to targeted users. These fabricated texts are anonymous, and they are occasionally linked to possible URLs. A precise statistical classification for a piece of news is not possible with the current systems because of data limitations and communication types. We will look at a variety of research publications that employ a variety of strategies for master training in the prediction and detection of harmful material on social media websites and networks. In this study, we attempted to identify spam tweets from a large collection of tweets by utilizing TVC algorithm to identify it. When a content paper is incorporated in this manner, a brief summary produced by utilizing the most important keywords from the first document is known as summarization. It is necessary to have a dynamic approach of dealing with the condensed information that is supplied through Twitter feeds. This research presents a novel way for producing a major substance-based summary in a shorter period of time than previously available. We also propose to detect harmful tweets both offline and online and to do so in real time. Most notably, when compared to the other current frameworks, the proposed framework performs multi-subject summarization on an online dataset, which results in a reduction in the amount of time required.

Keywords Tweet clustering · Tweet stream · Malicious tweet · Timeline · Summary · Continuous summarization

P. Mane (✉) · S. Sonekar · S. Kausar
JDCOEM, An Autonomous Institute, Nagpur, Maharashtra, India
e-mail: pranoti.mane23@gmail.com

© The Author(s), under exclusive license to Springer Nature Singapore Pte Ltd. 2023
Y.-D. Zhang et al. (eds.), *Smart Trends in Computing and Communications*, Lecture Notes in Networks and Systems 396, https://doi.org/10.1007/978-981-16-9967-2_1

1

1 Introduction

Due to technological advancements, an online community has evolved into a critical network for sharing and exchanging information. Twitter is among the most popular online social networking services because it allows users to send and read a large number of text-based messages. Thousands of millions of individuals share their information with their contacts on Twitter, and around 400 million tweets are sent every day. It is a frequent feature of social networks that multiple different individuals have a strong influence on one another. Because of the huge amount of available individuals and the volume of data shared on Twitter social networks, spammers engage in a slew of criminal operations aimed at spreading spam messages via connected website URLs.

Many e-commerce applications have made use of the social networking site Twitter to promote their products. Bot software automates functions in social media networks such as Twitter. Robotic process automation (RPA) is a type of software that automates the execution of operations over the online platform and in other web-based applications. In terms of computerization, Twitter is not particularly picky. Twitter just includes the definition of a CAPTCHA graphic as part of the registration procedure. Using Twitter APIs, bots conduct automation tasks for users after they have completed the sign-up procedure and received all of their personal information. Cyborgs are in charge of controlling the interface between humans and bots software.

In the case of large-scale tweets, conventional document summarizing algorithms are insufficient, and they are also insufficient in the case of tweets that are received quickly and continuously. To completely overcome this problem, Twitter summarization is required, which should have a new level of utility that is fundamentally different from ordinary summary. When summarizing tweets, it is important to consider the worldwide element of the tweets that have arrived.

Stemming is a phrase that is used in the field of etymological morphological and data recovery to describe the procedure of reducing curved (or occasionally determinate) words to their basic stem, base, or root frame, which is by and large a constructed word shape. The stem does not have to be identical to the grammatical underpinning of the word; it is often sufficient that related terms point to a comparable stem, independent of whether or not this stem is a legal root in and of itself.

During the registration process, collocations will be words that are sorted through prior to or after the preparation of ordinary dialect information is completed (content). Despite the way that stop words are frequently used to refer to the most well-known words in some kind of a language, there is no single widely used list of stop words that is utilized by all normal dialect preparation equipment, and to be sure, not all gadgets even use such a list. A few instruments, in particular, refrain from emptying particular stop phrases in order to aid in state seeking efforts.

2 Literature Survey

The process of summarizing a tweet is divided into two stages. The introduction advance necessitates the clustering of tweet information, followed by the summarizing of the information.

Twitter posts are limited to 140 characters or less in length, and we are only considering English posts in this case. All of the tweets are written in a casual manner with non-standard spelling and, as much as feasible, without the use of accentuation. Nonlinear and non-descriptions of Twitter posts were created using the crossovers TF-IDF-based calculation [1] as a starting point. The following are some ways to document summarizing that have been clarified. Using the arbitrary summarizer approach, k posts or even every subject is randomly selected for inclusion as a summary. This technique was beneficial in terms of achieving the final aim of providing the most exceedingly terrible execution and, in addition, setting the lowest bound of scenario execution. The latest summarizer technique selects the most recent k posts from the determination pool to serve as a summary. It has the ability to select the first portion of a news story to serve as a summary. This strategy is implemented in light of the fact that smart summarizers are unable to outperform a plain summarizer in terms of performance. This summarizer just makes use of the first segment of the document to provide a summary.

The computation for streaming informational aggregation has been extensively researched and written about by a variety of authors. BIRCH is an improved iterative decreasing and clustering algorithm that makes use of progressions' calculation. An unstructured machine learning technique [2] was used to perform this calculation.

An authentic and realistic result can be developed through mathematical approach for selecting the cluster based on distance and energy. Through this, the cluster can be easily identified and can take up the responsibility of running the entire network [3]. It is used to achieve multiple levels of clustering over large informational sets of varying sizes and shapes. The advantage of BIRCH is that it has the ability to form groups in response to incoming information and to respond in a dynamic manner to those groups.

People can share their opinions on Twitter, which is a good platform for doing so. Because of the sheer volume of tweets, it takes a lot of effort to keep up with what is going on inside the events. Presented here is a new technique for outlining events that makes use of excellent journalists and creates live game updates based on Twitter posts about the events. Great journalists selected illustrative tweets from the majority of non-educational tweets [4] to use as examples.

TextRank summarizer [5] is yet another positioning computation that is based on a chart. It is this strategy that makes use of the PageRank computation. This resulted in yet another chart-based summarizer that merges potentially even more data than LexRank. This occurs as a result of the fact that it modifies the weights of postings in a recursive manner. The final score assigned to each post is determined by how well it is associated with other posts that are swiftly affiliated with it, as well as the method in which distinct presentations are associated on different articles. Content ranking does not necessitate any further semantic learning. TextRank computation is a diagram-based approach that is used to discover the sentences that are in the top place.

In the web mining industry, differential evolutionary timeline Summarization (ETS) [6] is a tool that builds timelines for large amounts of information. ETS provides developmental instructions for a given time period. ETS provides outlines that are based on a score of time characteristic attributes. The most popular point of view is that it stimulates speedy news perusal and learning grasp of the issues. Iterative substitution is used to transform the ETS errand into an adjusted enhanced issue.

This summarizer employs a chart-based method, as demonstrated in LexRank [7]. It distinguishes between pairwise similarities between two phrases either between two blog posts, respectively. It has an impact on the proximity to score, which is the weight of the edge between both the two sentences. According to the weights of the connections that are connected with one another, the last rating of postings is processed in the following way: In contrast to coordinate recurrence summarization, this summarizer is excellent for providing summarizing in light of standard to chart. In spite of the fact that it is predicated on repetition, this framework makes use of the links between phrases to contain more information. In comparison with the recurrence-based method, this is a more complicated computation.

It is also possible to use Twitter streams for event summaries, which allows you to communicate with data in real time. When it comes to occasion summarization, the participant technique is used. Summary tweet extraction is accomplished through the use of three essential segments: Participant detection, sub-occasion detection, and summary tweet extraction [8]. Members of an occasion are distinguished by their member identifier. Individuals who participate in events are referred to as members. Sub-occasion discovery is a feature that summarizing sub-occasions that have been identified with members. By applying the summary tweet extraction section [9], the tweets are extracted from the sub-occasions.

Zhenhua Wang and colleagues have developed a summarizing structure known as Sumblr. Continuous summarization using stream clustering is what we are talking about here. Because of the vast amount of noisy and excessive tweets that are contained in it, continuous summarization is a challenging task to accomplish. The notion in question was taken into consideration when the idea of continuous tweet streaming summarization was being discussed. A tweet stream clustering component, a high-level summarization module, and a timeline generation module are the three primary components of this system. The social media platform Sumblr can also be used to experiment with dynamic, rapidly incoming, and large-scale tweet streams [10], among several other things.

3 Proposed Approach

Realizing continuous Twitter stream summarization is not an easy task, given that countless people are excellent for nothing, immaterial, and raucous in nature as a result of the social concept of tweeting, among other factors. Tweets are inextricably linked to the time at which they are posted, and fresh tweets have such a tendency to reach their target audience at a rapid pace. Since the volume of tweets is always significant, the summary calculation must be extremely efficient. It should be able to provide Twitter summaries of conscience spans of time. It should be able to automatically recognize subtheme modifications as well as the minutes in which they occur. A multi-subject reproduction of a continuously tweet stream summarizing structure, to be particular Sumblr, is proposed in this paper. It is tested on more comprehensive and significant scale informative collections, and the results are discussed in this study.

Figure 1 depicts the multi-theme variant of the Sumblr structure, which is composed of three core components: the tweet stream clustering module, the abnormal state summary module, and the timeline age module. Figure 1 shows the multi-theme form of the Sumblr structure. The Twitter stream clustering module is responsible for keeping the online truthful information current. The incremental clustering technique is used to keep tweets in their web version current. When a theme-based Twitter stream is provided, it can be used to group tweets in a productive manner and maintain reduced group data. The different leveled summarization module provides two types of synopses: Online rundowns and rundowns that have been recorded. In a brief web summary, it is depicted what is now being investigated among the general public. As a result, the contribution for building online synopses is retrieved precisely from the current bunches that have been kept up in memory in this manner. A verifiable summary, on the other hand, helps individuals to perceive the most important events that occurred during a specific period, which means we must

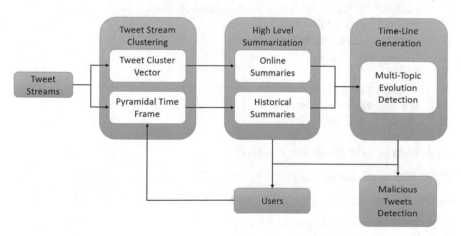

Fig. 1 System overview

disregard the influence of tweet content that occurred from outside that period in order to do this. As a result, the recovery of the information essential for the creation of genuine designs is more difficult. The development recognition computation at the heart of the chronology age module gives both ongoing and runtime timelines in the same calculation.

4 Implementation

Algorithm 1: Multi-topic Version of Sumblr

Input: a cluster set CL_set

1. While! Stream.end() do
2. Tweet t = stream.next();
3. Choose Cpt in CL_set whose centroid closet to it.
4. If MaxSim(t) < MBS then
5. Create a new cluster CLnew = {t};
6. CL_set.add(CLnew);
7. Else
8. Update Cpt with t;
9. If TScurrent % (αi) == 0 then
10. Store CL_set into PTF;

Algorithm 2: Tweet Stream Clustering

Input: CL_set represent a set of cluster

1. While! stream.end () do
2. Tweet t = stream.next ();
3. Choose Cpt in CL_ set whose centroid closest to t;
4. If MaxSim (t) < MBS then
5. Generate new Cluster CLnew = {t}
6. CL_set.add (CLnew)
7. Else
8. update Cpt with t
9. If TScurrent%(αi) = = 0 then
10. Store CL-set into PTF.

Algorithm 3: TCV-Rank Summarization

Input: The set of clusters D(c)

Output: SS represent the set of summaries

1. SS = Φ, T = {represent number of tweets in set D(c)};
2. Comparability graph to be built on T;

3.   Calculate LR i.e. LexRank scores;
4.   Tc represent the tweets having highest LR score in each cluster;
5.   while |SS| < L do
6.   for each tweet ti in Tc_SS do
7.   calculating vi in accordance the Equation given in (2);
8.   select tmax with the highest vi;
9.   SS.add (tmax);
10.   while |SS| < L do
11.   for each tweet t'i in T _ SS do
12.   calculate v'i according to Equation
13.   select t'max with the highest v'i;

      1.   SS.add (t'max);
      2.   returns SS;

Algorithm 4: Topic Evolution Detection

Input; tweet stream binned by time units

Output: timeline node set TLN

1.   TLN = ∅;
2.   Whilestream.end() do
3.   Bin Ci = stream.next();
4.   If hasLargeVariation() then
5.   TLN.add(i);
6.   Return TLN;

5 Result Analysis

Using Sumblr with a single topic against Sumblr with many topics, the time comparison in Fig. 2 is shown. In comparison with Sumblr with several topics, Sumblr with a single topic took more time to complete.

Fig. 2 Time comparison

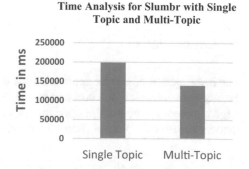

Time Analysis for Slumbr with Single Topic and Multi-Topic

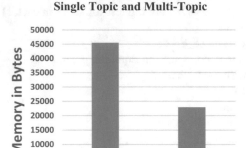

Fig. 3 Memory comparison

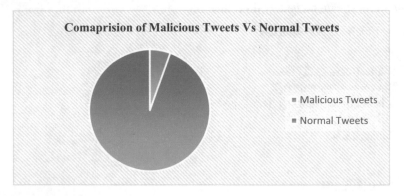

Fig. 4 Comparison of malicious and non-malicious tweets

Figure 3 depicts a comparison of memory. When compared to the suggested solution, the existing system needed greater CPU consumption. The storage comparison between different systems is depicted in Fig. 4.

6 Conclusions

The investigation of several approaches for document summary, such as sifting and Twitter summarization, has come to an end. These approaches are used to monitor a large number of tweets at the same time. Due to the fact that Twitter information is raucous and excessive, sifting is not an effective strategy. As a result, the summarizing technique is used to compress the information included in tweets. Traditional document summarizing algorithms are not practical for large-scale tweets, and they are also not adequately applicable for tweets that are received quickly and continually; in addition, they do not focus on static and small-scale informational indexes. An adjustment of a constant Twitter post live feed summarization system, specifically Sumbler, has been developed to address this issue by producing simple summaries and timelines with respect to streams and assessing it on more comprehensive and considerable magnitude information examination, which manage dynamic tweet streams with rapid arrival and a large scale. During the process of continuous summarization, our technique will identify the altering dates and timeframes in a progressive manner. Furthermore, evolutionary timeline summarization (ETS) somehow does not pay attention to concerns such as proficiency and adaptability, which are critical in our streaming-based environment.

References

1. S. Anjali, N.M. Meera, M.G. Thushara, A graph based approach for keyword extraction from documents. in *2019 Second International Conference on Advanced Computational and Communication Paradigms (ICACCP)* (2019), pp. 1–4. https://doi.org/10.1109/ICACCP.2019. 8882946
2. R. Yan, X. Wan, J. Otterbacher, L. Kong, X. Li, Y. Zhang, Evolutionary timeline summarization: A balanced optimization framework via iterative substitution. in *Proceedings of the 34th International ACM SIGIR Conference on Research and Development in Information Retrieval* (2011), pp. 745–754
3. S.V. Sonekar, M.M. Kshirsagar, L. Malik, Cluster head selection and malicious node detection in wireless Ad Hoc networks. In: *Next-Generation Networks*, ed. by D. Lobiyal, V. Mansotra, U. Singh. Advances in Intelligent Systems and Computing, vol 638 (Springer, Singapore, 2018). https://doi.org/10.1007/978-981-10-6005-2_55
4. M. Kubo, R. Sasano, H. Takamura, M. Okumura, Generating live sports updates from Twitter by finding good reporters. in *2013 IEEE/WIC/ACM International Joint Conferences on Web Intelligence (WI) and Intelligent Agent Technologies (IAT)* (2013), pp. 527–534. https://doi. org/10.1109/WI-IAT.2013.74
5. T. Chen, D. Miao, Y. Zhang, A graph-based keyphrase extraction model with three-way decision. in *Rough Sets. IJCRS 2020*, ed. by R. Bello, D. Miao, R. Falcon, M. Nakata, A. Rosete, D. Ciucci. Lecture Notes in Computer Science, vol 12179 (Springer, Cham, 2020). https://doi. org/10.1007/978-3-030-52705-1_8
6. S. Dutta, V. Chandra, K. Mehra, S. Ghatak, A.K. Das, S. Ghosh, *Summarizing Microblogs during Emergency Events: A Comparison of Extractive Summarization Algorithms* (2019)
7. C.-F. Tsai, Z.-C. Chen, C.-W. Tsai, MSGKA: an efficient clustering algorithm for large databases. in *IEEE International Conference on Systems, Man and Cybernetics*, vol. 5 (2002), p. 6. https://doi.org/10.1109/ICSMC.2002.1176400
8. C. Shen, F. Liu, F. Weng, T. Li, A participant-based approach for event summarization using twitter streams. in *Proceedings of the Human Language Technology Annual Conferene North America Chapter Association for Computer Linguistics* (2013), pp. 1152–1162
9. P. Nerurkar, A. Shirke, M. Chandane, S.G. Bhirud, Empirical analysis of data clustering algorithms. Procedia Comput. Sci. **125**, 770–779 (2018)
10. D. Rudrapal, A. Das, B. Bhattacharya, A survey on automatic Twitter event summarization. J. Inf. Process. Syst. **14**, 79–100 (2018)

Comparative Analysis of a Secure Authentication Protocol for 5G-Enabled IoT Network Using Public and Shared Secret Key

Shamshuddin K and Vishal B. Pattanashetty

Abstract Due to advancements in technologies and to achieve high efficiency, the world is shifting from 4G to a 5G network. The advantages of 5G network are enormous in every field and thus are catching speed to be a better version of what exists in today's world. The features like increased bandwidth availability, massive network capacity, and ultra-low latency are keeping 5G out of the box and giving raw network engineers to explore and invent high standards. As the world goes with pros and cons hand in hand, so is the advent of 5G technology. Finding its applications in communication, massive IoT, and mobile broadband, it also faces some challenges which affects the day-to-day life of humans. 5G applications face security risks due to the new technology used and the performance requirements of the specific application scenario. To avail the 5G services, the IoT devices need to communicate with an intermediate network known as an access network, which is usually publicly accessible and thus prone to attacks. In this paper, implementation and results of a 5G-enabled secure authentication protocol using symmetric and asymmetric key are compared to identify the best technique for privacy preservation in 5G-enabled IoT network.

Keywords Cooja · Cryptography · Privacy · Security · IoT

1 Introduction

The Internet of Things (IoT) is evolving as a platform for Internet-based service and comprises smart devices, such as smart phones and sensor nodes [1]. The IoT devices consist of communication system, RFID, and complex computational system. The most challenging aspects in IoT network are security and privacy [2].

S. K (✉) · V. B. Pattanashetty
School of Electronics and Communication Engineering, KLE Technological University, Hubballi, Karnataka, India
e-mail: shamshuddin@kletech.ac.in

V. B. Pattanashetty
e-mail: vishalbps@kletech.ac.in

© The Author(s), under exclusive license to Springer Nature Singapore Pte Ltd. 2023
Y.-D. Zhang et al. (eds.), *Smart Trends in Computing and Communications*, Lecture Notes in Networks and Systems 396, https://doi.org/10.1007/978-981-16-9967-2_2

In the present networking system, the corrupted devices can corrupt the functioning of network. If an IoT device is corrupted, then the entire connected physical world will get influenced dangerously. The advancements in the technologies in the communication sector occurring like a whirlwind, from first generation (1G) of cellular technologies to 5G in the expanse of about 40 years, has opened up so many different ways of communication and data transfer not only between two individuals but also between different devices like sensors, servers, IoT, and Internet of Vehicles (IoV). These provide platforms for the emanating technologies that are transforming into an essential component of our economy as well as our community. The 5G technology is becoming an indispensable technology as it is taking a humongous leap in performance when compared to the 4G networks. In the IoT network, the communication system will be broadcast in nature which makes wireless transmission system susceptible to eavesdropping and endangering the secrecy of data [3]. So there arises an increased need to protect and safeguard the data we send and receive. In today's world, due to increase in online data transfer such as online banking transaction and payment, security and privacy of data are gaining more importance and also area of concern.

2 Literature Survey

In 5G-enabled IoT network, the number IoT node will be enormous in number. Authentication of device is very important for providing privacy and security in 5G and IoT network. Liyanage et al. [2] describe the various privacy issues due to the new technologies involved in 5G network. The various 5G security and privacy issues, their defense, and also the road map for future research are described by the author of the paper [3]. Using Cooja simulator, the authors simulate IoE-based home automation system to analyze IEEE 802.15.4 protocol to explore IoT environment [1]. In another paper, authors show simulation results of energy consumption by a node using Cooja simulator [4]. The author has conducted experiment on energy constrained IoT devices using Cooja simulator and also carried out comparison of two protocols: constrained application protocol (CoAP) and hypertext transfer protocol (HTTP) [5].

Many researchers have proposed various authentication protocols. In the paper [6], authors proposed a secure authentication protocol for 5G-enabled IoT network. The proposed protocol is verified using Scyther tool [7] for various security parameters. In this paper, the proposed protocol in paper [6] is simulated using Cooja simulator [8] and further analyzed using symmetric and asymmetric key.

The proposed system is designed by using algorithms like the Diffie–Hellman for key exchange and RSA, and other cryptographic techniques [9] to authenticate the user equipment (UE) and to authenticate the access and mobility management (AMF) in 5G network [10], to check the integrity of the device that requests for the service, and finally to acknowledge the service request [11]. Typically, a user device requests for service to core access and mobility management function and is responsible for service allocation after the verification of network slice selection association information (NSSAI) [12].

'

3 System Design

The system design can be best explained with the help of Fig. 1. The black box on the Fig. 1 shows the focus on the area we are working on. The main goal is to authenticate and preserve security between the user equipment and AMF. User equipment such as mobile phones, various IoT devices like cars, or a fitness wearables, send a request to the 5G network [13], the request reaches to the AMF for the services, and the AMF is accessible by any IoT node, hence, it is prone to attacks by the hackers. That being the case, the proposed protocol presents a scenario where attackers try to attack, but attack failed, because of the authentication protocol checks for the integrity of the device requesting service, the attacks are stopped at the root itself.

3.1 Functional Block Diagram

Figures 2 and 3 represent the functional block diagram of the protocol [6]. The block diagram consists of the extreme ends between which the communication takes place that is user equipment and the AMF. The different blocks that are required to carry

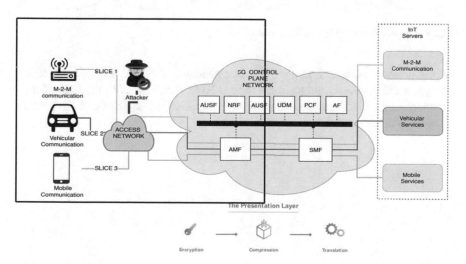

Fig. 1 System design [6]

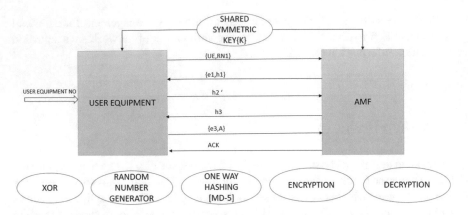

Fig. 2 Functional block diagram for symmetric key

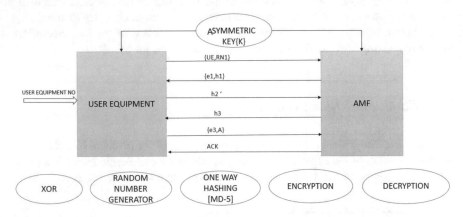

Fig. 3 Functional block diagram for asymmetric key

on the protocol include a random number generator (nonce), a hashing function, encryption, decryption, and a simple logical XOR function. With the help of these functions and a UDP protocol running at the application layer in end devices, the protocol is implemented.

3.2 Algorithm

Algorithm 1:

STEP 1: Select RandomNo : rn and User id :
UEid
Send (UEi to AMF) : UEid, rn
STEP 2: Ki = retrievekey from UEid
Choose RandomNo : rn2
Skey = rn2 exor Ki
e1 = Eki(rn, Skey)
h1 = H(Skey — e1 — rn)
send (AMF to UEid) : e1, h1
STEP 3: (rn, Skey) = Dki(e1)
rn2 = Skey exor Ki
e2 = Eki(rn, Skey)
h2 = H(Skey — e2 — rn)
if(e1 == e2 and h1 == h2)
Authenticated AMF
else (Terminate the protocol and communication)
h2d= H(h1 — Skey — e2 — rn)
send (UE to AMF) : h2d
STEP 4: h3 = H(h1 — Skey — e1 — rn)
if(h3 == h2d)
authenticated User Equipment UEid
send (AMF to UE) : h3
STEP 5: e3 = Eki(ServiceRequest)
h4 = H(e3)
A = h4 ĥ2d
send (UEid to AMF) : e3, A
STEP 6: h4 = A exor h2d
h4= H(e3)
if(h4d== h4)
then integrity is preserved
ServiceRequest = Dki(e3)
execute the ServiceRequest
send (AMF to UE): Ack

---

4 Results and Discussion

The analysis of results is based on the following parameters, which are average power consumption, radio duty cycle, node information, network graph, and sensor map.

- Average power consumption: The sum of the power contributions of the node's electronic components, which are in turn dependent on the component state and

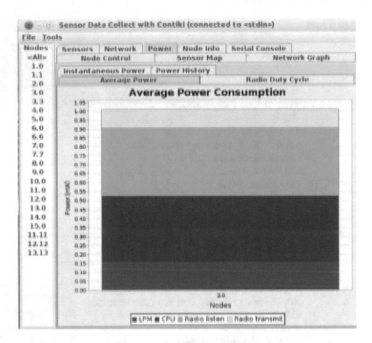

Fig. 4 Average power consumption

the actual operation executed, determines the node's power consumption as shown in Fig. 4.

- Radio duty cycle: The ratio of time when transmission of signal is ON to the time when transmission of signal is OFF, as shown in Fig. 5.
- Node information: Figures 6 and 7 give information about various parameters when simulated using symmetric key and asymmetric key.
- Network graph: Figure 9 shows interconnections between a group of elements that are depicted in network diagrams. A node is the representation of each entity. Links are used to represent connections between nodes.
- Sensor map: Figure 8 shows sensor map, and device manager allows users to examine real-time graphs of sensor data over several date and time periods using the logged sensor readings. The graphs can be useful in identifying power or cooling patterns, as well as possible trouble locations in the data center environment.

The comparison of symmetric key and asymmetric key is listed in Table 1. From node analysis, we found that asymmetric key encryption can be used to perform simple operations like exchanging passwords, etc., because it is more secure than symmetric key encryption. Public key generation will take more time, and its database has to be maintained. Symmetric key encryption is suitable if there is bulk data transfer. If more nodes are present, symmetric key encryption is the go-to method, and it is less secure.

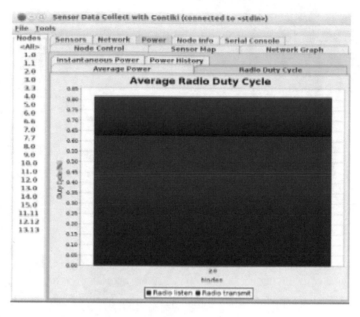

Fig. 5 Radio duty cycle

Nodes <All>	ETX	Churn	Beacon Interval	Reboots	CPU Power	LPM Power	Listen Power	Transmit Power	Power	On-time	Listen Duty Cycle	Transmit Duty Cycle
1.0	0.000	0		0	0.000	0.000	0.000	0.000	0.000		0.000	0.000
2.0	16....	0	34 min, 57 sec	0	0.019	0.163	60.000	0.002	60.1...	0 min,...	100.000	0.004
3.0	0.000	0		0	0.000	0.000	0.000	0.000	0.000		0.000	0.000
4.0	16....	0	34 min, 57 sec	0	0.019	0.163	59.995	0.003	60.1...	0 min,...	99.992	0.005

Fig. 6 Node information for asymmetric key

Nodes <All>	ETX	Churn	Beacon Interval	Reboots	CPU Power	LPM Power	Listen Power	Transmit Power	Power	On-time	Listen Duty Cycle	Transmit Duty Cycle
1.0	16....	0	34 min, 01 sec	0	0.024	0.163	59.996	0.003	60.1...	3 min,...	99.994	0.006
2.0	0.000	0		0	0.000	0.000	0.000	0.000	0.000		0.000	0.000
3.0	16....	0	33 min, 06 sec	0	0.017	0.163	59.997	0.003	60.1...	3 min,...	99.995	0.006

Fig. 7 Node information for symmetric key

4.1 Conclusion

An efficient application layer security protocol between user equipment and AMF in 5G network is implemented and verified. The protocol is tested for real time parameters using Cooja simulator on Contiki OS. Based on the observation, asymmetric encryption can be used to perform operations like exchanging passwords, etc. Public key generation will take more time, its database has to be maintained, and it is more secure. Whereas symmetric encryption is suitable for bulk data transfer. If more nodes are present, symmetric encryption is the preferred method, and it is less secure compared to asymmetric encryption.

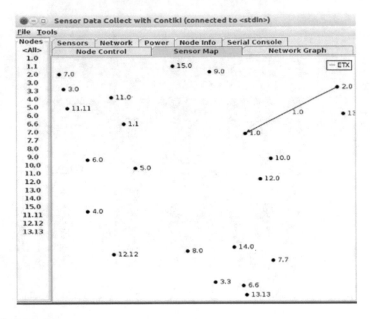

Fig. 8 Sensor map

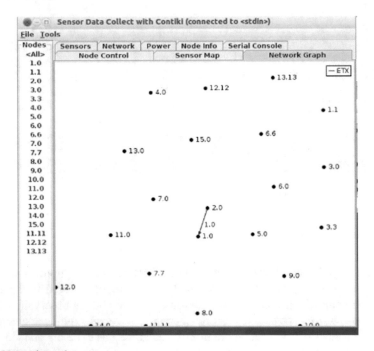

Fig. 9 Network graph

Table 1 Comparison of encryption algorithms

Parameter	Asymmetric key	Symmetric key
Average power consumption (mW)	60.185	60.182
Listen duty cycle (%)	99.996	99.994
Average transmission time (ms)	600	450

4.2 Future Scope

The secured authentication protocol can be further analyzed by implementing a key management system for public and private key, also sharing secret or session key using public key, and analyzing the performance of the both techniques. In this paper, the results obtained for single user equipment (UE) node is described; in the future, the number of UE nodes can be increased, and performance can be analyzed.

References

1. G.P. Dave, N. Jaisankar, Analysis and design of energy efficient protocols for wireless sensor networks. in *2017 Innovative Power Advanced Computing Technology* (2017), pp. 1–7
2. M. Liyanage, J. Salo, A. Braeken, T. Kumar, S. Seneviratne, M. Ylianttila, 5G privacy: scenarios and solutions. in *IEEE 5G World Forum, 5GWF 2018—Conference Proceedings* (2018), pp. 197–203
3. E. Bertino, S.R. Hussain, O. Chowdhury, 5G Security and Privacy: A Research Roadmap. *arXiv*, (2020)
4. M. Tutunoić, P. Wuttidittachotti, Discovery of suitable node number for wireless sensor networks based on energy consumption using Cooja. in *2019 21st International Conference on Advanced Communication Technology* (2019), pp. 168–172
5. K.P. Naik, U.R. Joshi, Performance analysis of constrained application protocol using Cooja simulator in Contiki OS. in *2017 International Conference on Intelligence Computer Instrumentation and Control Technology* (2017), pp. 547 550
6. S. Sharma et al. Secure authentication protocol for 5G enabled IoT network. in *2018 Fifth International Conference on Parallel, Distributed and Grid Computing (PDGC)* (2018), pp. 621–626
7. C. Cremers, *Scyther: Semantics and Verification of Security Protocols* (2006)
8. A. Dunkels, B. Grönvall, T. Voigt, Contiki—a lightweight and flexible operating system for tiny networked sensors. in *Proceedings of Conference on Local Computer Networks, LCN* (2004), pp. 455–462
9. J.H. Anajemba, C. Iwendi, M. Mittal, T. Yue, *Improved Advance Encryption Standard with a Privacy Database Structure for IoT Nodes* (2020), pp. 201–206. https://doi.org/10.1109/csn t48778.2020.9115741
10. F. Pan, H. Wen, H. Song, T. Jie, L. Wang, 5G security architecture and light weight security authentication. in *2015 IEEE/CIC International Conference on Communication China—Work. CIC/ICCC 2015* (2017), pp. 94–98. https://doi.org/10.1109/ICCChinaW.2015.7961587
11. Y. Li, L. Cai, Security aware privacy preserving in D2D communication in 5G. IEEE Netw. **31**, 56–61 (2017)

12. D. Fang, Y. Qian, 5G wireless security and privacy: architecture and flexible mechanisms. IEEE Veh. Technol. Mag. **15**(2), 58–64 (2020). https://doi.org/10.1109/MVT.2020.2979261
13. N. Alsaffar, H. Ali, W. Elmedany, Smart transportation system: a review of security and privacy issues. in *2018 International Conference on Innovative Intelligence Informatics, Computer Technology 3ICT 2018* (2018), pp. 2018–2021. https://doi.org/10.1109/3ICT.2018.8855737

DDoS Botnet Attack Detection in IoT Devices

Bandari Pranay Kumar, Gautham Rampalli, Pille Kamakshi, and T. Senthil Murugan

Abstract Distributed denial of service (DDoS) attacks are most harmful threats on present Internet. The Internet of things (IoT) weaknesses give an optimal objective to botnets, making them a significant supporter of the expanded number of DDoS attacks. The expansion in DDoS attacks made it essential to address the results as it suggests toward the IoT business as one of the significant causes. The main motive of this paper is to provide an analysis of the attempts to prevent DDoS attacks, which mainly at the network level. The reasonableness of these arrangements is removed from their effect in settling IoT vulnerabilities. It is obvious from this survey that there is no ideal arrangement yet for IoT security, but this field has large chances for innovative work.

Keywords Distributed denial of service · Internet of things · Botnet

1 Introduction

DDoS attacks are subclass for denial of service (DoS) attacks. A DDoS attack involves multiple connected online devices, collectively referred to as a botnet, which are wont to flood a target Website with fake traffic. The most commonly DDoS attack involves the flooding of a huge amount of traffic to deplete network resources, bandwidth, target CPU time, etc. The DDoS most attacks are HTTP flood, NTP amplification, SYN flood, DDoS flood, hoping flood, UDP flood, and ICMP (PING) flood. In this paper, we use DDoS attacks-based network flow type of testing in IoT botnet a categorical specific dataset.

The IoT botnet works for launching DDoS attacks on target IoT structures to interrupt the services and operations of IoT. This type of malware can interrupt the functions of connected devices after executing the code. Sometimes hackers can create the botnet to target specific IoT devices. The botnet works as an attacker. Firstly, they will create the bots for getting their tools and scripts for the DDoS

B. P. Kumar (✉) · G. Rampalli · P. Kamakshi · T. Senthil Murugan
Kakatiya Institute of Technology and Science, Warangal 506015, India
e-mail: bandaripranay766@gmail.com

attack and they remotely send commands to active IoT devices to activate the botnet. After activation, the botnets are sending requests to many targeting IoT devices. After attacking all the target IoT devices, the attacked can crash the entire network.

The Internet of things also being called as the Internet of objects, which refers to the wireless networks present between the objects. The devices are connected through the Internet on stands on the IPV6 allows the unique address for them. There some various security services that are mandatory for IoT, and this build for information for security, they are

- Confidentiality: Messages passing from source to destination can be easily intercepted by the attacker, and the content can be hacked. So this message must be hidden from all relay nodes, meaning the message will pass securely end to end is required in the Internet of things. The same can also be applied to devise storage. A simple solution to this is the encryption/decryption mechanism.
- Integrity: The message passing from source to destination should not be changed; it must be received by the recipient. The same is sent by the sender. No medium must change the content of the message as it is being passed or passed device.
- Availability: For the continuous operation of the Internet of things and access to data when necessary, it is also important. The services provided by the devices shall always be available and continuously in working order. So it is important to intrusion detection and intrusion prevention to ensure availability.
- Authenticity: The end-user must be able to learn the identity of the other to ensure that he or she interacts with them. The same entities they claim.

2 Literature Survey

The process of IoT botnet attack detection by the acceptance of IoT technology is spread strongly because the aptitude to produce a far better service. Security additionally changing into a significant difficult role with this development, and IoT contains a resource low procedure and memory resource constraints of IoT. Ali et al. [1] performed a comprehensive systematic literature review on IoT-based botnet attacks. The prevailing state of the art among the realm of study was given and mentioned intimately. A systematic methodology was adopted to create positive the coverage refers of all vital studies. It have been terminated methodology and duplicable. The review made public the prevailing planned contributions, datasets handled, network rhetorical ways used, and analysis focus of the primary handpicked studies. The results of this review unconcealed that analysis throughout this domain gains momentum, notably among past three years (2018–2020). Nine key contributions were additionally known, with analysis on system, model is that the foremost conducted. This method concluded solvation to detect the botnet activity corresponding to a consumer of IoT devices and networks related.

McDermott et al. [2] proposed deep learning method to detect model based on a bidirectional long short-term memory-based recurrent neural network (BLSTM-RNN). The Mirai botnet attack was detected by the BLSTM-RNN detection model was helped by the four attack vectors. Results for DNS, UDP, and Mirai were highly supportive with 98, 98, 99% validation accuracy and validation loss metrics of 0.116453, 0.125630, 0.000809 accordingly.

IoTs known as a device-to-device communication, communicate with each other without human presence, such as a small room lamp to the remote controller. Sonar et al. [3] conduct a survey on DDoS attacks on IoTs. Authentic user restricted, network resources unavailable, and bandwidth are consumed, when DDoS is attacked. Upgraded and high-power capacity devices must be developed to find and shut out the attack. Sonar et al. [3] reveled by few DDoS attacks, which are capable of malfunctioning or complete shutdown of an entire IoT network.

Thousands of IoT devices are vulnerable to even a sings DDoS attacks and IoT botnets group of thousands of IoT devices causing huge threats to Internet security in health, home, social, defense, etc. Al-Duwairi et al. [4] proposed a security information and event management-based IoT botnet DDoS attack detection and helped system. Monitoring packets are TCP SYN, ICMP, DNS packets that were used by the system to detects and block DDoS attack traffic from the compromised IoT devices.

In late 2016, Internet was stormed by large high-profile targets with DDoS attacks Antonakakis et al. [5] provided a quick analysis of the genesis of a botnet, and the type of classes related to the devices were being affected and the way its variants evolved to compete for vulnerable devices provided alongside a seven-month revised analysis done on the Mirai's growth to a peak of 600K infections and also analysis done on the history of its DDoS victims. Poor security practices were the most reason for the emergence of Mirai and have become a serious concern for safety, security, and the privacy of IoT devices.

The proposed algorithm experiment results expressed a good performance to strengthen the safety and security of the IoT with multiple unprotected devices. Yin et al. [6] proposed a structure for software-defined IoT with the help of an SDx pattern. Elaborated an algorithm for detection and interception of DDoS attacks based on SD-IoT. Boundary switches of SD-Iot are used to confirm whether the DDoS attack happened, finding the source of attack, and attacker behind the DDoS attack.

3 Methodology

In this paper, we are implementing our model on the DDoS dataset. The DDoS dataset consists of 1,048,555. We observed that attacks are used in a single dataset. In this paper, we consider only 1000 rows and 47 columns to implement our model. In our model building, we used Python language code implementation to specify IoT attacks of the dataset. We started our implementation by import libraries and defining functions for plotting. We using matplotlib for graphical representation of data as shown in Fig. 1.

Counting on the data, not all plots are getting to be made. An Axes3D object is formed a touch just like the other axes using the projection = "3D" keyword. Create

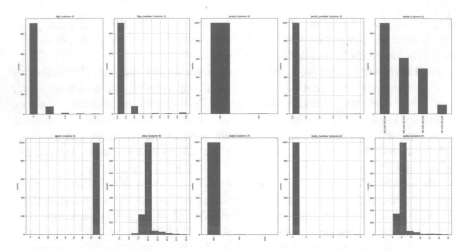

Fig. 1 Graphical representation of the column distribution

a replacement matplotlib. Figure and add replacement axes thereto of type Axes3D. Standardize the related type of dataset along any axis. Center to the mean and the component-wise scale to the unit variance the axis wants to compute the means and standard deviations along.

After importing the plot, they a graphical technique for knowledge of the set, usually as a graph showing the connection between either two or more variables. The design is usually drawn in both the cases either hand or by a computer. Within the past, they used sometimes in mechanical, or electronic plotters were commonly used. Reading the info from CSV could even be an indispensable essential in data science. Repeatedly, we get data from to varied sources which may we get exported to comma separated values (CSV) format so as that they are getting to be employed by the opposite systems. The Pandas library provides features using which we will read the CSV enter full also as in parts for fewer than a specific group of columns and rows. The read_csv province of the pandas library is employed by reading the content of a CSV file into the Python environment thanks to the pandas DataFrame. The function can read the files from the OS by using the proper path to the file. After showing, there is a 1 CSV enter the present version of the datasets. Subsequent hidden code cells define functions for plotting data after the code was connected to the importing libraries within the dataset. For the representation of the whole data within the graphical format, we use histogram representation.

IoT botnet is a network where devices are connected via the Internet of things (IoT). The DDoS attacks commonly attempt a malicious script to spoil the normal target on the target server. They have the types of attacks, and they are reflection attack and exploitation attack. The reflection attack has three types of attacks, such as TCP-based attack, TCP/UDP-based attack, and UDP-based attack. TCP-based attack has used Microsoft Structured Query Language MSSQL and simple service discovery protocol (SSDP). TCP/UDP-based attacks may happen with the help of

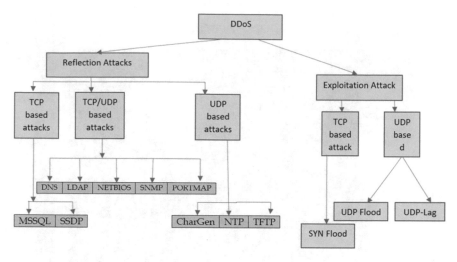

Fig. 2 DDoS taxonomy

Domain Name System (DNS), Lightweight Directory Access Protocol (LDAP), Network Basic Input/Output System (NetBIOS), Simple Network Management Protocol (SNMP), or PORTMAP. UDP-based attacks may be using network time protocol, the trivial file transfer protocol, the character generator protocol (CharGen) to attack the system. Exploitation attack is also divided into two classes—TCP-based attack which is using SYN flood and UDP-based attacks which is using UDP flood and UDP-lag. (Fig. 2).

4 Experiment and Result Analysis

Figure 3 shows results are detected dataset abnormal traffic and DDoS attack. We can also see the Fig. 3 anomaly detected the data correlation matrix. It calculates the Pearson coefficient of correlation (abbreviated the r), with each pair of quantitative features within the data receiving one number from -1 to 0 to $+1$, indicating that their linear dependence is strongly negative, absent, or strongly positive, respectively.

We can see that within the Fig. 4, each numerical attribute is plotted in rows and in columns. The diagonals are showing the histograms (if it is an equivalent column on the x and y-axis). The other charts are showing the scatter plots for every combination of the numerical variables within the dataset.

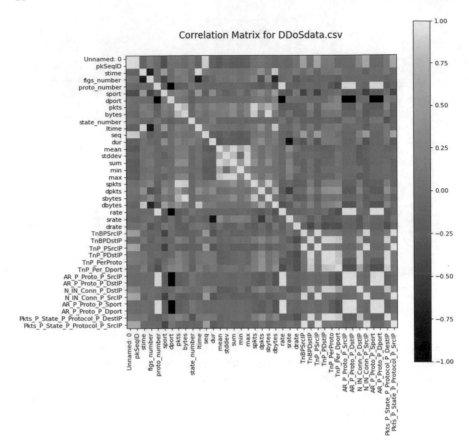

Fig. 3 Correlation matrix representation using DDoS attacks

5 Conclusion

In this paper, the DDoS botnet attack data enters detection algorithm based on the pre-processing of a dataset predicted method has been investigated. We implemented the plat graphs also can detecting an inconsistency caused any by a DDoS flooding attack. We achieving high detection on showing results.

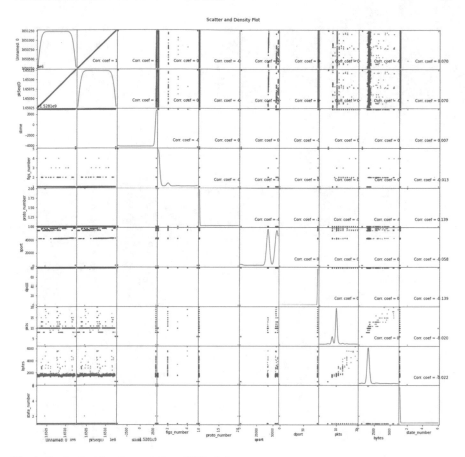

Fig. 4 Scatter matrix representation of DDoS dataset

References

1. I. Ali, et al., Systematic literature review on IoT-based botnet attack. IEEE Access (2020)
2. C.D. McDermott, F. Majdani, A.V. Petrovski, Botnet detection in the internet of things using deep learning approaches. in *2018 International Joint Conference on Neural Networks (IJCNN)* (IEEE, 2018)
3. K. Sonar, H. Upadhyay, A survey: DDOS attack on Internet of things. Int. J. Eng. Res. Dev. **10**(11), 58–63 (2014)
4. B. Al-Duwairi et al., SIEM-based detection and mitigation of IoT-botnet DDoS attacks. Int. J. Electr. Comput. Eng. (2088-8708) **10**(2) (2020)
5. M. Antonakakis et al., Understanding the Mirai botnet. in *26th {USENIX} Security Symposium ({USENIX} Security 17)* (2017)
6. Da. Yin, L. Zhang, K. Yang, A DDoS attack detection and mitigation with software-defined Internet of things framework. IEEE Access **6**, 24694–24705 (2018)

Text/Sign Board Reading Aid for Visually Challenged People

Srikhakollu Pandu Ranga Avinash, V. Krishna Sree, M. Saahith Reddy, K. Sravani Kumari, and A. Gaurav

Abstract The technological advancements in today's world are focussing on designing smart devices that improve lifestyle of human beings by making things easier and simple. Based on this, a productive approach is proposed for detecting the text from the natural scenes and converting it into voice output. The aim of this project is to overcome the reading problems of visually impaired people and their complete dependence on Braille code. The project is developed using MATLAB software in which the algorithms used are maximally stable extremely regions (MSER), stroke width transform, optical character recognition (OCR), and speech synthesizer of MATLAB. By utilizing these techniques, the text from the images can be identified effectively, and it can be converted into speech output.

Keywords Maximally stable extremal regions · Optical character recognition · Stroke width transform · Bounding boxes · Speech synthesis

1 Introduction

Communication plays an important role in the life of a human being whether it is in audio or written format. Most importantly, text that is present visually carries important and valuable information that guides people when they go out. Hence, the advancements in the technology can be applied to digitalize the images to extract specific data from the images. As a part of this, we can extract the data from the images that contain useful information using specific text extraction techniques. Text extraction can be combined with text-to-speech synthesis (TTS) which can convert the visible text in audio output [1]. A model developed in this process can be utilized by the visually impaired people to read the text from the natural scenes.

S. Pandu Ranga Avinash (✉) · V. Krishna Sree · M. Saahith Reddy · K. Sravani Kumari · A. Gaurav
Vallurupalli Nageswara Rao Vignana Jyothi Institute of Engineering and Technology, Bachupally, Nizampet, Hyderabad, Telangana 500090, India
e-mail: avinashkrishna37@gmail.com

V. Krishna Sree
e-mail: krishnasree_v@vnrvjiet.in

© The Author(s), under exclusive license to Springer Nature Singapore Pte Ltd. 2023
Y.-D. Zhang et al. (eds.), *Smart Trends in Computing and Communications*, Lecture Notes in Networks and Systems 396, https://doi.org/10.1007/978-981-16-9967-2_4

2   Literature Review

The literature survey for this project was carried out by studying and analyzing a variety of paper publications published by different authors at national and international level based on the preferred subject. A detailed survey of the papers includes

(1)   Zafar Khan, Ismail Taibani, and Soheil Sayed M. H. Sabo of Siddiq College of Engineering have proposed a finger reader system where the text will be scanned by the device, and the person will be able to hear those words as synthesized speech. This device lacks in providing accuracy under low-light conditions.
(2)   Professor Priya U. Thakre, Kote Shubham, and Shelke Om developed an assistance system for the blind people that puts forward a system which allows blind people to detect and avoid hurdles/obstacles was implemented as an android app. The limitations of the system are that the stability of glasses and using an android phone was weak.
(3)   Rupali D. Dharmale proposed a text recognition model for visually challenged person in Int. Journal of Engineering Research and Applications. Through this paper, they briefly discussed about the object identification techniques and image processing that are used in fields of projects for visually impaired people.
(4)   K.G. Krishnan, C.M. Propodi, and K. Kanimozhi have proposed an image to speech converter in which they have described a model using which a blind person can get information about the shape of a given image input.
(5)   Hangrong Pan, Chucai Yi, and Yingli Tian have developed a AI model that is capable to recognize the information of an incoming bus. This system notifies the blind person about the bus details with the help of a speech signal.

3   Proposed System

The proposed model makes use of text which is present in physically available images and converts it into speech that can be heard by visually impaired person. In order to achieve this goal, various objectives are needed to be accomplished as listed below.

1.   Taking the input image and converting the image into greyscale format from RGB format.
2.   Detection of the text regions from the greyscale image by implementing maximally stable external regions (MSER).
3.   Removal of non-text regions based on geometrical properties.
4.   Removal of the non-text regions present in the image by using stroke width transform (SWT).
5.   Expanding the bounding boxes around the identified letter candidates.
6.   Applying optical character recognition (OCR) to recognize the identified areas.

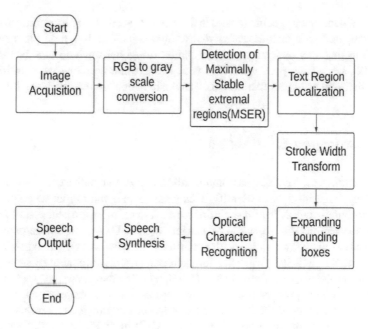

Fig. 1 Block diagram of proposed method

7.   Implement the speech synthesizer to receive a speech signal as voice output of the identified text.

Various stages that are involved in the implementation of the proposed model are shown in Fig. 1.

4   Implementation

The implementation of the above model starts by giving an image from the natural scene as the input to the system and converting the image to a greyscale format for increasing the processing efficiency. After the RGB image is converted to greyscale image, character identification techniques are applied for identifying and detecting the text from the given image. As a part of this, maximally stable extremal regions and stroke width transform are applied.

4.1   Maximally Stable Extremal Region

Maximally stable extremal regions or MSERs are a blob detection technique that is used to identify the text regions from the given input image. This technique uses the strategy of identifying specific locales in the input image that vary in certain properties when compared with encompassing district locations. It is based on identifying all the regions that stay constant or nearly same throughout a wide range of threshold values [2], i.e. all the pixels present inside the extremal regions have an intensity value either low or higher than the pixel intensities outside the given region.

4.2 Text Region Localization

In general, images have different components that resemble with a text or alphabetical format which the model can identify. This leads to an invalid input when compared with the original output. To eliminate this kind of errors, there is a necessary to remove the non-text regions that appear to be as a text region. These non-text regions can be eliminated based on few geometrical properties which measures the maximally stable extremal region (MSER) properties for each connected component or object in the binary image and removes them. In MATLAB, by using the command "regionprops", we can measure the properties of the image region. This command measures a set of properties for each connected component in the binary image. The properties used in this thesis are "bounding box", "eccentricity", "solidity", "extent", "Euler", and "image".

4.3 Stroke Width Transform

Stroke width transform (SWT) is an algorithm that is designed to compare the stroke of a regions with the strokes of the other regions contained in the background. Here, a stroke refers to group of values that show uniform width throughout. The image that is produced as an output of the stroke width transform has a size which is equal to the input image where each element has the value of the width of the pixel associated. The main purpose of using stroke width transform is to eliminate all the non-text regions that are detected as a part of text detection [3].

The calculation gets greyscale picture as input and returns comparable picture of a similar size, where the areas of suspected content are checked. It has 3 significant advances: the stroke width change, gathering the pixels into letter competitor's dependent on their stroke width, and lastly, gathering letter applicants into districts of text. Here, Fig. 2a shows the input image. The stroke width of the input image is calculated as shown in the Fig. 2b.

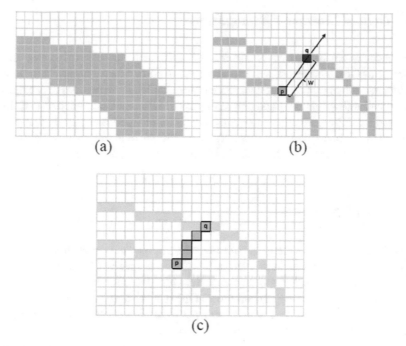

Fig. 2 Example of stroke width transform

4.4 *Expanding Bounding Boxes*

After recognizing the text and elimination of non-text regions, we consolidate the text regions for definite detection result. For this a rectangular box is plotted around each character for object identification [9]. This is done by expanding bounding boxes around each letter candidates and grouping them into words and sentences. By defining the X and Y coordinates, a bounding box is drawn around the letter which acts as a point of reference for the process of text detection. After creating the bounding boxes, the individual letter candidates have to be merged together to form group of words or a meaningful sentence. This is done by calculating the overlap ratio. In MATLB, we have a function "bboxOverlapRatio" which gives the overlap ratio that is present between every two bounding boxes that are considered. The overlap ratio has a value between 0 and 1 where 1 represents that the overlap is perfectly done. In the next step, we combine the bounding boxes according to the overlap ratio as well as maximum and minimum dimensions.

4.5 Optical Character Recognition

After identifying the text areas and removing the non-text areas, the next step is to detect the text characters that are to be received as voice output. This done by implementation of optical character recognition (OCR). It is a process of recognizing printed or hand-written text and converting them into machine-encoded text. In the proposed technique, we utilize an optical character recognition function that accepts the input as text boxes that are previously made through bounding boxes. This process produces the extracted text from the input image as a text document [4].

4.6 Speech Synthesis

After the text output is received, we need to implement a speech synthesizer to convert the identified text into voice output. Text-to-speech (TTS) is an algorithm that is designed to provide a speech signal from the given text data. By implementing text to speech, we can easily convert the text into speech and audio signal. MATLAB consists of an audio tool box which can used to implement the speech synthesizer that can convert the input text data to audio format.

5 Results

The system takes a maximum time of 4–5 s in processing all the output images and giving an audio output. The results of the above implemented algorithms in MATLAB are as follows:

1. The output image after transforming the input image to greyscale format is given in the Fig. 3.
2. The output image of the second stage is an image with maximum stable extremal regions (MSER) identified. Figure 4 shows maximum stable extremal regions that are identified in the given input image.

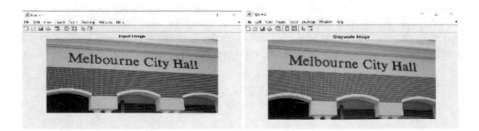

Fig. 3 Input image and greyscale image

Fig. 4 Image with identified MSER

3. Figure 5 shows the image with non-text regions removed based on geometrical properties. This image had removed maximum number of non-text regions from the set of regions identified by MSER.
4. Few of the non-text regions that are still visible in the input image are further removed by employing stroke width variation. This output related to stroke with variation is shown in the Fig. 6.
5. The output image with bounding boxes expanded around the letter candidates is shown the Fig. 7.
6. The output image with detected text after the letter candidates are merged into words and sentences is shown in Fig. 8.

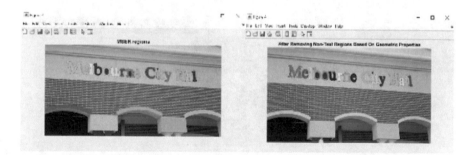

Fig. 5 Removing non-text regions based on geometrical properties

Fig. 6 Removing non-text regions based on stroke width variation (SWT)

Fig. 7 Image with expanded bounding boxes

Fig. 8 Image with detected text after merging of bounding boxes

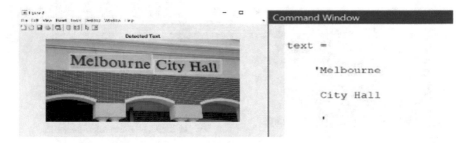

Fig. 9 Identified text output

7. The recognized text from the input image by using optical character recognition is shown in the Fig. 9.
8. Also, with the help to speech synthesizer used in the programme, the identified text output is converted into a speech signal, and an audio output is generated.

6 Conclusion

In this project, we developed a system which can convert text or image into speech by using MATLAB software. To the best of researcher's knowledge, the first step was to take an input image and perform image pre-processing. Segmentation of text images is done using maximally stable extremal regions. They are made unique by taking various thresholds. Then, the individual letter candidates are recognized. Later on, these individual letters are grouped to form the words. For word recognition, optical character recognition (OCR) is implemented, and speech processing toolboxes are used to convert text into speech.

As a part of future scope, this concept of text-to-speech conversion for blind people can be extended by integrating the model with Internet of things (IoT) so that it can be connected to various devices such as a smart watch.

References

1. C. Yi, Y. Tian, Scene text recognition in mobile applications by character descriptor and structure configuration. IEEE Trans. Image Process. 7 (2014)
2. S.M. Qaisar, A computationally efficient EEG signals segmentation and de-noising based on an adaptive rate acquisition and processing. In: 2018 IEEE 3rd International Conference on Signal and Image Processing (ICSIP) (IEEE, 2018), pp. 182–186
3. A. Kumar, S. Gupta, Detection and recognition of text from image using contrast and edge enhanced segmentation and Ocr. IJOSCIENCE (Int. J. Online Sci.) Impact Factor **3**(3), 3 (2017)
4. Y. Zhu, C. Yao, X. Bai, Scene Text Detection in Technological Advancements, 10(1), 19–36 (2016)

Automatic Segmentation of Red Blood Cells from Microscopic Blood Smear Images Using Image Processing Techniques

K. T. Navya⬭, Subhraneil Das, and Keerthana Prasad⬭

Abstract Human blood is a very effective parameter to detect, diagnose and rectify ailments of the human body. Complete blood count (CBC) is a method to clinically obtain a statistical measure of blood and its related parameters, i.e., red blood cells (RBCs), white blood cells (WBCs), platelets, hemoglobin concentration to name a few. This helps to determine the physical state of the subject. For further diagnosis, peripheral blood smear, a thin layer of blood smeared on a microscope slide and stained using various staining methods is examined for the morphology of the cells by the pathologists. However, manual inspection of smear images is tedious, time-consuming, and laboratorian-dependent. Although there are certain software-based approaches to tackle the problem, most of them are not robust for all staining methods. Thus, the need is to create an automated algorithm that will work for different staining types, thereby alleviating both the aforementioned drawbacks. This work aims to create an automatic method of segmenting and counting RBCs from blood smear images using image processing techniques to help diagnose RBC-related disorders. In the proposed method, the images are first preprocessed, i.e., standardized to a uniform color and illumination profile using contrast enhancement, adaptive histogram equalization followed by Reinhard stain normalization algorithms. WBCs and platelets are extracted in HSI color space and subtracted from the original image to retain only RBCs. Thereafter using morphological operations and active contour segmentation algorithms, a count of total RBCs were obtained even for overlapped cells in the microscopic blood smear image. The proposed method achieved counting accuracy of 89.6% for 150 images.

K. T. Navya · S. Das
Department of Electronics and Communication, Manipal Institute of Technology, Manipal, India

K. Prasad
Manipal School of Information Sciences, Manipal, India
e-mail: keerthana.prasad@manipal.edu

K. T. Navya (✉) · S. Das · K. Prasad
Manipal Academy of Higher Education, Manipal 576104, India
e-mail: navya.kt@manipal.edu

Y.-D. Zhang et al. (eds.), *Smart Trends in Computing and Communications*, Lecture Notes in Networks and Systems 396, https://doi.org/10.1007/978-981-16-9967-2_5

Keywords Red blood cells · Color normalization · Image processing · Segmentation · Active contour

1 Introduction

Blood sampling and diagnosis of diseases based on the blood count and characteristics are an important aspect of medicinal science and engineering. Human blood is an essential source for developing diagnosis patterns and antigens for numerous diseases. Blood is a specialized body fluid and has four main components [1].

- *Plasma*, liquid component produced in the bone marrow to transport blood cells throughout the body
- *Erythrocytes or red blood cells* is a biconcave disk with a central pallor, 6–8μm in diameter with pink stain helps carry oxygen from the lungs to the rest of the body.
- *Leukocytes or white blood cells*, largest cells, varies between 8 and 20μm in diameter, a dark purple nucleus with pale cytoplasm produce antibodies to fight off infection.
- *Thrombocytes or platelets*, small cell fragments without nuclei, 2–3μm in diameter, bluish in color helps in blood clotting.

Figure 1 shows a typical blood film or peripheral blood smear stained using the Romanowsky staining method to examine the blood cells microscopically. Staining plays a very important role in the analysis of blood smear images as the color profile, contrast, and feature details in a microscopic blood smear image are determined initially by this aspect. Popular staining methods under Romanowsky staining include Giemsa, Wright-Giemsa, Leishman stain, etc. The blood smear thus taken is analyzed in the feather edge of the slide, and hence, subsequent conclusion helps in knowing the current state of health of the patient as well as diagnosing a specific ailment. Considering the manual techniques followed previously for analyzing blood smear images, it was very tedious and not reliable. Thus, this work is aimed at developing

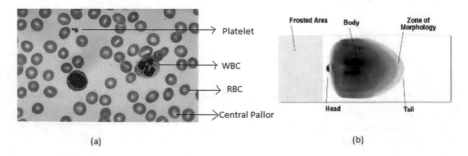

(a)   (b)

Fig. 1 Blood smear images [2] **a** microscopic view of blood smear **b** blood smear slide

an autonomous and robust system for image preprocessing followed by segmentation of red blood cells in a given blood smear image using image processing techniques in MATLAB to help diagnose RBC-related disorders.

2　Related Work

This section attempts at giving a brief overview of various past developments in the field of automated blood counting and its related parameters such as color normalization, preprocessing, RBC segmentation, background, and foreground differentiation using various segmentation methods based on underlying principles of image processing. Sharif et al. [3] elaborated a robust methodology for RBC segmentation using masking and morphological operation. This paper has a combination of pixel-based, region-based, and morphological segmentation. YCbCr color has been chosen for illumination issues. A marker-controlled watershed algorithm was used to handle overlapping cells from 20 images. Mazalan et al.[4] proposed an automated method to count RBCs in microscopic images using circular Hough transform (CHT) for 10 sample images and obtained 91.9% accuracy. This method is cost-effective and provides an alternative way to recognize and count circular cells. Alomari et al. [5] proposed an iterative structured circle method to segment WBC and RBC, and average accuracy of 95.3% for RBCs and 98.4% for WBCs was achieved for 100 images. However, this method depends on the number of iterations. Abbas et al. [6] presented a method for RBC segmentation in YCbCr color space. K-means clustering was used to identify cells from 90 Giemsa stained images. Tomari et al. [7] proposed a Hough transform (HT) method to count overlapped RBCs and obtained 94% average accuracy for four sample images. However, this method involved many parameters. Alam et al. [8] presented the YOLO algorithm to identify and count blood cells and obtained accuracy 96.09% RBC, 86.89% WBC, and 96.36% for 364 annotated images. Wei et al. [9] proposed K-means clustering-based method to segment and count overlapped RBCs. The method used the H and S components to differentiate between WBC and RBC. The author of the work obtained 92.9% accuracy for 100 Wright-Giemsa stained images. Acharya et al. [10] presented a method to identify and count RBC using K-Medoids and geometric features. This method achieved 98% accuracy for 1000 Wright stain images. Ejaz et al. [11] proposed HT-based method to segment and count RBCs and obtained 94.9% accuracy for 500 subjects. Berge et al. [12] proposed RBC segmentation method using boundary extraction and curvature calculation. The Delaunay triangulation method was used to split overlapped RBCs of any shapes. This method obtained 2.8% absolute error for 49 Giemsa stained images. Hegde et al. [13] proposed active contour method to segment WBCs. G'G'B channel representation for handling illumination and color variations and obtained 96% sensitivity for 54 images from the ALL-IDB2 dataset. Adagale et al. [14] proposed an overlapped red blood cell counting algorithm using template matching and pulse coupled neural network and achieved 90% average accuracy for 40 images. However, accuracy decreases due to overlapped RBCs. Cruz et al. [15] presented RBC counting method using blob analysis and watershed transform in the HSV component and obtained 96% average accuracy for 10 blood samples. Loddo et al. [16] proposed a blood cell counting method using nearest neighbor and SVM techniques by cropping each cell manually. Clumped cells were counted using CHT. This method used 368 images from the ALL-IDB dataset and obtained an average accuracy of 99.2% for WBCs and 98% for RBCs. Yeldhos et al. [17] implemented FPGA-based RBC counting system. Watershed transform and CHT segmentation method were used in YCbCr color space and obtained 90.98% accuracy for 108 blood smear images from the ALL-IDB database. Tran et al. [18] presented deep learning semantic segmentation method for RBC and WBC segmentation and counting. SegNet architecture was utilized to segment blood cells by labeling each pixel. The segmentation accuracy for WBCs and RBCs was 94.93% and 91.11%, respectively. For cell counting, Euclidean distance transform and binary dilation are used and obtained 93.3% for RBC and 97.29% for WBC for 42 ALL-IDB database images. From the past studies, different methods such as circular Hough transform, watershed transform, morphological operations, thresholding-based methods, K-means clustering, ANN, DNN have been used for RBC segmentation and counting. However, these methods lack robustness in handling blood smear images with multiple stains. Hence, there is a need for developing a robust system to handle images taken from various laboratory settings.

3 Methodology

This section gives a sequence of steps that are to be followed to obtain the desired results. A methodology of the work is depicted in Fig. 2.

3.1 Image Acquisition

The required Leishman stained blood smear images are acquired from Kasturba Medical College (KMC), Hematology department, Manipal, with a 100x lens objective. Also, smear images from Isfahan MISP online database[19] are gathered for the process.

Fig. 2 Methodology of proposed work

3.2   Preprocessing

Due to the varied color profile and contrast of the microscopic smear images, there is a need to standardize the image to obtain a consistent color profile. To make an image ready for the automatic segmentation under various image settings and stains, illumination correction and color normalization methods have been implemented. The various preprocessing methods used are discussed further [20].

- *Linear Contrast Enhancement using Adaptive Histogram Equalization*
  This method is helpful to reduce illumination variation. The images are first converted from RGB to LAB color space where a consistent normalization of the L parameter (Luminosity) is performed followed by grayscale conversion and subsequent adaptive histogram equalization.

- *Gamma Correction*

  is a nonlinear method used to correct the image's luminance; i.e., it amplifies the shadows or the bright regions of the image as per the requirement. It is used to correct uneven illumination by encoding luminance in video or still image systems. Gamma correction is defined as:

$$V_{out} = \Lambda V_{in}^{\Gamma} \tag{1}$$

  Powers larger than 1 make the shadows darker and smaller than 1 make dark regions lighter.

- *Gray World Assumption method*

  It is a color correction function based on the principle that on average the scene is neutral gray. The algorithm produces an estimate of illumination by computing the mean of each channel of the image. The average pixel value of an unsigned 8-bit integer image is 127.5, so by calculating the real average pixel value, the scaling value is computed to scale the entire image linearly. In practice, the average of each individual channel is used to calculate a separate scaling value for each channel.

- *Histogram Matching*

  It adjusts histogram of the 2D image to match histogram of reference image in order to normalize the color. The reference RGB image is converted into HSV, and hue component is extracted and used for histogram matching with the input image to segment WBC and platelets. The HSV image is then converted into RGB image for further processing.

- *Reinhard's Stain Normalization*

The algorithm thus maps the color distribution of an image to that of a well-stained target image thereby solving the problem of inconsistency. Technically it matches the mean and the standard deviation of each color channel in the two images in that color space [21]. The statistical approach to color mapping can be shown as in equations

$$l_{mapped} = \frac{l_{original} - \tilde{l}_{original}}{\hat{l}_{original}} \hat{l}_{target} + \tilde{l}_{target}$$

$$\alpha_{mapped} = \frac{\alpha_{original} - \tilde{\alpha}_{original}}{\hat{\alpha}_{original}} \hat{\alpha}_{target} + \tilde{\alpha}_{target}$$

$$\beta_{mapped} = \frac{\beta_{original} - \tilde{\beta}_{original}}{\hat{\beta}_{original}} \hat{\beta}_{target} + \tilde{\beta}_{target}$$

$$(2)$$

where the superscripted variables $\tilde{l}$ and $\hat{l}$ represent the mean and standard deviation of luminosity parameter, respectively, in the LAB color space. Similarly, the subscripted variables $l_{original}$, l_{target}, and l_{mapped} indicate whether the parameter concerned is either of the original, target or the mapped image of the LAB color space parameters. An image with acceptable color profile, contrast, and illumination was chosen as the target image, and the source image was subjected to Reinhard staining normalization.

- *Noise Removal*

The blood smear image post color normalization is subjected to noise removal and deblurring using Wiener filter. The filter processes a two-dimensional image and eliminates granular noise from the image. It is also useful in removing motion blur from the image, thereby making the image ready for further processing.

The image enhancement algorithms on being implemented on a set of images gave a series of results based on their resolution, contrast standardization, and processing method as stated in this section. In Fig. 3, the original faded Leishman stained microscopic image of a normal blood smear is shown which is almost devoid of sharpness, contrast, and color and the reference image for color normalization. The contrast-enhanced reference image is exposed to gamma correction for illumination correction and then passed through a Wiener filter with a filtering window of [10]. The resultant images are smoothened and exhibit a much less extent of jagged edges in the features of the image. The input image is subjected to grayworld color normalization in order to correct uniformity of color, contrast, and illumination variation. However, results obtained show consistency in image regions with proper contrast and deviation for other regions thereby proving to be inefficient for the required purpose. The images are then subjected to the statistical means of color matching in order to achieve a fixed level of uniform tone and contrast which can then be applied and fixated to any image which is fed as an input to the algorithm. The target image ideally should possess uniformity in regard to the factors of illumination, color tone, contrast, warmth,

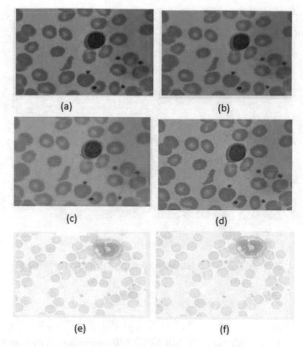

Fig. 3 Preprocessed blood smear images **a** Leishman reference image **b** contrast-enhanced image **c** gamma correction for image **d** Wiener filtered image **e** Leishman stained input image **f** grayworld normalized image

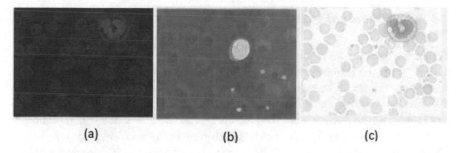

Fig. 4 **a** H component of input image **b** H component of preprocessed reference image **c** histogram matched RGB image

and sharpness. Histogram matching technique is used for color normalization and illumination correction where hue component of the reference image is matched to hue component of the source image to extract WBCs and platelets from blood smear images, and the matched RGB image is as shown in Fig. 4. In order to compare the histogram matched image with another color normalization method, Reinhard stain normalization was applied for source image with reference to the target image. The normalized results obtained with different stain input and target images taken in 100x

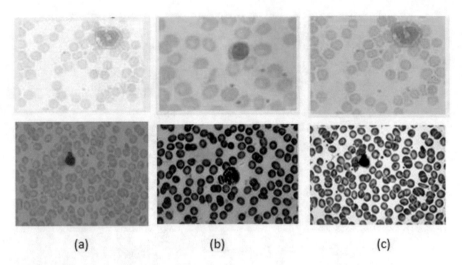

(a)   (b)   (c)

Fig. 5 **a** Leishman and Wright stain input image **b** 100× and 40× reference image **c** color normalized image

and 40X magnification are shown in Fig. 5. On visualization, Reinhard normalized image combat both illumination and color variation and show a prominent difference between the blood cells.

3.3   Segmentation of Blood Cells

The normalized image is processed further for WBC, platelets, and RBC differentiation. Thus, the Reinhard normalized image is subjected to unsharp masking for sharpening the object boundaries to distinct with respect to the background, especially at the edges of the RBCs. WBCs are extracted using the S-channel of HSV that highlights the nucleus and green channel of RGB color space which highlights RBCs and cytoplasm of WBC. Then by combining both the channel outputs, WBCs are removed. Further, traces of WBCs are removed using morphological opening via a disk-shaped morphological element with the radius range of 80–90, and then similarly, platelets are removed, subtracted, and smoothened to obtain the image with only RBCs. The appropriate results are shown in Fig. 6.

We can observe from the RBC segmented image that due to morphological operations, boundaries of RBCs are altered. For further RBC disorder analysis, the morphology of the cell is very important. So to improve on this diverging active contour segmentation algorithm is used. The basic premise of active contours is energy minimizing models; i.e., the snake tends to minimize its energy, thereby shrinking onto the boundaries of the image objects[22]. The entire process of fitting onto the bound-

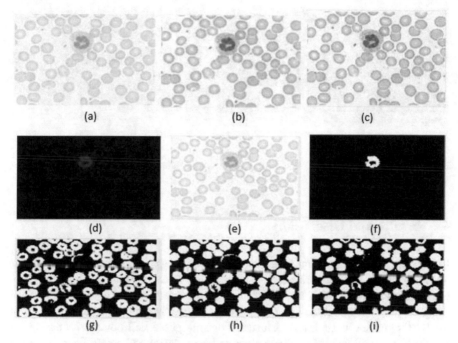

Fig. 6 **a** Input image **b** reinhard normalized image **c** sharpened image **d** S-channel of original image **e** green channel extraction **f** extracted WBC image **g** subtracted image **h** image smoothened by morphological dilation **i** resultant image after erosion

aries occurs in several iterations to get the best fit possible. Now, considering the snake as a continuous parametric variable, we can define its position in the image as

$$v(t) = (x(t), y(t)) \tag{2}$$

where $v(t)$ is the active contour and $x(t)$, $y(t)$ are the continuous contour coordinates. Thus, the energy equation associated with active contours is as follows:

$$E = \int_0^1 [E_{int}(v(t)) + E_{img}(v(t)) + E_{con}(v(t))]dt \tag{3}$$

where E_{int} is internal energy, E_{img} is image forces and E_{con} is external constraint force. The divergence of the contour was achieved by exercising a certain degree of *contraction bias*. This parameter ranges from -1 to 1, and a negative value indicates expanding contour, whereas a positive value indicates a shrinking contour. For the initial contour mask, centroids of the morphologically segmented RBCs are taken as a seed point to diverge. Figure 7 depicts an initial mask and contour detected RBC binary image as an overlay mask to present the detected cells clearly.

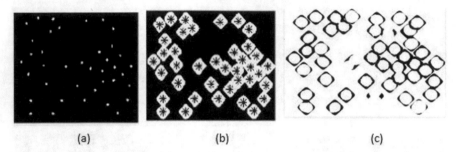

(a)   (b)   (c)

Fig. 7 **a** Initial contour mask **b** RBC segmented image **c** Detected RBCs along with their shape enclosed within binary mask

4 Results and Discussion

In the case of any algorithm, a reference or ground truth is required to measure the accuracy of the used algorithm. The segmented RBCs are counted manually and compared with the ground truth. Here for ground truth formation, we make use of ImageJ Tool to segment the microscopic blood smear images to get a clear count of the RBCs present in the image. Clustered or ambiguous cell boundaries are drawn using a freehand line tool to obtain clear cell boundaries and count. Now using the count obtained from the ground truth of the images and the count obtained via our proposed algorithm we determine the accuracy of the method. In this work, accuracy is determined as the percentage of detected count divided by the actual count of RBCs present in that image. A total of 150 images, 75 Leishman stained images from KMC and 75 images from the online database [19] were used in the study. The active contour algorithm yields an overall accuracy of 89.6% for 150 images. Though it resolves a significant amount of overlapped cells present in an image, some overlapped cells and clusters still remain unresolved. Still, there is a need for a robust segmentation method and accurate ground truth for RBC count to achieve higher accuracy.

5 Conclusion

The methods implemented to meet the objective of the work has been achieved so far. Preprocessing of the images proved to be a cumbersome task and after several trials, Reinhard's method demonstrated to be the most effective method of color normalization. Similarly, post-WBC and platelet removal, the active contour model provided an RBC count closest to the actual values as obtained from the ground truth and achieved an overall accuracy of 89.6%. However, to overcome overlapping and densely clustered RBCs, a robust segmentation algorithm has to be implemented to get a much more accurate count. Further, segmentation processes using convolutional neural networks and artificial intelligence toolboxes can be used to create a shape classifier for identifying cell edges, overlapped cells, and cell clusters accurately in order to detect specific diseases and anomalies.

References

1. H. Mohan, *Textbook of Pathology* (Medical Publishers Pvt. Limited, Jaypee Brothers, 2018)
2. K.W. Jones, Evaluation of cell morphology and introduction to platelet and white blood cell morphology. Clin. Hematol. Fundam. Hemost. 93–116 (2009)
3. J.M. Sharif, M. Miswan, M. Ngadi, M.S.H. Salam, M.M. bin Abdul Jamil, Red blood cell segmentation using masking and watershed algorithm: a preliminary study, in *2012 International Conference on Biomedical Engineering (ICoBE)* (IEEE, 2012), pp. 258–262
4. S.M. Mazalan, N.H. Mahmood, M.A.A. Razak, Automated red blood cells counting in peripheral blood smear image using circular Hough transform, in *2013 1st International Conference on Artificial Intelligence, Modelling and Simulation* (IEEE, 2013), pp. 320–324
5. Y.M. Alomari, S. Abdullah, S.N. Huda, R. Zaharatul Azma, K. Omar, Automatic detection and quantification of WBCs and RBCs using iterative structured circle detection algorithm. Comput. Math. Methods Med. **2014** (2014)
6. N. Abbas, D. Mohamad et al., Microscopic RGB color images enhancement for blood cells segmentation in YCBCR color space for k-means clustering. J. Theor. Appl. Inf. Technol. **55**(1), 117–125 (2013)
7. R. Tomari, W.N.W. Zakaria, R. Ngadengon, M.H.A. Wahab, Red blood cell counting analysis by considering an overlapping constraint 2006–2015. Asian Res. Publishing Netw. (ARPN) **10**(3) (2015)
8. M.M. Alam, M.T. Islam, Machine learning approach of automatic identification and counting of blood cells. Healthcare Technol. Lett. **6**(4), 103–108 (2019)
9. X. Wei, Y. Cao, G. Fu, Y. Wang, A counting method for complex overlapping erythrocytes-based microscopic imaging. J. Innov. Optical Health Sci. **8**(06), 1550033 (2015)
10. V. Acharya, P. Kumar, Identification and red blood cell automated counting from blood smear images using computer-aided system. Med. Biol. Eng. Comput. **56**(3), 483–489 (2018)
11. Z. Ejaz, A. Hassan, H. Aslam, Automatic red blood cell detection and counting system using Hough transform. Indo Am. J. Pharm. Sci. **5**(7), 7104–7110 (2018)
12. H. Berge, D. Taylor, S. Krishnan, T.S. Douglas, Improved red blood cell counting in thin blood smears, in *2011 IEEE International Symposium on Biomedical Imaging: From Nano to Macro* (IEEE, 2011), pp. 204–207
13. R.B. Hegde, K. Prasad, H. Hebbar, B.M.K. Singh, Image processing approach for detection of leukocytes in peripheral blood smears. J. Med. Syst. **43**(5), 114 (2019)
14. S. Adagale, S. Pawar: Image segmentation using PCNN and template matching for blood cell counting, in *2013 IEEE International Conference on Computational Intelligence and Computing Research* (IEEE, 2013), pp. 1–5
15. D. Cruz, C. Jennifer, L.C. Castor, C.M.T. Mendoza, B.A. Jay, L.S.C. Jane, P.T.B. Brian et al., Determination of blood components (WBCs, RBCs, and platelets) count in microscopic images using image processing and analysis, in *2017 IEEE 9th International Conference on Humanoid, Nanotechnology, Information Technology, Communication and Control, Environment and Management (HNICEM)* (IEEE, 2017), pp. 1–7
16. A. Loddo, L. Putzu, C. Di Ruberto, G. Fenu, A computer-aided system for differential count from peripheral blood cell images, in *2016 12th International Conference on Signal-Image Technology & Internet-Based Systems (SITIS)* (IEEE, 2016), pp. 112–118
17. M. Yeldhos, Red blood cell counter using embedded image processing techniques. Res. Rep. **2** (2018)

18. T. Tran, O.H. Kwon, K.R. Kwon, S.H., Lee, K.W. Kang, Blood cell images segmentation using deep learning semantic segmentation, in *2018 IEEE International Conference on Electronics and Communication Engineering (ICECE)* (IEEE, 2018), pp. 13–16
19. M. Kashefpur, R. Kafieh, S. Jorjandi, H. Golmohammadi, Z. Khodabande, M. Abbasi, N. Teifuri, A.A. Fakharzadeh, M. Kashefpoor, H. Rabbani, Isfahan MISP dataset. J. Med. Signals Sens. **7**(1), 43 (2017)
20. Image enhancement techniques. https://in.mathworks.com. Accessed May 2020
21. E. Reinhard, M. Adhikhmin, B. Gooch, P. Shirley, Color transfer between images. IEEE Comput. Graph. Appl. **21**(5), 34–41 (2001)
22. M. Kass, A. Witkin, D. Terzopoulos, Snakes: active contour models. Int. J. Comput. Vis. **1**(4), 321–331 (1988)

Design and Simulation of Microstrip Bandpass Filter Techniques at X-Band for RADAR Applications

Shraddha S. Shinde, D. P. Rathod, and ArunKumar Heddallikar

Abstract A new approach to compact bandpass filter techniques such as end-coupled, parallel-coupled, hairpin, interdigital, and combline filters with their enhancement in size and frequency performance is introduced in this paper. Designed fifth-order filters that are simulated in the compact narrowband frequency is targeted for high selectivity, better return loss, and insertion loss with a passband ripple using Chebyshev response. To minimize the dimension of filters, we implemented an optimized design by determining its parameters, coupling resonators, and connecting elements. The results are simulated using Genesys Keysight software and iterated for the desired specimen. The prototype filter consists of Rogers RT/duroid 5880 substrate with a thickness of 0.508 mm. Experimental result of hairpin filter structure shows a reduced design (20.2 mm × 10 mm) and improvises the stopband characteristics with better harmonic suppression and narrower frequency bandwidth than the other conventional filter for future development in RADAR applications.

Keywords Bandpass filter · Microstrip · End coupled · Parallel coupled · Hairpin · Interdigital · Combline · Chebyshev response · RADAR

1 Introduction

Evolution in communication took place from the telegraphy in the Second World War toward Radio Frequencies (RF) in remote sensing and navigation. RF filters are required in the telecommunication system to reduce noise and interference of signal

S. S. Shinde (✉) · D. P. Rathod
Department EnTC, VJTI (Veermata Jijabai Institute of Technology), Mumbai, India
e-mail: ssshinde_m18@et.vjti.ac.in

D. P. Rathod
e-mail: dprathod@el.vjti.ac.in

A. Heddallikar
RADAR Division, SAMEER, IIT B CAMPUS, Mumbai, India
e-mail: arunkumar@sameer.gov.in

© The Author(s), under exclusive license to Springer Nature Singapore Pte Ltd. 2023
Y.-D. Zhang et al. (eds.), *Smart Trends in Computing and Communications*, Lecture Notes in Networks and Systems 396, https://doi.org/10.1007/978-981-16-9967-2_6

[1]. X-band (8–10 GHz) frequencies are used because it allows smaller RF components compared to other bands of frequencies. Microstrip is unique and contains a Radio Frequency (RF) filter used in many applications such as RADAR, satellite, and wireless communications [5, 7]. A filter is made of two-port networks that are used to regulate frequency response according to the requirement by transmission lines at frequencies within passband and attenuate frequency in stopband [1, 8]. This paper illustrates the X-band filter (8–12 GHz), designed by combining two or more resonators to make a more compact size structure terminated with 50 Ω characteristics impedance [9, 10]. Chebyshev response is selected in this paper as compared to Butterworth response as it is having better selectivity and steeper roll-off with flatter passband magnitude response [3]. Some of the X-band filters such as end-coupled, parallel-coupled, hairpin, interdigital, and combline filters were reported [3, 4]. The end-coupled filter is half-wavelength resonators at mid-frequency of bandpass filter design, a gap between the two adjacent resonators. In a parallel-coupled filter, parallel arrangements of resonators along the half of wavelength resonators are specified [11]. To minimize its dimension of the half-wavelength filter, it required a hairpin microstrip line structure. A hairpin filter is compact and can be folded from the open line $\frac{\lambda}{2}$ wavelength microstrip resonator having stepped impedance to form 'U'-shaped small structure that consists of parallel-coupled lines on both sides ends for miniaturization, which is optimal space utilization [12, 13]. The interdigital filter has properties that consist of arithmetical symmetry, considered as advantageous as it correlates with combline and other filters [14–16]. The drawback of a filter is that the frequency response gets affected by coupling among non-adjacent resonators [14]. Combline filters having large lumped capacitance and shorter resonator lines form a compact structure, and the filter size is relatively reduced. To achieve this further, all five filters are compared with analysis and simulation to determine their performance [14, 15]. To miniaturize the dimension, our work focused on the conventional bandpass filter technique. In this way, we have designed filters on RT/duroid 5880 substrates of thickness 0.508 mm, (dielectric constant) $\varepsilon_{re} = 2.2$, and loss tangent = 0.0009. The design of filter techniques is performed on Genesys Keysight software. The filter design is chosen in the X-band frequency such that it can be used for various RADAR applications.

2 Material and Method

The Chebyshev filter illustrates a good performance compared to the Butterworth filter for frequency response and its cutoff [3, 17]. We considered Chebyshev prototype, and element parameters are given as $g_0 = g_6 = 1$, $g_2 = g_4 = 1.3712$, $g_1 = g_5 = 1.1468$, and $g_3 = 1.9750$ [2, 3].

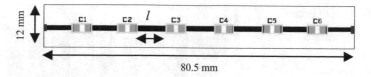

Fig. 1 Structure of end-coupled bandpass filter

Fig. 2 Structure of parallel-coupled bandpass filter

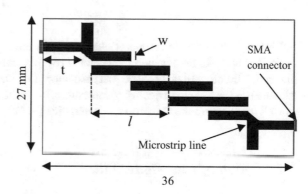

2.1 Design of End-Coupled Bandpass Filter

End-coupled bandpass filter Fig. 1 is half guided wavelength resonator designed in Genesys Keysight software. Figure 1 shows width $= 12$ mm, length $= 80.5$ mm, and l $= 10$ mm, where 'l' is the length of the resonator. SubMiniature version A connector (SMA) is connected at both ends of the microstrip line [3, 4]. Using Genesys Keysight software, width, length, and line spacing can be determined [3, 11].

2.2 Design of Parallel-Coupled Bandpass Filter

Figure 2 shows a parallel-coupled filter having a width of 27 mm and a length of 36 mm, t = 6 mm, $l = 5.25$ mm, and w = 1.5 mm, where 't' is the tapping length, 'l' is the length of a resonator, and 'w' width of each resonator. A parallel-coupled filter is a cascaded-line section with open ends. Electrical length is specified by even and odd mode characteristics impedance [6, 11].

2.3 Design of Hairpin Filter

The tapped input and output hairpin structure is shown in Fig. 3, with width $= 10$ mm, length $= 20.2$ mm, w $= 1.5$ mm, t $= 3.6$ mm, S1 $= 0.2$ mm, and S2 $= 0.4$ mm, where

Fig. 3 Structure of hairpin
bandpass filter

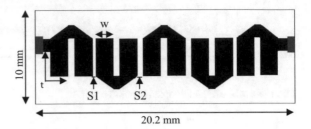

'w' is the width of each resonator element, 't' is the tapping length of the hairpin
filter, 'S1' and 'S2'are spacing between the two resonators. The input transmission
line is not directly attached to the first resonator [12]. Decreasing the tapping length
't' increases the unloaded quality of resonators, hence narrowing the 3 dB bandwidth
as well as changing the response characteristics [3, 12].

2.4   Design of Interdigital Filter

Figure 4 shows the structure of the interdigital bandpass filter modeled in Genesys
software of width = 10 mm, length = 21.5 mm, l = 6 mm, w = 1.7 mm, S1 =
0.2 mm, and S2 = 0.3 mm, where 'l' is tapped line, 'w' is the width of the resonator,
and 'S1' and 'S2' are spacing between the two resonators having resonator length
of $\frac{\lambda}{4}$ long with tapped line at first and last resonator [15, 16]. Each resonator has an
electrical length of 90° at the mid-band frequency which is short-circuited at one end
and open-circuited at another end. It is shorted to ground through the via-holes [3,
15].

Fig. 4 Structure of
interdigital bandpass filter

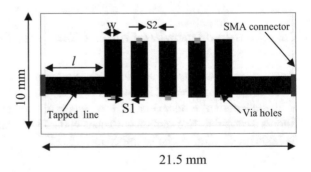

Fig. 5 Structure of combline bandpass filter

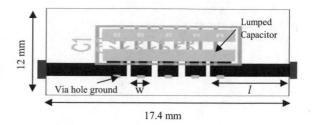

2.5 Design of Combline Filter

Figure 5 shows combline filter with width = 12 mm, length = 17.4 mm, l = 5 mm, C1 = C2 = C3 = C4 = C5 = 0.25pf, w = 1.8 mm, where 'l' is the length of the tapped line, 'C1….Cn' are the capacitance values, and 'w' is the width of the resonator. It consists of lumped capacitance at one end and short-circuited at another end. The lumped capacitors are usually used for narrowband filters [17].

3 Proposed Design

The microstrip bandpass filter design techniques designed in this paper are based on miniaturized structure and better frequency response [3, 4]. Chebyshev response is chosen for its passband characteristics. Considering these values, the dimensions of the structure are modeled based on the capacitances and inductances of the equivalent circuits [3]. The step-by-step discussion structure and simulation and its tuning of filter techniques are presented in Fig. 6. After determining the specifications of the filter, mathematical calculations are carried out using empirical formulae [2, 3]. Electromagnetic simulation is done in Genesys Keysight software. Genesys software is used for accurate frequency response of parameters S11 (return loss) and S21 (insertion loss) with good comparison performance of all microstrip bandpass filter techniques. In this context, some modifications were performed such as tunning and spacing between the resonators to get the exact required response. The required filter frequency response is done in two iterations. After iterations, the fractional bandwidth of the filter is in the range of 7.3–15.8%. After determining the prototype, the filter design is finalized.

4 Result and Discussion

Figure 7 above shows the simulated result in Genesys software. The filter structure is realized using a dielectric material RT/duroid 5880. Using Genesys software, the design is simulated, and the length, width, and spacing were calculated. At the

Fig. 6 Flowchart for
microstrip bandpass filter
techniques

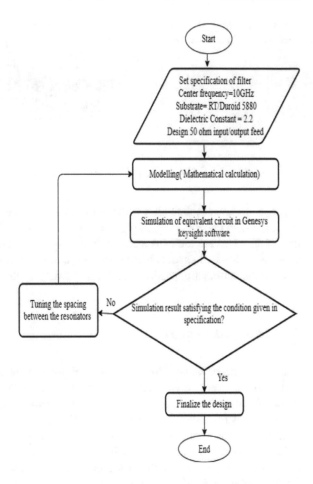

center frequency, the insertion loss (S21) is −0.73 dB, and the return loss (S11) is −39.523 dB.

The characteristics of the proposed parallel-coupled microwave bandpass filter are shown in Fig. 8. The bandpass filter is having an insertion loss of −1.496 dB across the passband, while the passband return loss is −42.837 dB across the required band 9.3–10.7 GHz.

Figure 9 shows the simulated result of the fifth-order hairpin bandpass filter. Comparing these results with end coupled and parallel coupled above in Fig. 7 and Fig. 8, it is noticeable that results are better for the following parameters. The insertion loss (S21) is −0.676 dB, and the return loss (S11) is −47.986 dB with a dimension reduction. Improved response characteristics are the reason why tapped structures are used. Hence, the hairpin bandpass filter is more preferred as compared to other filters.

Based on the above simulation, there is a minimum insertion loss of −0.795 dB and return loss of −41.275 dB at 10 GHz at the center frequency. It is observed that the

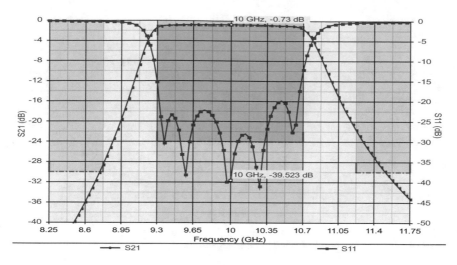

Fig. 7 Simulated result of fifth-order end-coupled bandpass filter

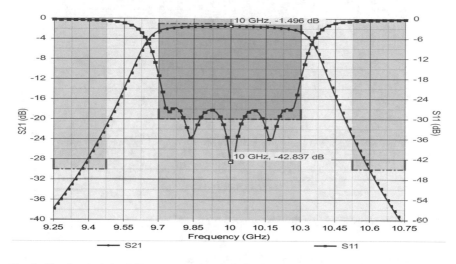

Fig. 8 Simulated result of fifth-order parallel-coupled bandpass filter

filter dimension is a bit larger than the hairpin filter. The frequency response increases slightly juxtapose to the hairpin filter, but it is much better than the end-coupled and parallel-coupled filter with a minimized dimension.

Figure 11 shows the simulation result that presents good S21 and S11 curves for the bandpass filter response. The center frequency is 10 GHz with an insertion loss of −0.597 dB and a return loss of −33.431 dB. It is observed that return loss and insertion loss have improved considerably with the miniaturized design when compared to other filters.

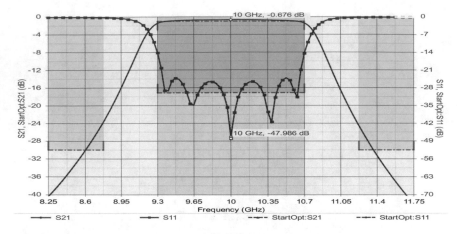

Fig. 9 Simulated result of fifth-order hairpin bandpass filter

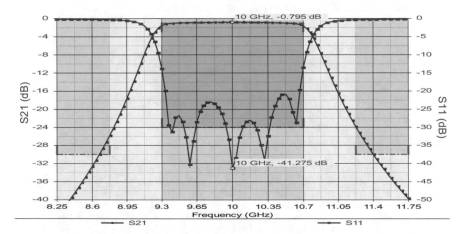

Fig. 10 Simulated result of fifth-order interdigital bandpass filter

Table 1 shows the comparison of techniques such as end-coupled, hairpin, interdigital, and combline bandpass filters. It is observed that hairpin filter structure and combline filter are more compact with a great improvement in frequency response and dimensions. This shows that the design can be easily developed to handle and permit configurability which can be easily integrated with various applications.

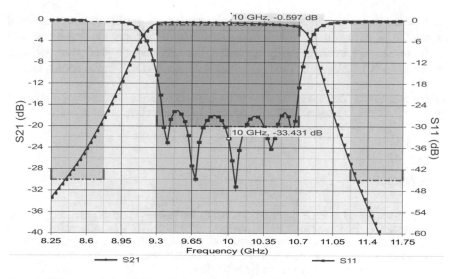

Fig. 11 Simulated result of fifth-order combline bandpass filter

Table 1 Comparison of end-coupled, parallel-coupled, hairpin, interdigital, and combline bandpass filter techniques from Genesys software simulation

Type of filter	Center frequency (f0) GHz	Fractional bandwidth (FBW)(%)	Insertion loss (S_{21}) In dB	Return loss (S_{11}) In dB	Dimension (mm) L × W
End coupled	10	15.8	0.73	39.523	80.5 × 12
Parallel coupled	10	7.3	1.496	42.837	36 × 27
Hairpin	10	15	0.676	47.986	20.2 × 10
Interdigital	10	15.5	0.795	41.275	21.5 × 10
Combline	10	15.7	0.597	33.431	17.4 × 12

5 Conclusion

This paper presents novel techniques to design microstrip bandpass filters for the X-band with RT/duroid 5880 substrate. The simulated results from Genesys Keysight software show the best outcome with improved fractional bandwidth, good narrow-band response, lower insertion loss, and higher return loss. In this paper, we have introduced simulated results of different types of microstrip bandpass filters with improved steeper roll-off and high selectivity. We observed that fractional band-widths of the filters in the range from 7.3% to 15.8% with average insertion loss and return loss of 0.8588 dB and 41.010 dB, respectively. Comparative analysis of the existing microstrip bandpass filter with the designed filter shows that combline filter with dimension 17.4 mm × 12 mm is the smallest among other filters with the least insertion loss of 0.597 dB and fractional bandwidth of 15.7%, but it requires lumped capacitance to reduce the resonator lines. However, hairpin filters have fractional bandwidth of 15% with an insertion loss of 0.676 dB without the requirement of via-holes. So, hairpin can be easily integrated, has relatively simple design, and is of low cost. Such a filter can be used for possible future RADAR-based applications.

Acknowledgements The author acknowledges the Keysight technology for providing student trial versions of Genesys Keysight software and SAMEER for precious support and guidance.

References

1. J. J. Jijesh, Shivashankar, S. K. Anusha, K. M. Guna, Rashmi, "Design and development of bandpass filter for X-band RADAR receiver system," 2017 2nd IEEE International Conference on Recent Trends in Electronics, Information & Communication Technology (RTEICT) (2017), pp. 2065–2070. https://doi.org/10.1109/RTEICT.2017.8256963.
2. D.M. Pozar, *Microwave Engineering* (Wiley, 2012)
3. W. Jia-Sheng, M. J. Lancaster, *Microstrip Filters for RF/Microwave Applications* (Wiley, 2001)
4. I.C. Hunter, *Theory and Design of Microwave Filter*. IET (2001)
5. I. Hunter, L. Billonet, B. Jarry, P. Guillon, Microwave filters-applications and technology. Microwave Theory Tech. IEEE Trans **50**, 794–805 (2002). https://doi.org/10.1109/22.989963
6. A. Grebennikov, *RF and Microwave Transmitter Design* (Wiley, 2011)
7. B. Adli, R. Mardiati, Y. Y. Maulana, "Design of microstrip hairpin bandpass filter for x-band radar navigation," 2018 4th International Conference on Wireless and Telematics (ICWT) (2018), pp. 1–6, DOI: https://doi.org/10.1109/ICWT.2018.8527781
8. C. Ping, Y. T. Li, Z. Y. Chuan, Y. W. Sheng, "Design of X-band receiver on airborne SAR/GMTI multi-model reconnaissance radar," 2011 3rd International Asia-Pacific Conference on Synthetic Aperture Radar (APSAR) (2011), pp. 1–4
9. H. Xu, W. Sheng, "The X-band microstrip filter design," 2017 7th IEEE International Symposium on Microwave, Antenna, Propagation, and EMC Technologies (MAPE) (2017), pp. 351–355. https://doi.org/10.1109/MAPE.2017.8250872
10. F.T. Ladani, S. Jam, R. Safian, A novel X-band bandpass filter using substrate integrated waveguide resonators. IEEE Asia-Pacific Conf. Appl. Electromagn. (APACE) **2010**, 1–5 (2010). https://doi.org/10.1109/APACE.2010.5720093
11. A. O. Lindo et al., Parallel and end coupled microstrip bandpass filters at W-band. 2009 Asia Pacific Microwave Conference (Singapore, 2009), pp. 345–348. https://doi.org/10.1109/APMC.2009.5385376
12. R. Ahmed, S. Emiri, Ş. T. İmeci, Design and analysis of a bandpass hairpin filter. 2018 International Applied Computational Electromagnetics Society Symposium (ACES) (Denver, CO, 2018), pp.1–2. https://doi.org/10.23919/ROPACES.2018.8364310
13. S. M. Kayser Azam, M. I. Ibrahimy, S. M. A. Motakabber, A. K. M. Zakir Hossain, "A Compact Bandpass Filter Using Microstrip Hairpin Resonator for WLAN Applications," 2018 7th International Conference on Computer and Communication Engineering (ICCCE), (2018), pp. 313–316. https://doi.org/10.1109/ICCCE.2018.8539247
14. B. Mohajer-Iravani, M. A. El Sabbagh, Ultra-wideband and compact novel combline filters. 2009 IEEE International Symposium on Electromagnetic Compatibility (Austin, TX, 2009), pp. 176–179. https://doi.org/10.1109/ISEMC.2009.5284644
15. D. R. Basavaraju, H. V. Kumaraswamy, M. Kothari, S. Kamat, In-terdigital bandpass filters for duplexer realization for LTE band 28. 2017 International Conference on Intelligent Computing and Control (I2C2). Coimbatore (2017), pp. 1–4. https://doi.org/10.1109/I2C2.2017.8321919

16. A. Makrariya, P. K. Khare, Design and analysis of a 5-pole inter-digital band-pass filter at 2.4GHz. 2015 International Conference on Communication Networks (ICCN) (Gwalior, 2015), pp. 93–96, https://doi.org/10.1109/ICCN.2015.19
17. M. Yuceer, A reconfigurable microwave combline filter. IEEE Trans. Circuits Syst. II Express Briefs 63(1), 84–88 (2016). https://doi.org/10.1109/TCSII.2015.2504010

A Novel Approach to Detect Plant Disease Using DenseNet-121 Neural Network

Nilesh Dubey, Esha Bhagat, Swapnil Rana, and Kirtan Pathak

Abstract The disease of crops is a major risk to food security and can incur a makeable loss to the people. But, the latest development in deep learning for solving this problem surpasses all the traditional methods in terms of efficiency, time period for detection and accuracy. In this paper, we came up with a rapid identification of leaf image and classify the image to correct class by using classical deep neural network architecture, DenseNet-121. This deep learning model has the ability to recognize 15 types of different plant disease, three of which are healthy ones, for better accurate results. The algorithm is highly optimized to produce results in less than 5 s after being fed into the system. The model's total testing accuracy for plant disease detection is 99%.

Keywords Plant disease · DenseNet-121 · Neural network · CNN · Agriculture

1 Introduction

India is the world's second largest agricultural producer, producing more than 280 million tons and accounting for more than 15% of India's GDP. Though the use of new technologies and equipment replaced nearly all the traditional methods of farming, but there are plenty of problems that farmers are been faced in his field of work like unpredictable climatic conditions and lack of support and awareness. Due to all of these issues, farmers' usage of chemical pesticides has increased, resulting in chemical build up in our soil, water, air, environment, animals, and even our own bodies.

One of the most important aspects of plant maintenance is disease detection because if disease is not discovered, the plant will not grow as it should. Farmers often do this by sight, but due to the rapid increase in complexity and variance in crop cultivation, it has become difficult for farmers, as well as agricultural professionals

N. Dubey (✉) · E. Bhagat · S. Rana · K. Pathak
Department of Computer Science and Engineering, Devang Patel Institute of Advance
Technology and Research, Faculty of Technology Engineering (FTE), Charotar University of
Science and Technology (CHARUSAT), Changa, India
e-mail: nileshdubey.ce@charusat.ac.in

and pathologists, to diagnose the disease, and it is also time-consuming. Tomatoes, potatoes, and peppers are the veggies that everyone eats because no cuisine can be imagined without them. However, these crops are now threatened by a variety of diseases, including early blight, septoria leaf spot in potatoes and tomatoes, and bacterial spot in peppers. Moreover, early detection of these agricultural diseases has the potential to reduce waste and control expenses, as well as improve product quality. Many classical technologies [1] such as biosensors have been proposed to enable simultaneous detection of many DNA or RNA sequences in a single reaction. Similarly, another thermography system catches emitted infrared radiation with thermo-graphic cameras and analyzes color differences. However, these systems have several drawbacks, such as the fact that it takes a long time to collect plant samples and test them physiologically in order to precisely diagnose the illness. This is where deep learning comes into play. It teaches a computer to filter inputs such as images, video, text, and so on through layers in order to learn how to predict and classify information without the need for human interaction. Though specific network topologies are utilized in deep learning, convolutional neural networks (CNNs) are commonly used in image recognition. CNNs are made up of four hidden layers:

(1)   Convolutional
(2)   Rectified linear unit (ReLU)
(3)   Pooling
(4)   Fully connected.

The principle and crucial features of images are extracted via convolution and then passed to the ReLU layer, where element-wise operations are conducted. The rectified feature map is then sent via a pooling layer, which decreases the dimensionality of the feature map and extracts prominent features, before the model learns and classifies the picture into the label using a fully connected layer and outputs a label. DenseNet-121 outperforms them all for image classification (leaf) because it is pretrained on the ImageNet dataset, which is the best training dataset for image classification.

2 Literature Review

The technology, which is currently used, includes spectroscopic, image-based, and volatile segregation-based methods for the detection of plant disease for the objective of generating a ground-based sensor system that helps to monitor the plant conditions [2].

Some procedure includes features retrieved from particle swarm optimization (PSO) [3, 4] and determine the affected cotton leaf spot and providing better final accuracy of 95%.

The apple (Malus domestica) disease detection [5] technique is implemented by incorporating k-means clustering and the Bayes classifying algorithm. The breakdown of texture is accomplished easily by using k-means clustering algorithm and co-occurrence matrix. The author carried out the PCA analysis, and the Bayes classification algorithm does the disease detection. PNN algorithms are used for the classification of plant diseases that use layers like competitive and radial. The leaves images are preprocessed and converted to gray scale before feeding into the system [6]. Another paper [7] addresses specific thresholding methodologies. The threshold techniques such as triangle and simple are used to differentiate the affected area and the leaf segment. The disease is classified in the final step using percentage calculation of the area of the lesion and leaf.

Soft computing techniques [8] are used to present generic image segmentation algorithms. The grayscale leaf image, SVM classifier, and minimal distance criteria are used for image classification. One paper [9] proposes a framework for rapid and precise recognition of plant diseases using artificial neural networks and with other techniques involving image processing. The end system provides better, accurate results of up to 91%. The detection of plant disease by large deep neural network structures, such as AlexNet and GoogleNet [10], is carried out on a dataset with 38 class labels. In the case of AlexNet trained from scratch, the accuracy varies considerably from 85.53% to that of GoogLeNet by transfer learning, which has accuracy 99.34%, obtained from the Plant Village dataset.

The PlantVillage image dataset was used to train the model to classify 14 species of plants and 26 diseases. They carried out an evaluation of two new sets of data of 121 and 119 images, found on the Internet, after getting a testing accuracy of 99.35% on the PlantVillage dataset [11]. They got 31.40% accuracy for the first dataset and 31.69% for the second dataset. Challenges and inadequacies of works that used convolutional neural networks (CNNs) to detect specific crop diseases [12].

CNN architectures such as InceptionV3, VGG16, DenseNet-121, and ResNet50 for image processing are grounded on various characteristics that are present mainly on the leaves infected with anthracnose compared to the non-diseased leaves. The accuracy ranges from 92.4% to 98.7% for identifying brown spots on leaves [13]. One proposed approach implemented the k-means clustering algorithm for grape disease detection. It configured an SVM classifier based on 31 viable particular features to recognize grape powdery and grape downy mildew diseases with 90 and 93.33% as testing accuracies [14].

3 Methodology

3.1 Experimental and Technical Design

The full process of developing the model for plant disease detection using CNN is detailed. The entire operation is divided into various necessary stages in the subsections below, beginning with obtaining photographs from the Internet for classification.

Dataset. Proper datasets are required at all stages of object detection and recognition research, from the training phase to the evaluation of recognition algorithms' performance. All of the photos from the dataset were obtained from the GitHub repository [15]. The images in the dataset are categorized into 21 different classes, and we have chosen 15 of them. To make the model function, we included these 15 classes in our dataset. There are two types of classes in the dataset: 13 for disease classification and 2 for healthy detection. As a result, 20,638 photos have been collected for the model.

Next step was to make the images in the dataset randomize using NumPy and pandas with a SEED of 42 and adding two more attributes "Disease Type" and "Disease ID." After this, every class has given an ID for better classification and ease of programming. By this, every image will have two attributes.

(i) Disease ID.
(ii) Disease Type.

The dataset can be visualized using a histogram (Fig. 1). Here is the diagrammatic representation of whole dataset is shown.

Displaying the images of particular class using Python script (Fig. 2). The script will detect all the images of specific class and disease present in that crop. Here, we have displayed "Tomato Plant," and the disease selected was "Tomato Bacterial Spot."

Image Processing. Image processing is always required for every model in machine learning and deep learning. It depends on the system which one is using; our system can handle above size very smoothly and efficiently. Following will be classes for the disease detection:

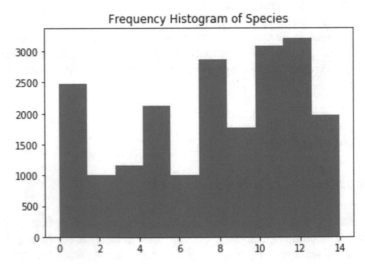

Fig. 1 X-axis = Disease ID, Y-axis = Frequency of disease. Bins of the chart represent the number of times a particular disease ID occurs in the dataset

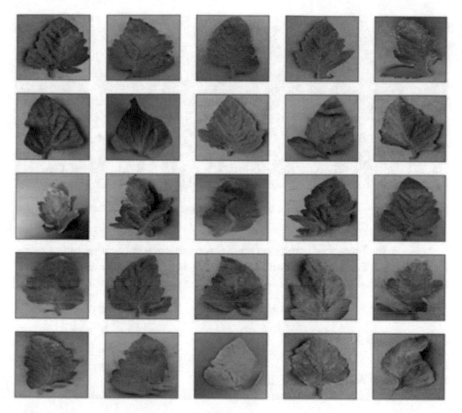

Fig. 2 Plant tomato having tomato bacterial spot disease

(1)   Pepper bell bacterial spot.
(2)   Pepper bell healthy.
(3)   Potato early blight.
(4)   Potato late blight.
(5)   Tomato bacterial spot.
(6)   Potato healthy.
(7)   Tomato Early Blight.
(8)   Tomato Late Blight.
(9)   Tomato Leaf Mold.
(10)   Tomato Spetoria Leaf Spot.
(11)   Tomato Spider mites two Spotted spider mites.
(12)   Tomato Target Spot.
(13)   Tomato Tomato YellowLeaf Curl Virus.
(14)   Tomato mosaic Virus.
(15)   Tomato Healthy.

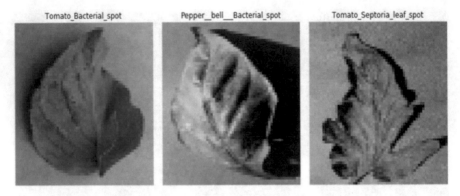

Fig. 3 Images converted to 64 × 64 resolution using image processing techniques

In Fig 3, the images of resolution are 64 × 64 that are to be fitted in the model. We can see that the images are quite blurry which might affect the training of model, but due to limited computation resource, we decided to make it 64 × 64 (Fig. 3).

4 Experimental Results

4.1 Neural Network Training

Dataset pictures were divided into two sets, training and validation datasets with a ratio of 8:2. Using the image instances, we developed a convolution neural network (CNN) based on DenseNet-121 which accepts a three channels input image of 64 × 64 × 3 resolution and returns a 15-dimensional vector where this vector is generated using soft-max activation function. Also, we had added global max pooling which extracts best possible features of the image. Using the global max pooling costs us less in computation level as it only takes the maximum values from a square matrix. Batch normalization is being used because it solves the problem called "internal covariate shift" and to the speed of training. We had used three kernels of size 1 × 1 to convolute the input layer image. We decided to use this network architecture because its rationalize the pattern of relationship between layers which was present in introduced architectures and pretrained on the ImageNet dataset which is one of the most preferable dataset when it comes to object detection and recognition:

(1) Highway networks
(2) Residual networks
(3) Fractal networks.

We solved the problem by assuring maximal information (and gradient) flow. DenseNets utilizes the network's potential through feature reuse by logically connecting every layer directly with each other. Instead of pulling representational power from exceptionally deep or wide topologies, we rationally connect every layer directly with each other. It provides benefits such as strong gradient flow, parameter and computational efficiency, maintains low complexity features, and so on. The architecture of DenseNet-121 is depicted in the image below.

Dense block and transition block. Due to heavy number of connection in the DenseNets, the conceive gets less simple compared to other CNN like VGG and ResNets. The below figure demonstrates a straightforward schematics on the detailed architecture of DenseNet-121. It is the simplest DenseNet in terms of complexity among those designed over the ImageNet dataset. In Fig. 4, the measures under each volume represent the size of the width and depth. Figure 5 will tell more about upcoming structure in the architecture.

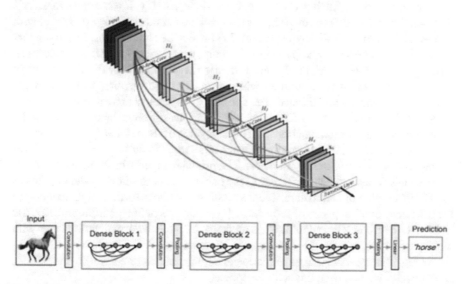

Fig. 4 Schematic diagram of DenseNet-121 architecture with dense block, convolution block, and transition layer

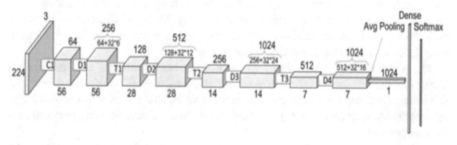

Fig. 5 Detailed structure of DenseNet-121

In Fig. 5 D1, D2 up to D4 represents dense block, and T1, T2 up to T3 represents transition block. There are bypass connections all over the network.

Moreover, image augmentation was used for better classify the unseen image and to increase the training accuracy. Convolution training was done using two Python libraries, namely Keras and TensorFlow back end, which is a framework, especially designed for deep learning purposes. Pixel values of every input images were normalized by dividing each and every value by 255 so that they lie within [0.0,1.0]. Moreover, the network was initialized with weights derived from ImageNet dataset which is world's largest collection of images. However, with the help of a categorical cross-entropy, loss metric model is compiled, and weights of the layers were optimized using optimizer called the Adam optimization algorithm having a learning rate of 0.02. A collection of 64 images, each having size of 64 × 64, were provide to the network as single batch per iteration. Considering timing, it was approximately 1 to 2 min for each epoch out of 70 in our experimental condition (Kaggle kernel with GPU-enabled mode). These hyper-parameters are decided after tuning the parameters with range of values and comparisons of accuracy and precision. After a successful training of the CNN model, the feature extraction layers which are from convolute to global max pooling layer were improvised to detect major features from the image in order to diagnose plant disease. The summary of model is shown in Fig. 6.

As the Fig. 6 elicits the information regarding the different layers in the model with respective activation functions, the input layer is of shape 64 × 64 × 3; after that hidden layer is consisting of multiple layers of neurons including convolution layer of shape 64 × 64 × 3 with 12 parameters. The principle layer in the hidden layer is the DenseNet-121 layer which can be say as the backbone of the whole model consists of 7,037,504 number of parameters. Then, a pooling and normalization layer to increase the efficiency and last part is of fully dense layer consists of 300 neurons and 307,500 number of neurons. The output layer is preceded by a batch normalization layer with 1200 parameters, and itself has a shape of 15 neurons and 4515 parameters.

As Fig. 7 suggests that final validation accuracy and loss are 99.00% and 0.02%, respectively. We achieved this with 70 epochs.

Performance metrics and testing. Essentially, the matrix was used to assess the performance of a classifier that had been pretrained with new images from the testing dataset. These results are then used to assess the performance metrics that are commonly employed in AI classification algorithms. Table 1 shows the performance indicators utilized in our study to evaluate the performance of our classifier, along with their descriptions and mathematical formula.

The confusion matrix is a table that displays the abovementioned numbers in such a way that the number of correctly categorized cases, as well as false positives and false negatives, can be easily seen. In our example, the matrix will correctly and incorrectly identify diseases. Figure 8 depicts the matrix created using performance metric approaches.

Some more details about the model insights are given in the following two graphs (Fig. 9).

```
Downloading data from https://github.com/keras-team/keras-applicat
29089792/29084464 [==============================] - 0s 0us/step
Model: "model_1"
_________________________________________________________________
Layer (type)                 Output Shape              Param #
=================================================================
input_2 (InputLayer)         (None, 64, 64, 3)         0
_________________________________________________________________
conv2d_1 (Conv2D)            (None, 64, 64, 3)         12
_________________________________________________________________
densenet121 (Model)          multiple                  7037504
_________________________________________________________________
global_max_pooling2d_1 (Glob (None, 1024)              0
_________________________________________________________________
batch_normalization_1 (Batch (None, 1024)              4096
_________________________________________________________________
dropout_1 (Dropout)          (None, 1024)              0
_________________________________________________________________
dense_1 (Dense)              (None, 300)               307500
_________________________________________________________________
batch_normalization_2 (Batch (None, 300)               1200
_________________________________________________________________
dropout_2 (Dropout)          (None, 300)               0
_________________________________________________________________
root (Dense)                 (None, 15)                4515
=================================================================
Total params: 7,354,827
Trainable params: 7,268,531
Non-trainable params: 86,296
```

Fig. 6 Model summary with parameter evaluation and segmentation

The left graph shows representation of epoch's vs loss in training as well as testing phase. It can be clearly see that the loss during training was very high and gets decreased constantly with increase in epochs. At the maximum epochs of 70, losses are very near to 0.0. Considering the testing phase, loss was more than loss occurred in training phase, and it also gets decreased by oscillating as number of epochs increase. The decrement of loss is because of the Adam optimizer we have applied; it tries to improve the network better and better while learning it. Similarly, with the right graph, it can be clearly see that the accuracy during training was very low, approx. 0.4 and gets increased constantly with increase in epochs. At the maximum epochs of 70, the accuracy is very near to 0.0. Same optimizer was used in this phase.

5 Conclusion

Plant diseases are major food threats that ought to be overcome before it results in further loss of the whole field. But, often farmers confuse between diseases and that leads to destruction of the whole field. Due to this dilemma, farmers apply small

```
Downloading data from https://github.com/keras-team/keras-applicat
29089792/29084464 [==============================] - 0s 0us/step
Model: "model_1"
_________________________________________________________________
Layer (type)                 Output Shape              Param #
=================================================================
input_2 (InputLayer)         (None, 64, 64, 3)         0
_________________________________________________________________
conv2d_1 (Conv2D)            (None, 64, 64, 3)         12
_________________________________________________________________
densenet121 (Model)          multiple                  7037504
_________________________________________________________________
global_max_pooling2d_1 (Glob (None, 1024)              0
_________________________________________________________________
batch_normalization_1 (Batch (None, 1024)              4096
_________________________________________________________________
dropout_1 (Dropout)          (None, 1024)              0
_________________________________________________________________
dense_1 (Dense)              (None, 300)               307500
_________________________________________________________________
batch_normalization_2 (Batch (None, 300)               1200
_________________________________________________________________
dropout_2 (Dropout)          (None, 300)               0
_________________________________________________________________
root (Dense)                 (None, 15)                4515
=================================================================
Total params: 7,354,827
Trainable params: 7,268,531
Non-trainable params: 86,296
```

Fig. 7 Final loss and validation accuracy

Table 1 Performance matrix formula

Measure	Formula
Accuracy	(TP + TN)/(TP + TN + FP + FN)
Misclassification rate (1—accuracy)	(FP + FN)/(TP + TN + FP + FN)
Sensitivity (or recall)	TP/(TP + FN)
Specificity	TN/(TN + FP)
Precision (or positive predictive value	TP/(TP + FP)

or excess amounts of pesticides and other chemicals. Here, we employ a convolution neural network (CNN) having multiple layers of ANN called deep learning algorithms to scale back this loss and guide farmers with suggestions.

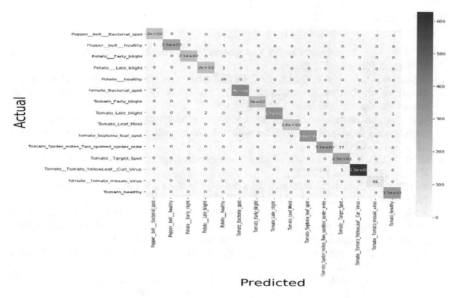

Fig. 8 Confusion matrix of model

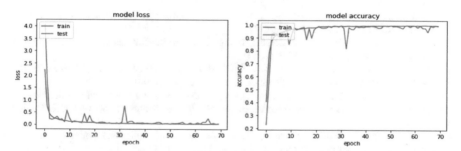

Fig. 9 Model loss (left) and model accuracy (right) produced at each epoch in training as well as testing

References

1. Y. Fang, R.P. Ramasamy, Current and prospective methods for plant disease detection. Biosensors **5**(3), 537–561 (2015)
2. S. Sankaran, A. Mishra, R. Ehsani, and C. Davis, "A review of advanced techniques for detecting plant diseases," Comput. Electron. Agricult. **72**(1), 1–13 (2010); F. Author, S. Author, T. Author, Book Title. 2nd edn. Publisher, Location (1999)
3. P. Revathi, M. Hemalatha, Identification of cotton diseases based on cross information gain deep forward neural network classifier with PSO feature selection. Int. J. Eng. Technol. **5**(6), 4637–4642 (2014)
4. C. Zhou, H.B. Gao, L. Gao, W.G. Zhang, Particles warm optimization (PSO) algorithm. Appl. Res. Comput. **12**, 7–11 (2003)
5. B. Sabah, N. Sharma, "Remote area plant disease detection using image processing". IOSR J. Electron. Commun. Eng. **2**(6), 31–34 (2012), ISSN: 2278–2834
6. S. G. Wu, F. S. Bao, E. Y. Xu, Y.-X. Wang, Y.-F. Chang, Q.-L. Xiang, "A leaf recognition algorithm for plant classification using probabilistic neural network". 2007 IEEE International Symposium on Signal Processing and Information Technology (2007)
7. S.B. Patil et al., Leaf disease severity measurement using image processinG. Int. J. Eng. Technol. **3**(5), 297–301 (2011)
8. V. Singh, A.K. Misra, Detection of plant leaf diseases using image segmentation and soft computing techniques. Inf. Process. Agricult. **4**(1), 41–49 (2017)
9. H. Kulkarni Anand, R.K. Ashwin Patil, Applying image processing technique to detect plant diseases. Int. J. Mod. Eng. Res. **2**(5), 3661–3664 (2012)
10. S. P. Mohanty, D. Hughes, M. Salathe, *"Using Deep Learning for Image-Based Plant Disease Detection,"* in arXiv:1604.03169[cs.CV]
11. S.P. Mohanty, D.P. Hughes, M. Salathé, Using deep learning for image-based plant disease detection. Front. Plant Sci. **7**, 1419 (2016)
12. J. Boulent, S. Foucher, J. Théau, P.-L. St-Charles, Convolutional neural networks for the automatic identification of plant diseases. Front. Plant Sci. **10**, 941 (2019)
13. A. Anagnostis, G. Asiminari, E. Papageorgiou, D. Bochtis, A convolutional neural networks based method for anthracnose infected walnut tree leaves identification. Appl. Sci. **10**, 469 (2020)
14. M. Ji, L. Zhang, Q. Wu, "Automatic grape leaf diseases identification via UnitedModel based on multiple convolutional neural networks", Inf. Process. Agricult.
15. https://github.com/spMohanty/PlantVillage-Dataset/tree/master/raw/color

Offline Handwritten Signature Forgery Verification Using Deep Learning Methods

Phan Duy Hung, Pham Son Bach, Bui Trong Vinh, Nguyen Huu Tien, and Vu Thu Diep

Abstract Offline signature verification is one of the most challenging tasks in biometric authentication. Despite recent advances in this field using image recognition and deep learning, there are many remaining things to be explored. The most recent technique, which is Siamese convolutional neural network, has been used a lot in this field and has achieved great results. This paper presents an architecture that combines the power of Siamese Triplet CNN and a fully-connected neural network for binary classification to automatically verify genuine and forgery signatures even if the forged signature is highly skilled. On the challenging public dataset for signature verification BHSig260, the proposed model can achieve a low False Acceptance Rate = 13.66, which is slightly better than the reference model. Based on this approach, the one-shot learning should make it possible to determine if the input image is genuine or fraudulent just from one base image. Therefore, our model is expected to be extremely suitable for practical problems, such as banking systems or mobile authentication applications, in which the amount of data for each identity is limited in quantity and variety.

Keywords One-shot learning · Offline signature verification · Siamese convolutional neural network · Triplet loss

P. D. Hung · P. S. Bach · N. H. Tien
FPT University, Hanoi, Vietnam
e-mail: hungpd2@fe.edu.vn

P. S. Bach
e-mail: bachpshe130246@fpt.edu.vn

N. H. Tien
e-mail: tien19mse13047@fsb.edu.vn

B. T. Vinh
Hanoi Procuratorate University, Hanoi, Vietnam
e-mail: vinhbt@tks.edu.vn

V. T. Diep (✉)
Hanoi University of Science and Technology, Hanoi, Vietnam
e-mail: diep.vuthu@hust.edu.vn

© The Author(s), under exclusive license to Springer Nature Singapore Pte Ltd. 2023
Y.-D. Zhang et al. (eds.), *Smart Trends in Computing and Communications*, Lecture Notes in Networks and Systems 396, https://doi.org/10.1007/978-981-16-9967-2_8

1 Introduction

A signature is a handwritten depiction of a person's name, nick name or symbol. The traditional function of signature is to permanently affix to a document, which plays a role as a physical evidence of the author's personal witness and certification of the content [1]. One of the explanations for its widespread use is that it is simple, fast, non-invasive and other people are acquainted with it in everyday life. Signature verification system aims to automatically discriminate if a biometrics sample is indeed of a claimed individual. In other words, it is wont to classify if a signature is genius or fraud. The fraud signatures, which are called forgeries, are commonly classified in three types: random (blind) forgery, simple forgery and skilled forgery. Based on the signature acquisition method, most recent signature verification systems are categorized in two types: online (dynamic) signature verification and offline (static) signature verification [2]. In the online method, the user's signature is acquired by using an acquisition device like a digitizing table. The online signatures are collected as a sequence over time and contain numerous information, such as pen position, pen inclination, pressure. On the other hand, signatures acquired from the offline method are the digital images of the user's signatures signed in the document after the writing process has completed. When the online method can contribute more diverse features for the verified process, it requires significant devices and techniques to acquire data, which makes the system more expensive and cumbersome. On the other hand, the data from offline methods are easier to obtain, and the image digital signatures are more suitable for practical problems. However, the lack of informative features compared to the other method and the quality instability of image data make the verification process more challenging.

The problem of offline signature verification is often modeled as follows: given a group of genuine signatures of users, a model is trained to extract meaningful features from them. Then, the model is employed for verification: a user claims their identity and provides one or some new signatures, which can be employed by the model to classify those signatures as genuine (belong to the claimed individual) or forgery (created by someone else).

In signature verification, one of the most important and familiar challenges for this task is that handwritten signatures have high intra-class variability, which means there are always differences between each handwritten signature from the same person [2]. This issue leads to an aggravation when considering skilled forgeries, especially with the presence of low inter-class variability in the dataset. In short, when there are large differences between each signature sample (high intra-class variability) and the forgeries are very skillful that they are nearly the same with the genuine (low inter-class variability), the possibility of confusing the forgeries with the genuine is higher.

The second challenge when training an automatic signature verification system comes from the presence of particle knowledge during training. In the realistic scenario, we have got only to access genuine signatures for users enrolled within the system during the training phase, in line with [3]. During operations, however, we would like the system to not only be able to accept the real, but also to reject forgeries. To solve that the forgery signatures for each user should be required for a better classification. But in general, it is not reasonable to require users to provide signatures that forge their own. Even if they can be collected and created from the service provider or from a third party, it would be a challenge to create a good forgeries data which is good enough to be absolutely distinguishable from the genuine signatures.

Thirdly, the amount of training data is always being considered. During the enrollment process in the real application, users are often required to produce only some samples of their signatures. In the meantime, some approaches like the Writer Independent approach need a large enough amount of data to perform well. That's why it is hard to create an efficient system with the very limited data collected from enrolled users in a very real case scenario. Whether or not there's an outsized number of users supplying their signatures to the system during the training phase, the performance of the classifier has to be superb for the new user, for whom supply only few samples of signatures.

Last but not least, like the face recognition system, an effective real-world signature verification system should be able to deal with the one-shot learning problem. Which means the verification application should instantly classify the genuine and fraudulent signatures of a user from just one genuine signature of that individual. However, deep learning algorithms do not work well with a very small dataset, especially if you only have one training example.

In general, signature verification systems have two phases: feature extraction and classification. In the first phase, the feature obtained from the extraction must be meaningful enough to represent the distinctive features of each individual (Writer Depend approaches) or show the clear difference between the genuine and forged signatures (Writer Independent approaches). Then, based on those features, a classifier will be used to make the final prediction.

In 2008, Impedovo et al. [1] provided a summary in automatic signature verification, with specific attention to most outstanding advancements in machine learning. Then, Shah et al. [2] presented a survey about the critical evaluation of 15 techniques applied on offline signature verification systems, which classify each work according to the feature extraction methods, classifiers and overall strengths and limitations of the systems. However, these reviews do not update to capture more recent trends, specifically the usage of deep learning methods, which have demonstrated superior results in multiple benchmarks.

Comprehensive surveys and state-of-the-art reviews of the recent literature are often found within the works of [3, 4] in step with these reviews, most deep learning methods for this task can archive results with high accuracy on an outsized dataset. Beside, among numerous researchers, a growing number of researchers have realized that "learning with small datasets" are a key factor for the success of signature verification in real practical scenarios. Authors like Bouamra et al. [5] attempt to train the model with a tiny low number of signature samples through a one-class support vector machine (OC-SVM) classifier. Hafemann et al. [6] propose an answer supported a meta-learning approach.

In recent years, many interesting techniques do not depend on hand-engineered feature extraction anymore. With the increase of the many deep learning approaches, the power to execute more meaningful feature representation from data (like pixels, just in case of images) by itself can promote faster and efficient learning, which helps researchers to save lots of significant amounts of work. In this field, Khalajzadeh et al. [7] applied CNNs for Persian signature verification, but only considered random forgeries in their tests.

Koch et al. [8] firstly propose an approach about Siamese network and one-shot learning for image recognition. This approach outperformed many available methods by previous authors. Moreover, they have argued that this approach can extend in other domains, especially for image classification.

Schoff et al. [9] came up with the same idea of applying the Siamese network, but in a different way. In this research, they propose a new system called FaceNet, which learns a way to map face images to a compact Euclidean space where distances directly correspond to a measure of face similarity. They achieved high accuracy (95.12% ± 0.39) in their published dataset, and also introduced new concepts about harmonic embeddings and harmonic triplet loss to deal with the face classification task.

The above approaches [8, 9] both apply Siamese structure including deep CNNs for getting a new representation of image feature, which will be used to classify tasks by comparing them over their distance between each other in new vector space. Those techniques are very flexible and do not need a dataset with millions of samples to make them work.

Dey et al. [10] applied convolutional Siamese network with contrastive loss function for writer independent offline signature verification and their system performs very well on cross dataset for this task. In addition, their approach has surpassed the state-of-the-art result on most of the benchmark signature datasets, which is encouraging this technique for further research. However, when performing evaluation across different datasets, the accuracy of their system gradually decreases on the datasets which have more distinctive features compared to the training dataset.

Chhabra et al. [11] have constructed an interesting method based on the convolutional Siamese net architecture from FaceNet [9] and triplet loss concept to solve the one-shot problem for offline signature verification. Their model has high accuracy and generalizability on the public signature database. However, they do not show the evaluation result in various databases and the performance comparison with other state-of-the-art methods.

Inspired by the work of Chhabra et al. [11], this study improves the network architecture of Deep Triplet Ranking CNN Network by applying more robust transfer learning models and more efficient classification methods. The work focuses mainly on forgery detection in signature verification systems with one-shot learning strategy, which means our model may not perform well on recognition identity from signature but to eliminate forgeries as much as possible. The model is evaluated on the dataset BHSig260, the challenging public dataset for signature verification.

The rest of this paper is organized as follows: In Sect. 2, we are going to give an outline of the general public dataset used, and present the steps of preprocessing.

Section 3 describes the research methodology. Section 4 includes experimental results and comparing them with other approaches. Finally, in Sect. 5, we summarize the result and define the longer term work.

2 Data Preparation

In this work, the signature dataset BHSig260 is used. These data are public, free to access, and contain skilled forgeries. BHSig260 dataset contains the offline signatures of 260 persons, in which 100 persons were signed in Bengali and the remaining were signed in Hindi. For each of the signers, 24 genuine and 30 forged signatures are available. This results in 2400 genuine signatures and 3000 forged signatures in Bengali, and 3840 genuine and 4800 forged signatures in Hindi. Our experiment is only in the Bengali signatures datas from this dataset.

Through analyzing the characteristics of the signature image, we have carried out the following preprocessing steps to improve the efficiency of the recognition process:

- The color factor is not important when comparing the dissimilarity of genuine and fraudulent signatures. That is why we convert all signature's image to grayscale.
- Using a median filter with kernel size 3×3, most salt and pepper noise are easily eliminated.
- After removing noise, using the Otsu threshold to convert the gray images to be the binary image. In addition, invert to the negative image so the background become zero and only the signature's area has pixel value 255.
- Finally, resizing all signature images to a fixed size for feeding into the network 128×128.

3 Research Methodology

The architecture proposed in Fig. 1 is modified from the architecture introduced by Chhabra et al. [11]. We uses 3 sysmetrics deep convolutional neural networks played as the subnetworks in the Siamese architecture. *Instead of building the new deep CNNs and training it from scratch, the Xception model, a pre-trained model in ImageNet is used for three branches.* The transfer learning strategy is used to reduce time for the training process: firstly, the model with the weights trained on ImageNet dataset will be removed the fully-connected layers on top, then all layers are frozen except the exit flow for train on the new prepared dataset. This strategy helps our model have the ability to extract the high feature representation of the image (from the freezed layers) while still being able to learn the important features from the signature image in new dataset. In this case, the Xception was chosen because of its superiority over models VGG-16, ResNet and Inception V3 and has been summarized

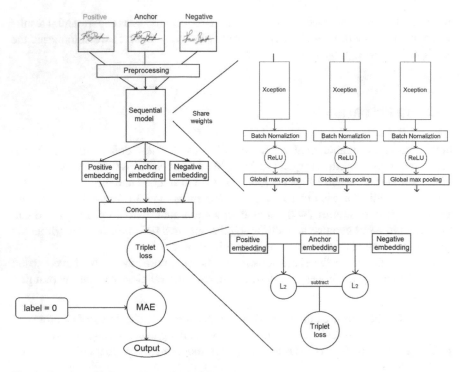

Fig. 1 Our Deep triplet ranking CNN architecture with Xception

in [12]. Moreover, it has fewer parameters than Inception V3, which can save more memories on the training and deployment processes.

In this design, our model accepts the input as three signature images categorized as: anchor (an individual's genuine signature), positive (another genuine signature of the same individual) and negative (a forged signature of the same individual). After the preprocessing, those triplet images are fed through the triple subnetworks in which each is constructed by transfer learning Xception top-up with a stack of a Batch Normalization, ReLU activation function and a global max pooling layer.

After being concatenated, our network uses a Lambda layer to compute two distances between embedding vectors: "positive-anchor", "anchor-negative". And the last calculates the triplet loss. Of course triplet loss optimization will continue to update the weights of the sequential part so that the embedding images have adjusted distance to become closer if they are genuine, or become further away from the forgeries.

In the next phase of genuine-forgery classification, we continue to use the idea of using distance vectors for the classifiers of weighted authors in [11] (Fig. 2), but replace the logistic regression classifier with a new classifier stack. The new binary classification includes a fully-connected layer and ReLU activation function, which is top-up with sigmoid function to calculate the probability that a signature is forged or genuine.

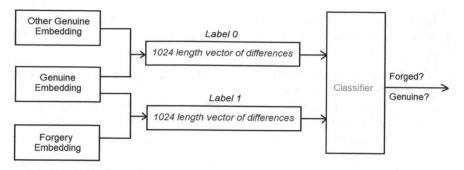

Fig. 2 Classification flow illustration

Table 1 Pairs labeling method for train and test

Set(Train/Test)	Pair	Label
Train users	Genuine—Genuine	0
	Genuine—Forgery	1
Test users	Base genuine—Genuine	0
	Base genuine—Forgery	1

4   Research Experiment

4.1   Experimental Setup

First of all, we split 20% of users and take their signature images because the test data. When forming the triplet combinations to train the triplet net, we take an anchor image (genuine signature of a person) and placing it in conjunction with both a positive sample (another genuine signature of the identical person) and a negative sample (a forged signature by some other person of the identical person). By that strategy, from dataset BHSig260 Bengali, we have 654,120 triplet combinations for training. The training pairs for the classifier are formed after step of embedding of all signatures using the train triplet net. We have the vector pairs, label them (see Table 1) and feed them to the classifier.

4.2   Results

The triplet net is trained by using Adam Optimizer with the learning rate equal to 1e-5 and mean absolute error for backpropagation to get the final encoding. The triplet combination is randomly split into a train set and validation set with ratio 2:1, respectively.

In this work, the positive class is the forged signature and the negative class is the genuine signature. We assess or compare the performance of system based on the following values:

- Area Under the Curve (AUC): the measure of the power of a classifier to differentiate between classes.
- False Acceptance Rate (FAR): the percentage of identification instances within which unauthorized persons are incorrectly accepted. FAR = FNR = FN/(FN + TP) = 1—recall. This parameter is that the most important during this study, the smaller the better.
- False Rejection Rate (FRR): the percentage of identification instances within which authorized persons are incorrectly rejected. FRR = FPR = FP / (FP + TN).
- Equal Error Rate (EER): The intersect of lines FAR, FRR.

The Tables 2 and 3 show the performance of the final model for detecting forgery and genuine signatures:

Beside, we have the performance on the test set in case of the pair Base Genuine— Forgeries is no longer depend on identity anymore, which means we construct the pair from the base genuine signature with all forgery signature in the dataset to get more positive label (the forged signature) for testing (Table 4).

The result from Table 4 can be used to compare with some state-of-the-art and the recent studies with Siamese network in the following Table 5:

When the solution is used in real applications, only one genuine signature image is necessary for new user. The image will be preprocessed and feed through the encoder to get the base embedding vector. This vector will be stored in the database and be marked by an id to query in future.

When a new signature image is attempted on the system, if the owner of the new signature claims to be anyone in the system by promoting the id, the system will use that id to track and query the base embedding out for comparison. The system then uses the encoder to get the embedding of the new signature image, and get

Table 2 Performance on validation set

AUC (%)	ERR	FAR	FRR
99.99	0.13	0.00	0.00

Table 3 Performance on test set

AUC (%)	ERR	FAR	FRR
86.16	22.75	18.57	27.97

Table 4 Performance on the test set with random forgeries by other user forged signatures

AUC (%)	ERR	FAR	FRR
99.34	14.18	13.66	27.97

Table 5 State-of-the-art performance on BHSig260 dataset (WD = Writer Dependent, WI = Writer Independent)

Language	Type	Features and algorithm	FRR	FAR
Bengali	WI [10]	SigNet	13.89	13.89
	WI [13]	Dutta et al	14.43	15.78
	WI [14]	Pal et al	33.82	33.82
	–	Our model	27.97	13.66

the vector's difference between the embedding of the base signature and the new signature. Lastly, the classifier will use this vector's difference as an input to predict the probability if the new image is a forged signature or not. The higher the value of this predicted result, the higher the probability that the new signature is a fake one.

5 Conclusions and Perspectives

The paper proposes an effective method for verifying the offline handwritten forgery signatures based on the improvement of a highly applicable study. The article has performed image preprocessing to improve recognition efficiency, improve in processing steps such as using transfer learning strategy and choosing a suitable machine learning architecture model for the problem.

From the result, we can achieve a good FAR result on the BHSig260 dataset which is slightly better than the reference results.

In the future, the study need to find more suitable preprocessing methods for different kinds of datasets. The model also need to be tuned hyper parameters of more powerful triplet ranking loss, and may need to apply triplet online mining techniques to let the learning process be better.

References

1. D. Impedovo, D. Pirlo, Automatic signature verification: the state of the art. IEEE Trans. Syst. Man Cybernetics Part C (Appl. Rev.), **38**(5), 609–635 (2008)
2. A.S. Shah, M.N.A. Khan, A. Shah, An appraisal of off-line signature verification techniques. Int. J. Modern Educ. Comput. Sci. **4**, 67–75 (2015)
3. L.G. Hafemann, R. Sabourin, L.S. Oliveira, Offline handwritten signature verification - Literature review. In: Proceedings of the Seventh International Conference on Image Processing Theory, Tools and Applications (IPTA), Institute of Electrical and Electronics Engineers (IEEE) (Montreal, QC, Canada, 2017), pp. 1–8
4. M. Diaz, M.A. Ferrer, D. Impedovo, M.I. Malik, G. Pirlo, R.A. Plamondon, Perspective analysis of handwritten signature technology. ACM Comput. Surv. **51**, 1–39 (2019). https://doi.org/10.1145/3274658

5. W. Bouamra, C. Djeddi, B. Nini, M. Diaz, I. Siddiqi, Towards the design of an offline signature verifier based on a small number of genuine samples for training. Expert Syst. Appl. **107**, 182–195 (2018). https://doi.org/10.1016/j.eswa.2018.04.035

6. L.G. Hafemann, R. Sabourin, L.S. Oliveira, Meta-Learning for fast classifier adaptation to new users of signature verification systems. IEEE Trans. Inf. Forensics Secur. **15**, 1735–1745 (2020). https://doi.org/10.1109/tifs.2019.2949425

7. H. Khalajzadeh, M. Mansouri, M. Teshnehlab, Persian signature verification using convolutional neural networks. Int. J. Eng. Res. Technol. **1** (2012)

8. G. Koch, R. Zemel, R. Salakhutdinov, Siamese neural networks for oneshot image recognition. In: ICML, pp. 1–8 (2015)

9. F. Schroff, D. Kalenichenko, J. Philbin, Facenet: A unified embedding for face recognition and clustering. In: Proceedings of the IEEE Conference on Computer Vision and Pattern Recognition (CVPR) (2015), pp. 815–823

10. S. Dey, A. Dutta, J.I. Toledo, S.K. Ghosh, J. Llados, U. Pal, *SigNet: Convolutional Siamese Network for Writer Independent Offline Signature Verification*. arXiv:1707.02131v2 (2017)

11. O. Chhabra, S. Chakraborty, Siamese triple ranking convolution network in signature forgery detection. In: Proceedings of the Alliance International Conference on Artificial Intelligence and Machine Learning (AICAAM) (2019)

12. F. Chollet, *Xception: Deep learning with Depthwise Separable Convolutions*. arXiv preprint arXiv:1610.02357 (2016)

13. A. Dutta, U. Pal, J. Llados, Compact correlated features for writer independent signature verification. In: Proceedings of the 23rd International Conference on Pattern Recognition (ICPR) (2016), pp. 3411–3416

14. S. Pal, A. Alaei, U. Pal, M. Blumenstein, Performance of an off-line signature verification method based on texture features on a large indic-script signature dataset. In: Proceedings of the 12th IAPR Workshop on Document Analysis Systems (DAS) (2016), pp. 72–77

Potential Applications of Advanced Control System Strategies in a Process industry—A Review

Pradeep Kumar Juneja, Sandeep Kumar Sunori, Kavita Ajay Joshi, Shweta Arora, Somesh Sharma, Prakash Garia, Sudhanshu Maurya, and Amit Mittal

Abstract Present paper attempts to review the applications of advanced control strategies based on artificial intelligence techniques and its hybrid counterparts applicable in process industry. This chemical process industry may be textile, paper, water purification plant, sugar mill, leather, steel, or any sub-process which may be common in all these industries. It covers an exhaustive literature review.

Keywords Process control · Control system · Controller tuning · Time delayed process

1 Introduction

The process modeling involves the development of dynamic mathematic models with state space forms describing the thermal state in the heated and cooled objects. Dynamic model-based optimal control strategies for both batch and continuous thermal processes in terms of the maximum principle, dynamic programming, heuristic search, etc. are also proposed in this study. The target of the development of optimal control for the heating processes is to provide the optimal heating patterns based on the given criteria and constraints associated with dynamic models and others. A complete hierarchical computer control structure for heating processes is proposed [1]. Nixon discussed recent advances in the use of programmable logic controllers for batch house control systems [2]. Jaggers explored that more accurate sensors make possible tighter control loops [3]. Taniguchi et al., applied advanced technology for cold strip mill installed in Japan that improved gage accuracy. The specifications and features of the mill are set forth, and the gage control system is examined in more detail [4] (Table 1).

P. K. Juneja (✉)
Graphic Era University, Dehradun, India
e-mail: mailjuneja@gmail.com

S. K. Sunori · K. A. Joshi · S. Arora · S. Sharma · P. Garia · S. Maurya · A. Mittal
Graphic Era Hill University, Bhimtal Campus, India

© The Author(s), under exclusive license to Springer Nature Singapore Pte Ltd. 2023
Y.-D. Zhang et al. (eds.), *Smart Trends in Computing and Communications*, Lecture Notes in Networks and Systems 396, https://doi.org/10.1007/978-981-16-9967-2_9

Table 1 Literature review

Sr. no	Authors	Year	Technology	Application
1	S. W. Nixon	1988	PLC	Batch house control systems
2	T. Taniguchi et al	1989	Advanced gage control system	Cold strip mill
3	C. Constantinescu et al	1990	DCS	Industrial applications
4	P. P. Aslin	1992	ACS	pH control in sugar mill
5	E. M. Heaven	1993	Parametric techniques	Paper machine
6	G Lightbody et al	1994	System identification techniques	Polymerization reactor
7	G. Rigler, et al	1994	Simulation and control	Multiple stand hot strip mill
8	Zhang Qiping et al	2003	APC	Industrial application
9	H. Ren-Chu et al	2013	MVC controller design	Ammonia synthesis process
10	P. K. Juneja et al	2013	MPC strategy	Lime kiln process
11	P. K. Juneja, et al	2018	Modeling and simulation	Paper machine headbox
12	P. K. Juneja et al	2020	Modeling, control and instrumentation	Review on lime kiln process

Constantinescu et al., presented the hierarchical structure of an integrated control and management system for industrial applications. Its main levels are: distributed process control, process operations support and plant management. This structure has been specialized in the case of a cement plant. The architecture of a gracefully degrading advanced control station is given which supports the process operations in the case of medium size applications, such as cement factories. Result of the global and local repair actions, effect of the spare processing elements and influence of different types of faults have been emphasized. The main advantages of the presented advanced control station are that only the redundancy manager and the application software have been especially designed. This has led to a low cost implementation of the station [5].

Lewis discussed a practical process industry problem of considerable difficulty and defeated conventional approaches [6]. The chemical company Du Pont uses advanced process control to improve operability of plants. This enhances safety, protecting people and the environment, and defines product quality within narrow limits. It reduces costs and increases profits. It is shown that the use of advanced process control enhances safety and increases efficiency [7]. Aslin presented advanced control process for pH control in a sugar industry [8]. Jones, discussed the application of dynamic simulators in industry. The use of training simulators is quite widespread within certain process industry sectors such as refining, petrochemical and oil and gas. Due to advances in technology and a wider awareness of the benefits these simulators are increasingly used for engineering and operations purposes. In the future the distinction between training and engineering simulators can be expected to be less clear. It is thought that soon a single tool will be available to enable a simulator to be used from conceptual design through to operations support [9]. Gough shown that the use of orthogonal filters allows transfer function identification [10]. Vachtsevanos et al., presented a new technique to optimize the slasher control parameters [11].

Heaven et al., examined some of the traditional parametric identification techniques [12]. Cameron presented the results of field applications in mercury reclaim and sulfur recovery [13]. Model and rule-based controllers can provide adaptive feature within a robust controller design framework [14].

Lightbody et al., proposed to utilize model-based approaches to improve the control performance of an industrial polymerization reactor. This involves the development of a process model using system identification techniques, the simulation of the plant within the Simulink environment to allow for the design and validation of control strategies. From these studies a Smith predictor was implemented to significantly improve the polymer viscosity control. Finally, a hardware platform is developed to facilitate the implementation of sophisticated algorithms such as recursive least squares that could not be accommodated on the present DCS [15]. Rigler et al., attempted the simulation and control of a multiple stand hot strip mill [16]. Arruda et al., proposed integrating different software resources with an application example of an oil industry [17]. Graebe Goodwin et al., described three different central strategies that were implemented and evaluated viz. a dithering controller, a linear cascade controller, and a nonlinear cascade controller [18]. Wilkinson et al., reviewed chronologically the status of fuzzy logic from the start to the present scenario [19]. Zhang Qiping et al., applied APC technique on an industrial process and shown that APC technique is capable of mastering and improving the key process targets [20]. Xiaoming Jin et al., performed advanced process control techniques [21]. Li-hong Dai et al., applied a method based on human machine intelligence [22]. Ren-Chu et al., designed six MVC controllers for the ammonia synthesis process. The final industrial application resulted in good control performance and economic benefit [23]. Juneja et al., applied MPC strategy on a lime kiln process. The lime kiln model is perturbed and the responses achieved are compared for controller designed based on MPC strategy [24]. Satisfactory system performances have been achieved by implementing MPC on real plant [25]. Juneja et al., showed that FOPDT model resembles consistency parameter [26, 27]. MOR and MIMO control system analysis techniques are depicted with the aid of flow chart [28, 29]. Modeling and control features of lime kiln process are attempted [30].

Yunhui Luo et al. [31] gives an improved version for the identification of FOPDT model when the response data is very less. So to approximate the step responses a B-spline series expansion are used which in turn provide more operative interpolation values for modeling computation. Least squares method diminishes the eccentricity of response between identified model and actual process by adjusting the error weight coefficients. Anindo Roy et al., [32] used the solidity framework of Hermite-Biehler theorem for FOPDT process model. The resultant simulation results illustrates that this frame work can be efficiently used for the synthesis of PID controller of the FOPDT model. The proposed method gives comparably more superior results over traditional PID tuning approaches.

Qiang Bi et al. [33] projected a robust identification technique that has been derived from a step test. This method gives improved identification result then that of the prevailing method under step testing and can be easily applied to PID auto tuning. IMC-based control system displays better control action in comparison to Ziegler Nichol's based control system [34].

2 Conclusion

An attempt has been done to review the applications of advanced control strategies which are applicable in process industry. It covers many control strategies viz. programmable logic control, distributed control system, system identification, multivariable control system, modeling and simulation, model predictive control, etc. And many application areas such as, cold strip mill, sugar mill, paper machine head box, polymerization reactor, multiple stand hot strip mill, ammonia synthesis process, limekiln process.

References

1. Y. Lu, Application of modern control strategies to thermal processes in metal industry. 1987 American Control Conference (1987), pp. 1053–1058
2. S. W. Nixon, Advancements in batching control systems [glass industry]. Conference Record of the 1988 IEEE Industry Applications Society Annual Meeting, vol 2 (1988), pp. 1073–1075
3. H. T. Jaggers, A new measurement and control system for rubber calendaring. IEEE Conference Record of 1988 Fortieth Annual Conference of Electrical Engineering Problems in the Rubber and Plastics Industries (1988), pp. 28–34
4. T. Taniguchi, H. Tanaka, T. Kawabata, E. Yasui, T. Ooi, A new control system for reversing cold strip mill. Conf. Record IEEE Indus. Appl. Soc. Ann. Meet. **2**, 1472–1477 (1989)
5. C. Constantinescu, C. Sandovici, Towards the fault-tolerant advanced control of a cement plant. 1990 Second International Conference on Factory 2001-Integrating Information and Material Flow (1990), pp. 168–172
6. D. G. Lewis, Experiences in the application of advanced control engineering in the process industry. IEE Colloquium on Case Studies in Industrial Control (1990), pp. 3/1–3/3
7. J. P. McCormick, K. J. Kelly, Computer based process control applications. IEE Colloquium on Case Studies in Industrial Control (1990), pp. 4/1–4/4
8. P. P. Aslin, Connoisseur applications in the food industry. IEE Colloquium on Automation and Control in Food Processing (1992), pp. 8/1–8/4
9. D. R. Jones, Current application of simulators in the process industries and future trends. IEE Colloquium on Operator Training Simulators (1992), pp. 3/1–3/4
10. B. Gough, Advanced adaptive control applications (in industry). Conference Record on Pulp and Paper Industry Technical Conference (1992), pp. 122–132

11. G. Vachtsevanos, J. L. Dorrity, A. Kumar, S. S. Kim, Advanced application of statistical and fuzzy control to textile processes. [Proceedings] IEEE 1993 Annual Textile, Fiber and Film Industry Technical Conference (1993), pp. 6/1–6/8

12. E.M. Heaven, T.M. Kean, I.M. Jonsson, M.A. Manness, K.M. Vu, R.N. Vyse, Applications of system identification to paper machine model development and controller design. Proc. IEEE Int. Conf. Control Appl. **1**, 227–233 (1993)

13. M. M. Cameron, Use of advanced controls in environmental remediation. Industry Applications Society 40th Annual Petroleum and Chemical Industry Conference (1993), pp. 177–183

14. P. J. King, K. J. Burnham, D. J. G. James, Combined model-based and rule-based controller for process control. IEE Colloquium on Advances in Control in the Process Industries: An Exercise in Technology Transfer (Digest No. 1994/081) (1994), pp. 4/1–4/5

15. G. Lightbody, G. W. Irwin, A. Taylor, K. Kelly, J. McCormick, Advanced control of a polymerisation reactor. IEE Colloquium on Advances in Control in the Process Industries: An Exercise in Technology Transfer (Digest No. 1994/081) (1994), pp. 7/1–7/3

16. G. Rigler, H. Aberl, W. Staufer, K. Aistleitner, K. H. Weinberger, Improved rolling mill automation by means of advanced control techniques and dynamic simulation. Proceedings of 1994 IEEE Industry Applications Society Annual Meeting, vol 3 (1994), pp. 2030–2037

17. L. Arruda, Amaral, Gomide, An object-oriented environment for control systems in oil industry. Proc. IEEE Int. Conf. Control Appl. **2**, 1353–1358 (1994)

18. Graehe, Goodwin, West, Stepien, An application of advanced control to steel casting. Proc. IEEE Int. Conf. Control Appl. **3**, 1533–1538 (1994)

19. J. Wilkinson, Additional advances in fuzzy logic temperature control. IAS '95. Conference Record of the 1995 IEEE Industry Applications Conference Thirtieth IAS Annual Meeting, vol 3 (1995), pp. 2721–2725

20. Z. Qiping, G. Jinbiao, W. Xiangyu, W. Youhua, An industrial application of APC technique in fluid catalytic cracking process control. SICE 2003 Annual Conference (IEEE Cat. No.03TH8734), vol 1 (2003), pp. 530–534

21. X. Jin, G. Rong, S. Wang, Advanced process control and its application to industrial distillation chain. Fifth World Congress on Intelligent Control and Automation (IEEE Cat. No.04EX788), vol 4 (2004), pp. 3400–3404

22. D. Li-Hong, C. Xue-Bo, Y. Zheng-Jun, F. Yong-Jun, Method of intelligent object-oriented process control design and application on wire cooling system. 2009 Chinese Control and Decision Conference (2009), pp. 5775–5780

23. H. Ren-Chu, W. Hao, G. Xiao-Jing, Advanced process control technology implementation in ammonia plant. 2013 25th Chinese Control and Decision Conference (CCDC) (2013), pp. 1200–1204

24. P. K. Juneja, A. Ray, Robustness analysis using prediction based control strategy for an industrial process. 2013 IEEE International Conference on Signal Processing, Computing and Control (ISPCC) (2013), pp. 1–3

25. S. M. Zanoli, C. Pepe, M. Rocchi, G. Astolfi, Application of advanced process control techniques for a cement rotary kiln," 2015 19th International Conference on System Theory, Control and Computing (ICSTCC) (2015), pp. 723–729

26. P. K. Juneja, M. Chaturvedi, S. Suman, K. Antil, Modeling of Stock Consistency in the Approach Flow System of the Headbox," 2018 3rd International Conference On Internet of Things: Smart Innovation and Usages (IoT-SIU) (2018), pp. 1–4

27. P. K. Juneja, M. Chaturvedi, A. K. Ray, V. Joshi, N. Belwal, Control of Stock Consistency in Head Box Approach Flow System. 2019 International Conference on Innovative Sustainable Computational Technologies (CISCT) (2019), pp. 1–5

28. P. K. Juneja, A. Sharma, A. Sharma, R. R. Mishra, F. S. Gill, A Review on Model Order Reduction Techniques for Reducing Order of Industrial Process Transfer Function Model. 2020 International Conference on Advances in Computing, Communication & Materials (ICACCM) (2020), pp. 346–350

29. P. Juneja, A. Sharma, V. Joshi, H. Pathak, S. K. Sunori, A. Sharma, Delayed complex multivariable constrained process control—a review. 2020 2nd International Conference on Advances in Computing, Communication Control and Networking (ICACCCN) (2020), pp. 569–573

30. P. K. Juneja, S. Sunori, A. Sharma, A. Sharma, V. Joshi, Modeling, control and instrumentation of lime kiln process: a review. 2020 International Conference on Advances in Computing, Communication and Materials (ICACCM) (2020), pp. 399–403
31. Y. Luo, W. Cai, H. Liu, R. Song, Identification of first-order plus dead-time model from less step response data. 9th IEEE Conference on Industrial Electronics and Applications (2014), pp. 1410–1415
32. A. Roy, K. Iqbal, PID controller tuning for the first-order-plus-dead-time process model via Hermite-Biehler theorem. ISA Trans. **44**(3), 363–378 (2005)
33. Q. Bi, W.-J. Cai, E.-L. Lee, Q.-G. Wang, C.-C. Hang, Y. Zhang, Robust identification of first-order plus dead-time model from step response. Control. Eng. Pract. **7**(1), 71–77 (1999)
34. S. K. Sunori, P. K. Juneja, M. Chaturvedi, N. Agarwal, Design and analysis of control systems for heat exchanger system of sugar mill. 8th IEEE International Conference on CICN 2016 (2016)

Analysis of Algorithms for Effective Cryptography for Enhancement of IoT Security

Valerie David⬤, Harini Ragu⬤, Vemu Nikhil⬤, and P. Sasikumar⬤

Abstract The Internet of Things has emerged as one of the most prevalent technologies of the current times, finding its place in a myriad of applications and is widely used for digitization and automation applications such as smart city development, automated monitoring systems, healthcare, energy management and much more. With more devices being connected to the Internet, one of the biggest challenges faced by the Internet of Things surfaces privacy and security risks. The vulnerability of IoT devices and networks have been brought to light, presenting a threat to the integrity of data. Cryptography has proved itself as a method to secure communication channels and data, as a way to ensure IoT security. In this paper, we aim to compare cryptographic algorithms, namely—AES, DES, RSA and lightweight cryptographic algorithm Fernet, to determine which cryptographic algorithm is the most efficient and secure, and can thereby minimize the risk to data integrity and security in IoT applications.

Keywords IoT · Security · Lightweight · Cyberattack · Cryptography · DES · AES · Fernet · RSA

1 Introduction

The Internet of Things (IoT) is a network that connects physical entities or "things" to the Internet through a wireless network to exchange data without the need for human intervention. The Internet of Things is bringing about a transformation in our physical world by introducing a dynamic system of devices that are connected at an unparalleled scale. Technological advancements are allowing IoT to be used in many fields to enhance and automate processes, be it smart sensors that could be

V. David (✉) · H. Ragu · V. Nikhil · P. Sasikumar
School of Electronics Engineering, Vellore Institute of Technology, Vellore, India
e-mail: valeriedavid2101@gmail.com

P. Sasikumar
e-mail: sasikumar.p@vit.ac.in

used in home automation or even healthcare applications to help provide remote aid and medical developments such as smart inhalers.

Although the Internet of Things has helped make lives easier by connecting devices and performing tasks that would otherwise require manual labour and human intervention, it is also susceptible to security threats, as there is a deficit of built-in security. IoT devices are often vulnerable to targeting malware and hacking, which can then set off botnet attacks at a large scale and cause a threat to the stability and efficiency of networks and the associated devices. According to reports by Kaspersky, as of early 2019, 276,000 IP addresses launched around 105 million attacks on IoT endpoints. The number of attacks of IoT devices increased by around nine times in comparison to early 2018. Since IoT connects devices through networks, continued attacks could cause a threat to developed smart cities and industries as well. These cyberattacks are known to be frequent and are expected to escalate in the near future. Hence there arises a dire need to ensure that all data exchanged over the Internet of Things can be secure and their integrity remain unharmed. Security has an important role to play in order to prevent unauthorized access and misuse of data. In order to protect the data exchanged over IoT from cyberattacks, cryptography has emerged as a method to encrypt and thereby protect data. Cryptography is the process of making data unrecognizable to users that are unauthorized, hence providing confidentiality to authorized users. Encryption and data integrity are necessary requirements for authentication, secure communication and protection of firmware. Ideally, a preferred cryptographic algorithm is one that combines high performance with low cost. However, in most cases it can be seen that there are often performance-cost trade-offs when it comes to choosing an algorithm.

Lightweight cryptography, a section of classical cryptographic algorithms, is often used for IoT devices and applications. This is because lightweight algorithms consume less memory and resources for computation, and is pertinent for resource-constrained conditions whilst delivering adequate security. In cryptography, encryption is the process of converting plaintext sent by the user into an unintelligible ciphertext whereas decryption performs the reverse operation, where the encrypted ciphertext is translated back into an intelligible plaintext. Cryptographic algorithms are largely split up into types—symmetric and asymmetric key cryptography. Symmetric Key Cryptography—This kind of encryption involves a single secret key that is used to encrypt and decrypt a message. It uses a secret key made up of various characters and it is necessary for both the authorized sender and recipient to be aware of the key in order to access the messages. Asymmetric Key Cryptography—This kind of encryption uses two keys in the process of encryption. A message that has been encrypted making use of a public key can be translated back to plaintext only by using a private key. The symmetric key cryptographic algorithms compared in this paper are AES, Fernet and DES whereas the asymmetric key algorithm is RSA.

2 Literature Review

Abu-Tair, Mamun et al. recognized the security threats associated with IoT-enabled smart homes such as data being hacked or a large amount of data being erroneous, and have highlighted a few lightweight cryptographic algorithms such as TRIVIUM and CLEFIA that could be applied in order to prevent the same [1]. Similarly, Saraiva, Daniel AF, et al. analysed cryptographic algorithms such as AES, Twofish, SPECK128, RC6, and many more in order to find the most efficient algorithm for devices connected to the Internet that are constrained by resources. These algorithms were compared specifically to test execution times, power consumed and throughput for IoT devices using them [2]. Datta, Debajit, et al., propose a system in order to combat unprecedented cyberattacks by secure authenticated communication between devices via sound, as opposed to the conventional QR or pin-based authentication systems. The system involves encryption of a random signal before it is transmitted through sound. Many encryption algorithms were combined in order to determine what the most accurate and efficient algorithm would be [3].

Ismail et al., identified that the Message Queue Telemetry Transport (MQTT) protocol, popularly used in IoT environments and machine to machine interactions owing to its small footprint and efficiency, and consuming less memory, energy and time. However, the protocol is susceptible to exploitation and hence implemented the lightweight Fernet algorithm, based on AES-128-CBC to ensure security of IoT devices and the messages sent through them [4]. Umesh V. Nikam et al. also worked towards finding methods to ensure secure communication in an IoT network through the MQTT protocol through standard techniques of cryptography such as end-to-end payload encryption and digital signature implementation [5]. Effy Raja Naru, et al., have compared various lightweight cryptographic algorithms for resource-constrained IoT devices for secure transmission of data. The comparison of algorithms was based primarily on computation and required storage space in order to protect and ensure the privacy, integrity and security of the data being transmitted [6].

Pradeep Semwal et al., studied various cryptographic algorithms for data security applications in cloud computing. The algorithms were compared based on criteria including computation time for encryption and decryption, memory consumption and Avalanche effect. The algorithms compared in this study were DES, 3DES, RSA, AES, Blowfish, IDEA and CAST-128 [7]. Similarly, Priyadarshini Patil et al., studied the strengths, weaknesses, performance and cost of the same cryptographic algorithms on the basis of the time it takes for the algorithm to perform encryption and decryption operations on the data, the amount of memory consumed, entropy, and the requirement of certain bits to encode data flawlessly. The comparison was implemented using Java cryptography [8]. A survey was performed on cryptographic techniques by Padmavathi et al., on algorithms such as AES, DES and RSA with an added LSB substitution methodology on the basis the time it takes for the algorithm to perform encryption and decryption operations on the data, and buffer size given data of multiple packet sizes. The data transmitted is in the form of images [9]. Along

Table 1 Comparison of algorithms

Factors	RSA	DES	Fernet	AES
Cipher type	Asymmetric	Symmetric	Symmetric	Symmetric
Key length (bits)	>1024	56	256	128/192/256
Block size (bits)	>=521	64	128	128
Speed	Slowest	Slow	Very fast	Fast
Scalability	Not scalable	Scalable	Scalable	Not scalable
Power consumption	High	Low	Low	Low
Security	Least secure	Medium	Excellent	Excellent

similar lines, Prerna Mahajan et al., analysed encryption algorithms for security purposes. The algorithms analysed were RSA, AES and DES. The analysis was on the basis of evaluation parameters of the time it takes for the algorithm to perform encryption and decryption operations on the data packed into multiple sizes for comparison purposes [10]. Gurpreet Singh et al., evaluated encryption algorithms for ensuring the security of information being communicated via systems on the internet. Four algorithms were analysed during this study, namely RSA, AES, DES and 3DES. On comparing these algorithms in real-time, it was found that on the basis of speed, avalanche effect and throughput, AES resulted in being the most efficient algorithm for applications in information security [11].

3 Proposed System and Implementation

Through our work, we aim to compare a few lightweight cryptographic algorithms that can be used in resource-constrained IoT environments in order to determine which algorithm best suits the criteria of efficiency and security. The cryptographic algorithms explored within the scope of this paper are RSA, AES, Fernet and DES. Table 1 shows a comparative study between various parameters for the four cryptographic algorithms.

4 Results

The algorithms are evaluated based on five evaluation parameters namely: encryption time, decryption time, throughput of encryption, key lengths and average entropy per byte. The comparison is performed against multiple packet sizes for each algorithm.

On analysing Table 2 and Figs. 1, 2 and 3, it can be concluded that the Fernet algorithm takes the least amount of time for both encryption and decryption of data of all packet sizes, followed by AES and DES. The RSA algorithm takes the most time for

Table 2 Comparison of parameters

S. No	Algorithm	Packet size (KB)	Encryption time (sec)	Decryption time (Sec)	Throughput (KB/sec)
1	RSA	118	10.0	5.0	11.8
	DES		3.2	1.2	36.88
	Fernet		0.066	0.065	1787.8
	AES		1.7	1.2	69
2	RSA	153	7.3	4.9	20.9
	DES		3.0	1.0	51
	Fernet		0.066	0.066	2318.18
	AES		1.6	1.1	95.63
3	RSA	196	8.5	5.9	23.058
	DES		2.0	1.4	98
	Fernet		0.066	0.063	2969.7
	AES		1.7	1.24	115.29
4	RSA	312	7.8	5.1	40
	DES		3.0	1.6	104
	Fernet		0.076	0.070	4105.26
	AES		1.8	1.3	173.33
5	RSA	868	8.2	5.1	105.85
	DES		4.0	1.8	217
	Fernet		0.077	0.079	11,272.72
	AES		2.0	1.2	434

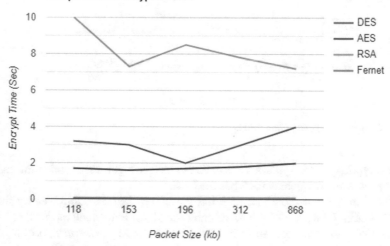

Fig. 1 Time for encryption of all the algorithms

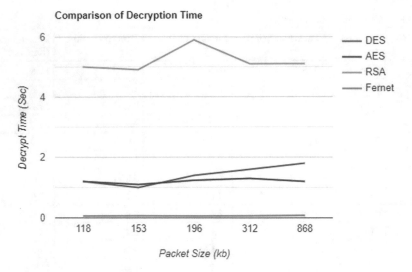

Fig. 2 Time for decryption of all the algorithms

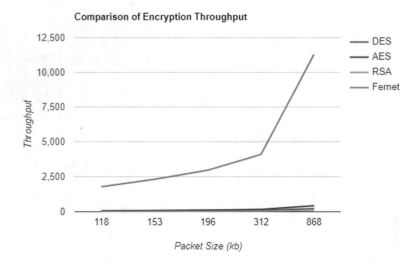

Fig. 3 Comparison of encryption throughput

both encryption and decryption. Hence, Fernet has the highest encryption throughput for all packet sizes, whereas RSA has the least.

Table 3 and Fig. 4 show that RSA has the largest key size, followed by Fernet, AES and DES. However, RSA is an asymmetric algorithm, which means that its key size of 1024-bit is analogous to an 80-bit key length for a symmetric algorithm, when algorithm strength is compared. This means that amongst the algorithms compared, Fernet is considered the strongest depending on the key size.

Table 3 Comparison of key lengths

Algorithm	RSA	DES	Fernet	AES
Key length (bits)	1024	56	256	128

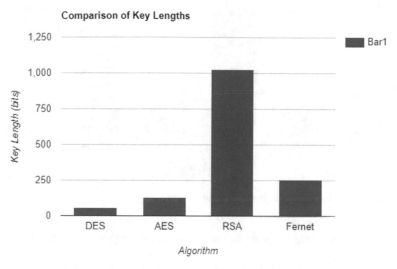

Fig. 4 Comparison of key lengths

Table 4 Comparison of entropy per byte

Algorithm	RSA	DES	Fernet	AES
Avg. Entropy/byte	3.0958	2.9477	3.47248	3.84024

On analysing Table 4 and Fig. 5, it can be seen that AES has the highest entropy per byte, followed by Fernet, RSA and lastly, DES. This shows that each time the algorithm is run, AES tends to produce a more randomized key, hence making it more difficult to crack, in comparison to the other algorithms compared within the study.

5 Conclusion

Through this paper, we have aimed to find a solution to the security challenges faced by IoT applications. Cryptographic algorithms, including the lightweight Fernet algorithm, have been compared based on evaluation parameters of the time it takes for the algorithm to perform encryption and decryption operations on the data, throughput of encryption, key length and average entropy per byte in order to determine the

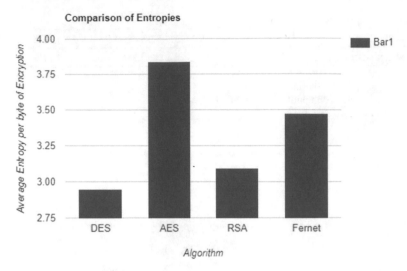

Fig. 5 Comparison of average entropy per byte

most secure and efficient algorithm which can alleviate the risk to data security in IoT applications.

Since Fernet is a lightweight algorithm, it emerges as the most efficient algorithm, with the least encryption and decryption time and the highest encryption throughput. AES has the highest average entropy per byte, which means that it is the most secure algorithm amongst the algorithms compared, followed by the Fernet algorithm. As IoT devices require cryptographic security provided by lightweight sources in order to reduce latency as well as the consumption of resources, yet maintain the authenticity of data, which can be determined by the difficulty in cracking the key, it can be concluded that out of the algorithms compared, the Fernet algorithm emerges as a better fit for security applications in IoT.

In the future, more lightweight cryptographic algorithms can be compared for IoT device security and authentication, in order to determine the strongest algorithm that can be used.

References

1. M. Abu-Tair et al., "Towards secure and privacy-preserving IoT enabled smart home: architecture and experimental study." Sensors **20**(21), 6131 (2020)
2. D.A.F. Saraiva et al., Prisec: comparison of symmetric key algorithms for iot devices. Sensors **19**(19), 4312 (2019)
3. D. Datta et al., An efficient sound and data steganography based secure authentication system. CMC-Comput. Mater. Continua **67**, 723–751 (2020)
4. E.L.G. Ismail, A. Chahboun, N. Raissouni. *"Fernet Symmetric Encryption Method to Gather MQTT E2E Secure Communications for IoT Devices."* (2020)

5. U. V. Nikam, H. D. Misalkar, A. W. Burange, *"Securing MQTT protocol in IoT by payload Encryption Technique and Digital Signature"* (2018)
6. E. R. Naru, H. Saini, M. Sharma, "A recent review on lightweight cryptography in IoT." *2017 international conference on I-SMAC (IoT in social, mobile, analytics and cloud)(I-SMAC)* (IEEE, 2017)
7. P. Semwal, M. K. Sharma, "Comparative study of different cryptographic algorithms for data security in cloud computing." *2017 3rd International Conference on Advances in Computing, Communication and Automation (ICACCA)(Fall)*. IEEE (2017)
8. P. Patil et al., "A comprehensive evaluation of cryptographic algorithms: DES, 3DES, AES, RSA and Blowfish." Procedia Comput. Sci. **78**, 617–624 (2016)
9. B. Padmavathi, S. Ranjitha Kumari, "A survey on performance analysis of DES, AES and RSA algorithm along with LSB substitution." IJSR, India (2013)
10. P. Mahajan, A. Sachdeva, "A study of encryption algorithms AES, DES and RSA for security." Global J. Comput. Sci. Technol. (2013)
11. G. Singh, "A study of encryption algorithms (RSA, DES, 3DES and AES) for information security." Int. J. Comput. Appl. **67**(19) (2013)

Relevance of Artificial Intelligence in the Hospitality and Tourism Industry

Smrutirekha, **Priti Ranjan Sahoo**, **and Ravi Shankar Jha**

Abstract The authors provide a crisp yet in-depth summary of the relevance of artificial intelligence in the hospitality and tourism industry. Focusing on artificial intelligence, the chapter draws the attention of the reader toward its usage and role in the lives of the customers and service providers in the hospitality sector. It also focuses on the costs involved in the incorporation of artificial intelligence in the hospitality industry. Some of the salient features of artificial intelligence in the hospitality sector have been discussed in the chapter by elucidating the benefits of artificial intelligence like financial benefits, technological benefits, and resource management benefits. This chapter sought to project the various factors affecting the role of artificial intelligence in the industry and how the hospitality and tourism industry adopts the technology with time, demand, and expectations of the customer. Both perspectives have been highlighted, i.e., from the customer's point of view as well as the managerial insights. The overall intention and thrust were to provide the reader with perspectives of the importance of advancing stages of artificial intelligence in the chosen industry along with a few contrasting thoughts as artificial intelligence cannot replace the essence of human touch completely as it is one of a kind of a servicescape involving humans as the most crucial and critical link.

Keywords Artificial intelligence · Hospitality · Tourism · Technology

1 Introduction

Technology is a necessary evil. It has evolved and become more sophisticated with time, but it has also disrupted many industries with its pros and cons to consider. The hospitality and tourism industry is also one sector that used to be related to humans with a human touch and experience. Enhanced guests' experiences were

Smrutirekha (✉) · P. R. Sahoo · R. S. Jha
KIIT School of Management, KIIT University (Institution of Eminence), Bhubaneswar, India
e-mail: smrutirekha195@gmail.com

P. R. Sahoo
e-mail: prsahoo@ksom.ac.in

Y.-D. Zhang et al. (eds.), *Smart Trends in Computing and Communications*, Lecture Notes in Networks and Systems 396, https://doi.org/10.1007/978-981-16-9967-2_11

due to a high level of human service aspect. The service providers were involved in providing services to human consumers like travelers, passengers, tourists, guests, and event attendees, to name a few. But, with technologies like artificial intelligence, the industry has been revolutionized and has gained popularity in incorporating automated services, leading to a better time, revenue, and human resource management.

Hospitality may be defined as the process involving the guests and service providers where they are provided with a set of experiences, be it for leisure or business. Food, accommodation, and experience are the primary concerns of hospitality services, which can be provided in commercial and non-commercial settings. Customer service, particularly the personal relationship between the host and the visitors, distinguishes hospitality as a service sector centered on the people who use it.

Artificial intelligence (AI) in the travel, tourism, and hospitality industries is gaining momentum as time passes by. Hospitality and tourism are now using AI to revamp their operations, thereby lowering costs, increasing production, and improving the efficiency and dependability of the services they provide [7].

In response to the guests' expectations and as a means of enhancing the customer's experience management, numerous hospitality and tourism organizations have begun to employ service automation, mobile applications, and artificial intelligence for ease and smart work. AI is projected to have a significant impact on all aspects of life and society. Hence, this chapter discusses the usage of artificial intelligence in the hospitality industry and how the management has adopted it. The chapter begins with the evolving AI and gradually highlights the relevance of AI in the chosen sector of study, thereby paving a pathway for further research in this field.

2 Overview of Artificial Intelligence

Artificial intelligence is one such innovation of the human mind that puts science in a dilemma whether to be admired or feared. John McCarthy coined the term artificial intelligence and defined it as the ability of a computer's system to interpret external data, learn from the external data obtained, and finally put them to use to achieve specific goals and targets, thereby adopting itself at its flexibility.

The ability of computers or machines to perform seemingly intelligent actions is referred to as artificial intelligence. Although artificial intelligence has been around since the 1950s, technology has only recently advanced to the point where it can be considered reliable enough to be used for critical commercial activities. When addressing artificial intelligence, it is essential to believe that scientists' programs, computers, and models don't have exact human intelligence; they perform the desired act intelligently when needed as per the instructions fed.

Artificial intelligence (AI) is the ability of a machine to understand and use human language before continuing to function on its own. Reasoning, knowledge, learning,

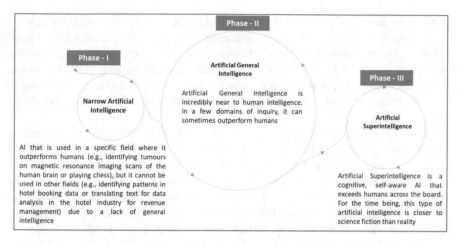

Fig. 1 Phases of artificial intelligence

communication, perception, planning, and other uses of modern AI are all used in society.

There are three types of artificial intelligence, which are as follows [8] (Fig. 1).

3 Adoption of Artificial Intelligence in the Hospitality Industry and Its Implications

The deployment of artificial intelligence significantly impacts corporate processes across all critical horizontals and verticals of the functional areas of hospitality and tourism organizations. AI adoption refers to how quickly hospitality businesses and consumers adopt AI. If consumers sense a financial benefit from employing AI, they will become more favorable. Consumers may be more willing to embrace AI solutions if the industry passes labor savings to them. Customers will eventually come around and embrace technologies after being exposed to them and become more normalized effective and efficient in hospitality processes. Still, to kick-start the process, it is expected that consumers will be informed about the technologies' benefits.

In recent times, AI has been applied in most business sectors, primarily those involving services like the hospitality industry. Tourism arrivals/demand/expenditure, hotel occupancy, waste generation rates in hotels, and energy demand have all been forecasted using AI (neural networks and machine learning) [12].

In the hospitality and tourism industries, AI offers substantial prospects by integrating with machine learning or big data or IoT to innovate products and services, improve decision-making, accelerate the rate of innovation, enhance business performance, obtain sustainable advantage, and remove business impediments [11]. These technologies are diverse in type and can be used at many stages of a tourism service encounter, such as booking, vacationing at the destination, and returning home. Personalization and customization of services in a unique and time-saving manner are one of the goals of incorporating artificial intelligence in the hospitality and tourism industry.

Intelligent agents, collaborative robotics (robots), biometrics, facial and gesture recognition, intelligent automation, recommendation systems, intelligent products, personalization, text, speech, image, and video recognition, and extended are all AI-enabled business applications. All these are relevant and applicable to the hospitality industry. Almost all jobs presently performed by humans will be performed by robots, AI, and natural language systems in the final stage, when AI is wholly developed and applied in the hospitality industry, and all systems are integrated and interoperable.

Researchers have employed AI for analytical reasons such as identifying destination features, sentiment analysis of Internet reviews, assessing hotel employee satisfaction, and market segmentation. Because it directs the evolution of a hotel firm, innovation can be characterized as the most significant component of its corporate strategy.

Robots, artificial intelligence, and service automation, among other emerging technology, are causing dramatic changes in the way hotels cater to their customers [15]. AI provides a significant opportunity for hotel companies to improve operations, increase productivity, and maintain a constant level of quality [6].

As per the customer expectations of the services, some of the new facets of measuring service quality have consistently induced the necessity for service automation and mobile technologies, mainly with the rise of readily available lodging and restaurant service providers. To meet these demands, an increasing number of hospitality businesses are investing in technology to improve the efficiency and reliability of their services. In a larger sense, numerous macro-environmental forces influence the adoption of AI technology in the hospitality and tourism industry (Fig. 2).

There are various implications for improving hotel customer experience utilizing AI before, during, and after the guest's stay. By making the hotel visible to potential clients, AI search platforms and VR promote the purchase. Virtual reality can be used to test a service that would otherwise be impossible to do. AI's application in consumer research provides limitless possibilities while still cost-effective [10].

This enables hotel firms to create personalized products and services by providing services that best match the needs and expectations of their clients. AI helps guests save time by consuming services when it is convenient for them [4]. This aids in the improvement of operational and service efficiency and effectiveness.

The human aspect of the hospitality and tourism industry operations has made the sector mostly a conventional labor-intensive one. This has shown a direction to the managers for using AI for innovative work instead of hard work, which leads to improving the lives of both the customer and the employees.

Two examples of product innovations include the expanding popularity of guest room automation and artificial intelligence. The Moon Pod, for example, is a

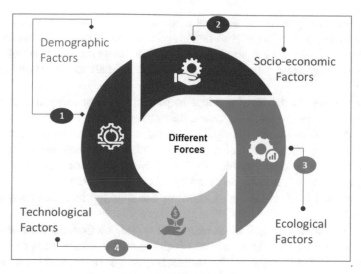

Fig. 2 Macro-environmental forces affecting artificial intelligence

room tablet designed by CitizenM in London that provides hotel guests complete control over their room's temperature, lighting, television, and even window blinds. When a guest is not there, the Shangri-La in Abu Dhabi has incorporated a guest automation system that transforms the room's status from occupied to vacant and adjusts the temperature while sending vital information to the front desk. In addition to technological developments, demographics tend to be the most important determinant.

Customers will be more inclined to employ automation technologies for specific activities if they believe they are appropriate for automation. Customers' perceptions toward AI are influenced and changed by their actual contacts with AI technologies and their participation in service processes. They may lead to a lack of willingness to use them [17].

Customers and managers will see AI as a tool to boost productivity, service quality, and, ultimately, the company's financial outcomes. They can do a cost–benefit analysis of AI adoption and choose whether to apply AI or not, which is then put into practice through an AI adoption program. On the other hand, employees may have more unfavorable attitudes toward AI since they see AI as a tool for their replacement as employees rather than a technique to improve their performance at work, leading to fear of losing their employment and increased turnover intentions.

The hospitality business's scale, market positioning, and corporate culture will all be important factors to examine. Another critical factor to consider is the technological complexity of AI solutions. The simplicity of use of new technology is one of the essential aspects of adopting new technology [14]. Customers' and service providers' cultural attributes may significantly impact how AI technologies are viewed and used. While some civilizations (like Japan, South Korea, and the United States) appear to be open to new ideas and technologies, other countries like Japan and South Korea are more conservative and dubious of or outright reject such advances. Some of the AI systems, particularly, those that can cause bodily harm to humans, should have safety qualities as a top priority (customers or employees). While technologies like chatbots may annoy some customers, they are unlikely to cause bodily harm to anyone.

When pricing AI-delivered services as part of the hospitality marketing mix, customers' perceptions of AI-provided services and their willingness to pay for non-human delivered tourism/hospitality services must be taken into account. AI technologies that are practical, effective, and economical to acquire will need to be developed. There is some opposition to AI's continuing growth, not only from the working class, who fear being replaced by robots and AI, but also from respected intellectuals and thinkers who see hazards in a technology that could become self-aware [16].

Innovative technologies, such as interactive social hubs, chatbots, in-room intelligent technologies, and robots are used during a stay to create a unique customer experience by providing guests with various helpful information and entertainment, saving them time, and being available to them 24 hours in the hospitality and tourism industry. Incorporating AI into the service process also eliminates the language barrier, resulting in a more efficient service [13]. For hotel visitors, innovative technology solutions are not only helpful but also amusing. Guests are frequently willing to use a service that involves a robot to observe how it works. Innovative technologies are a source of advertising, sales growth, and cost savings for the hotel industry.

Inhouse entertainment, keyless room access, and the ease with which AI-enabled virtual agents can be used are increasingly seen as the new adoptive technologies of guest satisfaction and innovation. Not only this, but also AI is empowered with recognition of the likes, dislikes, interests, and passion through the observation of behavioral patterns and gestures of the guest in a servicescape.

Current revenue management software, which is mainly used in lodging properties, will eventually spread to the inventory management systems of intermediaries and eventually cover the entire industry [5]. Intelligent systems will estimate, forecast, plan, and control operations across channels and suppliers, maximizing resource efficiency, revenue growth, and customer satisfaction. AIs will be the primary tool for improving operations, redesigning, and reengineering business processes, focusing on the customer while leveraging revenues and costs.

While present systems rely primarily on previously acquired data (Big data) to extract any implications and recommendations to tourists, predictive analytics solutions to anticipate customer behavior and project industry success will see a significant advancement in the future [9]. There are few risks involved in using AI, which directs toward loss of privacy, human bias, and the threat of technology guiding the society leading to a mechanized community rather than an ethical one.

4 Benefits and Costs Associated with Incorporation of AI

The adoption of AI technologies has enormous non-financial benefits for hospitality businesses and indirect financial consequences. AI technologies, above all, improve the quality of an employee's job [1]. AI assists staff in avoiding or reducing errors in the service process, such as incorrectly logging orders at a restaurant or incorrectly processing customer data in a hotel's property management system, lowering customer service expenses. At times, AI technologies can make the service process hilarious and enjoyable, therefore stimulating customer purchases.

Thanks to AI technologies, employees can save time, which frees them from tedious duties and focuses on more revenue-generating activities [3]. Reducing time, energy, and costs associated with human resource management is one of the most critical non-financial benefits of employing AI technologies.

The most obvious financial benefit is labor cost reductions due to AI technologies' benefits—robots, kiosks, or chatbots work around the clock and have incredibly high service capacity, meaning they can serve multiple consumers at once or for a set period [18].

The reduction of time, energy, and costs associated with human resource management is a non-financial benefit of employing AI technologies [2]. AI can assist in resolving some of the issues related to hiring and terminating personnel, particularly seasonal workers.

Through novel appealing and interactive ways of delivering services, connecting with clients, and engaging with them, AI could improve the perceived quality of service.

Figure 3 gives a bird' eye view of the benefits of artificial intelligence in the hospitality industry.

AI has gifted the hospitality industry with some very unique and innovative services like:

Fig. 3 Benefits of AI in the hospitality industry

- Dynamic pricing solutions using the market data, i.e., demand-based pricing of products and services, which depends on the current market demands
- Customized and personalized services as per the customer requirements and demands
- Intelligent and instant assistance concerning travel and tourism
- Guest rating and review analysis for generation of the accurate feedback interpretations
- Demand forecasting in terms of hotel occupancy predictions
- Designing of mobile business solutions to the management of the hospitality sector
- Modernization of the existing services and guest experiences leading to smart work
- Resource recruitment and management.

Costs associated with incorporation of artificial intelligence:

- *Acquisition costs*—these include the cost of buying a robot or a kiosk and buying a chatbot, or paying for AI software development.
- *Installation costs*—a robot or a kiosk must be physically delivered to the location where it will be used and installed.
- *Maintenance costs*—these expenditures include the robot/electricity kiosk's usage, spare components, routine maintenance, and machine repair, among other things.
- *Software update costs*—these costs include installing new and updated software for robots, kiosks, AI packages, chatbots, and other devices.
- *Costs for adapting the premises to facilitate robot's mobility.*
- *Costs for hiring specialist*—to operate and maintain the robots/kiosks/chatbots.
- *Costs for staff training*—who needs to know how to use the new technology efficiently and adequately.
- *Insurance costs*—they include insurance for robots/kiosks as assets and insurance for damages caused by a robot, while it is in use.

5 Conclusion

Advances in AI technology have allowed them to be used in various industries and societies, including manufacturing and smart factories, warehousing and supply chain management, and autonomous cars, to name a few. AI is also employed in the service industries, such as education, journalism, financial trading, and legal services.

Businesses use AI to reduce costs, eliminate waste, improve productivity, economic efficiency, and financial bottom lines, streamline operations, design service experiences, and increase revenues, resulting in significant changes to their business models and the nature of work.

It is undeniable that technological growth will continue unabated. Robots, AI algorithms, chatbots, and kiosks will enhance their technical properties and become more affordable, allowing them to be deployed more broadly in the hospitality industry. The introduction of AI will inevitably result in the loss of jobs, but it will also result in numerous new jobs that do not currently exist. We are unlikely to feel sorry for dishwashers and cleaners if robots do their jobs or if we utilize chatbots and kiosks to deliver faster, cheaper, and more efficient service than human staff.

While we are still in the early phases of significantly integrating AI into the industry, we can only expect an increase in the adoption of such technology. Human technological skills, a lack of substantial and organized opposition to the technology's adoption into the industry, and the desire to employ human labor more efficiently in industrialized countries are all driving factors in the technology's rise in the industry. While some externalities are expected, numerous issues will arise regarding how they affect the labor market, workplace culture, and international relations.

However, we must remember that technology is a tool, not an end in itself. As a result, the economic evaluation of the costs and advantages of AI adoption must go beyond traditional financial measurements and take a more holistic approach.

Innovative technologies bring value and personalize the stay, but allowing them to take precedence may jeopardize the human ties that visitors seek and enjoy. Finally, AI will never replace the human touch, which is critical for the hospitality industry to stay hospitable.

References

1. E. Brynjolfsson, A. McAfee, Will humans go the way of horses. Foreign Aff. **94**, 8 (2015)
2. P.K. Chathoth, The impact of information technology on hotel operations, service management and transaction costs: a conceptual framework for full-service hotel firms. Int. J. Hosp. Manag. **26**(2), 395–408 (2007)
3. T.H. Davenport, *The AI Advantage: How to Put the Artificial Intelligence Revolution to Work* (MIT Press, 2018)
4. S.J. DeCanio, Robots and humans-complements or substitutes? J. Macroecon. **49**, 280–291 (2016)
5. S. Ivanov, *Hotel Revenue Management: From Theory to Practice* (Zangador, 2014)
6. S. Ivanov, U. Gretzel, K. Berezina, M. Sigala, C. Webster, Progress on robotics in hospitality and tourism: a review of the literature. J. Hospitality Tourism Technol. (2019)
7. S.H. Ivanov, Tourism beyond humans–robots, pets and Teddy Bears, in *Paper to Be Presented at the International Scientific Conference "Tourism and Innovations"* (2018)
8. S.H. Ivanov, C. Webster, K. Berezina, Adoption of robots and service automation by tourism and hospitality companies. Rev. Turismo Desenvolvimento **27**(28), 1501–1517 (2017)
9. S. Ivanov, C. Webster, Economic fundamentals of the use of robots, artificial intelligence, and service automation in travel, tourism, and hospitality, in *Robots, Artificial Intelligence, and Service Automation in Travel, Tourism and Hospitality* (Emerald Publishing limited, 2019)
10. S. Ivanov, V. Zhechev, Hotel revenue management–a critical literature review. Tourism: An Int. Interdisc. J. **60**(2), 175–197 (2012)
11. R.S. Jha, P.R. Sahoo, Internet of Things (IOT)–enabler for connecting world, in *ICT for Competitive Strategies: Proceedings of 4th International Conference on Information and Communication Technology for Competitive Strategies (ICTCS 2019), December 13–14, 2019, Udaipur, India*, 1 (CRC Press, 2020)
12. J. Kim, S. Wei, H. Ruys, Segmenting the market of West Australian senior tourists using an artificial neural network. Tour. Manage. **24**(1), 25–34 (2003)

13. C. H. Ko, Exploring how hotel guests choose self-service technologies over service staff. Int. J. Organ. Innov. **9**(3), 16–27 (2017)
14. V. Liljander, F. Gillberg, J. Gummerus, A. Van Riel, Technology readiness and the evaluation and adoption of self-service technologies. J. Retail. Consum. Serv. **13**(3), 177–191 (2006)
15. C. Liu, K. Hung, Understanding self-service technology in hotels in China: technology affordances and constraints, in *Information and Communication Technologies in Tourism 2019*, 225–236 (Springer, 2019)
16. A. Parasuraman, C.L. Colby, An updated and streamlined technology readiness index: TRI 2.0. J. Serv. Res. **18**(1), 59–74 (2015)
17. D.C. Ukpabi, H. Karjaluoto, Consumers' acceptance of information and communications technology in tourism: a review. Telematics Inform. **34**(5), 618–644 (2017)
18. C. Waxer, Get ready for the BOT revolution. Comput. World (2016)

Deep Learning-Based Smart Surveillance System

G. Sreenivasulu, N. Thulasi Chitra, S. Viswanadha Raju,
and Venu Madhav Kuthadi

Abstract The role of CCTV cameras has been overgrown in this generation. CCTV cameras are installed all over the places for surveillance and security. Many surveillance systems still require human supervision. Recent advances in computer vision are, thus, seen as an important trend in video surveillance that could lead to dramatic efficiency gains. Various public places like shopping malls, supermarkets, ATMs, banks, and other places, where CCTV cameras are available, are the places we should concentrate on. Security can be characterized in various terms in various settings like robbery distinguishing proof, brutality recognition, odds of a blast, and so on. In jam-packed public places, the term security covers practically a wide range of strange occasions. So, it is important and challenging to build a model which detects these abnormal activities and generates some kind of alert. We used a combination of convolutional neural networks (CNNs) and long short-term memory (LSTM) which involve the concept of deep neural networks. It extracts the spatial–temporal features of the images and calculates the Euclidean distance between the original and reconstructed batch of images. We converted the training videos into images to train the model and calculated the loss between the images to identify the abnormality. To validate the proposed algorithm, 4 datasets as HOLLYWOOD, UCF101, HMDB51, and WEIZMANN are used for action recognition. The proposed technique performs better than the existing one. We made use of Jupiter notebook and Python frameworks.

Keywords Smart surveillance · Deep learning · CNN · LSTM and UCF101

G. Sreenivasulu (✉)
Computer Science and Engineering, ACE Engineering College, Hyderabad, India
e-mail: aceaimldept@gmail.com

N. Thulasi Chitra
Department of CSE, MLR Institute of Technology, Secunderabad, Telangana, India

S. Viswanadha Raju
CSE, JNTUHCEJ, Jagityal, Karimnagar, Telangana, India

V. M. Kuthadi
Department of CS&IS, School of Sciences, BIUST, Palapye, Botswana

Y.-D. Zhang et al. (eds.), *Smart Trends in Computing and Communications*, Lecture Notes in Networks and Systems 396, https://doi.org/10.1007/978-981-16-9967-2_12

1 Introduction

In the recent past, there is a sizable growth in multimedia content material, and this is growing everyday. Surveillance way regulation or maintaining an eye fixed on a few get-togethers or unique events. So, it is far taken into consideration that video investigation is the satisfactory alternative for tracking and detecting. "Manual surveillance" is just too stressful and inefficient. By the use of this observation system, we will easily stumble on what is occurring at a selected area, and remotely we will screen many locations within side the meantime [1]. Also, there has been fantastic development with inside the video evaluating strategies and methods.

Many scholars have endeavored to expand correct clever observation structures that can apprehend any person thru extraordinary methods. Correctness and competence are the principle issues as 100% efficiency is not always reached [2]. There are a whole lot of "video surveillance systems" and the point of interest of every of them is to yield its area with inside the arcade. Video observation entails investigation that includes a listing of steps like "video preprocessing, object detection, movement detection, recognition, and classification" of actions [3].

Deep neural learning methods are greater appropriate for conduct and studying such datasets [4]. These methods can carry out an evaluation of the picture and video datasets that are to be had openly. These skilled fashions of deep mastering can acquire an precision of greater than percent of 95 in a few cases (Fig. 1).

"Support vector machine (SVM)" is the rendition of administered dominating. SVMs utilize carefully an immense wide assortment of capacities for dominating denied of the utilization of extra computational force. These are effective to represent nonlinear highlights and fit for utilize green strategy for dominating.

There are several foremost levels of our method gadget, e.g., attainment of image, pre-processing of image, selection feature, train our data, and test our data classification [5]. Due to brand new capabilities and picture dispensation used. This paper is comprising of six sections. Section 1 is enabling introduction, Sect. 2 is talking about literature survey, Sect. 3 enables system methods, Sect. 4 is focusing on working model, Sect. 5 is results, and Sect. 6 is conclusion and future scope.

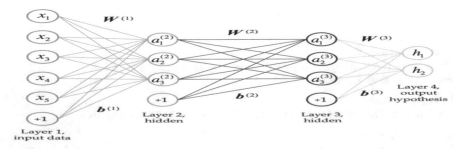

Fig. 1 Illustration of deep neural networks

2 Literature Survey

A major work has previously been carried out on smart investigation structures; many researchers have tried to increase right clever surveillance structures that can apprehend any human hobby via specific methodologies. The below are the specific strategies recycled.

A. **Recognition on Human Action (RHA) Methods**

Majority of scholars labored with inside the area of RHA. In the place of computer vision (CV), members are the use of a spread and dynamic surroundings for the overall presentation assessment of visual dataset. Prof. Ionut [2] future a scheme for indoctrination of functions and their withdrawal to get original dispensation of body charge for motion popularity structures.

A technique is future to get the movement material inside the taken video. The descriptor is proposed. The proposed approach has progressed with good fit.

B. **Features for shape primarily-based totally methods**

Azher [1] added a singular method to apprehend social movements. A new function description is added specifically. ALMD with the aid of using thinking about movement and presence. This method is an allowance of the local ternary pattern (LTP) that is used for stationary textual content evaluation. Spark machine learning library and random forest approach are hired to apprehend human movements. Dataset of six hundred movies such as six human motion training is used for trying out functions. UCF sports motion and UCF-50 records units also are castoff for end outcome evaluation.

The studies explain the methods of temporal units of nearby functions. Aggregation of extracted functions is carried out with VLAD algorithm. Supervised method is used to deliver efficiency in consequences. MSR-Action three-D, UT-Kinect Action 3D, and the Florence 3D.

The consequences indirect that efficiency in out of doors images changed into smaller than the managed surroundings. Zhong et al. added a singular method to apprehend cross-view movements completed with the aid of using the use of digital paths. Virtual view kernel is castoff for locating similarities among multi-dimensional model functions.

C. **Point features grade points**

Gao delivered a method for the gratitude of a couple of fused physical actions. The preliminary stage and wage functions are removed after which the multimedia bag of words version is organized. The graph version is likewise carried outs eliminates the intersecting arguments of the hobby from the data.

Yang supplied a method for RHA in multimedia datasets. For the removal of forefront movement, background motion is remunerated via way of means of the usage of a movement version. The wished foreground patch is over segmented in spatiotemporal patches the usage of methods are more efficient.

Zhang supplied a coding version which can examine greater correct and pictures then different present models. The proposed methodology brings strong consequences in opposition to obstruction and multi-dimensional views. For type of functions, classifiers of supervised learning methods are used.

3 System Approach

In our strategy, first, we put into impact the preprocessing technique that is the combination of various procedures. In the essential advance, we select the channel, practice the formal hat channel, changing the profundity esteems, and difference extending through method of methods for limit esteems executed to embellish the nature of the picture. After the pre-preparing, a weight-basically based absolutely the division approach is applied for discovery to figure body qualification the utilization of combined propose and to supplant the set of experiences through method of methods for the utilization of loads and also to distinguish the closer view regions, and correspondingly, a crossover trademark extraction procedure is utilized for notoriety of human activity. The removed capacities are melded essentially dependent on sequential fundamentally-based absolutely combination, and in some time, the intertwined trademark is applied for arrangement.

A. **Proposed Method**

Our proposed method will analyzes the video feed in real-time and identify any abnormal activities like violence or theft. We are using combination of convolution and LSTM networks (ConvLSTM). We will have to extend deep neural networks to three dimensions for learning spatiotemporal features of the video feed. The abnormal events are identified by computing the **reconstruction loss using Euclidian distance between original and reconstructed batch. With minimal** manpower, abnormal activities are detected, and an alarm will be activated.

B. **CNN and LSTM Architecture**

The CNN and LSTM engineering include utilizing convolutional neural network (CNN) layers for highlight extraction on input information joined with LSTMs to help succession forecast. A CNN LSTM can be characterized by adding CNN layers toward the front followed by LSTM layers with a dense layer on the yield. It is useful to consider this design characterizing two sub-models: the CNN model for highlight extraction and the LSTM model for deciphering the highlights across time steps (Fig. 2).

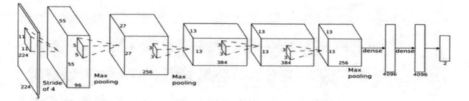

Fig. 2 CNN and LSTM architecture

C. Code Implementation

This section is divided into seven sections. Each section is considered to be equal importance and having its own importance. Each section is discussed in detailed below.

C.1: Importing Packages

A segment can encompass executable declarations in addition to characteristic definitions. These statements are meant to initialize the module. They are performed handiest the first time, and the module call is encountered in an import statement.

Each module has its very own non-public image desk that is used as the worldwide image desk through all features described with inside the module. In this manner, the author of a module can utilize overall factors with inside the module without irritating around inadvertent conflicts with a client's overall factors. On the contrary hand, on the off chance that you understand what you are doing, you could contact a module's overall factors with the indistinguishable documentation used to counsel its highlights, modname, itemname.

Modules can import various modules. It is ordinary yet at this point not, at this point needed to region all import proclamations toward the beginning of a module (or content, so far as that is concerned). The imported module names are situated with inside the transferring module's worldwide image work area.

C.2: Path Creation

Training videos are placed in train directory, creating a new directory named frames for storing converted image frames

```
❖  store_image=[]                              ❖  os.makedirs(train_images_path)
❖  train_path='./train'                        ❖  def store_inarray(image_path):
❖  fps=5                                        ❖  image=load_img(image_path)
❖  train_videos=os.listdir(train_path)         ❖  image=img_to_array(image)
❖  train_images_path=train_path+'./frames'     ❖  store_image.append(gray)
❖  from playsound import playsound             ❖  text=gTTS("Abnormal event detected!")
❖  from gtts import gTTS                        ❖  text.save("alert.mp3") .
```

Fig. 3 Conversion of video to image (for every 0.5 s is the frequency of frame)

C.3: Conversion of Videos to Images

Capturing an image for every 0.5 s, creating path for every image (Fig. 3).

```
✓ for i in range(1,17):
✓ vidcap =
    cv2.VideoCapture('./train/0{}.avi'.format(i))
✓ def getFrame(sec):
✓ vidcap.set(cv2.CAP_PROP_POS_MSEC,sec*1000)
✓ hasFrames,image = vidcap.read()
```

```
✓ if hasFrames:
✓ cv2.imwrite("./train/frames/image"+str(count)+".
    jpg", image)
✓ return hasFrames
✓ sec = 0
```

C.4: Saving Array of Images into .npy file

Image frames are converted into an array. After resizing and clipping, it is stored in training .npy file in a separate file.

Allow us to comprehend the interaction of convolution utilizing a basic model. Consider that we have a picture of size 3×3 and a channel of size 2×2 (Fig. 4; Table 1).

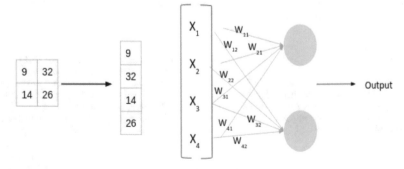

Fig. 4 CNN working model

Table 1 Convolution matrix

1	1	2	1	7
11	1	23	0	1
2	2	2		

C.5: Loading the Model

```
o  training_data=training_data[:,:,:frames]
o  training_data=training_data.reshape(-
   1,227,227,10)
   training_data=np.expand_dims(training_data
   , axis=4)
o  target_data=training_data.copy()
o  epochs=5
o  batch_size=1
o  callback_save
   =ModelCheckpoint("saved_model.h5",monito
   r="mean_squared_error",
   save_best_only=True)
o  stae_model.fit (training_data, target_data,
   batch_size=batch_size, epochs=epochs,

   callbacks = [callback_save,
   callback_early_stopping])
o  stae_model.save("saved_model.h5")
o  def mean_squared_loss(x1,x2):
o  difference=x1-x2
▪  a, b, c, d, e=difference.shape
▪  n_samples=a*b*c*d*e
▪  sq_difference=difference**2
▪  Sum=sq_difference.sum()
▪  distance=np.sqrt(Sum)
o  mean_distance=distance/n_samples
                return mean_distance
```

C.6: Running the model with static video

```
✓  model=load_model("saved_model.h5")
✓  cap =
   cv2.VideoCapture("test/testing_video.mp4")
✓  print(cap.isOpened())
✓  while cap.isOpened():
✓  x=0
✓  imagedump = []
✓  ret, frame = cap. read ()
```

See Fig. 5.

```
Model: "sequential"

Layer (type)                    Output Shape                 Param #
=================================================================
conv3d (Conv3D)                 (None, 55, 55, 10, 128)      15616

conv3d_1 (Conv3D)               (None, 26, 26, 10, 64)       204864

conv_lst_m2d (ConvLSTM2D)       (None, 26, 26, 10, 64)       295168

conv_lst_m2d_1 (ConvLSTM2D)     (None, 26, 26, 10, 32)       110720

conv_lst_m2d_2 (ConvLSTM2D)     (None, 26, 26, 10, 64)       221440

conv3d_transpose (Conv3DTran    (None, 55, 55, 10, 128)      204928

conv3d_transpose_1 (Conv3DTr    (None, 227, 227, 10, 1)      15489
=================================================================
Total params: 1,068,225
Trainable params: 1,068,225
Non-trainable params: 0
```

Fig. 5 Model for "sequential" to add convolution layers

✓ for i in range (10):
✓ ret, frame = cap. read ()
✓ try:
✓ image = imutils.resize(frame, width=700, height=600)
✓ except AttributeError:
✓ x=1
✓ break
✓ gray=(gray-gray.mean())/gray.std()

✓ cv2.putText(image, "AbnormalEvent", (100,80), cv2.FONT_HERSHEY_SIMPLEX,2, (0,0,255),4)

✓ imagedump.resize(227,227,10)
✓ imagedump=nn.expand_dims(imagedump, ...; .... axis=0) in ....=......vand dims (imagedump, axis=4)
✓ output=model.predict (imagedump)
✓ loss=mean_squared_loss (imagedump, output)
✓ if frame.anv () ==None:
✓ print("none")

✓ cv2.imshow("video", image) cap.release() cv2.destroyAllWindows()

✓ stae_model.summary()

C.7: Running the Model with Dynamic Video as an Input

LSTM needs one measurement, yet in the event that you get MobileNetV2 without "top layers" (that is the classification yields, we need not bother with them) so you will have a "convolution" layer. We need to add one layer to have the one measurement shape viable with GRU. Also, MobileNet is not at first intended to identify "activities," however, just to make picture acknowledgment. In this way, on the off chance that you change the whole layers as not "teachable," you will not have precision upgrades or "misfortune" improvement. For instance, I make the last 9 layers teachable.

```
➢   screen_size = (1366, 768)
    fourcc = cv2.VideoWriter_fourcc(*"XVID")
    video = cv2.VideoWriter("captured_video.avi", fourcc, 20.0, (screen_size))
➢   def min():
                 M=3Minimize = win32gui.GetForegroundWindow()

    while M >0:
              time.sleep(1)
              M -=1
         time.sleep(1)
         os.startfile("C:/Users/mpsre/test/test_video.mp4")
         time.sleep(2)

    win32gui.GetForegroundWindow()

         while M >0:
                  time.sleep(1)
                  M -=1
         time.sleep(1)

         os.startfile("C:/Users/mpsre/test/test_vide
    o.mp4")
         time.sleep(2)

         win32gui.ShowWindow(Minimize,
    win32con.SW_MINIMIZE)

➢   for i in range(20):
         img = pyautogui.screenshot()
         s=str(c)
         img.save("./images/image"+s+".jpg")
         c+=1
         fram = np.array(img)
         fram = cv2.cvtColor(fram,
    cv2.COLOR_BGR2RGB) video.write(fram)

➢   model=load_model("saved_model.h5")
    cap = cv2.VideoCapture("captured_video.avi")
    print(cap.isOpened())
    min()
    while cap.isOpened():
         img = pyautogui.screenshot()
         s=str(c)
```

```
    for i in range(10):
    img = pyautogui.screenshot()
         s=str(c)
         img.save("./images/image"+s+".jpg")
         c+=1
         fram = np.array(img)
    fram = cv2.cvtColor(fram,
    cv2.COLOR_BGR2RGB) video.write(fram)
    ret,frame=cap.read()
         try:
             image =
    imutils.resize(frame.width=700,height=600)
                 except AttributeError:
                     x=1
    break
    imagedump=np.array(imagedump)
    imagedump.resize(227,227,10)
    imagedump=np.expand_dims(imagedump,axis
    =0)
    imagedump=np.expand_dims(imagedump,axis
    =4)
    output=model.predict(imagedump)
    loss=mean_squared_loss(imagedump,output)

             if frame.any()==None:
                 print("none")
             if cv2.waitKey(10) & 0xFF==ord('q'):
    break
             if loss>0.00068:
```

In the event that you need to utilize a non-prepared model (setting the loads boundary to none), so you will change shape. However, you should leave every one of the layers to be "teachable." In the event that you as of now have an opened scratch pad where you prepared the past model, if it is not too much trouble, slaughter the bit first.

4   Results and Screenshots

See Figs. 6, 7, 8, 9, and 10.

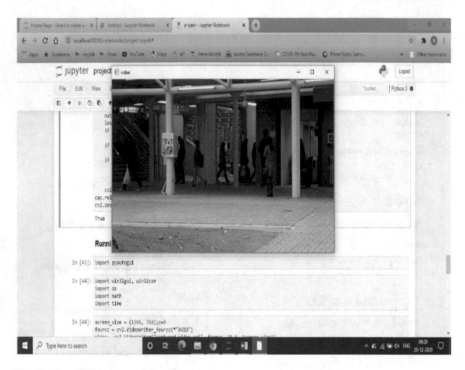

Fig. 6 Normal video

5 Conclusion and Future Scope

For detecting abnormal activities, we used combination of convolutional neural networks and LSTM (ConvLSTM2D) which is a concept of deep neural networks.

As we can see that the accuracy is 0.7327 which means 73.27% of total abnormal activities will be detected correctly. So, majority of the abnormal activities will be detected and appropriate action can be taken immediately. As the dataset has limited training videos, the accuracy is low. If we add more training videos to the model, the accuracy will increase, and the model will be able to differentiate between normal and abnormal events more accurately.

We have tested the model by giving input videos in both static and dynamic (live) ways. This project overcomes the drawbacks of the existing system which involves constant monitoring and decreases man power. So, any abnormal activities like theft, violence, and others at public places will be detected and produce a display and voice messages as alert.

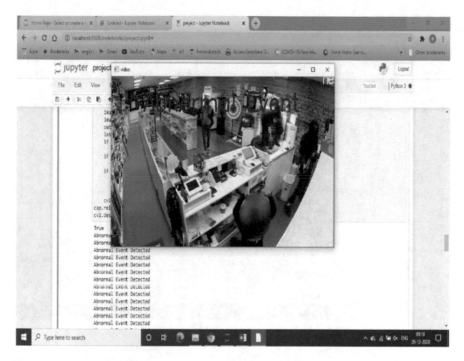

Fig. 7 Common moment video

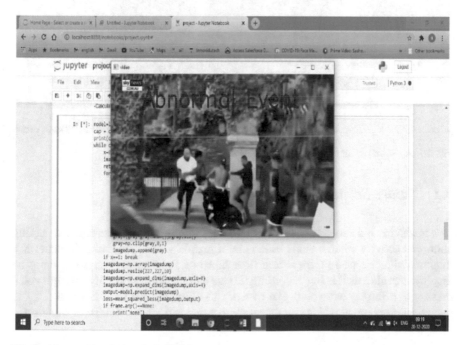

Fig. 8 Abnormality during theft (ATM)

Fig. 9 Abnomality detected

Fig. 10 Abnormal event detected (during street fight)

5.1 Future Scope

Our method is built to detect abnormal activities. Our model is currently detecting abnormal activity when a static video and dynamic video are passed. Whenever a dynamic video is passed, there is a time lapse of 5–10 s, and the text cannot be displayed as alert as it is live recording. So we can generate alert only through voice message. In future, we would like to work to eliminate the time lapse and try to increase the accuracy which can detect almost majority of the abnormal situations.

Rerferences

1. B. Janakiramaiah, G. Kalyani, A. Jayalakshmi, Automatic alert generation in a surveillance systems for smart city environment using deep learning algorithm. Evol. Intel. (2020)
2. R. Slama, H. Wannous, M. Daoudi, A. Srivastava, Accurate 3D action recognition using learning on the Grassmann manifold, Pattern Recognit. **48**, 556–567 (2015)
3. G. Sreenu, M.A. Saleem Durai, Intelligent video surveillance: a review through deep learning techniques for crowd analysis. J Big Data **6**, 48 (2019)
4. M. Baba, V. Gui, C. Cernazanu, D. Pescaru, A sensor network approach for violence detection in smart cities using deep learning. Sensors (2019)
5. G. Sreenivasulu, S. Viswanadha Raju, et al., A Review of clustering techniques, in *International Conference on Data Engineering and Communication Technology (ICDECT)*, Springer, March-2016.J ISSN: 2250-3439)

Pronunciation-Based Language Processing Tool for Tai Ahom

Dhrubajyoti Baruah, Joydip Sarmah, Kamal Gayan,
and Satya Ranjan Phukon

Abstract Words with similar pronunciation may not be of same in its written form. Such similarities can be measured by means of computer algorithms. For English language, an algorithm is available, which may be utilized as a database function "Soundex". This algorithm returns codes on the base of pronunciation of words. Same code signifies similar pronunciation, and dissimilar code indicates different pronunciation. This paper focuses on development of Soundex for Tai Ahom language. In some cases, a single word carries several different meanings. For such matters, exact meaning may be guessed on the basis of other words in the sentence. For this purpose, we have focused on developing another algorithm backed by sentiment analysis.

Keywords Text mining · Soundex · Natural language processing · Sentiment analysis · Tai Ahom

1 Introduction

Soundex is an algorithm used to index words of English language as pronounced. RC Russell and MK Odell had developed the English Soundex algorithm and was patented in the year 1918. Soundex is a phonetic algorithm that returns a specific code against words that sounds similar. The code is a combination of the starting alphabet of the word along with three digits fetched from a mapping table. Words having similar pronunciation get same code from Soundex. Initially, it was used in the US. census from 1890 to 1920. Soundex was used by US census to index people and also for other official purposes. Soundex system helps us to avoid problems like misspellings or alternate spellings of any word. Let us take two naming words "Rima" and "Reema". Though these words are written in different forms, the Soundex code will be the same. The words "Rima" and "Reema" both will produce the same Soundex code "R500". The Soundex system is also useful for searching the ancestor's data. It is common that family names are misspelled in official records. Also it is

D. Baruah (✉) · J. Sarmah · K. Gayan · S. R. Phukon
Department of Computer Application, Jorhat Engineering College, Jorhat, Assam, India
e-mail: dhrubaghy@gmail.com

© The Author(s), under exclusive license to Springer Nature Singapore Pte Ltd. 2023   125
Y.-D. Zhang et al. (eds.), *Smart Trends in Computing and Communications*, Lecture Notes in Networks and Systems 396, https://doi.org/10.1007/978-981-16-9967-2_13

quite obvious that written form of family names could change with time. National Archives and Records Administration (NARA) of USA applies Soundex system to search citizen data for different administration purposes. Soundex is also used to check the spelling of words in large databases. For example, in a dictionary, we can use Soundex to find out words on the basis of the pronunciation. Soundex algorithm for English language has some limitations, like the algorithm depends on the first letter of the word and, hence, cannot detect similarity in pronunciation of words with different starting letters. For example, "college" and "kollege", although similar in pronunciation, Soundex yields different codes that incorrectly indicates dissimilarity. We are focusing on design and development of Tai Ahom language Soundex. The other focus of this paper is based on sentiment analysis. In Tai Ahom language, a single word may carry several different meanings, which meaning fits best in a sentence depends on the other present words of the sentence. A word is likely to carry the positively sensed meaning within a sentence with optimistic, positive flavour. The same word may carry the negative meaning (if any) within a pessimistic, negative flavoured sentence.

2 Related Work

A lot of projects are going on Soundex for different languages. "The Soundex Phonetic Algorithm Revisited for SMS Text Representation" [1] work was carried out by David Pinto and his team at BUAP, University of Mexico. Gunma University, Japan, and RMIT University, Australia, are developing a set of phonetic matching functions for the Japanese language [2]. University of Science and Technology, Algeria, researchers have designed Soundex for Arabic language [3]. India, having several different languages, is also focusing deeply on phonetic research. Odia language Soundex is designed by Indian Institute of Information Technology, Bhubaneswar, Orissa [4]. One project work was carried out for the Assamese language by Jorhat Engineering College, Assam [5]. This project is a web dictionary, where similar pronunciation words are displayed from database according to the Soundex code. Soundex algorithm improvement for Indian language on the basis of phonetic matching [6] work was carried out by Rima Shah and Dheeraj Kumar Singh of Gujarat Technological University. For Sindhi language, a phonetic algorithm is developed, to be implemented for Sindhi Spell Checker [7].

3 Tai Ahom Language

The Tai Ahom language is an obsolete, extinct language which was spoken in the North East part of India, by the Ahom people [8]. The Tai Ahom are descendants of Tai-speaking people, who came to North East India in 1228AD [9], led by King Siukapha, and they settled on the Brahmaputra Valley. The Ahoms ruled in Assam from 1228 to 1826AD. The Ahom people came from Yunnan province of South East Asia [9]. Ahom was used as the prime language of the Ahom kingdom but was gradually replaced by Assamese language. Although Ahom language had lost its use as first language, it continued to be used in marriage and other religious ceremonies, and recently many efforts are made to re-establish the Ahom language [10]. Therefore, although this language is usually regarded as a dead language, it survives in the form of large number of documents written in the heritage Ahom script and exists as a ritual language in Ahom religious ceremonies. The old Ahom manuscripts have somehow helped to understand this dead language and also all the other ways in which it survives today. The derivation of Ahom script was probably from Brahmi origin [10]. The Tai Ahom language is tonal, but tones were not written; so it is not known properly as how tones were arranged and exact number of tones used in the language [11]. The Ahom language now has a place on Unicode [12].

4 Our Approach

Following list is prepared according to their pronunciation. The letters having same pronunciation are grouped together and assigned with a mapping code. The list is prepared based on pronunciation similarity, instead of human articulation base:

Now lets us take " ꏁ" and " ꏃ". Though they are not same letter, both are pronounced as "SA". Therefore, we group them and assign to mapping code "SA". Below is a graph showing the intensity while pronouncing " ꏁ" and " ꏃ" (Fig. 1).

Now let us take " ꏁꯛ" and " ꏃꯛ" Ahom words as example which are pronounced as "sang" and "shang". Now we will generate Soundex code for these two Ahom words according to the mapping codes mentioned in Table 1.

Mapping code for " ꏁꯛ":

$$ꏁ \rightarrow SA$$
$$ꯛ \rightarrow NA$$
$$\frown \rightarrow O$$

Therefore, the Soundex code will be SANAO.
Likewise mapping code for " ꏃꯛ":

$$ꏃ \rightarrow SA$$
$$ꯛ \rightarrow NA$$
$$\frown \rightarrow O$$

Therefore, the Soundex code will be SANAO.

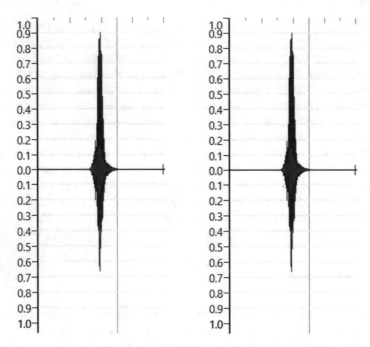

Fig. 1 Graph generated for " vo" and " w"

Table 1 Soundex mapping code of Tai Ahom language

Ahom letter	Code	Ahom letter	Code
vo, w	SA	ɯ	PHA
w, ψ	JA	ʋ	BA
ɕ, ʋɕ, ʁ	NA	ƌ	MA
ɞ, ', C, ⌣	RA	ɯ	THA
w	LA	ʍ	HA
ɯ̈, ɟ, ɭ	AA	Ƌ	DA
῀, ₒ	O	ʍ	DHA
ᶜ, ᵴ	U	∩	GA
ₒ ᵦ,	I	Ƴₒ	GHA
m	KA	Ƴᵖ	BHA
ᴎ	KHA	˧	E
ɑᴎ	TA	ᵣ	AW
ʋ	PA	ˀ	AI
		˴	AM

Now we observe that the words with similar pronunciation will have same Soundex code. On the basis of the grouping, an algorithm is developed to generate Soundex codes. Algorithm to generate Soundex for Ahom words is given below:

1. Take a variable name as word;
2. Take another variable name as soundex_code to store the generated code.
3. For i = 0;i < word.length;i++

If word[i] == ꪫ or ꪉ

Then soundex _code = soundex _code + SA

Else if word[i] == ꪥ or ꪤ

Then soundex _code = soundex _code + JA

Else if word[i] == ꪘ or ꪙ or ꪚ

Then soundex _code = soundex _code + NA

Else if word[i] == ꪦ or or ꪝ or ꪜ

Then soundex _code = soundex _code + RA

Else if word[i] == ꪪ

Then soundex _code = soundex _code + LA

Else if word[i] == ꪢ or ꪣ

Then soundex _code = soundex _code + AA

Else if word[i] == ꪶ or ꪳ

Then soundex _code = soundex _code + PHA

Else if word[i] == ꪚ

Then soundex _code = soundex _code + BA

Else if word[i] == ꪛ

Then soundex _code = soundex _code + MA

Else if word[i] == ꪒ

Then soundex _code = soundex _code + THA

Else if word[i] == ꪏ

Then soundex _code = soundex _code + HA

Else if word[i] == ꪕ

Then soundex _code = soundex _code + DA

Else if word[i] == ꪞ

Then soundex _code = soundex _code + DHA

Else if word[i] == ꪁ

Then soundex _code = soundex _code + O
Else if word[i] == 𑜤 or 𑜥
Then soundex _code = soundex _code + U
Else if word[i] == 𑜢 or 𑜣
Then soundex _code = soundex _code + I
Else if word[i] == 𑜀
Then soundex _code = soundex _code + KA
Else if word[i] == 𑜁
Then soundex _code = soundex _code + KHA
Else if word[i] == 𑜄
Then soundex _code = soundex _code + TA
Else if word[i] == 𑜆
Then soundex _code = soundex _code + PA

Then soundex _code = soundex _code + GA
Else if word[i] == 𑜖
Then soundex _code = soundex _code + GHA
Else if word[i] == 𑜗
Then soundex _code = soundex _code + BHA
Else if word[i] == 𑜦
Then soundex _code = soundex _code + E
Else if word[i] == 𑜧
Then soundex _code = soundex _code + AW
Else if word[i] == 𑜩
Then soundex _code = soundex _code + AI
Else if word[i] == 𑜪
Then soundex _code = soundex _code + AM

4. Soundex code generated and stored in variable Soundex_code.

Algorithm 1: Tai Ahom Soundex

Experimented Results for algorithm 1:

See Table 2.

This way, the solution path has been framed for getting correct word among different probable alternative words with variant spelling. In another scenario, a single word may have several meanings. Soundex algorithm is of no use for such cases. In this case, we can consider "sentiment analysis and subjectivity" [13] concept

Table 2 Tai Ahom Soundex output

Word	Soundex Code
𑜃𑜤𑜃𑜦𑜨, 𑜃𑜤𑜃𑜦𑜨, 𑜃𑜤𑜃𑜦𑜨	NAUNAO, NAUNAO, NAUNAO
𑜊𑜃𑜦𑜨, 𑜊𑜃𑜦𑜨	JANAO, JANAO
𑜏𑜤𑜆𑜦𑜨, 𑜏𑜤𑜆𑜦𑜨	SAUPAO, SAUPAO
𑜏𑜃𑜦𑜨, 𑜏𑜃𑜦𑜨	SANAO, SANAO

to get the proper meaning of word. Sentiment analysis, also termed as opinion mining, is the computational research of sentiments, opinions and emotions expressed in text. Using this concept, we can mine or extract emotions and opinions in a given sentence.

In our approach, categorization of words, according to positive and negative sense, is done. The word having positive sense is assigned a key value "positive", and the word having negative sense is assigned a key value "negative". On the basis of the key value, computation for meaning fetch of targeted word is done. A complete sentence is needed to work for detection of meaning of targeted word within it. The targeted word carries more than one meaning, and to fetch the actual meaning, the sentence should have at least one negative or positive word in it. If majority of the non-targeted words in a sentence is positive, then the positively sensed meaning of the targeted word is returned by an algorithm. Likewise, if the counter for negative word holds a bigger value, it indicates the negatively sensed meaning of the targeted word in the sentence.

As an illustration, some targeted words are as follows:

1. 𑜀 (Ka , meaning : to dance, suffering)
2. 𑜍 (Ra, meaning : epidemic, rain)
3. 𑜊 (Ja, meaning: medicine, abandoned)
4. 𑜉 (Ma, meaning: to come, to not eat)

Considered sentences are as follows:

1. 𑜒𑜧 𑜀 𑜃𑜫 𑜈𑜄	5. 𑜒𑜧 𑜉𑜦 𑜊𑜊𑜫𑜈𑜧 𑜎
2. 𑜒𑜧 𑜃𑜫𑜒 𑜀 𑜂𑜫𑜍 𑜀 𑜉𑜦	6. 𑜒𑜧 𑜉𑜦 𑜊 𑜊𑜫𑜂𑜫 𑜊𑜢
3. 𑜉𑜦 𑜂𑜫𑜍 𑜂𑜫𑜍	7. 𑜒𑜧 𑜀 𑜊𑜫 𑜉
4. 𑜒𑜨𑜍𑜂𑜫𑜍 𑜊𑜫𑜃𑜫𑜃𑜫𑜊𑜢	8. 𑜉 𑜨 𑜁𑜒𑜧 𑜉 𑜉𑜦 𑜈𑜄

The first sentence listed above is as follows:

𑜒𑜧 𑜀 𑜃𑜫𑜈𑜄 (She dances well)

Now we will try to find the meaning of the word " 𑜀 " which has both negative and positive sense, from the above sentence. Leaving our targeted Ahom word " 𑜀 " aside, we categorize the other words as follows:

ৰঙি -> Neutral word (Neither positive nor negative)

ৰ্চ্ভ -> Positive word

ৱৱৗ -> Neutral word

Now we can see that, among the three words, there is a positive word " ৰ্চ্ভ ". Therefore, count of positive is more than negative, and we will take " ক্ৰ" as a positive word and will return the positive meaning of it. So, " ক্ৰ" will mean "to dance" (positive) here instead of "suffering" (negative).

The second Ahom sentence is as follows:

ৰঙি ৰ্চ্ভ্যক্ৰ ষ্ম্ভচ্ৰ ম (She got hurt in quarrel).

The Ahom word " ক্ৰ" has both negative and positive senses. Keeping the targeted Ahom word " ক্ৰ" aside, we categorize the other words as follows:

ৰঙি -> Neutral word

ৰ্চ্ভ -> Neutral word

ষ্ম্ভ -> Neutral word

চ্ৰ -> Negative word

মঙি -> Neutral word

Now we can see that among the five words, there is a negative word " চ্ৰ". Therefore, count of negative is more than positive, and we will take " ক্ৰ" as a negative word and will return the negative meaning of it. So, " ক্ৰ" here will mean "suffering" or "hurt" instead of "dance". On the basis of this concept, an algorithm is developed to return the probable meaning of a word within a sentence. The algorithm is as follows:

1. Take a string variable containing the sentence as 'sentence_str'.
2. Take an integer variable containing the position of the targeted word in the sentence as 'pos'.
3. Take a variable name as 'Counter' to count the occurrence of positive and negative word and initialize it to zero
4. Now split the sentence into an array
5. For(i=0;i<sentence_str.length;i++)
6. If sentence_str[i] != sentence_str[pos]
   If sentence_str[i] is positive
   Counter++
   Else If sentence_str[i] is negative
   Counter - -
7. If(counter > 0)
   Then return positive meaning of sentence_str[pos]
   Else If(counter < 0)
   Then return negative meaning of sentence_str[pos]
   Else
   Return "No result found"

Algorithm 2: Meaning Mining
The algorithm designed was tested, and satisfactory results are found. Some of the test results are given in Table 3.

Table 3 Experimented results of algorithm 2 (Meaning Mining)

Sentence	Targeted Word	Meaning
ꯃꯑ꯰ꯑ꯱꯰꯱	꯱	Epidemic
꯰꯰꯰ꯑ꯰꯱ꯥ꯰꯵꯵꯰꯰	꯱	Rain
꯰꯰ꯃ꯰ꯠꯠ꯰꯰꯰	ꯠ	Medicine
꯰꯰ ꯠ꯱꯰꯰ꯃ꯰	ꯠ	Abandoned
꯰꯰ꯃ꯰꯰꯰	꯰	To come
꯰ ꯰ꯠ꯰꯰꯰ꯃ꯰꯰꯰	꯰	To not eat

5 Conclusion and Future Scope

Algorithm of Soundex for Tai Ahom language can be implemented using any programming language. This algorithm works flawlessly and still is open for any needed modification to improve its performance. The programme developed using this algorithm can be used in many fields, for example, in linguistic and computational task ranging from word suggestion to plagiarism [14]. In future, this algorithm can be used in further development of technology for Tai Ahom language which is an extinct language struggling for revival. This algorithm will be a boon to research and development of Tai Ahom language. The other algorithm based on sentiment analysis for meaning mining may be integrated as future scope for heritage Tai Ahom language research in web computing platform [15].

Acknowledgements This work is carried out under the project "Web Technology Development For Heritage Tai Ahom Manuscripts" which is funded by Dept. Of Science and Technology (DST), Govt. of India, under Science and Heritage Research Initiative (SHRI) Scheme (sanction no DST/TDT/SHRI-11/2018).

References

1. D Pinto, D Vilariño, Y Alemán, H Gómez N Loya, and HJ Salazar, "The Soundex Phonetic Algorithm Revisited for SMS Text Representation"
2. Michiko Yasukawa, J. Shane Culpeppery, Falk Scholery, "Phonetic Matching in Japanese", SIGIR 2012 Workshop on Open Source Information Retrieval. August 16, 2012, Portland, Oregon, USA
3. ND Ousidhoum, A Bensalah, and N Bensaou, "A New Classical Arabic Soundex Algorithm", Proc. of Int. Conf. on Advances in Communication and Information Technology 2012, ACEEE
4. Rakesh Chandra Balabantaray, Bibhuprasad Sahoo, Sanjaya Kumar Lenka, Deepak Kumar Sahoo, Monalisa Swain, IIIT Bhubaneswar, "An Automatic Approximate Matching Technique Based on Phonetic Encoding for Odia Query", IJCSI International Journal of Computer Science Issues, Vol. 9, Issue 3, No 3, May 2012
5. D Baruah, Jorhat Engineering College, AK Mahanta, Gauhati University, "Design and Development of Soundex for Assamese Language"
6. R Shah, Department of Computer Science And Engineering, Parul Institute of Engineering and Technology, Gujarat Technological University,Gujarat, Vadodara, DKSingh, Department of Information and Technology, Parul Institute of Engineering and Technology, Gujarat Technological University,Gujarat, Vadodara, "Improvement of Soundex algorithm for Indian language based on phonetic matching
7. Z Bhatti, A Waqas, IA Ismaili, DN Hakro, WJ Soomro, "Phonetic based Sou x & ShapeEx algorithm for Sindhi Spell Checker System"
8. https://en.wikipedia.org/wiki/Ahom_language
9. Stephen Morey, La Trobe Univeristy, "Ahom and Tangsa: Case studies of language maintenance and loss in North East India"
10. https://omniglot.com/writing/ahom.htm
11. https://scriptsource.org/cms/scripts/page.php?item_id=script_detail&key=Ahom
12. http://unicode.org/L2/L2012/12309-ahom-rev.pdf
13. Bing Liu, University of Illinois at Chicago, "Sentiment Analysis and Subjectivity"

14. D Baruah, AK Mahanta, "A New Similarity Measure with Length Factor for Plagiarism Detection", International Journal of Computer Applications (0975 – 8887) Volume 72– No.14, May 2013
15. www.ahomweb.in

Improving Ecommerce Performance by Dynamically Predicting the Purchased Items Using FUP Incremental Algorithm

K. Kalaiselvi, K. Deepa Thilak, S. Saranya, T. Rajeshkumar, M. Malathi, M. Vijay Anand, and K. Kumaresan

Abstract Data mining or knowledge discovery is the way toward examining data according to substitute perspectives and summarizing it into accommodating information. This information can be then used to fabricate a pay, decreases costs, or both. Programming made with web mining as its key subject ought to permit clients to isolate information from a wide extent of assessments or centers, demand it, and sum up the affiliations perceived. Taking everything into account, information mining is the way toward discovering affiliations or models among many fields in colossal social instructive assortments. This paper effectively tracks down the rehashed bought things by clients. This proposed algorithm is having a higher running time than the existing FUP incremental algorithm. This algorithm efficiently finds the frequent items, and dynamically the items can be added. The entire history of the frequent item database was added and put into separate clusters. At last, we compare and choose the best-purchased items of the customer and also predict the past purchased items in the history. Based on the output, we can easily find the current status of the customer purchase.

K. Kalaiselvi (✉) · K. Deepa Thilak
Department of Networking and Communications, SRM Institute of Science and Technology, Kattankulathur, Chennai, Tamil Nadu, India
e-mail: kalaisek2@srmist.edu.in

S. Saranya
Department of ECE, SRM Easwari Engineering College, Chennai, India

T. Rajeshkumar
Department of CSE, Saveetha School of Engineering, Saveetha Institute of Medical and Technical Sciences, Chennai, Tamil Nadu, India

M. Malathi
Department of ECE, Vivekananda College of Engineering for Women (Autonomous), Namakkal, Tamil Nadu, India

M. Vijay Anand
Department of Computer Science and Engineering, Saveetha Engineering College, Chennai, Tamil Nadu, India

K. Kumaresan
Department of CSE, K.S.R. College of Engineering, Tiruchengode, Tamil Nadu, India

Keywords Data mining · FUP · Intra-cluster · Cost

1 Introduction

This undertaking is an expansion of one of the eminent sub-classes of Data Mining:—"market basket analysis (MBA)," which is an appearance framework giving data into the client buying plans [1, 2]. A market compartment is made out of the thing sets which are bought in a solitary outing to the store. MBA from an overall perspective desires to discover the relationship between the things bought in this compartment [3–5]. As a showing mechanical get-together, it is utilized to mine out the ceaseless thing sets in an epic number of exchanges. Hence, it is likewise called "steady item set mining."

2 Proposed System

The proposed paper intends to achieve an improved foreseeing calculation to view the incessant things liable to be bought by the client. This calculation has preferable running time over FUP gradual calculation. It assists with finding continuous things in a powerfully added exchange. Disintegrate the exchange history data set into intentional example isolated bunches. It maps the current client to the most appropriate bunch and performs sequencing of past acquisition of the clients. It additionally predicts the buy succession of the current client and separates the regular thing from the exchanges [6].

This paper intends to achieve an anticipating calculation to discover the incessant things liable to be bought by the client. This calculation has preferred running time over FUP steady calculation. It assists with finding regular things in a progressively added exchange. Disintegrate the exchange history data set into intentional example isolated groups. It maps the current client to the most appropriate bunch and performs sequencing of past acquisition of the clients. It additionally predicts the buy arrangement of the current client, extricate the successive thing from the exchanges [7, 8].

3 Implementation

3.1 Allocation Phase

Allocation phase the total cost can be minimized for every transaction constant variable T assigned to newly created cluster. A similar will be written in the information base. The choice of whether to remember the exchange for one of the current bunches or to make another one is made by working out the expense of grouping. The expense comprises of intra-bunch disparity and between group closeness which is determined as follows. The decision of whether to recall the trade for one of the current gatherings or to make another is made by determining the cost of batching. The cost contains intra-bunch dissimilarity and between pack likenesses still up in the air as follows. Dispersion stage the outright cost can be restricted for each trade predictable variable T given out as of late made gathering. a comparable will be written in the database allocation phase the total cost can be minimized for every transaction constant variable T assigned to newly created cluster. The same will be written in the database.

Dissemination stage the supreme cost can be restricted for each trade reliable variable T given out as of late made gathering. A comparative will be written in the database Allotment stage the complete expense can be limited for each exchange steady factor T appointed recently made group. A similar will be written in the information base

3.1.1   (a) Intra-Cluster Dissimilarity

$$\text{Intra}(U) = |U_{kj} = 1\ \text{Sm}(C_j, E)|$$

S stands for little items. E—Maximum ceiling 5 C_j—jth cluster. To be known as a little thing, the most extreme roof is the greatest number of exchanges that can contain it. Intra-group uniqueness is characterized as the mix of one of a kind minuscule articles found in all bunches. The most extreme amount of exchanges that a roof can contain is called the greatest roof. It is a small thing, but it is a big deal. The association of distinct small elements existing throughout the bunches is the intra-group variation in this way.

3.1.2   (B) Inter-Cluster Dissimilarity

$$|U_{kj} = 1|\text{Inter}(U) = kj = 1|\text{La}(C_j, S)| - \text{Inter}(U)$$
$$- \text{Inter cluster dissimilarity La}(C_j, S)|\text{Large objects(la)}$$

S—Minimum support C_j—jth cluster.

The minimum support indicates the number of transactions in which an item must appear in order to be classified as a major item. The entire cost is computed using the formula below.

The least help indicates the minimum number of exchanges that a thing should be provided to ensure that it is a significant object. The equation that follows determines the total cost.

$$\text{COST} = w * \text{Intra}(U) + \text{Internal}(U)$$

where w stands for weight.

Intra(U)—Dissimilarity within a cluster. Inter(U)—Cluster similarity between clusters

Another exchange is first placed in every one of the current bunches, and the expense is determined for each group. Then, at that point, another bunch is made to oblige the exchange and the expense is determined. The exchange is then at last appointed to the bunch with the most reduced expense esteem as follows.

- For this reason (each new non-bunched exchange)
- Assign the transaction to the group c
- Calculate the cost for (each group, c)
- Compare the cost to the most cost-effective cost
- Assign existing expense to best cost if (new expense is superior)
- Assign the present bunch to the most suitable group
- Calculate the cost of a new bunch for the current exchange
- If the new expense is better than the best expense so far, assign the current expense to the best cost
- Assign the present bunch to the most suitable group
- End.

One more trade is first positioned in all of the current bundles and the still up in the air for each gathering. By then one more bundle is made to oblige the trade what's more, the not really settled. The trade is then finally given out to the bundle with the most minimal cost an impetus as follows. For this reason (each new non-bundled trade)

- In order to (each bundle, c)
- Assign the bundle c to the trade.
- Estimate the cost
- Compare the price to the best price
- In the event that (new cost is better)
- Assign the best cost to the existing cost
- Assign the present bundle to the most appropriate gathering
- Add another bundle to the current trade
- Estimate the cost
- Compare the price to the best price you've found thus far
- In the event that (new cost is better)
- Assign the best cost to the existing cost
- Assign the present bundle to the most appropriate gathering
- End.

3.2   Refinement Phase

In the refinement stage, the little enormous proportion (SL proportion) of the multitude of exchanges is determined as follows.

$$SLR = \text{(no. of little things)}/\text{(no. of huge things)}$$

The SL proportion of every exchange accordingly determined is then contrasted and the SLR limit. Assuming the SLR of the exchange surpasses the edge, the exchanges are moved to the overabundance pool. An endeavor is then made to oblige these exchanges are an alternate bunch, if the SLR of these exchanges in the new group does not surpass the edge. If not these exchanges are considered anomalies and are dispensed with from thought.

The interaction is clarified as follows.

- Calculate S-L proportion of each exchange.
- Move every one of the exchanges whose S-L proportion surpasses the edge to the overabundance pool.
- Shuffle the exchanges in the overabundance pool to various groups with the end goal that the S-L proportion esteem stays beneath the limit.
- Delete the leftover exchanges from the abundance pool. An illustrative illustration of the interaction is as per the following.

In the refinement stage, the little immense extent (SL extent) of all the not really set in stone as follows.

$$SLR = \text{(no. of easily overlooked details)}/\text{(no. of colossal things)}$$

The SL extent of each not really set in stone is then differentiated and the SLR edge. In case the SLR of the trade outperforms the breaking point, the trades are moved to the excess pool. An undertaking is then made to oblige these trades is a substitute pack, if the SLR of these trades in the new gathering does not outperform the edge. If not these trades are viewed as exemptions and are cleared out from thought.

The cycle is explained as follows.

- Calculate S-L extent of each trade.
- Move all of the trades whose S-L extent outperforms the edge to the excess pool.
- Shuffle the trades in the excess pool to different bundles with the ultimate objective that the S-L extent regard stays under the edge.
- Delete the extra trades from the bounty pool.

Initial Clustering (Allocation Phase) (Tables 1, 2, 3, and 4; Fig. 1):

The bunching system is consequently finished, consolidating both the portion and refinement stages. The gathering collaboration is thusly all out, intertwining both the task and refinement stages.

Table 1 Clustered transaction

Cluster 1		Cluster 2		Cluster 3	
TID	Item Set	TID	Item Set	TID	Item Set
110	B,D	210	B,I	310	B,I
120	A,B,D	220	A,B,I	320	A,B,I
130	B,C,D	230	B,E,I	330	B,E,I
140	D,F,H	240	B,C,E,I	340	B,C,E,I
150	B,G,I	250	C,I	350	C,I

Table 2 Excess transaction and elimination

Minimum Support = 60% Maximum Ceiling = 30%			
Cluster	Large	Middle	Small
1	B,D	A,C	
2	B,I	C,E	A
3	D,H		F,G
Intra-cluster dissimilarity = 7 Inter-cluster similarity = 2 Total cost = 9			

Table 3 Excess pool

TID	Item Set
140	D,F,H
150	B,G,I
330	B,C,D,F

Table 4 Final cluster

Minimum support = 60% Maximum Ceiling = 30%			
Cluster	Large	Middle	Small
C1	B,D	C	A,F
C2	B,I	C,E	A,G
C3	D,H	F	G
Intra(U1) = 3 Inter(U1) = 2 Cost(U1) = 5			

3.3   Incremental Association Rule Mining

Info: Transaction history information base

   Yield: Successive thing sets and guaranteed incessant thing sets.

   The exchange history information base contains the past exchanges made by the clients. The subtleties incorporate client id, the arrangement of things purchased

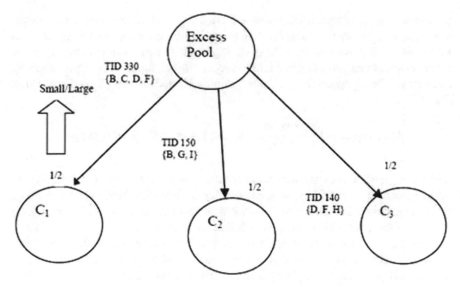

Fig. 1 Refinement

alongside the exchange id. This stage has two sub stages, viz. original information base discovery updating successive and promising incessant thing sets. Data:
Transaction history database yield: perpetual thing sets and ensured customary thing
sets. The trade history database contains the past trades made by the customers. The
nuances consolidate customer id, the plan of things bought close by the trade id. This
stage has two sub-stages, viz. unique informational collection discovery refreshing
nonstop and promising unremitting thing sets.

3.4   Original Database Discovery

A unique database may make it possible to embed new exchanges. Existing affiliation
controls may be invalidated, and new affiliation rules may be enacted as a result.
Maintaining affiliation rules for a big database is a serious challenge. As a result,
this paper provides a new computation to deal with such a refreshing situation. The
new approach is based on the assumption that the measurements of new exchanges
eventually diverge from those of unique exchanges. As the presumption suggests,
measurements of old exchanges obtained through previous mining can be used to
approximate measurements of new exchanges. As a result, the support count of object
sets obtained from previous mining may differ from the help count. After embedding
new exchanges into a unique data set that comprises prior exchanges, a set of item
sets is created. When new exchanges are embedded into the first data set, the new
calculation uses the greatest help count of 1-itemsets acquired from previous mining
to assess inconsistent thing sets of a unique information base that will be equipped

to become regular thing sets when new exchanges are embedded in the first data set. Support count for rare thing sets that will be equipped for continuous thing sets, such as min pL, is displayed in Eq. 1: With the highest help count and the largest size of new exchanges that permit embed into a unique data set, support count for rare thing sets that will be equipped for continuous thing sets, such as min pL, is displayed in Eq. 1:

$$\text{min\_sup}_{\text{DB}} - \left(\frac{\text{max supp}}{\text{total size}}\text{Xinc\_size}\right) \leq \text{min\_PL} < \text{min\_sup}_{\text{DB}}$$

where min_sup(DB) is least help count for a unique information base, max supp is the greatest help count of thing sets, current size is various exchange of a unique data set, and inc_size is the most extreme number of new transactions. Here, a promising successive thing set is characterized as following definition:

A rare thing set that meets requirement 1 is a promising succeeding thing set. Apriori calculation is used in this research to find all possible sequential k-thing sets as well as promising ordinary k-thing sets. For each emphasis, Apriori filters all exchanges of a unique information base using two phases of measures: join and prune. Things in both incessant k-thing sets and promising successive k-thing sets can be merged in the join step, unlike in usual Apriori calculations. Its help count should be higher than a client indicated least help count edge for a persistent item, and it should be higher than min PL but not higher than min PL for a promising successive thing.

A remarkable database might allow install new trades. This might invalidate existing association leads just as start new alliance rules. Keeping up alliance rules for an incredible informational collection is a huge issue. Thusly, this paper proposes one more computation to oversee such reviving situation. Our assumption for the new computation is that the experiences of new trades continuously change from special trades. As demonstrated by the assumption, the estimations of old trades, gained from past mining, can be utilized for approximating that of new trades. Thus, support check of thing sets got from past mining may imperceptibly not equivalent to help count of thing sets ensuing to embedding new trades into an exceptional database that contains old trades. The new estimation uses the most outrageous assistance count of 1-itemsets obtained from past mining to evaluate uncommon thing sets of a remarkable informational collection that will be prepared for being progressive thing sets when new trades are inserted into the principal informational collection. With the most outrageous assistance count and most prominent size of new trades that grant implant into a one of a kind informational index, maintain mean uncommon thing sets that will be prepared for ceaseless thing sets, for instance min_pL , is shown in Eq. 1, where min_sup(DB) is least assist with importance a novel database, max supp is most noteworthy assistance check of thing sets, current size is different trade of a remarkable informational collection and inc_size is a most outrageous number of new trades. Here, a promising standard thing set is described as keeping definition: A promising progressive thing set is something uncommon set that satisfies the condition 1. In this paper, Apriori estimation is applied to find all possible unending k-thing sets and promising progressive k-thing sets. Apriori channels all trades of a one of a kind informational index for each accentuation with two phases measures are joined and prune step. Unlike normal Apriori estimation, things in both nonstop k-thing sets and promising ceaseless k-thing sets can be solidified in the join step. For something standard, its assist count with being higher than not really settled least assist count with edging and for a promising persistent thing, its assist check with being higher than min_PL yet not actually the customer demonstrated least assistance count.

3.5   Updating Frequent and Promising Frequent Items

An old incessant k-thing may transform into a rare k-thing, and an old promising continuous k-thing could turn into a consecutive k-thing, when new exchanges are added to a unique knowledge base. As a result, new affiliation rules are introduced, and certain existing affiliation rules are rendered invalid. When new exchanges are introduced to a unique data set, all k-things should be refreshed to address this issue. In this segment, we'll show you how to spruce up any old item. When new exchanges are embedded into a unique data set, the size of the refreshed data set grows. Min PL should be updated in this fashion to correspond to the new size of a refreshed data collection. min PL (update) is processed in the same way as the following Eq. 2:

Then, at that point, if any k-thing has support count more prominent than or equivalent to min_sup(DBUdb), this thing set is moved to a regular k-thing of a refreshed information base. In the other case, if any k-thing has support count not exactly min_sup(DBUdb); however, it is more noteworthy or equivalent to min_PL(update), and this k-thing is moved to a guarantee continuous thing set of a refreshed data set. The accompanying calculations are created to refresh continuous and promising regular k-items of a refreshed data set. Right when new trades are added to a one of a kind database, an old relentless k-thing could transform into an uncommon k-thing and an old promising consistent k-thing could transform into a normal k-thing. This presents new alliance rules, and some current connection rules would get invalid. To deal with this issue, all k-things ought to be invigorated when new trades are added to a remarkable informational collection. In this part, we reveal how to revive each and every old thing. The size of a revived database additions when new trades are installed into a one of a kind informational index. Thusly, min_PL ought to be recalculated to associate with the new size of an invigorated database. min_PL (update) is handled as the follows:

Condition 2:

By then, if any k-thing has maintained check more conspicuous than or comparable to min_sup(DBUdb), this thing set is moved to a persistent k-thing of a revived informational collection. In the other case, if any k-thing has maintained check not actually min_sup(DBUdb) yet it is more unmistakable or comparable to min_PL(update), and this k-thing is moved to an assurance persistent thing set of an invigorated database.

Table 5 Transaction data and candidate 1-item sets

TID	Item Set	Item Set	Support
1	A,B,E	A	7
2	B,D	B	7
3	B,C	C	6
4	A,B,D	D	2
5	A,C	E	3
6	B,C		
7	A,C		
8	A,B,C,E		
9	A,B,E		
10	A,C		

Table 6 Candidate, frequent, and promise frequent 2-item sets

C2	Support	L2	Support
AB	4	AB	4
AC	4	AC	4
AD	1	PL2	Support
AE	3	AE	3
BC	3	BC	3
BD	2	BD	2
BE	3	BE	3
CD	0		
CE	1		
DE	0		

The going with estimations are made to invigorate relentless and promising persistent k-items of a revived informational collection (Tables 5, 6, and 7).

Table 7 Candidate, frequent, and promise frequent 3-itemsets

C3	Support	PL3	Support
ABC	1	ABE	3
ABE	3		
ACE	1		
BCD	0		
BCE	0		
BDE	0		

4 Conclusion

This proposed calculation has higher running time than existing FUP steady algorithm. This calculation productively tracks down the incessant things and powerfully the things can be added. The whole history of continuous thing information base added and put it into independent bunches. Finally, we look at and pick the best-bought things of the client and furthermore predict the past bought things in the set of experiences. Based on the yield, we can undoubtedly track down the current status of the client bought.

References

1. Frequent Itemset Mining Dataset Repository, http://fimi.ua.ac.be/data (2004)
2. Apache Hadoop, http://hadoop.apache.org/ (2013)
3. Apache Mahout, http://mahout.apache.org/ (2013)
4. R. Agrawal, J. Shafer, Parallel mining of association rules. IEEE Trans. Knowl. Data Eng., pp. 962–969 (1996)
5. R. Agrawal, R. Srikant, Fast algorithms for mining association rules in large databases, in *Proceedings of VLDB*, pp. 487–499 (1994)
6. G.A. Andrews, *Foundations of Multithreaded, Parallel, and Distributed Programming* (Addison-Wesley, 2000)
7. R.J. Bayardo, Jr., Efficiently mining long patterns from databases. SIGMOD Rec., pp. 85–93 (1998)
8. M. Boley, H. Grosskreutz, Approximating the number of frequent sets in dense data. Knowl. Inf. Syst., pp. 65–89 (2009)

A Novel Approach to Privacy Preservation on E-Healthcare Data in a Cloud Environment

Kirtirajsinh Zala and Madhu Shukla

Abstract Cloud environment enables healthcare professionals to work together. Despite its many benefits, it faces several difficulties: technical, legal, and managerial. Cloud computing can enhance healthcare facilities by utilizing its skills to assist information transfer within the health monitoring system, and without any geographical limitations, users such as patients, physicians, pharmacists, and health insurance agents can obtain health-associated information at any moment. With the widespread use of healthcare information and communication technology (ICT), creating a stable and sustainable data sharing scenario has attracted increasing interest in both academic research and the healthcare sector. This paper evaluates and compare the present situation with the security criteria for cloud-based medical e-health record. In addition, this paper outlines proposed cloud model and comparison of our proposed cloud model with different frameworks listed in different research papers are discussed through which how we can safeguard patient e-healthcare record over cloud environment.

Keywords EHR · Cloud computing · Privacy preserving · Health · Security

1 Introduction

Cloud computing provides cost-effective alternatives through various services such as storage, computing resources, and much more. Since big quantities of data stored in the cloud contain sensitive information, storing this information with third-party service providers presents a major danger to information privacy. The information location in the cloud is vibrant and depends on different variables including network velocity and storage accessibility. In this situation, traditional mechanisms for data protection to safeguard the user information at a known place fail owing to uncertainty

K. Zala (✉) · M. Shukla
Department of Computer Engineering, Marwadi University, Rajkot, India
e-mail: Kirtirajsinh.zala@marwadieducation.edu.in

M. Shukla
e-mail: Madhu.shukla@marwadieducation.edu.in

© The Author(s), under exclusive license to Springer Nature Singapore Pte Ltd. 2023   149
Y.-D. Zhang et al. (eds.), *Smart Trends in Computing and Communications*, Lecture Notes in Networks and Systems 396, https://doi.org/10.1007/978-981-16-9967-2_15

with regard to user data place. In today's healthcare settings, there is a powerful need to build an infrastructure that minimizes time-consuming and expensive operations to obtain a patient's full medical record and uniformly integrates this diverse collection of medical data to deliver it to healthcare professionals. EHRs are becoming more widely used, allowing healthcare professionals, insurance companies, and patients to generate, handle, and access patient healthcare information from any location and at any time. Threat to individual privacy, inference of sensitive information including personal information, patient health records splits among different stakeholders like insurance company pharmacy doctor or even patterns from non-sensitive information need to be protected [1–6]. Privacy preservation in data mining (PPDM) is a relatively recent development. Many researchers are looking into how this technology can be used in the field of cloud computing to secure data in the cloud. Furthermore, plenty of techniques have been proposed. The two factors of PPDM are data hiding and rule hiding. Data hiding methods aim to keep individuals' sensitive data private and to modify data mining algorithms so that sensitive data cannot be presumed from the results of data mining algorithms. Rules are hidden to safeguard output privacy [7].

1.1   What Is Cloud Computing?

Cloud computing relates to the remote control, configuration, and access of resources to hardware and software. It provides Internet data storage, infrastructure facilities, and implementation.

1.2   Deployment Model

Cloud can have any of the four types of access: public, private, hybrid, and community [2, 6, 8].

1.3   Service Models

Cloud computing is based on service models. These are categorized into three basic service models which are.

Infrastructure-as-a-Service (IaaS): IaaS provides access to basic resources, including physical, virtual, virtual, and storage. IaaS quickly increases demand and allows you to only pay for what you use. It enables you prevent the cost of purchasing and handling your own physical servers and other information centers [2, 6, 8, 9].

Platform-as-a-Service (PaaS): PAAS offers the apps, development and deployment instruments, etc., at runtime environment. PaaS enables you to prevent software license purchasing and management costs and complexities, fundamental application, and middleware infrastructure or development instruments and other resources [2, 6, 8, 9].

Software-as-a-Service (SaaS): The SaaS framework enables end-user services to use software applications. Users can connect and use cloud-base applications over the Internet using the SAAS service [2, 6, 8, 9].

1.4 Security and Privacy Concerns in the Healthcare Cloud Computing Environment

There are several challenges to ensuring the level of data security and privacy over cloud in healthcare environment. Despite its vast possibility and dynamic process, cloud privacy, security, and trust continue to be areas of concern and uncertainty. The following issues have been identified in the cloud computing environment for health care.

Virtualization: This way many users can share the same physical resources by developing virtual ones. Multiple virtual machines are modeled on the same hardware in a multi-tenant setting to allow resource pooling [1, 2, 6, 8].

Information and Storage Issues: In order to meet client requirements, this fresh paradigm depends on distributed systems. To that purpose, physical servers are distributed in various geographical places across various data centers, so there are different threats to the privacy and accessibility of the information in the cloud environment like vulnerabilities for data recovery, unsanitary media, and data backup [1, 2, 6, 8].

Web Services on Internet: It provides enormously powerful access to remote data service. Despite the many advantages of the key component, Web technology use can compromise client data privacy and put them at risk of various security issues like session management, SQL injection, cross-site scripting, abbreviations, and acronyms [1, 2, 4, 6, 8].

Data Confidentiality and Integrity: The confidentiality of cloud computing medical image processing is the way medical content and patient data are being kept confidential. It therefore protects unauthorized users from disclosing medical information. Cloud computing medical image processing protects the confidentiality of medical contents and patient information. As a result, it prevents unauthorized users from accessing medical information. The medical image content in the health system must be preserved. The primary challenge for providers who offer cloud-based medical image processing as a service is thus ensuring the integrity of medical images. [1, 2, 4, 6, 8, 10].

Data Availability and Ownership: The diagnosis and clinical decisions of medical image processing tools are of major relevance. In reality, medical images provide instant access to critical data that assists physicians in diagnosing and treating patients. As a result, this software should always be available and accessible to authorized users. Load balancing methods are also used to improve accessibility and efficiency. The process of establishing a link between medical records and their owners is commonly referred to as data ownership. In this case, the owners should have legal rights, complete control, and unrestricted access to medical data [1, 2, 4, 6, 8, 10].

2 Related Work

In this section, we briefly review the existing work. Author proposed secure multiparty computing protocol (SMC). Pailler scheme based on homographic encryption techniques is used to avoid possible disclosure of information in a collaborative environment [1]. The other proposed technique is based on the fully homographic encryption (FHE) algorithm, the attribute-based encryption (ABE), and the indexed searched algorithm, and it allows healthcare professionals to handle encryption keys in the cloud while also splitting the encryption key, so that patient information are never made available to a cloud service provider [2]. The author uses Depsky cloud storage to store health data, and the DepSpace protocol is responsible for access control. The author proposes a solution to the problem of maintaining user access control and preserving trust while storing medical information in groups. Access control (sharing) and encrypted (secure) data-based work is used [3]. Here, author suggested technique is based on the residue number system (RNS) homomorphic encryption system. Technique presented enables direct operation over the encoded data and hence facilitates the complete privacy protection [11]. With diverse users of EHR with different access privileges and permissions, the study presents distributed clinical data exchange via dynamic access control privacy employing hybrid clouds, as well as access control policy transformation and cryptographic building blocks [4]. Medcloud platform is proposed for developers to use in application development, and Restlet, a Web portal, is presented to users, to access the Medcloud system. People from different places can easily access and use, as well as help the developers in creating their own healthcare applications sharing the EHR [12]. In this paper author evaluates performance of cryptographic based identity-based encryption (IBE), and newly identity based proxy re-encryption scheme (IBPRE) which is designed for securing privacy of EHR [13]. On privacy access control, author proposed a prototype which is a combination of RBAC, PBAC, MAC, and DAC access control models. The current prototype is capable of demonstrating the process of setting the standard of access for patients and the health authority and handling access requests by health professionals [5]. The author has planned and built a simulative method for patient doctor apps to link the patient with the doctor in order to secure the privacy of patients [14]. The paper using image steganography for providing improved medical data security proposed an improved image steganography for securing medical data that anonymized and secures compressed data using swapped Huffman tree coding [15]. In order to exchange health information in the cloud, the author explores essential resources related to health information sharing and incorporation in health care and examines emerging problems and issues. Discussion on a range of possible options and opportunities to incorporate EHR in the healthcare cloud [16]. In cloud patient data privacy and security in sensor-cloud infrastructure, author proposes generic framework steps for patient psychological parameters (PPPs) to achieve patient privacy and security in S-CI extracted from available literature [17]. For secure management of medical e-health record in cloud, author sets security functions like an encryption and authentication in the system; author suggested security steps and security process with digital rights management (DRM) to protect patient record in cloud [18]. The author proposes a cloud computing framework for EMR system collaborations to securely share health and medical records among federations of healthcare information systems, enabling for the information exchange of medical records among healthcare professionals with no preexisting relationship on a global scale. The patient-provided selective access grants govern the sharing [19]. In this paper, paper shows that using the attributes-based encryption method is a safe way to access health information in cloud computing [20]. Many authors have used cloud simulator in the future. Cloud simulator is a virtual environment framework that allows for the modeling, simulation, and experimentation of evolving cloud computing infrastructure and application services [21]. There are some research challenges in the cloud in healthcare environment that have been briefly described by certain authors. How do users secure and protect the safety of data stored in the cloud, and how do they use data storage protection to ensure the safety of medical records? Which authentication scheme will be more efficient for secure EHR transfer? What encryption scheme can be used to ensure healthcare data security? These issues must be addressed in the future work [1, 2, 4, 9, 12].

3 Proposed Model

The patient's medical history is essential for making the right decision and assisting healthcare staff in responding as quickly as possible. This information is extremely private and must be kept private for the patient's sake. Medical records should be kept accessible at the same time. To ensure that a patient's data is accessible to all healthcare institutions, it can be accessed anywhere. This is a problem because unauthorized users can access patient data because of the large number of users, associations, and patients and the frequent changes in privileges that must be made to ensure anonymity. This paper proposes a cloud framework for managing user access control in a complex universe for user data and ensuring confidentiality by storing user's healthcare data in the cloud (Fig. 1).

Step-1:   Every user has to register with the system
Step-2:   User information will be checked using the user management module, and the user registration details will be sent to the crypto module, which will store all data using the DES technique.

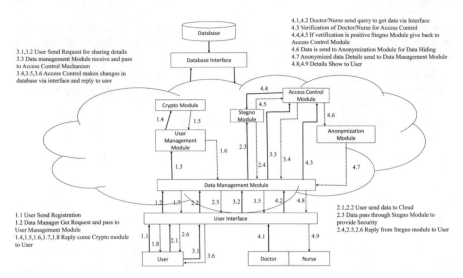

Fig. 1 Proposed model

Step-3:   If registered user, e.g., patient needs to store his/her data on the cloud, patient can transfer all data to the cloud after login via steganography module, which will hide all the EHR of the patient in one image and provide the EHR of the patient with protection.

Step-4:   Patient will provide access to his/her EHR data to doctor/hospital/user through access control module. At the time of defining access rights to its EHR, patient will define some sensitive EHR will not be seen to anybody which will be protected by anonymization module.

Step-5:   Now, registered doctor/hospitals/user wants to access patient EHR; then, they have to login with their credentials and will request EHR of patient through access control module.

Step-6:   Access control module checks Dr./hospital/user access rights/privileges and contact steganography module to provide the requested EHR of the patient stored in image.

Step-7:   Steganography module will reverse the process and will give EHR of patient to access control module.

Step-8:   Access control module will verify EHR with anonymization module.

Step-9:   Anonymization module will hide the specified critical EHR of the patient that cannot be exposed to anyone, and the rest of the EHR will be provided to the access control module.

Step-10:   Lastly, access control module will give requested EHR of patient to doctor/hospital/user.

So, here, author will be implementing below module in proposed model.

1. Cryptography encryption techniques to protect patient credentials in cloud.
2. Steganography to hide patient's EHR in image.

Table 1 Comparison with different frameworks

Paper Reference No.	Any approach?	Information hiding?	Data mining technique used?	Patient privacy maintained	Storing medical data using encryption technique?	Framework available?	Any techniques for sharing medical data over cloud available?
[12]	Yes	No	No	Yes	No	Yes	No
[1]	Yes	No	No	Yes	Yes	Yes	No
[3]	Yes	No	No	Yes	Yes	Yes	Yes
[2]	Yes	No	Yes	Yes	Yes	Yes	No
[4]	Yes	No	No	Yes	Yes	Yes	Yes
[20]	Yes	No	No	No	Yes	Yes	Yes
[19]	Yes	No	No	Yes	No	Yes	Yes
Our proposed cloud model	Yes	Yes	Yes	Yes	Yes	Yes	Yes

3. Access control module to set privileges and rights for patient's EHR to hospitals/doctor/user by patient itself.
4. Anonymized private field based on user choice provide data hiding mechanism.
5. Effectively mine the data to generate the required result.

4 Proposed Cloud Model Comparison with Different Frameworks

See Table 1.

5 Conclusion and Future Work

In this paper, we have discussed security techniques used in cloud environment to safeguard the information and comparison of different frameworks describe in different papers with the proposed cloud model presented in depth in this paper. Managing user access control, guarantee confidentiality of patient's data are complex task. So, our main aim in future publication is to mine the data efficiently, preserve privacy, data hiding, secure storing of sensitive data, and adequate access control for users. For future work, we intend to further investigate and implement the proposed model in the context. As the proposed cloud model is in stage of development, hence, actual results will be shared in future publication.

References

1. M. Marwan, A. Kartit, and H. Ouahmane, "A Cloud Based Solution for Collaborative and Secure Sharing of Medical Data," Cloud Security, pp. 1528– 1547, 2019.
2. H. Elmogazy and O. Bamasak, "Towards healthcare data security in cloud computing," 2013 IEEE Third International Conference on Information Scienceand Technology (ICIST), 2013.
3. D. R. Matos, M. L. Pardal, P. Adão, A. R. Silva, and M. Correia, "Securing Electronic Health Records in the Cloud,"Proceedings of the 1st Workshop onPrivacy by Design in Distributed Systems - W-P2DS18, Apr. 2018.
4. F. Rezaeibagha, Y. Mu, Distributed clinical data sharing via dynamic access-control policy transformation. Int. J. Med. Informatics **89**, 25–31 (2016)
5. Randike Gajanayake,Renato Iannella,Tony Sahema," Privacy Oriented Access Control for Electronic Health Records," InData Usage Management on the WebWorkshop at the Worldwide Web Conference, ACM, Lyon Convention Centre,Lyon, France2012.
6. N.A. Azeez, C.V.D. Vyver, Security and privacy issues in e-health cloud-based system: A comprehensive content analysis. Egyptian Informatics Journal **20**(2), 97–108 (2019)
7. H. Vaghashia and A. Ganatra, "Applications, vol. 119, no. 4, pp. 20–26, .A Survey: Privacy Preservation Techniques in Data Mining,"International Journal of Computer-2015.
8. M. Marwan, A. Kartit, and H. Ouahmane, "Using cloud solution for medical image processing: Issues and implementation efforts," 3rd International Conference of Cloud Computing Technologies and Applications (CloudTech), 2017.
9. Chenthara, Shekha, Khandakar Ahmed, Hua Wang et al. "Security and Privacy-Preserving Challenges of e-Health Solutions in Cloud Computing." IEEE Access, vol. 7, 30, pp. 74361–74382., doi:https://doi.org/10.1109/access.2019.2919982, May 2019.
10. Monica Adriana Dag, Emil Ioan ,Razvan Dobre," Data Hiding Using Steganography"IEEE 12th International Symposium on Parallel and DistributedComputing, 2013.
11. M. Gomathisankaran, X. Yuan, and P. Kamongi, "Ensure privacy and security in the process of medical image analysis,"IEEE International Conference on GranularComputing (GrC), 2013.
12. Dalia Sobhy, Yasser El-Sonbaty and Mohamad Abou Elnasr," MedCloud: Healthcare Cloud Computing System," The 7th International Conference forInternet Technology and Secured Transactions (ICITST-2012).
13. X.A. Wang, J. Ma, F. Xhafa, M. Zhang, X. Luo, Cost-effective secure E-health cloud system using identity based cryptographic techniques. FutureGeneration Computer Systems **67**, 242–254 (2017)
14. Farheen Pathan, Prof. Jadhav H. B.," Patient Privacy Control for Health Care in Cloud Computing System," International Research Journal of Engineering andTechnology (IRJET),vol.04,issue 07,july-2017.
15. M. A. Usman and M. R. Usman, "Using image steganography for providing enhanced medical data security," 2018 15th IEEE Annual ConsumerCommunications & Networking Conference (CCNC), 2018.
16. Y. Guo, M.-H. Kuo, and T. Sahama, "Cloud computing for healthcare research information sharing," 4th IEEE International Conference on Cloud ComputingTechnology and Science Proceedings, 2012.
17. I. Masood, Y. Wang, A. Daud, N.R. Aljohani, H. Dawood, Towards Smart Healthcare: Patient Data Privacy and Security in Sensor-Cloud Infrastructure. Wirel. Commun. Mob. Comput. **2018**, 1–23 (2018)
18. H. Ko, L. Mesicek, J. Choi, J. Choi, S. Hwang, A Study on Secure Contents Strategies for Applications With DRM on Cloud Computing. InternationalJournal of Cloud Applications and Computing **8**(1), 143–153 (2018)

19. S. Shenai and M. Aramudhan, "Cloud computing framework to securely share health & medical records among federations of healthcare information systems," Biomedical Research, Jun.2017.
20. K. Geetha, Impact of cloud database in medical healthcare records based on secure access. Int. J. Eng. Adv. Technol. Regular Issue **8**(6), 652–654 (2019). https://doi.org/10.35940/ijeat.f8097.088619
21. P. Humane, J. Varshapriya, Simulation of cloud infrastructure using CloudSim simulator: a practical approach for researchers, in *International Conference on SmartTechnologies and Management for Computing, Communication, Controls, Energy and Materials (ICSTM)*, pp. 207–211, May 2015

Temporal Reasoning of English Text Documents

Parul Patel

Abstract Temporal expression recognition and normalization (TERN) is promising research area in the field of natural language processing. Automatic temporal information extraction system is in great demand due to rapid growth of digitized information on the Internet. Lot of research has been done in temporal information extraction. But due to variety of temporal expression present in the textual data, interpretation needs more efforts. It is important to understand nature of expression, and then its interpretation ways can be defined. For example, "on the same day," it requires information about previous sentences in the text to define reference time rather than considering document creation time as reference time statically. In this paper, a normalization system is proposed that dynamically chose the reference time from text rather than relying on traditional approach of selecting a document creation date as reference time. The system is evaluated on gold standard dataset like WikiWar, AQUAINT, and TempEval-2 datasets.

Keywords Temporal expression · Temporal tagger · TIMEX · TERN

1 Introduction

Due to rapid increase in the digitization, amount of information has increased drastically on Internet. Time plays fundamental role in retrieving meaningful information from it. Recognizing time based information from text document becomes an emerging research area in natural language processing. For understating a text from time perspective, it is important to interpret temporal expressions that are available in various forms. Understanding temporal expressions require contextual information around text. For example, "On the same day, decision of making of power plant has been taken." In this temporal expression, "the same day" requires contextual information around the text for proper interpretation. Lot of work has been done in the temporal information extraction domain, but interpretation of implicit and relative

P. Patel (✉)
Department of ICT, VNSGU, Surat, India
e-mail: pjpatel@vnsgu.ac.in

temporal expression is done on the basis of fixed reference time like document publication time, document creation time, news date etc. In this paper, a novel approach is proposed to determine reference time dynamically based on the type of temporal expression rather than earlier static method. Rest of the paper is organized as follows: Sect. 2 is review of literature in the field of temporal reasoning. Section 3 describes research methodology adopted for novel approach. Section 4 shows evaluation and results of the implemented algorithm. Section 5 concludes the work and gives path to future work.

2 Literature Review

Temporal information can be expressed in variety of ways in natural language. Approaches to extract temporal information includes, machine learning-based, rule-based, data-driven-based etc. HeidelTime [1] is a rule-based temporal tagger based on UIMA (unstructured information management architecture). SUTime [2] is based on CFA (cascade of finite automata) with heuristics. Trips and Trios [3] is a system developed by UzZaman et al. by using conditional random field and hidden Markov models. Deep learning is also becoming very popular in the natural language processing. Kim et al. [4] have used recurrent neural network for recognizing temporal expressions. T-YAGO [5] uses knowledge base (KB) to extract temporal information. It extracts temporal information from Wikipedia, infoboxes, categories, and lists. ParsTime [6] is rule-based temporal tagger developed for Persian language. In [7], CRF-based approach is used to extract temporal expressions from English text document.

3 Research Methodology

Temporal information processing system also called temporal taggers involves two sub tasks:

(1) Temporal Expression Extraction
(2) Interpreting each temporal expression and converting into a standard format.

In this paper, main focus is on interpreting temporal expression by looking text around it in a sentence rather than considering document creation date, time as a static reference time. In this section, we introduce the normalization task that is used to convert the temporal expressions recognized by recognizer module into standard format. At first glance, it looks very easy to interpret temporal expressions by calculating the offset value from specific reference time, but it is not always true. There are some problematic cases that need very complex solutions. Normalization module depends upon external knowledge for interpreting implicit expression. We describe the fixed language specific knowledge. It is a rule format for temporal reasoning. Rules are operated on the basis of the matching pattern of the temporal expression. It is applicable only when time pattern matches. Rules are divided into two parts: constant and functions. Constant part is defined as follows:

%DCT, REFTIME, OFFSET, DCT_YEAR, DCT_MONTH, DCT_DAY

DCT is a document creation time, REFTIME is a reference time to for normalization, OFFSET defines the direction to be chosen for relative temporal expression, DCT_YEAR, DCT_MONTH, DCT_DAY defines date, month, and year of the document creation date. There are set of functions applied to calculate different normalization values.

For example:

ADD (date, granularity, number)

It calculates the date after adding the number in the corresponding granularity. For example Add (2017-04-01,day,-2) will output 2017-03-30. We have used SUTime's basic rules for interpreting explicit temporal expression. For interpreting implicit and relative temporal expression, we need to identify reference time correctly. Moreover, external knowledge need to be fed into the system to handle named temporal expressions. To impart such knowledge related to a festival or holidays, a special file is developed with the knowledge of it as an external information to be embedded in a system. Moreover, in a country like India, festival do not occur on a same date every year as it follows specific religion calendar. To deal with such situation, we have collected dates of such Indian festival through year with their synonyms in different regions of India along with their values. To generate a knowledge base for the holidays, first we have extracted name of festivals occurring on fixed day or on a variable day from Wikipedia. Data about festivals and special events are gathered from various sources like Wikipedia, different calendars available on Internet, and other resources available on Web that focuses on Indian festivals. Based on available data, rules are designed to interpret such expressions. Once the knowledge base is created, it can be extended for adding more information about other named temporal expressions and event. After generating knowledge base, we have generated algorithm to select the reference time dynamically. Generally, publication time of the document is referred as a reference time for all temporal expressions which is not always true. We control discourse-level information in order to find out reference time by using previously mentioned temporal expression and handling TIMEX that are affected by tenses of the verb. Consider following example

"News: 2020-04-18
*"A CEO of XYZ company has declared that they are going to purchase laptop, desktop and other required equipment for their organization for giving improved performance on **15th March 2020**, **After two days**, tender was published in local newspaper. **After one month**, a contract is given to ABC 'company for purchase and maintenance. Final delivery is on **this Thursday**".*

In above example, publication time of the document is "2020-04-18." But the expression after two days relies on the expression "*15th March 2020* of earlier sentence. However, another expression "this Thursday" depends on the publication time. In short, there cannot be only one logic to normalize temporal expressions. In the above example, following temporal expressions are extracted by recognition module:

(a)   2010-04-18,
(b)   March 8, 2010,
(c)   Week, (followed by modifier *after* and number $= 1$),
(d)   Thursday (followed by modifier *this*).

First two temporal expressions are explicit temporal expressions. Next two are context specific expressions which need a specific reference time to resolve it. First relative temporal expression depends upon the temporal expression of earlier sentence whereas last temporal expression depends on publication date of the news fragment. So, here, document publication date is treated as a publication time and temporal expressions occur in the sentences around are known as local temporal expressions. We have analyzed corpora like WikiWar, aquaint, and TempEval to get clear difference between two different type of relative temporal expression (1) expressions that depends upon publication time (2) expressions that depends upon the heuristic around sentence, and finally concluded various expressions in both the categories as shown in following table: We refer temporal expressions that depends upon the publication time as global temporal expressions, whereas temporal expressions that depends on its previous sentences are termed as local temporal expressions. Expressions like next month, this Friday, tonight, and this year depend on the publication time or a global reference time like document creation date, document publication date etc., and expressions like that year, on the same day, before one day, after one day etc., may depend on the TIMEX that present in the previous few sentences. Finally, all extracted temporal expressions have assigned their respective class for interpretation as: (1) Local and (2) Global. Normalization algorithm works as follows:

(1)   All explicit expression that do not require any other information for interpretation are converted into a TIMEX2 format yyyy-mm-dd. Example of such expression is shown in Table 1.

(2)   In this step, all relative temporal expressions are assigned their offset value based on their temporal modifier.

Table 1 Normalizing explicit temporal expressions

No.	Expression	After normalization
1	3rd January 2010	2010-01-03
2	Sunday, 14 November 2010	2010-11-14
3	15/03/1982	1982-03-15
4	November 2016	2016-11

Table 2 Offset direction from reference time

No.	Temporal expression	From reference time
1	Last Monday	< D1
2	Next Thursday	> D4
3	This coming Sunday	> D7
4	Next December	> M12
5	Last January	< M1

- I will see you next Friday
- They had a meeting last Wednesday.

In this example, temporal modifier like next, last are separated to decide the direction of offset in normalization (Table 2).

(3) Once offset is decided, then reference time for temporal expression is chosen. As shown in above example, relative temporal expression can be considered deictic whose reference time depends upon the publication time where anaphoric expression relies on the context of earlier sentence. Some of the expressions are event-based where event is an external knowledge to be fed into a system for its interpretation. Consider following examples:

(a) *"After meeting, they have decided to accept proposal."*
(b) *"He will visit china in next Christmas."*

In sentence (a), to resolve a temporal expression "after meeting," first the time of meeting needs to be resolved which is out of the scope of this work. In sentence (b), to resolve temporal expression "next Christmas," system has to externally add a knowledge about the date when Christmas occurs. For resolving such expressions, after detecting their offset direction, their year granule is identified by taking publication time as a reference time. For example, if the publication date of document is "20-02-2017" then, 2017 will be taken as year to decide final year of the expression. For example, if the expression is next Christmas, then it is incremented by 1 and so on. Once the granule is decided, then finally granule + festival name is searched in the data file to get final value.

Following is an algorithm to find out the reference time dynamically.

1: Load all temporal expression of Document d into a list Tlist
2: Assign label to each expression as local or global for choosing reference time.
3: Initialize GRT and LRT with either RT(reference time) or PT(publication Time)
//(Divide TE into two part: temporal modifier and temporal noun (it means after one week will be divided as temporal modifier: after, temporal noun =one week another e.g. previous month then temporal noun= month and //temporal modifier= previous)//
4: For each TE in a TList
      TE'=Segment(TE)
      If(TE'= explicit) then
           LRT= TE'
           Insert(TE',TList1)
      Else if(TE'=Local)
           TE'= Normalize(TE',LRT,offset)
      Else
           TE'=Normalize(TE',GRT,offset)
      Endif
           LRT=TE'
           Insert(TE',TList1)
      Endif

To deal with weekdays, straightforward rules are not always applicable. Consider following sentence.

(c)   Meera did not go to work on Tuesday.
(d)   Shlok will go to school on Monday.

In above two sentences, sentence (a) contains temporal expression "On Tuesday" and sentence (d) contains temporal expression "on Monday." During interpretation of such expression, selecting offset direction is quite complex as "On Tuesday" in sentence (c) focus on previous Tuesday, whereas "On Monday" in sentence (d) focus on next Monday. To deal with such weekdays expression with modifier that do not specify the direction, we have applied following algorithm (Fig. 1):

1. Insert TE (with Weekdays)
2. Move backwards in the sentence till the beginning of the sentence to find
    verb
3. If found(verb) =true then
        Find Tense of the Verb
    Else
        Move forward in the sentence till the end of the sentence to find
verb
            If found(verb)=true then
                Find Tense of the Verb
            Else
                Jump towards previous sentence.
                Goto step 2.
            End If
    End if
4. If Tense(Verb)= "PS" || " PP" || "PC"||"BPS"   then
        Offset.direction="Past"
    Else
        Offset.direction= "Future"
    End

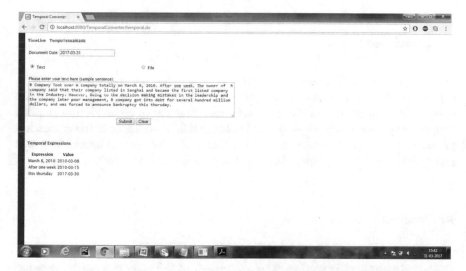

Fig. 1 Snapshot showing output of the system for choosing reference time dynamically

Table 3 Results on different
dataset

Dataset	Precision	Recall	F-measure
TempEval-2	0.94	0.91	0.92
Aquaint	0.88	0.9	0.88
IndiaTimes	0.89	0.85	0.86
TempEval-3	0.91	0.9	0.9
WikiWar	0.92	0.91	0.91

4 Evaluation and Results

The proposed algorithms are tested on the gold standard dataset TempEval-2,
Aquaint, TempEval-3 shared task, WikiWar and IndiaTimes. WikiWar contains 22
historical documents, Aquaint contains 73 documents, TempEval-2 and TempEval-3
contains 325 and 21 documents, respectively. The result is calculated in terms of
precision, recall, and F-measure. Precision, F-measure, and Recall are calculated as
follows (Table 3).

Precision = TRUE POSITIVE/(TRUE POSITIVE + FALSE POSITIVE)

Recall = TRUE POSITIVE/(TRUE POSITIVE + FALSE NEGATIVE)

F − Measure = 2 ∗ (Precision ∗ Recall)/(Precision + Recall)

5 Conclusion

In this paper, automatic temporal information recognition and normalization system
is proposed that focuses on all types of temporal expressions, explicit, implicit, and
relative. Two algorithm for choosing reference time for interpretation of implicit
temporal expressions are proposed and implemented. Major focus of this research is
to understand temporal semantics to avoid ambiguity that arise in natural language
processing. Finally, the proposed algorithms are evaluated on standard dataset.

References

1. J. Strötgen, M. Gertz, HeidelTime: high quality rule-based extraction and normalization of
   temporal expressions, Proceedings of the 5th International Workshop on Semantic Evaluation,
   ACL 2010, pages 321–324, Uppsala, Sweden, 15-16 July 2010.
2. A. Chang, C. Manning, SUTIME: a library for recognizing and normalizing time expressions
   (2012)
3. N. UzZaman, J. Allen, TRIPS and TRIOS system for TempEval-2: extracting temporal information from text, in *Proceedings of the 5th International Workshop on Semantic Evaluation*
   (2010)

4. Z.M. Kim, Y.-S. Jeong, TIMEX3 and event extraction using recurrent neural networks, pp. 450–453 (2016). https://doi.org/10.1109/BIGCOMP.2016.7425968
5. Y. Wang, M. Zhu, L. Qu, M. Spaniol, G. Weikum,. Timely YAGO: harvesting, querying, and visualizing temporal knowledge from Wikipedia. 697–700 (2010). https://doi.org/10.1145/173 9041.1739130
6. B. Mansouri, M. Zahedi, R. Campos, M. Farhoodi, M. Rahgozar, ParsTime: Rule-based extraction and normalization of Persian temporal expressions (2018). https://doi.org/10.1007/978-3-319-76941-7_67
7. P. Patel S.V .Patel, Article: CRF based approach for temporal information recognition from English text documents, in *IJCA Proceedings on International Conference and Workshop on Emerging Trends in Technology* ICWET 2015(1):1–4, May 2015

Design, Development, and Integration of DMA Controller for Open-Power-Based Processor SOC

Gannera Mamatha, M. N. Giriprasad, and Sharan Kumar

Abstract While the processor is momentarily disabled or busy performing other orders in parallel, the Direct Memory Access (DMA) technology allows direct access to peripherals and memory. DMA gets control of the buses to transfer the data directly to the I/O devices. DMA completes the data transfer to all peripherals without the interference of the processor. The DMA controller supports eight channels with 32-bit data transfer, and it has an interface toward user logic for data read and write The channel assignment is done based on priority. This project proposes to design, develop, and integrate DMA controller for open-power processor A2O core-based fabless SoC through AXI4 interface. The methodology used for designing is as follows: design state machines, develop Verilog HDL code, simulate using ModelSim Questa®, and synthesis using Vivado Design Suite-Xilinx®.

Keywords A2O core · AXI4 · DMA · Verilog HDL · Questasim

1 Introduction

Direct Memory Access (DMA) is a technique that allows an information/yield (I/O) device to send or receive data directly to or from the main memory, bypassing the CPU to speed up memory operations. A device called as a DMA regulator is in charge of the cycle (DMAC).

A DMA channel allows a peripheral to transfer data without putting undue strain on the CPU. Without the DMA channels, the CPU uses a fringe transport from the I/O peripheral to duplicate each piece of data. Using a fringe transport uses, the CPU during the read/compose step and prevents other work from being done until the activity is completed.

DMA allows the CPU to do several tasks at the same time, while data are being transferred. The CPU initiates the information exchange process. The controller can transfer data in one of three ways, which are listed below. Only, the new controller

G. Mamatha (✉) · M. N. Giriprasad · S. Kumar
Department of ECE, JNTUA College of Engineering Anantapur, Anantapur, Andhra Pradesh, India
e-mail: gmamatha.ece@jntua.ac.in

© The Author(s), under exclusive license to Springer Nature Singapore Pte Ltd. 2023
Y.-D. Zhang et al. (eds.), *Smart Trends in Computing and Communications*, Lecture Notes in Networks and Systems 396, https://doi.org/10.1007/978-981-16-9967-2_17

has operating modes based on these modes, which are essential for improving performance and data transfer time depending on the user's application.

In *burst mode*, the framework transport is delivered solely after the information move is finished.

The framework transport is relinquished for a few clock cycles during the exchange of information between the DMA channel and I/O peripheral in cycle stealing mode, so the CPU can conduct other activities. When the transfers are complete, the CPU receives an interrupt on demand from the DMA controller signaling that the data transfer is complete, and that the CPU is ready for the next transfer.

The DMAC can take over responsibility for framework transfer only when the CPU requires it in transparent mode.

If the CPU is accessing any external memories, data stored in internal memory, such as RAM, may not be sampled correctly when accessed by DMA controller. We can conclude that if sufficient care is not taken for the memories from which the DMA controller actually gets information, there may be a cache coherence problem with the controller as well.

2   DMA Features

In general, there are many design methods and techniques for the DMA controllers design based on the application. This DMA controller reference is taken from http://www.opencores.org.

This DMA consists of eight channels and supports the both read and write transactions for the address given by the user. The features of DMA controller are as follows:

1.   Eight DMA channels
2.   Three priority levels
3.   Support different modes
4.   Clock divider for slow channels
5.   Block transfer in a frame context
6.   Support FIFO buffer
7.   Windowed arbitration mode.

The AXI4 interfaces the master device and slave device interface. The slave device interface is used to complete DMA with the configured register operations. In some applications, the bus may not include master device, and the slave device interface of AXI is used.

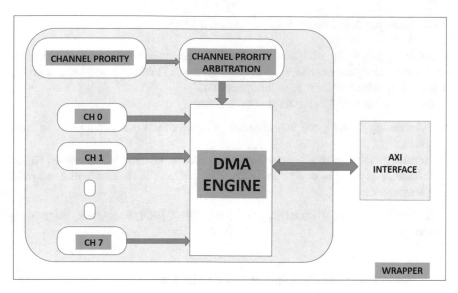

Fig. 1 This describes the all sub-modules in the architecture

2.1 Microarchitecture

The slave configuration is introduced in this DMA because whenever the DMA needs to send the data to the processor which is connected at that the core processor will be acting as the master and the DMA as slave.

The micro-architecture of the DMA controller is shown below (Fig. 1):

3 Modes of Operation

3.1 General Mode

The mode of operation determines the mode in which these channels is to use the AXI read and write buses. Each of the core has been configured to operate in an "Independent" or "Joint" mode. When using the independent mode, each of the channel can operate in the normal mode or the outstanding. When using the "Joint mode," each of the channel can work in either normal or joint mode.

3.1.1 Independent Mode: Normal Channel Mode

When using this mode, each core will have to make use of independent arbiter for read and write operations. Read operations on the AXI bus and will be carried out without taking into account of write operations.

Each channel works in the following manner:

1. According to AXI, the first channel calculates the next read and write burst sizes.
2. As long as the data buffer can contain the data, the channel will emit read bursts.
3. As long as there is write data in the data buffer, the channel will send write bursts.

In the CORE0 JOINT MODE and CORE1 JOINT MODE registers, the mode is turned on.

3.1.2 Independent Mode: Outstanding Channel Mode

If the core operates in independent mode, each channel can be set to "read outstanding," "write outstanding," or both. This mode works as a normal independent mode; the only *difference* is that the use of a read outstanding the read command to be issued *after* the *write* command has been issued to what are the items in the right way are recorded. With the "write outstanding," the write command will be issued *after the read* command is issued, as long as the data are read, it is actually a read from the data buffer. It works on the assumption that the AXI slave, it will respond quickly, else the data buffer is filled or overflowing.

This mode is configured with the help of the RD_OUTSTANDING and WR_OUTSTANDING registers in channel. This mode is efficient while working with the fast responding channels.

One advantage is that latency is improved and throughput for small data buffers.

3.2 Joint Mode

When using this mode, each core will have to make use of a *single arbitrator, both* for read and write operations. In the Joint Mode of operation, the channels are carried out at the same rate, for read and write operations. The present channel locks both the AXI read and write buses, and it is going to read all of the data in the write data. To the running of this channel configuration mode, this *will only happen* in the read registers, and they can have an impact on read operations, as well as the write operations.

This mode can be configured with help of CORE0_JOINT_MODE, CORE1_JOINT_MODE registers and JOINT_MODE register in the channel.

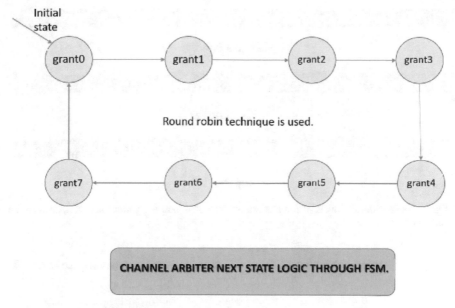

Fig. 2 This is the Round Robin model how channel is allocated based on request

4   Fsm Logic

4.1   Channel Arbiter FSM

This FSM implements the Round Robin logic in the next state logic (Fig. 2).

Initially, it will be in the grant0 state, and based on the valid request and priority, the algorithm is executed. For the channels having the same priority, then, the allocation of channel will be done with help of the Round Robin algorithm.

req and **advance** signals are used for the checking of the valid request for a channel, and **grant** signal will be used for the allocation of channel.

The priority arbitration has been depicted below with the three different levels.

4.2   Priority of the Channel

See Fig. 3.

Three different types of priority have been implemented based on the necessity of the channel usage. In the first priority mode (Normal priority), the channel is allocated to the peripherals based on the Round Robin algorithm. In the high priority mode, the preferred channel is allocated in alternate intervals of time, and in the

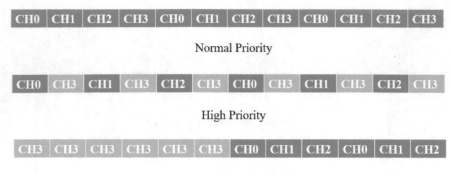

| CH0 | CH1 | CH2 | CH3 | CH0 | CH1 | CH2 | CH3 | CH0 | CH1 | CH2 | CH3 |

Normal Priority

| CH0 | CH3 | CH1 | CH3 | CH2 | CH3 | CH0 | CH3 | CH1 | CH3 | CH2 | CH3 |

High Priority

| CH3 | CH3 | CH3 | CH3 | CH3 | CH3 | CH0 | CH1 | CH2 | CH0 | CH1 | CH2 |

Top Priority

Fig. 3 Three different levels of priority are implemented, and the pattern is described with the yellow color

top priority mode, the preferred channel is allocated continuously till the end of its operation.

5 Block Diagram with AXI4

See Fig. 4.

The master and slave of the DMA controller have been connected to the AXI interconnect with the separate clock wizard and reset block. This can be packaged to

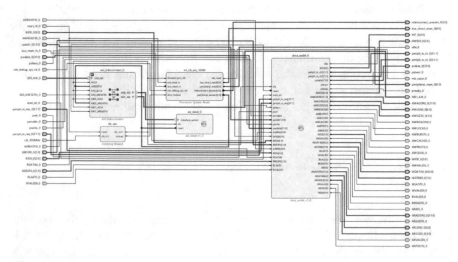

Fig. 4 With the on-chip bus, the pins are extended, and connections are shown

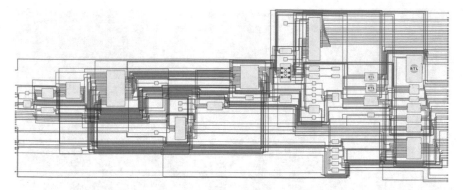

Fig. 5 Connection of the DMA controller with A2O core is depicted

use as an IP for the user. There are so many optional features for this DMA controller like outstanding mode, peripheral control, and multi-processor support.

5.1 A2O Core Interface

See Fig. 5.

This core is open sourced, and the developed controller has been integrated with the smart connect which is present in this by default.

6 Results

Unit testing of the design which is done in Questasim is attached below.

At the rising edge of the clock, the AXI interface signals AWVALID and AWREADY are getting asserted. These signals infer that the design has been checked correctly. 216 bits of data have been transferred total, so there are eight locations shift in the next, whenever the next transaction is occurring (Fig. 6).

At the rising edge of the clock, the AXI interface signals AWVALID and AWREADY are getting asserted. These signals are used to get the handshake correctly between the master as well as the slave. With the correct and accurate handshake, the transfers and the operation will be efficient. In the same way, for the read transaction, the interface signals ARVALID, ARREADY are also getting asserted at the rising edge of the clock only (Fig. 7).

WLAST and RLAST signals are asserted after the completion of the previous transaction. These AXI interface signals are useful in depicting the end of the transactions, while the current is going to end, and in other way, it is also depicting the information like in the next cycle the data being driven at the master, and for the read

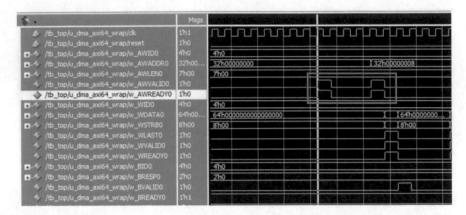

Fig. 6 WRITE transaction handshaking signals are highlighted. AWVALID and AWREADY are the two important signals which determine the correctness of the controller with the on-chip AXI4 bus

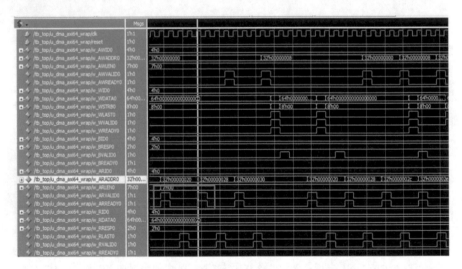

Fig. 7 READ transaction handshaking signals are highlighted and the address also same format with WRITE transaction

transactions in the next cycle, the driven data are being sampled and at the end again these signals getting asserted as shown in the waveform below (Fig. 8).

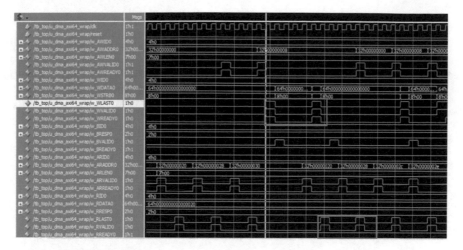

Fig. 8 At the end of each and every transaction, the WLAST and RLAST signals are asserted for WRITE and READ transactions, respectively

7 Conclusion

This work developed, simulated, and synthesized the AXI4 DMA controller, which was utilized in A20 processor-based fabless SoC. The controller achieves the optional features like cyclic buffer mode also. This proposed architecture has the potential to improve data transmission performance using AXI4. This design is elaborated and interfaced with A2O in Xilinx Vivado, and Simulation is done in Mentor Questa using slave-based function unit testing with the write and read operations. This controller supports the slow communication devices with the help of the different core and the clock divider.

The future scope of this work is mentioned as currently in this work DMA has a direct interface with the on-chip bus only. But this can be interfaced directly with the peripheral interfaces respectively.

References

1. G. Ma, H. He, Design and implementation of an advanced DMA controller on AMBA-based SoC, in *2009 IEEE 8th International Conference on ASIC* (2009), pp. 419–422. https://doi.org/10.1109/ASICON.2009.5351258
2. K. Chen, L. Qi, H. Yu, Design of two-dimension DMA controller in media multi-processor SoC, in *2008 Second International Symposium on Intelligent Information Technology Application* (2008), pp. 708–711. https://doi.org/10.1109/IITA.2008.493
3. C. Yu, C. Liu, C. Kang, T. Wang, C. Shen, S. Tseng, An efficient DMA controller for multimedia application in MPU based SOC, in *2007 IEEE International Conference on Multimedia and Expo* (2007), pp. 80–83. https://doi.org/10.1109/ICME.2007.4284591

4. Y.J.M. Shirur, K.M. Sharma, A. Aishwarya, Design and implementation of efficient direct memory access (DMA) controller in multiprocessor SoC, in *2018 International Conference on Networking, Embedded and Wireless Systems (ICNEWS)* (2018), pp. 1–6. https://doi.org/10.1109/ICNEWS.2018.8903991
5. H. Yuan, H. Chen, G. Bai, An improved DMA controller for high-speed data transfer in MPU based SOC, in *2004 Proceedings of the 7th International Conference on Solid-State and Integrated Circuits Technology*, vol 2 (2004), pp. 1372–1375. https://doi.org/10.1109/ICSICT.2004.1436811

A Comparative Study Using Numerical Simulations for Designing an Effective Cooling System for Electric Vehicle Batteries

Akshat Maheshwari, Vineet Singhal, Ajay Kumar Sood, and Meenakshi Agarwal

Abstract As technology is advancing rapidly in electric vehicle (EV), numerous automobile industries are focusing on shifting a part of their catalogue or have already started production of EVs. One of the most significant parts of an electric vehicle is battery which powers the vehicle, and it usually contains thousands of cells, which can be arranged in various combinations of series and parallel arrangements to form a module, many such modules together make a battery. Lithium-ion batteries are one of the preferred types for use in electric vehicles. The efficiency and life cycle of a lithium-ion battery depend on several factors. Battery temperature is one such critical factor and needs to be monitored to avoid early failure of the battery. Therefore, advanced cooling technology is to be incorporated into a battery to lower the temperature. This paper presents a comparative study of direct and tube cooling methods in a battery for effective cooling of an EV using ANSYS Fluent. These cooling systems are designed keeping an easier process of manufacturing in mind.

Keywords Electric vehicle · Lithium-ion battery · Battery cooling

1 Introduction

Lithium-ion batteries have been in extensive use in case of several electric vehicles because of their high energy density, operational temperature range being wider, less toxicity, and better charge cycles performance. In modern electric vehicles, there are around 5000–8000 cells (cell spec: 18,650), and these thousands of cells are divided into many modules of 200–500 cells [1]. These modules after long charge cycles and during acceleration of the vehicle produce large amount of heat which has to be removed to maintain the efficiency of the cells, retaining the capacity and number of

A. Maheshwari (✉) · V. Singhal · A. K. Sood · M. Agarwal
BML Munjal University, Gurugram, Haryana 122413, India

A. K. Sood
e-mail: ajay.sood@bmu.edu.in

M. Agarwal
e-mail: meenakshi.agarwal1@bmu.edu.in

© The Author(s), under exclusive license to Springer Nature Singapore Pte Ltd. 2023 179
Y.-D. Zhang et al. (eds.), *Smart Trends in Computing and Communications*, Lecture Notes in Networks and Systems 396, https://doi.org/10.1007/978-981-16-9967-2_18

life cycles. The optimum temperature for the well-functioning of the cell is around 15–35 °C [2]. Air has always been a good option when it comes to cooling, but as the demand for more power increases more heat is generally generated; as a result, air having a low heat capacity and thermal conductivity is then not a very good coolant [3, 4], and hence, we stick to water cooling in our cases of study. We need to implement a cooling system to remove the heat produced by the batteries as fast as possible using an efficient pump. Many small parts of such battery module are designed, and a cooling system is designed for this part of a module. This cooling system uses submerged water to cool down the cells and is simulated to find results and differences in flow and heat extraction. A fine mesh is generated in Ansys, and we use a micro-pump pumping 7 l/min of distilled water at temperature of 300 K and the outlet is an open outflow. The outer wall of cell is made of aluminum, and the entirety of the cell uses a custom-made Ansys material with a specific heat capacity of 998 J/Kg K and thermal conductivity in axial direction of 35 W/mK, radial direction 0.2 W/mK, and density of 2018 kg/m^3 [2, 5]. We compute steady flow, with energy equation and K epsilon standard viscosity equations on, each wall and interference are coupled to provide the heat exchange that is expected. The simulation is run for each design at two different discharge rates which supply us with different results to analyze.

2 Methodology

After exploring various design changes (distance between cells, distance from the shell), solution methods and boundary conditions two different cooling system models are designed. Each module is made using multiple lithium iron phosphate battery (LiFePO) [6]. This kind of battery uses graphite with metallic backing as anode and lithium iron phosphate as cathode.

The lithium cells referred in this research are of type 18,650, and each cell is created as a solid piece of a generalized matter named e-material which has generalized properties; this is done to not overcomplicate the individual design of a cell while avoid designing interior and exterior of the cell (standard size: 18 mm diameter and 65 mm length) with the following parameters [3, 7]:

Nominal voltage = 3.2 V, Density = 2018 kg/m^3, Reference temperature = 300 K, Specific heat capacity = 1018 J kg^{-1} K^{-1}, Thermal conductivity in radial direction = 1.5 Wm^{-1} K^{-1}, Thermal conductivity in axial direction = 25 Wm^{-1} K^{-1}, Energy density = 325 Wh/L (1200 kJ/L) and Cycle durability = 2000–12,000 cycles.

Pump properties are [8]: Max flow rate = 7, 9, and 12 L/min, Rated voltage = 12/24 V, Impeller type is closed impeller, Motor type is brushless DC motor, Liquid temperature = 0–100 °C, Ambient temperature = −40–70 °C, Media is pure water, antifreeze, working voltage range = 6–18 V or 12–28 V and power consumption = 1–32 W.

The cells are arranged in different orders to produce small modules, and each module has an outer shell made up of aluminum. Two different types of cooling

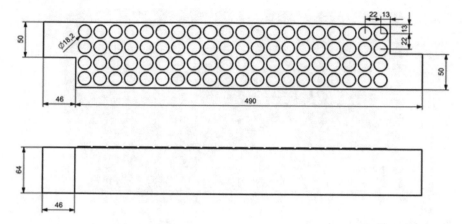

Fig. 1 20S 4P, 80-cells (direct cooling) (dimensions are in mm)

strategies are considered for simulations here in this paper, namely direct cooling and tube cooling which are presented in the following sections.

Boundary Conditions used for the simulations are:

- Inlet mass flow rate is 0.115 kg/s and 0.015 kg/s for direct cooling and tube cooling, respectively ([4, 9–11]).
- All cells are considered as a constant source of heat with heat generation rates corresponding for 1 and 2 °C discharge rates, (5318 W/m^3 and 19,452 W/m^3, respectively) ([12, 13]).
- All cell walls in contact with the fluid are thermally coupled with the system.

Simulation for two different discharge rates 1 and 2 °C was performed.

1. Discharge rate 1 °C is used to represent the ideal discharge rate during normal driving operation.
2. Discharge rate 2 °C is used to represent when there is a higher demand in energy during acceleration and hill climbing.

The heat generation rate is figured out using practical experiments on a set of twenty 18,650 lithium-ion cells at 1.35 A for 1 °C and 2.70 A for 2 °C, and the observed heat generation rate is 5318 W/m^3 and 19,452 W/m^3 [5] (Fig. 1).

2.1 Direct Cooling

Meshes for 20S 4P, 80-cells case, and straight tube case are as shown in Figs. 2 and 4, respectively. Orthogonal quality is >0.95 in all the cases. Quality of mesh was ensured by performing a grid independence test also to ensure that the final results

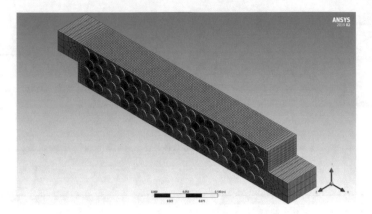

Fig. 2 Mesh for 20S 4P, 80-cells

are independent of the further refinement of the mesh. For direct cooling, a mass flow rate inlet is used with 0.115 kg/s, that is, 7 L/min, and for tube cooling, the inlet mass flow rate is 0.015 kg/s ([4] and [14]), the flow rate here is decreased because the area of cross section in case of tube cooling is very small, multiple tube-cooled modules can be cooled using the same pump to prevent heavy pressure differences in the tubes. The inlet temperature is maintained at 300 k. The outlet is made to be an outflow-type outlet. To achieve this flow rate, a readily available mini-DC water pump can be used with the specifications as shown in Table 2. We have used a pressure-based, steady-state solver with gravity set at 9.8 m/s^2, energy equations are turned on for the simulations.

2.2 Tube Cooling

80-cell module (20S 4P) straight tubes (see Fig. 3).

3 Results and Analysis

Velocity Contours for 1 °C discharge rate (see Figs. 5 and 6)

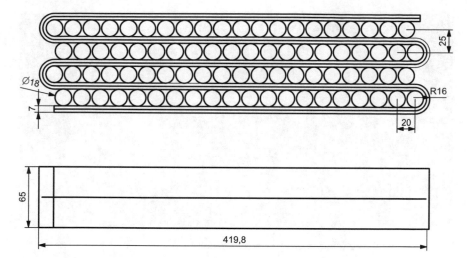

Fig. 3 20S 4P, straight tube

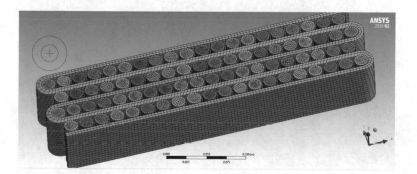

Fig. 4 Mesh for 20S 4P, straight tube

The velocity contours clearly show that straight tube model has a higher flow rate of liquid surrounding individual cells. Turbulence isn't very high in any of the cases. The velocity contours and types of packing are important because in real conditions, acceleration, and jerks have a huge impact on the actual flow rate, backflow is also a certain occurrence and can be prevented using backflow prevention valves at bottleneck regions.

Temperature Contours for 1 °C discharge rate (see Figs. 7 and 8)

A simple look at the temperature gradient contours shows that direct cooling is the superior form of cooling, these types of modules are easier to manufacture by using a punching press and boring process, seals can be made using rubber rings. The major drawback of this cooling method is the probability of a leakage, and this drawback is more observable when the entire cooling system is in an EV which is on move,

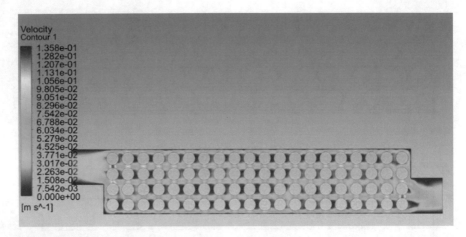

Fig. 5 20S 4P, 80-cells (direct cooling) velocity contour

Fig. 6 80-cell module (20S 4P) straight tubes (tube cooling) velocity contour

continuously accelerating and deaccelerating. Indirect cooling solutions like tube cooling when compared to direct cooling have moderate execution but have higher complexity when it comes to manufacturing. The temperature rises up drastically with the C rating and discharge amperage.

Tables 1 and 2 show the results for the two different charges, i.e., 1 °C and 2 °C, respectively. For indirect tube cooling method, the temperature rises much more as compared to the case with direct cooling with the increase in C rating from 1 to 2 °C.

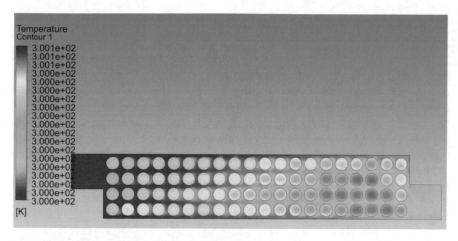

Fig. 7 20S 4P, 80-cells (direct cooling) temperature contour

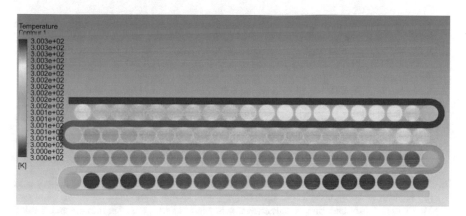

Fig. 8 80-cell module (20S 4P) straight tubes (tube cooling) temperature contour

Table 1 Results for 1C

Module	Maximum cell temperature (K)	Maximum fluid temperature (K)
80-cell module (20S 4P)	300.056	300.017
80-cell module (20S 4P) Straight tubes	300.3	300.13

Table 2 Results for 2C

Module	Maximum cell temperature (K)	Maximum fluid temperature (K)
80-cell module (20S 4P)	300.198	300.094
80-cell module (20S 4P) straight tubes	301.178	300.656

4 Conclusion

To recapitulate, two approaches are presented here. In case of indirect cooling, temperature rise is observed to be more with increase in C rating from 1 to 2 as compared to that in case of direct cooling. Direct cooling as observed from temperature contours seems to be a better option for providing more effective cooling as compared to the indirect cooling method. When it comes to designing a faction of a module, the results can help to fulfill the necessities of little and medium-sized companies. The exploration and approach prompt a decrease in the expense and the complexity of designing battery cooling systems thereafter improving the project lead time.

References

1. H. Liu, Z. Wei, W. He, J. Zhaoa, Thermal issues about Li-ion batteries and recent progress in battery thermal management systems: a review. Energy Convers. Manag., pp. 304–330 (2017)
2. L.H. Saw, Y. Ye, A.A.O. Tay, W.T. Chong, S.H. Kuan, M.C. Yew, Computational fluid dynamic and thermal analysis of Lithium-ion battery pack with air cooling. Appl. Energy **177**, pp. 783–792 (2016)
3. L.H. Saw, Y. Ye, A.A.O. Tay, Electrochemical-thermal analysis of 18650 lithium iron phosphate cell. Energy Convers. Manage. **75**, 162–174 (2013)
4. J.A. Brumley, A study of the energy consumption of a battery cooling system by different cooling, Graduate Theses, Dissertations, and Problem Reports, p. 5273 (2016)
5. Y. Huang, Y. Lu, R. Huang, J. Chen, Z. Liu, X. Yu, T. Roskilly, Study on the thermal interaction and heat dissipation of cylindrical Lithium-Ion Battery cells. Energy Procedia, pp. 4029–4036 (2017)
6. P. Cicconi, D. Landi, M. Germani, Thermal analysis and simulation of a Li-ion battery pack for a lightweight commercial EV. Appl. Energy, pp. 159–177 (2017)
7. W. Tong, K. Somasundaram, E. Birgersson, A.S. Mujumdar, C. Yap, Thermo-electrochemical model for forced convection air cooling of a lithium-ion battery module. Appl. Therm. Eng. **99**, 672–682 (2016)
8. X. Li, J. Zhao, J. Yuan, J. Duan, C. Liang, Simulation and analysis of air-cooling configurations for a lithium-ion battery pack, J. Energy Storage, pp. 102–270 (2021)
9. Q. Wang, B. Jiang, B. Li, Y. Yan, A critical review of thermal management models and solutions of lithium-ion batteries for the development of pure electric vehicles. Renew. Sustain. Energy Rev. **64**, 106–128 (2016)
10. S. Park, D. Jung, Numerical modeling and simulation of the vehicle cooling system for a heavy-duty series hybrid electric vehicle, SAE Technical Paper, p. 13 (2008)
11. A. Jarrett, I.Y. Kim, Design optimization of electric vehicle battery cooling plates for thermal performance. J. Power Sour. **196**(23), 10359–10368 (2011)
12. M. Rajadurai, S. Ananth, Battery thermal management in electrical vehicle-review article. IJISET—Int. J. Innov. Sci., Eng. Technol. **7**(2), (2020)
13. Z. Tang, X. Min, A. Song J. Cheng, Thermal management of a cylindrical lithium-ion battery module using a multichannel wavy tube. J. Energy Eng. **145**(1) (2019)
14. Y. Lyu, A.R.M. Siddique, S.H. Majid, M. Biglarbegian, S.A. Gadsden, S. Mahmud, Electric vehicle battery thermal management system with thermoelectric cooling. Energy Rep. **5**, 822–827 (2019)

Lexical Resource Creation and Evaluation: Sentiment Analysis in Marathi

Mahesh B. Shelke⊙, **Saleh Nagi Alsubari**⊙, **D. S. Panchal**⊙, and **Sachin N. Deshmukh**⊙

Abstract In India, the raise of regional language contents over social media, websites, blogs and news article are exponentially increasing because of ease of use of technology, and people are expressing their thoughts, opinion more conveniently and powerfully over Internet. In this paper, we evaluate the challenges of sentiment analysis in Marathi by setting up a baseline, where we produced an annotated dataset, however, initially, we created an annotated dataset consisting of Marathi news scraped from various newspaper/channel websites. Furthermore, domain experts annotated Marathi news with positive, negative and neutral polarity. And we used machine learning models such as logistic regression, Stochastic Gradient Decent (SGD), support vector machine (SVM), nearest neighbour, neural network, decision tree (DT), Naïve Bayes (NB) and proposed ensemble-based model for sentiment analysis to demonstrate effectiveness. In experimentation, the proposed ensemble classifier outperforms other classifiers with an accuracy of 94.16% and an F-score of 97.02% for fivefold validation. Also, for tenfold validation, the accuracy is 95.07%, and the F-score is 96.93%.

Keywords Marathi · Sentiment dataset · Machine learning · Indian language

1 Introduction

It is very easy at present to express thoughts on the web or on social media. After viewing films, using a product or visiting a locality, we can write movie reviews, product reviews or tourist reviews. This opinion-rich data will be interested both in the decision makers for the entities concerned and in enterprises seeking to improve their products or services. It provides people the opportunity to express themselves rather than media personalities speaking on behalf of the general majority of the

M. B. Shelke (✉) · S. N. Alsubari · D. S. Panchal · S. N. Deshmukh
Department of Computer Science & IT, Dr. Babasaheb Ambedkar, Marathwada University, Aurangabad, M.S., India
e-mail: mahesh_shelke21@hotmail.com

© The Author(s), under exclusive license to Springer Nature Singapore Pte Ltd. 2023   187
Y.-D. Zhang et al. (eds.), *Smart Trends in Computing and Communications*, Lecture Notes in Networks and Systems 396, https://doi.org/10.1007/978-981-16-9967-2_19

population. You can make your feelings heard by posting your thoughts on the Internet. As a result, vast amounts of information, including the author's point of view, are available on the Internet. It is now a challenge that these documents contain useful information. This enhances the application of sentiment analysis/ opinion mining.

Emotions play an important role in the communication and decision-making processes in our intellectual activities. Emotions are a succession of events made up of feedback loops. Feelings and behaviours, like cognition, may aspect cognition. In addition, the detection and interpretation of emotional data are vital in a wide range of fields, including human–computer interactions, e-learning, e-Health, automotive, security, user profiling and customisation. The analysis of textual sentiment is one of the popular computer linguistic methodologies.

Sentiment analysis is a process of recognising and categorising views expressed in a piece of text in a computational way, particularly to assess whether the person has a positive, negative and neutral opinion towards any certain topic, product, etc. Product reviews, Film reviews, blog and articles are the popular and available opinion-rich contents. There are three levels of sentiment analysis can be performed: Document level, Phrase/Sentence level and Entity/Aspect level. In document level sentimental analysis, the polarity is determined for the complete document. In sentence level sentiment analysis, polarity is determined for the individual sentences of the document. In the aspect level sentiment analysis, the polarity is determined for the entity/aspect of the document. Following methods can be used for sentiment analysis:

- **Using Lexicon-based Sentiment Analysis**: It is a polarity-based information dataset of phrases or word, in which scores are assigned to each word. This score describes the characteristics related to the term as positive, negative or neutral sentiment expressed in text.
- **Using N-Gram Model-based Sentiment Analysis**: It forms and uses the N-Gram model (unigram, bigram, trigram or mixed) using the categorization trainings data.
- **Using Machine Learning-based Sentiment Analysis:** Supervised and Unsupervised learning model can be used to perform prediction on data by extracting features from text.

The main objective of this paper is to develop and evaluate lexical resources for sentiment analysis in Marathi, as there are few lexical resources, libraries, Corpus and tools available in Marathi, which signifies that Marathi has not been explored in the field of sentiment analysis. The proposed approach is designed in combination with machine learning-based algorithm. The lack of Marathi preprocessing tools and annotated dataset for training the model was the main challenge during model development. As a result, we created a preprocessing tool using the Indic NLP Library and inltk tools [1, 2], as well as collecting Marathi news from various online newspapers/channels and manually annotating and validating it with domain experts. A comparison of all classification algorithms was also evaluated and discussed. In addition, an annotated dataset of Marathi news with sentiment orientations of positive, negative and neutral was developed [3, 4].

2 Related Work

Sentiment analysis (SA) and Opinion Mining are become popular in the field of natural language processing for Indian languages such as Hindi, Bengali, Tamil, Telugu and so on, but there are still some languages that remain unexplored due to a lack of lexical resources, such as Marathi, Gujarati, and Punjabi [5]. Three types of sentiment classification techniques are machine learning approaches, lexicon-based approaches and hybrid approaches. Machine learning approaches are classified into three types: supervised learning, semi-supervised and unsupervised learning. When there is annotated dataset unavailable to train the model then unsupervised learning is used, whereas supervised learning is used when there is a substantial annotated dataset available. The authors proposed a sentiment analysis (SA) method for Gujarati tweets that used POS tagging to extract features and SVM to classify the tweets. They have collected 40 tweets as dataset [6].

For performing sentiment analysis of Malayalam movie reviews, the authors devised a hybrid approach based on fuzzy logic. This consists of a tagging machine learning system and a fuzzy logic approach to determining review membership. TnT Tagger was also used to train datasets that had been manually tagged [7]. For Tamil and Bengali tweets, the authors developed machine learning-based sentiment analysis model based on probabilistic and decision tree classification. The amount of the dataset used, variations in writing style, and the incorrect use of punctuation marks all have an impact on the model's performance [8]. Using machine learning approaches such as support vector machine, maximum entropy, decision tree and Naïve Bayes, the author developed a system for predicting emotion in Tamil films. When it comes to accuracy, SVM outperforms other algorithms [9].

3 Proposed Methodology

This section describes the corpus creation process in detail, beginning with data collection, preprocessing, corpus annotation and inter-annotator agreement for measurement of agreement between annotators. And also, this section covers sentiment classification approach.

3.1 Lexical Corpus Creation

Web Scraping
We created a python-based web scrapper to collect Marathi news from various online newspapers/channels and collected 1649 news headlines from categories such as general news, current affairs, sports, entertainment, art, culture, science and technology, health and medicine and so on.

Data Preprocessing
We pre-processed the crawled data into the desired forms, following steps are carried out in preprocessing:

- Manually identify and remove duplicate and irrelevant news.
- Identify news items that contains English words and transliterated them.
- Remove improper punctuation marks, smileys, hashtags and photo tags.
- Remove complex sentences because they are unsuitable for sentiment analysis.

Data Annotation and Inter-annotator agreement
For manual data annotation, we chose three Marathi language domain experts who are academicians and researchers to annotate a Marathi news dataset with polarity scores -1, 0 and 1 (Negative, Neutral and Positive, respectively). For evaluation of manual data annotation, we have used the Fleiss' Kappa inter-annotator agreement score. Following formula is used to calculate Fleiss' kappa score [10].

$$k = \frac{\overline{P}_x - \overline{P}_x}{1 - \overline{P}_x}. \tag{1}$$

where the factor $1 - \overline{P}_x$. represents the degree of agreement that can be obtained other than by chance, The degree of agreement that was achieved above chance is given by $\overline{P}_x - \overline{P}_x$. and if the evaluators are totally in agreement, Kappa $k = 1$ and $k = 0$ if there is no agreement amongst the evaluators (other than what would expected by chance).

And the inter-annotator agreement score for Marathi news dataset is $k = 0.957$, which is almost perfect agreement. Above Table 1. Fleiss's Kappa Inter-annotator agreement score. And following Table 2 shows examples of data annotation. Table 3 shows the statistics for Marathi news dataset after preprocessing and data annotation.

Table 1 Fleiss's Kappa inter-annotator agreement score

Notator (i with j)	K-score
R_{12}	0.953
R_{23}	0.965
R_{13}	0.954
Fleiss's Kappa score (R_{123})	**0.957**

Table 2 Shows examples of data annotation

News	Polarity
पकिअप पलटी होऊन भीषण अपघात, ११ जणांचा मृत्यू ७ जखमी	−1
ऐतिहासकि अर्थसंकल्प, सुकरॅप पॉलसीिमुळे देशात ५० हजार नवे जॉब	1
वॉर्ड आरक्षणासंदर्भात औरंगाबाद महापालकिने शपथपत्र दलि्लीला पाठवलि	0

Table 3 Shows the statistics for Marathi news dataset after preprocessing and data annotation

S. No.	Statistics	No. of news
1	Initial	1649
2	After preprocessing	1321
3	Positive	538
4	Negative	536
5	Neutral	237

3.2 Feature Extraction

To train the model, supervised machine learning algorithms require a text document in the form of a feature vector. Feature extraction techniques reduce the length of the feature vector by transforming all of the features in a lower-dimensional feature vector. Unigram features are bag-of-words (BoW) features obtained by removing unnecessary spaces and noisy characters between two words.

3.3 Sentiment Classification

Experiment work is based on supervised machine learning algorithms such as logistic regression, Stochastic Gradient Decent (SGD), support vector machine (SVM), nearest neighbour, neural network, decision tree (DT), Naive Bayes (NB), and the proposed ensemble-based model.

Individual algorithms compose an ensemble-based sentiment analysis in order to develop a highly accurate predictive model that classifies Marathi news in terms of sentiment orientation. The ensemble classifier system uses the average predicted probability, which is a soft voting approach, to determine the sentiment orientation. Figure 1 shows sentiment classification approach for Marathi news.

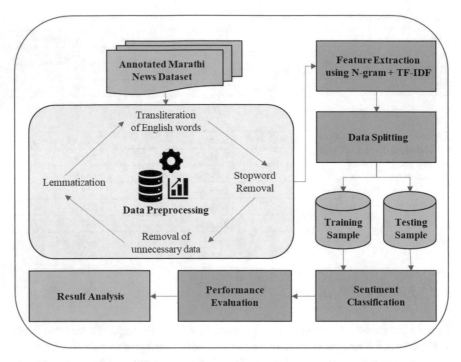

Fig. 1 Shows sentiment classification approach

4 Experimental Evaluation

In the experiment, we focused on the three types of class problems: positive, neutral and negative. We collected Marathi news headlines from a variety of newspaper and channel websites. In addition, the Marathi news dataset is divided into three categories based on the sentiment expressed in the sentences. If the expressed opinion is positive, it is labelled as 1, neutral, 0, and negative, it is labelled as −1. For training and testing samples, the dataset is divided into 80:20 ratios and various preprocessing techniques, such as data cleaning, URL and Hashtag removal, extra blank spaces, punctuation mark removal, emoticons, transliteration of English words into Marathi, Stopword removal and lemmatization, are used on the dataset. In experimentation validation, we used k-fold cross validation with $k = 5$ and $k = 10$. A classification model can be evaluated using a variety of measures, with accuracy being one of the most straightforward. The number of correctly classified examples divided by the total number of examples is the definition of accuracy. Accuracy is useful, but it ignores the complexity of class imbalances and the different costs of false negatives and false positives. Following Table 4. Shows performance evaluation of individual classifier with k-fold validation.

We performed fivefold cross validation on training dataset, and obtained highest accuracy for ensemble classifier as 95.19%, and also performed tenfold cross

Table 4 Shows performance evaluation of individual classifier with k-fold validation

S. No.	Classifier	$k = 5$		$k = 10$	
		Accuracy	F-score	Accuracy	F-score
1	Logistic regression	86.15	92.53	86.72	92.84
2	Stochastic Gradient Decent	94.12	96.84	94.44	95.32
3	SVM	88.78	93.87	90.95	95.01
4	Nearest neighbour	92.56	95.81	93.01	96.02
5	Neural network	94.16	97.02	95.07	96.93
6	Decision tree	93.12	94.16	94.21	95.16
7	Naïve Byes	87.30	93.11	88.44	93.71
8	Ensemble classifier	95.19	97.15	96.33	97.92

validation on training dataset, and we obtained highest accuracy for ensemble classifier as 96.33% for all Marathi news dataset. Figures 2 and 3 shows performance evaluation of individual classifier with fivefold and tenfold validation.

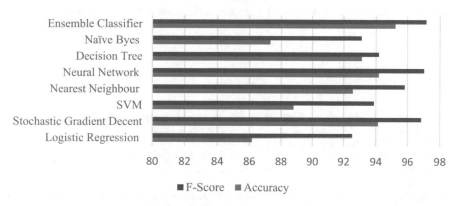

Fig. 2 Shows performance evaluation of individual classifier with fivefold cross validation

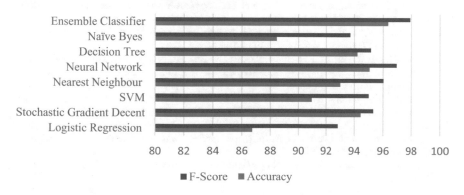

Fig. 3 Shows performance evaluation of individual classifier with tenfold cross validation

5 Conclusions and Future Work

This paper presents a baseline model for sentiment analysis in Marathi. We created an annotated dataset of Marathi news scraped from various newspaper/channel websites, and domain experts annotated Marathi news with positive, negative and neutral polarity. For sentiment analysis, we used machine learning models such as logistic regression, Stochastic Gradient Decent (SGD), support vector machine (SVM), nearest neighbour, neural network, decision tree (DT), Naive Bayes (NB) and the proposed ensemble-based model. The proposed ensemble classifier outperforms other classifiers in experiments, with an accuracy of 94.16% and an F-score of 97.02% for fivefold validation. In addition, the accuracy for tenfold validation is 95.07%, and the F-score is 96.93% and this dataset will be made available publicly for advancement in research.

In the future, we can create more domain-specific annotated lexical resources to expand resources for Indian languages, increase dataset size and implement using a deep learning-based model.

Acknowledgements Authors acknowledges the Chh. Shahu Maharaj National Research Fellowship (CSMNRF-2019), Pune, Maharashtra.

References

1. G. Arora, iNLTK: natural language toolkit for Indic languages, in *Proceedings of Second Workshop for NLP Open Source Software (NLP-OSS)*, 2020
2. A. Kunchukuttan, *The IndicNLP Library* (2020)
3. S. Badugu, Telugu movie review sentiment analysis using natural language processing approach, in *Data Engineering and Communication Technology. Advances in Intelligent Systems and Computing*, 2020
4. A. Rajan, A. Salgaonkar, Sentiment analysis for Konkani Language: Konkani Poetry, a case study, in *ICT Systems and Sustainability. Advances in Intelligent Systems and Computing*, 2020
5. S. Rani, P. Kumar, A journey of Indian languages over sentiment analysis: a systematic review, pp. 1415–1462 (2018)
6. V.C. Joshi, V.M. Vekariya, An approach to sentiment analysis on Gujarati Tweets. Adv. Comput. Sci. Technol. **10**(5),1487–1493 (2017)
7. M. Anagha, R.R. Kumar, K. Sreetha, P.C. Reghu Raj, Fuzzy logic based hybrid approach for sentiment analysis of Malayalam movie reviews, in *IEEE International Conference on Signal Processing, Informatics, Communication and Energy Systems (SPICES)*, Kozhikode, 2015
8. S. Se, R. Vinayakumar, M. Anand Kumar, K.P. Soman, Predicting the sentimental reviews in tamil movie using machine learning algorithms. Indian J. Sci. Technol. **9**(45), 1–5 (2016)
9. S.S. Prasad, J. Kumar, D.K. Prabhakar, S. Tripathi, Sentiment mining: an approach for Bengali and Tamil tweets, in *2016 Ninth International Conference on Contemporary Computing (IC3*, Noida, 2016
10. "Wikipedia," [Online]. Available: https://en.wikipedia.org/wiki/Fleiss%27_kappa. [Accessed May 2021]

11. M.G. Jhanwar, A. Das, An ensemble model for sentiment analysis of Hindi-English code-mixed data, in *Workshop on Humanizing AI (HAI)*, Stockholm, Sweden, 2018
12. M.A. Ansari, S. Govilkar, Sentiment analysis of transliterated Hindi and Marathi Script, in *Sixth International Conference on Computational Intelligence and Information*, Cochin, India

Artificial Neural Network Method for Appraising the Nephrotic Disease

K. Padmavathi, A. V. Senthilkumar, Ismail Bin Musirin, and Binod Kumar

Abstract In the shape of an affected person's proof, the medical report is an ever-developing supply of record for a medical institution. One of the complex issues that get up in the transplanted kidneys is glomerulonephritis. In AI, there are two methodologies: managed and solo mastering. Characterization is a method that falls underneath controlled learning. Out of numerous arrangement models, the maximum prevalently applied is the artificial neural community. While neural networks turn out tremendous in characterization and preparing a device, the precision of the outcome may also anyways be beneath inquiry. The enhancement of the artificial neural networks is completed by using the exactness and space of the result. For this, ANN may be hybridized with a metaheuristic algorithm referred to as the cat swarm optimization (CSO) set of rules. The benefits of optimization artificial neural community are normally the development in the precision of the order, translation of the statistics, and reduction in fee and time utilization for buying real outcomes and so forth within the prevailing study, a correlation between the aftereffects of an ANN decrease again propagation version and the proposed ANN-CSO version is carried out for medical assessment.

Keywords Machine learning · ANN · CSO algorithm

1 Introduction

The measure of information is developing dramatically and it is important to break down this immense measure of information to separate helpful data from it. This

K. Padmavathi (✉) · A. V. Senthilkumar
Computer Applications, Hindusthan College of Arts and Science, Coimbatore, India
e-mail: padhukarups3@gmail.com

I. Bin Musirin
Faculty of Electrical Engineering, Centre of Electrical Power Engineering, Universiti Teknologi, Skudai, Malaysia

B. Kumar
Department of MCA, Rajarshi Shahu College of Engineering, Pune, India

© The Author(s), under exclusive license to Springer Nature Singapore Pte Ltd. 2023   197
Y.-D. Zhang et al. (eds.), *Smart Trends in Computing and Communications*, Lecture Notes in Networks and Systems 396, https://doi.org/10.1007/978-981-16-9967-2_20

prompted the development field of information mining. Information mining alludes to removing information from such enormous measures of data sets. Information mining is the center of knowledge discovery data measures. Knowledge discovery data is a coordinated cycle of the recognized, substantial, novel, valuable and justifiable examples from huge and complex datasets [1]. Data mining assignments may be partitioned into two instructions: expressive and predictive. Prescient undertaking incorporate grouping, relapse, time arrangement examination and elucidating task incorporate bunching, affiliation rule [2]. These procedures can be utilized in explicit zones [3]. Discuss this process and their utility in a different area. Statistics mining applications contain special discipline, as an example, deals, medicine, account, advertising, clinical services, banking and safety [4, 5]. Order is information—digging method utilized for the expectation of a class of items. It is an illustration of directed learning. Grouping anticipates clear cut marks (discrete, requested). Information arrangement includes two stages. The preliminary step is the mastering step (getting ready step) wherein a classifier is labored to depict a foreordained association of data training. Within the 2nd step the model that's the inherent preliminary step is utilized for the characterization of obscure records as an example test records is applied for assessing the classifier exactness. There are various characterization calculations like preference tree, gullible classifier and artificial neural network gives the relative research of these order calculations [6, 7]. Synthetic neural community is a machine gaining knowledge of approach utilized for classification troubles.

2 Related Work

Various analysts executed different strategies to distinguish CKD and anticipate the endurance of patients. Various strategies, execution apparatuses and techniques give various outcomes and give alternate point of view on this specific sickness [8].

In [6], the creators thought the precision and execution time is based on the performance of the ocular networks. In light of the examination results, ocular network performs better than the other techniques.

In [7], artificial neural network (ANN) is executed with the back-propagation network, summed up generalized forward neural networks (GRNN) and modular neural network (MNN) (BPN) for early recognition of CKD. In light of the exhibition as far as precision, affectability and explicitness, the best model is picked to create of the smart framework. The framework is lined up with Google cloud stage by utilizing Google Application Engine. Subsequently, CKD can be recognized in the beginning phase all the more effectively. It is likewise valuable for general individuals to self-identify the danger of contact to CKD.

In [9], three information mining procedures have been analyzed for kidney sickness examination. These strategies are artificial neural network, decision tree and logical regression. From the computational outcome it is seen that ANN performs better and gives higher exactness.

An examination between Back spread fake neural network, PSO neural network and a Bat calculation streamlined neural network was made by Golmaryami et al. [10] on stock value forecast. The outcomes indicated that bat calculation changed the weight lattice more precisely than the other two utilized. The bat calculation was assessed alongside its the present variations and was introduced by Induja et al. [11]. Certain contextual investigations including bat calculation alongside its different applications were generally chosen and surveyed. The surveys and synopses dependent on these investigations have been quickly advanced in [11].

3 Optimization-Based Artificial Neural Network Classification

This segment depicts the fundamental commitment of this whole work. The subsection talks about the means remembered for the proposed mixture highlight choice strategy. As a rule, by decreasing the quantity of highlights, it will lessen the preparation time to assemble and limit the unpredictability of the classifier model. Highlight choice is to choose the most significant highlights before it adjusted as a contribution to the classifier to get upgrade arrangement. Thus, the less pertinent or less critical highlights will be dispensed with or taken out from the rundown. Half and half highlights determination technique utilizing highlight positioning procedure specifically alleviation f with ground-breaking transformative calculation improvement calculation is proposed. The detail clarification of our proposed strategy is depicted in the tracing after subsection [12].

3.1 Cat Swarm Optimization

The optimization technique unites two modes, looking for search mode and the tracing mode. Taking modes into the count, a mixture ratio of consolidating seeking for mode with tracing mode. While they might be resting, they move their position purposely and steadily, every now and then even live inside the main job. No matter what, for utilizing this lead into CSO, we utilize searching for mode to adapt to it.

The direct of looking for after focal points of cat is associated with tracing mode. for this reason, it is miles clear that MR have to be extremely a motivation with a selected final goal to guarantee that the cats make a contribution most of the strength in search of for mode, tons equal to this present day truth. The technique of CSO can be created into the following steps.

Step 1: Based on the identical time create N cats.
Step 2: Arbitrarily sprinkle the number of cats into the M-dimensional arrangement space and erratically pick qualities, which are in-degree of the most incredible speed, to the speeds of each cats. By utilizing then aimlessly

choose amount of felines and set them into following mode as demonstrated by means of MR, and the others set into looking for mode.

Step 3: Assess the well-being appraisal of each feline through utilizing the spots of felines into the wellness work, which tends to the proportions of our goal, and keep up with the quality feline into memory. Know that we most certainly need to think about the state of the quality feline (x best) because of it tends to the top notch plan as of as of now.

Step 4: Stream the felines as demonstrated by their flags, if feline k is looking for mode, apply the feline to the looking for mode measure, via and enormous use it on the following mode measure. The procedure steps are shown ahead.

Step 5: Again the select amount of felines and set them into following mode as affirmed through MR, with the guide of then set trade felines into searching for mode.

Step 6: test the give up circumstance, at whatever factor satisfied, prevent this machine, and for the maximum factor reiterate step 3 to step 5.

3.2 Artificial Neural Network

A network is prepared for appearing conceptual acknowledgment in la small amount of time. A ocular community abuses the non-linearity of an trouble to symbolize a group of wanted statistics sources. Ocular networks were massive in expertise a superior manner for grouping in AI and discovers utility in one of a kind fields, as an instance, statistics mining, design acknowledgment, crime scene investigation and so on.

An artificial neural network structure comprises of essentially 3 parts, they are input layer, hidden layer and output layer. There are the accompanying kinds of artificial neural network. Least difficult neural network structures is accompanied by feed—forward. It contains of a data film in which the records enters and is inferred out on the grounds that the yield from the yield film. The presence of shrouded layers is discretionary. It uses an incitation work for affiliation. Lower back-multiplication neural organization. A returned unfurl ocular local area is indistinguishable as the feed forward neural organization beside one little change.

4 Proposed Method

Proposed include determination method in CKD dataset. The cycle of proposed highlight determination methods in CKD dataset. In this cycle, we have utilized two covering techniques in particular Genetic Search and Greedy stepwise [13, 14]. In Genetic hunt approach, overlaying subset evaluator with development primarily based is applied to select the vast 18 highlights from CKD informational collection. Moreover, if there should be an prevalence of keen stepwise, covering subset evaluator with enhancement-based classifier is applied to pick out the good sized 15 highlights from CKD informational series.

5 Result and Discussion

UCI machine learning repository CKD dataset was used for this proposed approach. Attributes are displayed accordingly. Classification algorithms are applied to the dataset for performance testing and assessment training [15].

Table 1 presents the performance of the various classification techniques. The accuracy of the ANN peaks high (Fig. 1)

Table 2 shows the various levels of classification techniques using the Cat Swarm Optimization techniques.

Table 1 Classifiers KNN, SVM, ANN performing their classification levels during their performance

S. No.	Algorithm	Accuracy	Precision	Recall
1	KNN	89	86	87
2	SVM	92	90	91
3	ANN	94	92	94

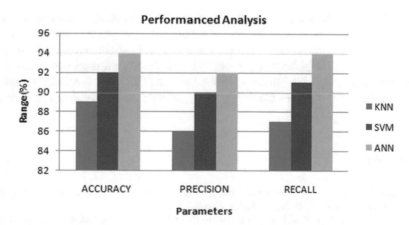

Fig. 1 Show the classification comparison of KNN, SVM, ANN classification algorithm ranges

Table 2 Performance of the classification techniques of Cat Swarm optimization techniques levels of KNN, SVM, ANN

Algorithm	Accuracy	Precision	Recall
CSKNN	92	90	91
CSSVM	95	93	95
CSANN	98	96	97

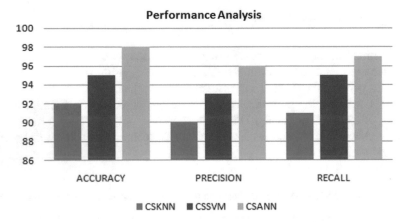

Fig. 2 Shows the classification comparison of optimization-based classifier algorithms. The comparison between existing methods (CSKNN, CSSVM) and proposed (CSANN) method

6 Conclusion

This examination built up an adjusted form of (CSO) to improve seeking through capacity by amassing look in the region of the best arrangement sets. We at that point consolidated this with an ANN to create the CSO-ANN strategy for include determination and ANN boundary streamlining to improve arrangement exactness. Assessment utilizing UCI datasets exhibited that the CSO-ANN technique requires less time than CSO-ANN to acquire grouping aftereffects of prevalent exactness.

References

1. J. Han and M. Kamber, *Data mining: concepts and techniques*, 2nd ed. Amsterdam; Boston: San Francisco, CA: Elsevier; Morgan Kaufmann, 2006.
2. P. Sondwale, Overview of Predictive and Descriptive Data Mining Techniques. Int. J. Adv. Res. Comput. Sci. Softw. Eng. **5**(4), 262–265 (2015)
3. B. Ramageri, "DATA MINING TECHNIQUES AND APPLICATIONS," *Indian J. Comput. Sci. Eng.*, vol. 1, no. 4, pp. 301–305.
4. P. Chouhan and M. Tiwari, "Image Retrieval Using Data Mining and Image Processing Techniques," *Image (IN)*, vol. 3, no. 12, 2015.
5. Simmi bagga and g. n. singh, "applications of data mining," *Int. J. Sci. Emerg. Technol. Latest Trends*, pp. 19–23, 2012.

6. N. Padhy, The Survey of Data Mining Applications and Feature Scope. Int. J. Comput. Sci. Eng. Inf. Technol. **2**(3), 43–58 (2012)
7. S.S. Nikam, A comparative study of classification techniques in data mining algorithms. Orient J Comput Sci Technol **8**(1), 13–19 (2015)
8. ArturQ.B.da Silva, Taina V. Delaware Sandes-Freitas, JulianaB.Mansur,Jose Osmar Medicina - Pestana,and Gianna Mastroianni-Kirsztajn."Clinical Presentation, Outcomes, and Treatment of Membranous renal disorder once Transplantation", International Journal of Nephrology, 2018.
9. Dr. S. Vijayarani and Mr. S. Dhayanand, "KIDNEY DISEASE PREDICTION USING SVM AND ANN ALGORITHMS," International Journal of Computing and Business Research (IJCBR), vol. 6, no. 2, 2015.
10. Ms. Sonali Sonavane, Prof. A. Khade and Prof. V. B. Gaikwad, "Novel Approach for Localization of Indian Car Number Plate Recognition System using Support Vector Machine," International Journal of Advanced Research in Computer Science and Software Engineering, vol. 3, no. 8, pp. 179–183, 2013
11. P.S. Devi, C.H. Sowjanya, K.V.N. Sunitha, A Review of Supervised Learning Based Classification for Text to Speech System. International Journal of Application or Innovation in Engineering & Management (IJAIEM) **3**(6), 79–86 (2014)
12. A.V. Senthilkumar, K. Padmavathi An efficient Meta classifier technique for Membranous Nephropathy Kidney disease. ISSN:227-8616, 2020
13. R. K. Chiu, R.Y. Chen, S. A. Wang "Intelligent systems on the cloud for the early detection of chronic kidney disease," in International Conference on Machine Learning and Cybernetics, Xian, 2012(July).
14. K.R. Lakshmi, Y. Nagesh, M.V. Krishna, Performance Comparison of Three Data Mining Techniques for Predicting Kidney Dialysis Survivability. International Journal of Advances in Engineering & Technology (IJAET) **7**(1), 242–254 (2014)
15. A.V. Senthilkumar, K. Padmavathi, A proposed method for Prediction of Membranous nephropathy Disease. International Journal of Innovative Technology and Exploring Engineering", ISSN:22783075, Volume -8 Issue-10,August 2019

Artificial Neural Network and Math Behind It

Harshini Pothina and K. V. Nagaraja

Abstract Article is structured in such a way that the reader could easily understand from the roots of an artificial neuron to its applications in between gaining knowledge on how an artificial neuron exists. It starts with the history of neuron, and the explanation goes on by describing the architecture of artificial neuron and on to the functioning of neuron. It also explains the basic math involved in constructing an artificial neuron by describing the activation functions used in the network, and the mathematical models helps in building the neural network. It also talks about the structure of feed forward network, loss function imposing an error on the output, also the gradient descent, and the back propagation method involving the computation of cost function with respect to the network parameters. As a conclusion, paper describes the importance of artificial neural networks in the upcoming artificial intelligence and many more applications behind it.

Keywords Artificial neural network · Back propagation · Feed forward network · Gradient descent · Perceptron · Sigmoid neuron

1 History

Neural networks that we are going to consider are artificial neural networks, which are based on human brains structure and its function. The first step toward artificial neural networks came in the year 1943; Warren McCulloch and Walter Pitts released a journal on working of neurons so as to depict its capacity in human brain. They demonstrated a basic neural system utilizing electric circuits.

H. Pothina (✉)
Department of Electronics and Communication Engineering, Amrita School of Engineering,
Amrita Vishwa Vidyapeetham, Bengaluru, India
e-mail: BLENU4EIE18031@bl.students.amrita.edu

K. V. Nagaraja
Department of Mathematics, Amrita School of Engineering, Amrita Vishwa Vidyapeetham,
Bengaluru, India
e-mail: kv_nagaraja@blr.amrita.edu

© The Author(s), under exclusive license to Springer Nature Singapore Pte Ltd. 2023
Y.-D. Zhang et al. (eds.), *Smart Trends in Computing and Communications*, Lecture Notes
in Networks and Systems 396, https://doi.org/10.1007/978-981-16-9967-2_21

Later, in 1949, Donald Heb composed a book named "The Organisation of Behaviour," which brought up the way that the pathways of neurons are reinforced each time they are utilized, an idea in general by which humans learn. He contended that if two nerves fire simultaneously, the association between them is upgraded.

In 1962, Widrow and Hoff assembled a studying approach that inspects the yield before the weight alters it (that is 0 or 1) as per the standard: change in weight = (pre-weight line yield)*(error/number of input values). It depends on the possibility that the major error of one perceptron can be modified by distributing it over the system while other perceptrons modify their weights. Even after applying this standard, despite everything brings about an error, however, it will run right independently [1].

2   Introduction

Neural networks process the data as like the human brain does. Network is structured with a huge number of interrelated processors called neurons that work along to deal with certain problem. *What are artificial neural networks?* ANNs are popular machine learning techniques that simulate the mechanism of learning in biological organisms. The cells present in the human nervous system are referred to as neurons. Axons and dendrites are used to connect these neurons to one another using a connecting region called synapses. The change in the response of external stimuli depends upon the strength of the synaptic connections, and due to these changes, learning in living organisms takes place. This biological procedure is replicated the same way in artificial neural networks, which contain processing elements called neurons. These processing elements are joined to each other by means of weights, which play the similar part as the power of synaptic connections in biological structures [2–4]. Each input to a neuron is associated with a weight which acts on the function operated at that subdivision.

3   Artificial Neuron

See Fig. 1.

An ANN computes a function of the input neurons to the output neurons using an intermediate parameters called weights and biases. The data from the input layer is taken into the hidden layer and the information is seen at the output layer after an amazing process performed in the hidden area. It is within the hidden area where all the process really happens through a system of connections specified by "weights" and "biases" (generally referred as W and b). Implementation and optimization of a neural network lies in understanding what exactly happens in the hidden area. The neuron receives the input and computes the weighted sum and by adding the bias it decides whether neuron should be "fired" or "activated" based on the result and a pre-set "activation function." After then, the neuron sends the data to the next

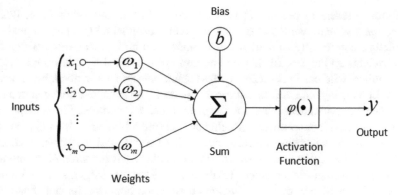

Fig. 1 Artificial neuron [5]

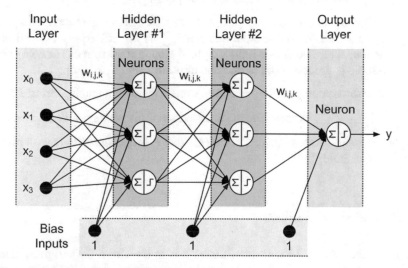

Fig. 2 Different layers of a network [9]

attached neuron in a process called "forward pass." At the termination of this action, output layer with one neuron for every one feasible output is connected to the hidden layer [6–8] (Fig. 2).

4 Perceptron

We have a general thought of the architecture of a neural network. Now, let's continue our discussion with the explanation of working of a neural system. *How does a neural network run?* To understand its working, we need to know the different kinds of neurons present in our neural systems. Perceptron is one of its kind. *What is perceptron and what can it do?* A basic artificial neuron with input and output layers is called perceptron. It is a mathematical model of a biological neuron which talks about weights and biases. In an actual neuron, the electrical signals are received by the dendrites, whereas in perceptron, these electrical signals are given by numerical values. In a biological neuron, electrical signals at the synapses between dendrites and axons are modulated in various amounts. It can be modulated in perceptron by multiplying each input value by weights and adding a bias to it. Bias is a constant value which helps in better performance of the neuron, and it is given a constant value 1 by default. A neuron discharges an output only if the strength of the input neuron exceeds a certain threshold. Total strength of the input signals can be modeled in a perceptron by computing its weighted sum and applying its step function to determine its output value. As in biological neural systems, this can be fed to the other perceptrons. Perceptron takes the following steps:

(i) Multiplying all the input samples with their weights w, (ii) Summation of the weighted sum of all the input samples: $\sum w_j x_j$, (iii) Apply the activation function, at the end decide if the weighted total is more than a threshold, where threshold is proportionate to bias and allocate 1 or 0 as an output.

We can likewise compose the perceptron work in the accompanying terms: $F(x) = \begin{cases} 0 \text{ if } w.X + b < 0 \\ 1 \text{ if } w.X + b \geq 0 \end{cases}$. Here, $w.X$ is the dot product of the vector with components of weight and input vector matrix, b is the bias and is proportional to the $-$threshold (bias $= -$threshold) [10].

5 Activation Function

The major purpose of the activation function is to transform the weighted sum of the inputs of every neuron at the input side to the output signal, and these output values are distributed as inputs to the next layer neurons. Any activation function that we consider must be differentiable in view of the fact that a back propagation technique is used to reduce the error and run the weights accordingly. One of the foremost advantages of the activation function [2] is that we can get various models of the output of our choice by varying the weights and bias. According to the mathematical equation of the perceptron that we have seen above, the output can be made more likely with the large positive value of bias, and a large negative value of bias makes the output more unlikely, and weights can be changed accordingly to get the desired output. A perceptron is capable of decision-making by analyzing the given data according to the training set. Complex networks with lot many layers of perceptrons can be made possible where the output from previous layer is taken by the next successive layer as input signal, and weights it up to yield. We moderately vary weights and bias to observe the performance of the neural network, but for a perceptron, small variations in the weights and bias lead output shift from 0 to 1 and vice-versa. Due to this drawback of perceptron, network is replaced with many other activation functions like tanh, rectified linear unit (ReLU), and Sigmoid functions, also called logistic function. The present time activation function which is used more is sigmoid function due to the fact that output can be any continuous value between 0 and 1.

Sigmoid function: It is demonstrated by the following equation: $\sigma(x) = \frac{1}{(1+e^{-x})}$, and the output:

$F(x) = \sigma(wX + b)$ So, the formula of output becomes $\sigma = \frac{1}{\left(1+e^{-\sum w_j X_j + b}\right)}$.

Artificial neurons receive multiple inputs as like biological neurons receives from pre-synaptic neurons. These artificial neurons take all the inputs, adds them up, and process using sigmoid function only if the summation exceeds the threshold limit, and then, the processed data is served as output.

Derivatives of common activation functions:

1. **Sigmoid activation function:** $\sigma(x) = \frac{1}{(1+e^{-x})}$

Quotient rule is used to calculate its derivative: $\sigma'(x) = \frac{\partial\left(\frac{1}{1+e^{-n}}\right)}{\partial(x)} = \text{Sigmoid}(x)(1 - \text{sigmoid}(x))$.

If we have a careful look at its derivative, we can observe that the outcome of its derivative is simply sigmoid(x) weighted by $1 - \text{sigmoid}(x)$. This is very convenient to compute effective gradients in a neural network [7, 11]. The output of the sigmoid function ranges between 0 and 1, but for very large negative or positive input sample, the output at the sigmoid function becomes saturated at the rear end of 0 or 1. Hence, the derivative of the function becomes less in value preventing the updates of the weight. Even though the sigmoid function has a very good interpretation, it can cause neural network to get stuck at their current state during training. Hyperbolic tangent function is the substitute to the sigmoid function.

2. **Hyperbolic Tangent function:** $\tanh(x) = \frac{\sinh(x)}{\cosh(x)} = \frac{e^x - e^{-x}}{(e^x + e^{-x})}$

Like sigmoid function, tanh is also a sigmoidal function which is "S" shaped. The output of tanh function ranges from -1 to $+1$, thus the very large negative input maps to the negative [2] output, very large positive input maps to the positive output, and the zero valued input maps to the near zero output. This makes the function less likely to get stuck while training [12].

Quotient rule is used to calculate its derivative:

$$\tan h'(x) = \frac{\partial \tanh(x)}{\partial x} = 1 - (\tanh(x))^2$$

3. **Identity Activation Function:** Linear $(x) = x$

The output values from this function ranges from $-\infty$ to $+\infty$. It maps the pre-activation to itself. Linear activation function helps to formulate a multilayered neural network with linear activation into a single-layered linear network. The derivative of linear activation function is unity in case of one-dimensional inputs.

Linear'$(x) = 1$, So after having a knowledge of how neurons work in a network, let us now know how such multiple of neurons can be brought together to form a feedforward neural network.

6 Structure of Feedforward Neural Network

Feedforward neural network consists of three different layers, namely input layer, hidden layer, and output layer. Data is first given at the input layer which is being sent to the hidden layer for further process. Hidden layer is present with actual neurons, and after a magic trick, output is shown at the output layer. The information flows only in one direction which is forward. Hence, it is known as feedforward neural network [13]. If the information is not passed in one direction and the output of neuron is feedback into previous layer in a cycle, then it is known as recurrent neural network called back propagated neural network. Let us now consider a three-layered network with sigmoid neurons in a training set. The following network has four input signals, two neurons at the hidden layer, and three at the output layer (Fig. 3).

We first have to assign weights and biases to the connections between neurons at each layer [6]. In general, weights and biases are all arranged randomly in a matrix form and vector form, respectively.

The total weighted sum of the parameters is given by: $Y = wX + b$

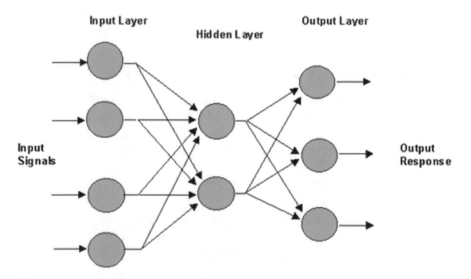

Fig. 3 Structure of feedforward neural network [14]

$$Y = \begin{bmatrix} w_{(1,1)} & w_{(1,2)} & w_{(1,3)} & \cdots & w_{(1,n)} \\ w_{(2,1)} & w_{(2,2)} & w_{(2,3)} & \cdots & w_{(2,n)} \\ w_{(3,1)} & w_{(3,2)} & w_{(3,3)} & \cdots & w_{(3,n)} \\ \vdots & \vdots & \vdots & \ddots & \vdots \\ w_{(m,1)} & w_{(m,2)} & w_{(m,3)} & \cdots & w_{(m,n)} \end{bmatrix} \begin{bmatrix} x_1 \\ x_2 \\ x_3 \\ \vdots \\ x_n \end{bmatrix} + \begin{bmatrix} b_1 \\ b_2 \\ b_3 \\ \vdots \\ b_m \end{bmatrix}$$

In a weight matrix, number of columns is given same as the number of neurons existing at the previous layer, and number of rows is given same as the number of neurons existing at the current layer. In a bias vector, the number of vector components is given same as the number of neurons existing at the present layer. Now, we need to build these different layers in a neural network by assigning weights and biases to the neurons starting with the feedforward step to predict the output [7]. First layer is the input layer which takes the input samples represented by the vector "X." This vector X is multiplied with the weighted matrix "w" and sums up with bias at the connections between the layer1 and layer2, and the result is passed to the second layer as input. Input to the layer1 is given by $X^{(4)}$ which represents four input samples from four neurons.

The input to the layer2 is given as: $Y_1 = w^{(2,4)} X^{(4)} + b^{(2)}$.

Here, $w^{(2,4)}$ represents weighted 2×4 matrix with two neurons at the second layer and four neurons at the first layer.

$b^{(2)}$ represents bias vector of two neurons at the 2nd layer.

In this layer, the weighted input is computed under sigmoid activation function to form an output, and it is weighted up with the weights and bias present at the connections between second and third layer. The final output is taken by the third layer as input. Output at layer2 is given as

$$F_1(x) = \sigma(Y_1)$$
$$F_1(x) = \sigma(w^{(2,4)} X^{(4)} + b^{(2)})$$
$$F_1(x) = \sigma\left(w^{(2,4)} X^{(4)} + b^{(2)}\right)$$

The input to the third layer is given as

$$Y_2 = w^{(3,2)} F_1(x) + b^{(3)}$$
$$Y_2 = w^{(3,2)} \left(\sigma(w^{(2,4)} X^{(4)} + b^{(2)}) + b^{(3)}\right)$$

Here, $w^{(3,2)}$ represents weighted 3×2 matrix with three neurons at the third layer and two neurons at the second layer.

$b^{(3)}$ represents the bias vector of three neurons at the third layer.

This input to the third layer to computed under sigmoid activation function to form the final result which is given as

$$F_2(x) = \sigma(Y_2)$$
$$F_2(x) = \sigma(w^{(3,2)}(\sigma(w^{(2,4)}X^{(4)} + b^{(2)}) + b^{(3)}))$$

In this concept of neural networks, weights and bias values are picked in such a way that the output value of the network is approximated to the desired value of the function $F(x)$ for all inputs in the training set. Things being what they are, how would we evaluate how far our expectation is from our desired output with the goal for us to know whether we have to continue scanning for more exact boundaries? For this point, we need to compute an error or such characterize a cost function also called as loss function [2, 7]. It is the deduction between the desired output and the actual output. The most ordinarily utilized one in our neural networks is the quadratic cost function, likewise called mean squared error, characterized by the equation:

$$C(w, b) = \frac{1}{N} \sum_{i=1}^{N} \left| F_{i(\text{desired})}(x) - F_{i(\text{actual})}(x) \right|^2$$

"w" refers to the weight, "b" refers to bias, "N" is the total number of input samples in the training set, "$F(x)$" is the output of the network, "i" refers to the sample number, and "$\sum$" refers to the summation over all the inputs in the training set.

In order to modify the variables, weights and biases to get the approximated output, mean squared error method is preferred over linear error because tiny modifications in weights and biases do not produce any change within the range of desired outputs; therefore, employing a quadratic function wherever massive variations have additional impact on the loss function than small ones helps in deciding the way to modify these variables. To minimize the error, we have a new algorithm called gradient descent. It is used to minimize the loss function by finding the new set of values of the input parameters at each layer as the loss function gets reduced whenever the output approaches the desired value of Y. Adjusting the weights by calculating the gradient of a loss function [11] helps in reducing the error of an output. In general, gradient is a measure of change in output value due to the change in input parameter values. This can be achieved by finding a global minimum of the loss function. Mathematically, gradient is nothing but the partial derivative of the function with respect to its input value (Fig. 4).

As we see from the above graph between cost function and weight, weight is changed to a certain value in a direction to the negative gradient to get the minimum value of a cost function. Now, we present a new parameter called "Learning rate" which has the greatness of explaining what number of units should we move from a current position to the heading that makes our work quicker. We have known how to calculate the error and the derivative of a loss function to update weights and biases. Now, we will move further to know how to compute the gradient descent in order to minimize the error [16–18]. Here is where a new method called "Back Propagation" arises. The word back propagation says that the gradient of the error at each layer is propagated back starting from the output side to the input side by making changes

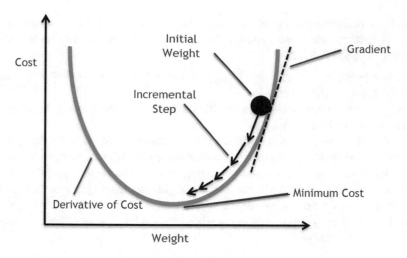

Fig. 4 Graph of gradient descent with respect to weight [15]

in the weights and bias values for each layer in a network. The gradient at each layer with respect to all the parameters is computed according to the chain rule by updating weights and bias values in the direction of negative gradient toward its minimum point.

Now, let's consider the equation of loss function or cost function to compute the gradient. Gradient with respect to weights is

$$\frac{\partial C}{\partial w_{(L)}} = \frac{\partial C}{\partial F_{L(\text{actual})}(x)} \frac{\partial F_{L(\text{actual})}(x)}{\partial Y_{(L)}} \frac{\partial Y_{(L)}}{\partial w_{(L)}}$$
$$= 2\big(F_{\text{desired}}(x) - F_{L(\text{actual})}(x)\big)\sigma'(Y_{(L)})(F_{L-1(\text{actual})}(x))$$

where "L" is the last layer and "$L - 1$" is the previous layer or second last layer, and we require three main equations to back propagate the error gradient with respect to weight and bias at current layer (L).

$$\frac{\partial C}{\partial w_{(L)}} = \frac{\partial C}{\partial F_{L(\text{actual})}(x)} \frac{\partial F_{L(\text{actual})}(x)}{\partial Y_{(L)}} \frac{\partial Y_{(L)}}{\partial w_{(L)}},$$
$$\frac{\partial C}{\partial b_{(L)}} = \frac{\partial C}{\partial F_{L(\text{actual})}(x)} \frac{\partial F_{L(\text{actual})}(x)}{\partial Y_{(L)}} \frac{\partial Y_{(L)}}{\partial b_{(L)}}.$$

Gradient with respect to the activation function at previous layer or second-last layer ($L - 1$)

$$\frac{\partial C}{\partial F_{L-1(\text{actual})}(x)} = \frac{\partial C}{\partial F_{L(\text{actual})}(x)} \frac{\partial F_{L(\text{actual})}(x)}{\partial Y_{(L)}} \frac{\partial Y_{(L)}}{\partial F_{L-1(\text{actual})}(x)}$$

The above gradient equations are to measure how a weight and bias parameters affect the cost function which we optimize. Now, learning rate comes into picture while updating weights and bias.

$w_{i(\text{new})} = w_{i(\text{old})} - \alpha\left(\frac{\partial C}{\partial w_i}\right)$, where $\frac{\partial C}{\partial w_i}$ is the gradient of loss function with respect to weight at a specific layer.

$w_{i(\text{new})}$ is the updated weight at a specific layer, $w_{i(\text{old})}$ is the old weight at a specific layer and "α" is the learning rate.

At the last layer or third layer in our neural network, refreshing the weights and biases relies on cost function, weights, and biases given to that layer, but when comes to hidden layer or second-last layer, updating weights and biases depends upon the computations involved in the last layer and weights and biases given to that second-last layer. These dependencies would add up if we have more hidden layers. Let's see how the dependency happens by considering the gradients with respect to weights and biases at third layer

$$\frac{\partial C}{\partial w_{(3)}} = \frac{\partial C}{\partial F_{3(\text{actual})}(x)} \frac{\partial F_{3(\text{actual})}(x)}{\partial Y_{(3)}} \frac{\partial Y_{(3)}}{\partial w_{(3)}},$$

$$\frac{\partial C}{\partial b_{(3)}} = \frac{\partial C}{\partial F_{3(\text{actual})}(x)} \frac{\partial F_{3(\text{actual})}(x)}{\partial Y_{(3)}} \frac{\partial Y_{(3)}}{\partial w_{(3)}}.$$

To compute the updates for weights and biases in the second-last layer, we would have to utilize some of the computations from last layer [7].

$$\frac{\partial C}{\partial w_{(2)}} = \frac{\partial C}{\partial F_{3(a)}(x)} \frac{\partial F_{3(a)}(x)}{\partial Y_{(3)}} \frac{\partial Y_{(3)}}{\partial F_{2(a)}(x)} \frac{\partial F_{2(a)}(x)}{\partial Y_{(2)}} \frac{\partial Y_{(2)}}{\partial w_{(2)}},$$

$$\frac{\partial C}{\partial b_{(2)}} = \frac{\partial C}{\partial F_{3(a)}(x)} \frac{\partial F_{3(a)}(x)}{\partial Y_{(3)}} \frac{\partial Y_{(3)}}{\partial F_{2(a)}(x)} \frac{\partial F_{2(a)}(x)}{\partial Y_{(2)}} \frac{\partial Y_{(2)}}{\partial b_{(2)}}.$$

$\frac{\partial C}{\partial F_{3(a)}(x)}, \frac{\partial F_{3(a)}(x)}{\partial Y_{(3)}}$ are the re-utilized partial differential terms from last layer or third layer.

7 Applications of Artificial Neural Network

1. **Neural Network approach to an Autonomous Car Driver**: In this scenario, neural network is trained under Gazelle system to determine the movement of the car. Network is trained with back propagation algorithm and uses five neurons to assign five different input values at input side, hidden layer, and output layer. Hyperbolic tangent activation function is used to train the neuron. The output yields the target angle for steering the car. Five input values are given by target angle, current angle, lateral offset of the car, and the distances which

are the difference between the distance ahead in the direction of movement and the distance ahead at a certain angle [19].

2. **Neural Networks for modeling Brain Wave Data:** Wisconsin Card Sorting test is used to record the signal waves produced by the brain. It uses feed forward neural network and Elman recurrent neural network to analyze the accurate brain wave forecasts with the help of linear activation function and logistic activation function. The signal waves are generated by varying number of input neurons and number of nodes at the hidden layer, and the best architecture of the neural network is established by the root mean square error (RMSE) and mean absolute percentage error (MAPE) [20].

3. **Artificial Neural Networks-based controllers for biped Robots:** Biped robots are the ones which have two legs, same like humans. Different kinds of artificial neural network controllers are used to control and for the stability of the robot. Recurrent neural network (RNN), self-recurrent wavelet neural network (SRWNN), and Recurrent wavelet Elman neural network (RWENN) are the three main ANN-based controllers [8] used to stabilize and improve the response of the robot. These controllers work the same like a neural network with input, hidden layer, and output layer neurons with a transfer function. These controllers are compared based on the result by computing root mean square error (RMSE) [21].

4. **Artificial Neural Network for predicting Crop Yield:** Earlier to understand the weather condition toward the crop yield, crop modeling was used as a tool which is based on linear methods. These methods were neither enough to know the interconnections between these external factors and crop yield. To get the better of these linear methods, nonlinear methods like neural networks, fuzzy logic, and many more are used. Application of mathematical models and artificial neural network approach are used to predict the crop response as they give accurate outputs with minute error. Multilayered feed forward neural network is chosen to know the interactions between the soil layer and the crop yield. Based on the crop, the input parameters given to the network changes like planting date, minimum and maximum temperature conditions at which crop grows, maturity group etc., to leave leaf area index, mass and temperature stress factor, relative growth rate etc., as outputs [22].

5. **Neural Networks to estimate participation in elections:** Artificial neural networks are widely used to anticipate polity issues with a best accuracy. To select the members in an election, multilayered neural network and tan sigmoid activation function are used to predict the selected people in elections. Data such as qualification, age, opinion about the government services, and political orientation of people are given as inputs to the network with hidden layers and gives the target members as output of the network. Weights and biases are chosen as number and percentages at each stage of the network. Best accuracy of the data is finalized based on the error graphs left by the network [23].

6. **Neural Networks in Electric Power Industry:** Optimization techniques are very much used to deal with problems related to power systems, controlling, and many such complex parameters. Economic load dispatch, load forecasting,

and security analysis are the three main constraints in electric power industry, which requires mathematical optimization technique methods to solve such complex and large problems. One such optimization tool to dispatch required number of units to the load of the system, to estimate the load and to supply the power with required number of volts and frequency limits is artificial neural network. Multilayered perceptron with back propagation algorithm is used by giving security indicators such as fuel mix diversity and energy intensity as inputs to the network [24].

7. **Neural Network to predict the yield of enzymatic synthesis of betulinic acid ester:** Synthesis of betulinic acid and phthalic anhydride gives a $3\beta - O -$ phthalic ester which acts as an anticancer agent. Esterification is carried out using feedforward neural network with sigmoid function as transfer function and gradient descent algorithm. Amount of ester, reaction time, reaction temperature, and number of moles of the substrate are given as data at the input neurons to the network, and the number of neurons at the hidden layer depends upon the complexity of the network. The output of the neuron gives the percentage of yield ester. Weights and biases values are assumed accordingly, and gradient descent back propagation technique is used to predict the yield by training the network [25].

8. **Neural Networks for Face Recognition:** Face recognition is a kind of pattern recognition problem which has face detection, face alignment, face extraction, and face matching. It uses three-layered feedforward network to detect the human faces and involves multilayered perceptron for face alignment and other models of artificial neural networks combined with geometric feature-based methods for face extraction and face matching. Length of the extracted image vector, that is, the number of pixel boxes of the image becomes the number of neurons at the input side, whereas number of neurons at the hidden layer depends upon the complexity of the network and the output neuron as one or many depend on the kind of task to be solved. Length of the pixel of the image is given at the input neurons, and the artificial neural network filters out the false negative images and yields out the true image in the form of 0 or 1 if it contains human face [26].

9. **Neural Networks in geotechnical engineering:** Artificial neural networks in this field are used to estimate the properties of soil and to predict the liquefaction which causes the ground failure during earthquakes. ANNs are also used to retain walls, mining, blasts, and dam. Many other mathematical models fail to encounter these complex problems in geotechnical areas, but ANN helps in solving those by providing an input and output architecture with appropriate yield as the desired output can be obtained by changing the input parameters using back propagation algorithm. Multilayered feedforward network with logistic activation function is used, and number of hidden layers is found by hit and miss method which depends on the complexity of the network, but, however, scientists had found that two hidden layers are sufficient to model the algorithm to obtain liquefaction potential and other such related parameters as output [27].

10. **Neural Networks in HIV Modeling:** Artificial neural networks are used for HIV modeling which includes stages like prediction, function approximation, and classification of the pattern. Multilayered perceptron is used for prediction and function approximation of HIV which takes HIV drug resistance mutations and treatment change episodes (TCEs) as input variables and for the prediction of HIV protease cleavage sites in proteins. It also uses feedforward neural network along with hyperbolic tangent and sigmoid activation functions. ANNs are also used in prediction of creatinine clearance in patients with input variables as their age, actual and ideal body weight, body surface area, and some blood chemistries like RBC count and WBC count, platelet count, albumin, blood urea nitrogen, and many more such variables. Multilayered perceptrons are also used for the classification of HIV patients into positive and negative result, which takes input data as their age, weight, CD4, CD8, gender, HB, and TB and also used to classify the health and symptomatic status, visual fields (VFs) between eyes of a normal person and HIV person [28].

11. **Neural Networks for Industrial drying:** Artificial neural networks are used to increase the quality in the drying process of grated coconut by controlling the product humidity level. Multilayered perceptron with back propagation method and hyperbolic tangent function are used to maintain the required humidity level. Number of neurons at the input and output layer depends upon the data given and the yield. Number of hidden layer neurons depends upon the selected architecture of the problem with less error. The seven temperatures of the product dryer, final temperature of the product, and the initial moistures of the crushed coconut and pasteurized grated coconut are considered as input data which yields the final moisture of the grated coconut [29].

12. **Neural Networks for analysis of path planning of Humanoid Robots:** Artificial neural networks are used for the path making of humanoids which reach the target place by avoiding the obstacles on their path. Feedforward neural network with an input, two hidden and output layers are considered, and a hyperbolic tangent activation function is used along with back propagation algorithm. The network takes heading angle, distance to the right, left, and front obstacles are considered as input data and gives the steering angle at the output layer. Simulation process using ANN and experimental process of a humanoid are compared to observe the errors in the length of the path and the time taken for the robot to reach the final point by avoiding the obstacles. Simulation process using ANNs shows less error compared to the experimental process. ANN works efficiently in the path planning of humanoid robots in a cluttered environment [30].

13. **Neural Networks to early Lung Cancer detection:** Artificial neural networks are used efficiently to predict the patients with and without lung cancer. Multilayered perceptron with gradient descent algorithm and other two such algorithms for error measurement are used. Hyperbolic tangent activation function and sigmoid activation function are used to train the network. Variables like age, weight, gender, pneumonia, cough, bone pains, lung cancer in the family,

WBC and RBC count, fever, cigarettes, hemoglobin, platelets, Na, K, vital capacity, creatinine, cardiovascular diseases, appetite etc. Such kind of 48 parameters count of the patients are taken as the input data and gives an output value as 0 or 1, indicating with and without lung cancer of the patient. The final architecture of the network is with 48 input neurons, 9 hidden layers, and 2 output neurons are considered based on the experiment [31, 32].

14. **Neural Networks for Medical Diagnosis:** Artificial neural networks are used in diagnosis of diseases like acute nephritis disease and heart disease. Feedforward neural network along with back propagation method is used to differentiate the infected and non-infected people. Disease symptoms, images, and signals related to the test part are used as input data; hidden layer consists of 20 neurons, and output layer has two neurons which shows 0 or 1 identifies as non-infected or infected person. Sigmoid activation function is used to process the data, and the performance of the network is increased by the back propagation method and the calculation of loss function (mean square error of the samples) [33].

15. **Neural Networks to simulate performance of Osmosis membrane:** Artificial neural networks are used in desalination process of seawater using reverse osmosis membrane method. It uses feedforward network with sigmoid activation function and back propagation algorithm to calculate the error. Desalination process is done for different sea water reverse osmosis membranes (SWRM) which calculates the fresh flow concentration and total suspended solids rate (TDS). Number of input layer and hidden layer neurons depends upon the membrane that we are selecting. Parameters like temperature (T), influent flow (q), percentage of recovery (Re), and saline water with influent concentration (So) are given at the input nodes to yield fresh flow concentration (Se) [34].

16. **Neural Networks to predict moisture in Halite:** Artificial neural networks are used to foresee the presence of water on the exterior of halite rocks. Multilayered perceptron with back propagation algorithm and sigmoid activation function are used to direct the neural network with the given input data. Presence or absence of water on the halite rocks is found by the yield of electrical conductivity of the surface of a halite rock using conductivity sensors. Relative humidity and temperature of the air are given as input parameters to yield electrical conductivity. Number of neurons at the hidden layer is estimated by trial and error method based on the complexity of the algorithm, weights and biases are assigned randomly. Error between the actual and desired value is calculated using mean square error method at the output layer to further proceed for the increments in weights to get the desired yield [35].

17. **Neural Networks for inverse design and simulation of Nano-photonic particle:** Artificial neural networks are used to approximate the simulation of problems on light scattered by the nanoparticle and to solve problems on inverse design in physics like quantum theory, photonic devices, and photovoltaic materials. Inverse design problems are related to the geometry of a spectrum closest to any arbitrary spectrum which was chosen. Lossless silica shells

and lossless TiO_2 were considered as nanoparticles, and the neural network is trained with 250 neurons at each layer. Thickness of the nanomaterial shells was given as the input data, and the output was the spectrum which is sampled at a certain range of wavelength. Training of network is made by changing the learning rate and number of neurons at each layer. Neural network approach was compared with the simulation model and observed that it works well in approximating nanoparticle scattering phenomena. Neural network produces an accurate geometry of an arbitrary spectrum for inverse design problems. Inputs to the network were given by the trainable variables with fixed weights, and the network is trained with the back propagation algorithm by fixing the output node with the desired output to propose the geometry of the spectrum [36].

18. **Finite element method based Neural Networks:** Finite element method is generally used to tackle mathematical problems related to optimization techniques, partial differential equations, fluid flow, heat analysis, electromagnetism, and many more such problems from different branches in engineering. This method serves as a tool in dealing complex algorithms like solving partial differential equations in back propagation algorithm in ANN. Finite element analysis is performed at the beginning for the parameters required for a network construction. These results are passed through the network layers to proceed with further actions. FEM is a powerful tool to deal with boundary values in a neural network by updating the necessary parameters like weights, inputs etc. This results in accurate output in less time with minimal error [37–41].

8 Conclusion

Artificial neural networks have a wide range of applications in artificial intelligence, cloud computing, cyber security, and many more areas. ANNs are widely used for face recognition, image processing, language processing, text classification, and so on. Nowadays, handling problems in medical science, financial analysis, agriculture, education, banking systems, marketing, and many more has become an easy task for neural networks. Applications of ANN have a great impact in the current trend because of its analysis factors like processing speed, tolerance, performance etc. ANNs take part in control systems and robotics. Mathematics is the key step behind all these applications to structure a neural network. Mathematical model provides the means to implement the application and to reach the goal. Some of the topics from mathematics like calculus, linear algebra, probability, statistics, vectors, matrices etc., play a dominant role to architect any type of neural network. This article on artificial neural networks helps the reader to get a basic idea of artificial neuron. It draws the reader's attention by giving the knowledge of math required to build a neural network and makes their task easy in solving such algorithms. This paper can also help the reader in implementing the network algorithm in a Python code, and finally, it can summarize its applications in different areas.

References

1. History of Neural Networks, Stanford computer science
2. C.C. Aggarwal, Neural Networks and Deep learning, IBM T. J. Watson Research Center International Business Machines Yorktown Heights, NY, USA
3. R. Singh, Basics of Artificial Neural Network: International Journal of Engineering Research and Technology (IJERT), Department of Mechanical Engineering, Ganga Institute of Technology and Management, Kablana, Jhajjar, Haryana, India
4. T.J. Griinke, Development of an artificial neuralnetwork(ANN) for predicting tribiological properties of KENAF FIBRE REINFORCED EPOXY COMPOSITES (KFRE)
5. A. Choudhary, D. Pandey, S. Bhardwaj, Artificial neural network based solar radiation estimation: a case study of Indian cities. Int. J. Emerg. Technol. **11**(4), 257–262 (2020)
6. S. Haykin, Neural Networks, A Comprehensive Foundation", 2nd edn, Mc Master University, Hamilton, Ontario, Canada, Published by Pearson Education (Singapore) Pte. Ltd., Indian Branch, 482 F.I.E. Patparganj, Delhi 110092, India
7. C.F. Higham, D.J. Higham, Deep Learning: An Introduction for Applied Mathematicians, January 19, 2018
8. O.I. Abiodun A. Jantan, A.E. Omolara, K.V. Dada, N.A. Mohamed, H. Arshad, State of the art in artificial neural network applications
9. M. Devlin, B.P. Hayes, Non-intrusive load monitoring using electricity smart meter data: a deep learning approach, November 2018. https://doi.org/10.13140/RG.2.2.29463.42402
10. M. Nielsen, Neural Networks and Deep learning
11. S.-H. Han, K.W. Kim, S.Y. Kim, Y.C. Youn, Artificial Neural Network: Understanding the basic concepts without mathematics: Dementia and Neurocognitive disorder(DND)
12. A.S. Jacobs, R.J. Pfitscher, R.L.D. Santos, M.F. Franco, E.J. Scheid, L.Z. Granville, Artificial neural network model to predict affinity for virtual network functions, in *NOMS 2018—2018 IEEE/IFIP Network Operations and Management Symposium*, Taipei, 2018, pp. 1–9. https://doi.org/10.1109/NOMS.2018.8406253
13. M.H. Sazli, *A Brief Review Of Feed Forward Neural Networks* (Ankara University, Faculty of Engineering, Department of Electronics Engineering 06100 Tandoğan, Ankara, Turkey)
14. http://www.geocomputation.org/1998/05/gc05_01.gif
15. https://www.oreilly.com/library/view/learn-arcore-/9781788830409/assets/f3899ca3-835e-4d3e-8e7f-fd1c5a9044fb.png
16. L. Fridman, Recurrent Neural Networks for Steering through time, MIT 6.S094: youtube video: https://youtu.be/nFTQ7kHQWtc
17. A. Karpathy, Stanford University: Back Propagation and Neural Networks1, youtube video: https://youtu.be/i94OvYb6noo
18. S. Muhammad, O. Tomio, K. Kazuhiko, N. Kunihiko, The Development of Anomaly Diagnosis Method Using Neuro-Expert for PWR Monitoring System
19. K. Albelihi, D. Vrajitoru, An Application of Neural Networks to an Autonomous Car Driver, Computer and Information Sciences Department, Indiana University South Bend, South Bend, IN, USA
20. C.H. Aladag, E. Egrioglu, C. Kadilar, Modeling Brain Wave Data by using Artificial neural networks
21. N.M. Mirza, Al Ain University and Al Ain, Comparison of Artificial Neural Networks based on Controllers for Biped Robots, United Arab Emirates
22. S. Khairunniza-Bejo, S. Mustaffha, W.I.W. Ismail, Application of Artificial Neural Network in Predicting Crop Yield
23. S.R. Khaze, M. Masdari, S. Hojjatkhah, Application of Artificial neural networks in estimating participation in elections
24. M. Mohatram, P. Tewari, Shahjahan, Applications of Artificial Neural Networks in Electric Power Industry

25. M.G. Moghaddam, F. Bin H. Ahmad, M. Basri, M. Basyaruddin Abdul Rahman, Artificial neural network modeling studies to predict the yield of enzymatic synthesis of betulinic acid ester
26. T.H. Le, *Applying Artificial Neural Networks for Face Recognition* (Department of Computer Science, Ho Chi Minh University of Science, Ho Chi Minh City 70000, Vietnam)
27. A. Mohamed Shahin, M.B. Jaksa, H.R. Maier, State of the Art of Artificial Neural Networks in Geotechnical Engineering
28. W. Sibanda, P. Pretorius, Artificial Neural Networks—A Review of Applications of Neural Networks in the Modeling of HIV Epidemic, North West University, Vaal Triangle Campus Van Eck Blvd, Vanderbijlpark, 1900, South Africa)
29. E. Assidjo, B. Yao, K. Kisselmina, D. Amane, Modleing of an industrial drying process by Artificial neural networks
30. B. Sahoo, D.R. Parhi, P. Biplab Kumar, Analysis of Path Planning of Humanoid Robots using Neural Network Methods and Study of Possible use of Other AI Techniques
31. K. Goryński, I. Safian, W. Grądzki, M.P. Marszałł, J. Krysiński, S. Goryński, A. Bitner, J. Romaszko, A. Buciński, Artificial neural networks approach to early lung cancer detection
32. L. Bertolaccini, P. Solli, A. Pardolesi, A. Pasini, An overview of the use of artificial neural networks in lung cancer research
33. Q. Kadhim Al-Shayea, MIS Department, Artificial Neural Networks in Medical Diagnosis, Al-Zaytoonah University of Jordan Amman, Jordan
34. E.S. Salami, M. Ehetshami, A. Karimi-Jashni, M. Salari, S. Nikbakht Sheibani, A. Ehteshami, A mathematical method and artificial neural network modeling to simulate osmosis membrane's performance
35. K. Wierzchos, J.C. Cancilla, J.S. Torrecilla, P. Díaz-Rodríguez, A.F. Davila, C. Ascaso, J. Nienow, C.P. McKay, J. Wierzchos, Application of artificial neural networks as a tool for moisture prediction in microbially colonized halite in the Atacama Desert
36. J. Peurifoy, Y. Shen, L. Jing, Y. Yang, F. Cano-Renteria, B.G. DeLacy, J.D. Joannopoulos, M. Tegmark, M. Soljacic, Nanophotonic particle simulation and inverse design using artificial neural networks
37. L. Neamt, O. Matei, O. Chiver, *Finite Element Method Combined with Neural Networks for Power System Grounding Investigation* (Electrical, Electronic and Computer Engineering Department Technical University of Cluj-Napoca Cluj-Napoca, Romania)
38. T. Guillod, P. Papamanolis, J.W. Kolar, Artificial neural network (ANN) based fast and accurate inductor modeling and design. IEEE Open J. of Power Electron. **1**, 284–299 (2020). https://doi.org/10.1109/OJPEL.2020.3012777
39. Padmasudha Kannan, K.V. Nagaraja, An efficient automatic mesh generator with parabolic arcs in Julia for computation of TE and TM models for waveguides. IEEE Access **8**, 109508–109521 (2020)
40. T.V. Smitha, K.V. Nagaraja, An efficient automated higher-order finite element computation technique using parabolic arcs for planar and multiply-connected energy problems. Energy **183**, 996–1011 (2019)
41. T.V. Smitha, K.V. Nagaraja, Application of automated cubic-order mesh generation for efficient energy transfer using parabolic arcs for microwave problems. Energy **168**, 1104–1118 (2019)

Machine Learning-Based Multi-temporal Image Classification Using Object-Based Image Analysis and Supervised Classification

Swasti Patel, Priya Swaminarayan, Simranjitsingh Pabla, Mandeepsingh Mandla, and Hardik Narendra

Abstract During last decade, there has been tremendous research related to the image-based technique in remote sensing; object-based classification is one of the popular techniques due to its capacity of promising results. This paper presents a novel approach where a hybrid method of object-based image analysis and supervised classification is used. The data used in this study is high-resolution multi-spectral 4-band images from 2017 to 2019 provided by the PlanetScope satellite of region Chandigarh, India. First, the data has been pre-processed through passing it in a pipeline of steps followed by a multi-resolution segmentation algorithm and classifying the image into seven classes based on the spectral signature using algorithms like maximum likelihood (ML), support vector machine (SVM), Mahalanobis distance (MD). Comparing the three algorithms, it was observed that SVM and ML have given the highest overall accuracy of 95.21% and kappa coefficient = 0.9159. Also, the overall accuracy 91.91% and kappa coefficient = 0.8860 were achieved.

Keywords Object-based approach · Machine learning · Land cover classification · Change detection · Remote sensing

1 Introduction

There are three techniques for image classification in remote sensing [3]: unsupervised [13] classification, supervised [3, 7, 14] classification, and object-based classification. Multi-temporal images are combined imagery of multiple images of the same place or location taken at different points of time. For research purpose,

S. Patel (✉)
Parul Institute of Technology, Parul University, Vadodara, India
e-mail: swasti.patel20247@paruluniversity.ac.in

P. Swaminarayan
Parul Institute of Computer Applications, Parul University, Vadodara, India
e-mail: priya.swaminarayan@paruluniversity.ac.in

S. Pabla · M. Mandla · H. Narendra
Parul Institute of Engineering and Technology, Parul University, Vadodara, India

© The Author(s), under exclusive license to Springer Nature Singapore Pte Ltd. 2023   223
Y.-D. Zhang et al. (eds.), *Smart Trends in Computing and Communications*, Lecture Notes in Networks and Systems 396, https://doi.org/10.1007/978-981-16-9967-2_22

two multi-temporal images of Chandigarh were taken into consideration, from April 2017 to May 2019. In the traditional approach, land classification [15] and changes were detected with pixel-based [12] image detection. This was replaced by object-based image detection due to higher accuracy compared to former method. Object-based image analysis (OBIA) includes pixels first being grouped into objects based on spectral similarity as geological unit. In OBIA, there are two main approaches segmentation and classification.

Considering all this, a new approach is introduced where OBIA [6] is used for segmentation and supervised learning is used for classification of the classes, instead of using OBIA for both purposes. For this research, the following classes were classified: road, urban area, barren land, water, wooden land, grassland, and fields. While doing this process, it was discovered that if two classes are very near to each other like grassland and fields, then it is recommended to make a new class. Kappa coefficient is used for calculating the accuracy of the classification done by supervised learning algorithms. Furthermore, to identify change detection over time, the thematic change technique is utilized. This new method gives promising results with high-resolution and medium-resolution images.

2   Study Area, Dataset, and Software Used

2.1   Study Area

This study focuses on Chandigarh City located in India which is a perfect suitable for because of its geo-location and well-structured planning. This region has vast variety due to its combination of urban area, rural area, water bodies, mountains, and agricultural areas.

2.2   Data

The high-quality data is provided by Planet [10] via Planet's Education and Research Program. For performing chance detections, two images were acquired; April 11, 2017 and May 4, 2019. All the images are of high resolution. Table 1 provides details for the image acquired. In order to perform the classification, the images need to be resized equally into a shapefile that is being used with all the coordinates of the location as shown in Fig. 1.

Table 1 Summary of the remotely sensed datasets along with the satellite used for this study

Satellite name	Sensor type	Acquisition date	Spectral band (nm)	Pixel size (m)
PlanetScope analytic ortho scene	4-band natural color (red, green, blue) and near infrared (NIR)	04/11/2017–05/04/2019	Blue 455–515 Green 500–590 Red 590–670 NIR 780–860	3 3 3 3

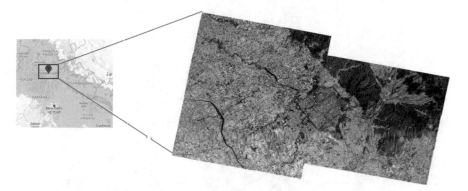

Fig. 1 Multispectral four-band images acquired from PlanetScope of Chandigarh

2.3 Software

For image pre-processing, GIS-software [11] is used, and for OBIA, eCognition Developer 9.0 software is used, and eCognition Developer 9.0 makes it easy for the region of interest (ROI) creation. ROI plays a vital role in training the machine learning algorithms; for thematic change Envi 5.1 software is used.

3 Methodology

The flow of the system is depicted in Fig. 2. The process starts with attaining the images from PlanetScope. This is followed by the process of combination and clipping of the images. Once this is done, the image segmentation is performed using OBIA. After the segmentation, it is essential to define ROI for classification purpose. ML algorithms run on the classes as training data. Final step includes determining the accuracy of the classification. This process is repeated until desired accuracy is achieved.

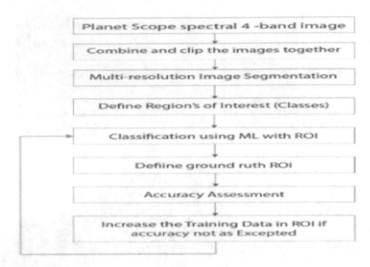

Fig. 2 Flow diagram that shows OBIA and classification workflow

Fig. 3 Pre-process of image shown here uses mosaicking to merge the image and shapefile to clip the image

3.1 Pre-processing Image

A pipeline is being created for pre-processing the image where initially the image is being merged using mosaicking and then resized using the shapefile [13, 16]. Furthermore, to make the shapefile, the coordinates of the desired location have been pin-pointed by user preference, so both the images are resized evenly shown in Fig. 3.

3.2 OBIA

In this research, segmentation [5, 13] is difficult to perform directly because of the processing that is done in order to get hand-like image layer mixing using equalizing histogram with three-layer mix. This is the pre-processing of the images in order to increase the color vibrancy. Defining the rule-set for segmentation where a process tree should be created is shown below.

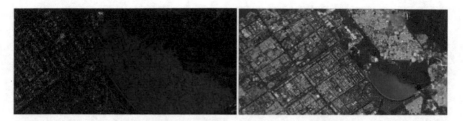

Fig. 4 Multi-resolution segmentation with hyper-parameter for composition of homogenous criterion where shape value is set to 0.2 and compactness to 0.8

In this process, selection of the multi-resolution algorithm where first image object domain is to be set to pixel level because there are not any objects for now. Secondly, in the segmentation settings, the hyper-parameter and image layer weights of the algorithm should be blue: green: NIR: red to 1:1:2:1. Nonetheless, the last hyper-parameter is composition of homogenous [1, 12] criterion where the shape is set to 0.2 and compactness [11] to 0.8. In the end, the image is exported with all the 4-band with a file extension of.tif file. This enables the process tree to run and obtain a segmented image as shown in Fig. 4.

3.3 Classification

In earlier studies, the classification part was done with OBIA; thus, machine learning [2, 14] has been incorporated to take care of this part. The segmented image is taken as an input and load it in Envi as a raster file after this makes the ROI (i.e., region of interest) file also referred to as training data in terms of ML in which seven specified classes are used as shown in Table 2. While specifying classes [4, 15], use of ROI type is chosen as a polygon to choose the pixels from the image to get better training

Table 2 Overview about the ROI classes with color of each class used for the classification and brief class description

ROI Classes	Classes Colour	Class Description
Wooden land		Non-wetland which contains large presence of medium to large trees.
Urban land		Area which has artificial surfaces such as houses and buildings
Mountain		Landform which rises above or area which is elevated from the surrounding land
Field		Non-wetland class where farming is present
Grassland		Non-wetland where small grass or bushes
Water		Exposed fresh or saline surface water
Roads		Artificial surface to travel on roads, foot-paths and Highways
Barren land		Non-wetland bare land with no or very less vegetation cover

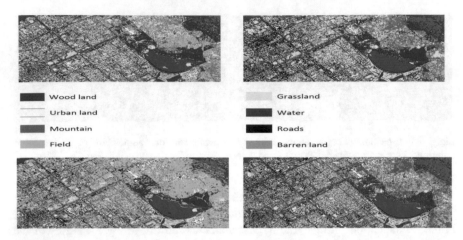

Fig. 5 Land cover classification results using maximum likelihood algorithm of subset image of April 2017 and May 2019 (i.e., upper half) and support vector machine algorithm of subset image of April 2017 and May 2019 (lower half)

data. The supervised algorithm on both the images has been performed that leads to mimic the classification that is being done by OBIA. Moreover, the input file and ROI are to be passed together when choosing any supervised algorithm.

If the user did not get the desired results from visualizing the image, then it is recommended to start with less training data at first and then increase it accordingly as it is completely trying and viewing method. One can see the figure after the classification with the desired output in Fig. 5, then move on to the next step.

Now to extract the data from the classified image, it has to be converted from raster to vector [7], so the database file will be created that is.vcf file.

3.4 Change Detection

The.vcf file which consists of class_name, class_id, parts, length, and area as headers makes new filtered database file. The details of the training data are given in Table 3. For research purpose, two databases have been used created from both the images. ROI is to be created for ground truth [13, 14] that is used for accuracy assessment. For identifying changes, it is required to provide both the classified images from the timeline and thus calculate the thematic change in order to find out the change.

Table 3 Selection of training data

ROI classes	Number of objects (i.e., pixels) classified	Actual objects (i.e., pixels) taken into consideration
Wooden land	5110	104
Urban land	1406	285
Mountain	21,243	118
Field	10,761	216
Grassland	642	65
Water	4051	193
Roads	1070	33
Barren land	989	193

4 Results

This novel approach for classification is promising because in the first place, the pre-processing of the data is done by setting certain parameters and weights in image bands [6] before proceeding it for segmentation. After this process, an overall accuracy for both the images with a kappa coefficient of 0.915 for support vector machine (SVM) (Ma et al., n.d.), [14] was achieved. For the classified images, the kappa coefficient 0.886 for maximum likelihood (ML) April image is as shown in Table 3. And, both the algorithms that are used in this research have provided the best results. Thus, SVM is great for urban classification and wooden land, while ML is good for classifying road and grassland. One can say that the data here is a delicious cake and the machine learning algorithm provides the icing on this cake. This approach takes less time than the general approach where the classification is done in OBIA. Not only it is taking less time but also gives great accuracy result. The output of change detection over a duration of time is shown in Table 4, and for thematic change, refer to Fig. 6.

5 Discussion

From the past studies, it has already been known that object-based approach provides superior results over pixel-based. The main reason behind this is it allows to incorporate the textual and spatial features [4]. However, the approach our team developed is not only giving better results than traditional pixel-based approach but also gives a competitive performance to object-based approach. Moreover, the multi-resolution (Fig. 4) algorithm cannot be ignored which plays an important role in the object-based approach. So, decided to keep the multi-resolution [5, 13] algorithm because of it distinct benefits, but the software that does it is quite expensive (i.e., eCognition Developer 9.0 [6], (Ma et al., n.d.)), although there are always other options

Table 4 Summary of results achieved by thematic change detection algorithm's used for land cover change detection from 2017 to 2019

	Water	Mountain	Field	Urban Land	Road	WoodenLand	Grassland	Barren Land	Row Total	Class Total
Unclassified	0	0	0	0	0	0	0	0	0	0
Water	246,567	87,581	23,776	5898	28,775	6589	2502	9446	411,134	411,134
Mountain	16,028	7,185,805	3,710,606	37,959	308,036	1,542,371	223,365	973,046	13,997,216	13,997,216
Field	1864	382,384	11,443,395	11,586	9293	639,630	378,427	784,651	13,651,230	13,651,230
Urban Land	4603	115,505	310,648	2,589,912	1,162,597	2882	28,978	889,623	5,104,748	5,104,748
Road	11,147	417,475	134,649	487,379	1,400,830	3421	11,868	423,399	2,890,168	2,890,168
WoodenLand	855	452,152	1,047,513	82	1068	2,393,803	51,187	29,023	3,975,683	3,975,683
Grassland	2238	232,317	2,639,939	34,063,768	14,125	135,545	738,764	679,254	38,505,952	38,505,948
Barren Land	3341	774,413	4,278,263	292,788	351,504	56,081	712,914	2,660,497	9,129,801	9,129,801
Class Total	286,643	9,647,632	23,588,788	37,489,372	3,276,228	4,780,322	2,148,005	6,448,939	0	0
Class Changes	40,076	2,461,827	12,145,393	34,899,460	1,875,398	2,386,519	1,409,241	3,788,442	0	0
Image difference	124,491	4,349,584	−9,937,558	−32,384,624	−386,060	−804,639	36,357,944	2,680,862	0	0

0: no change
1: from 'Unclassified' to 'Water'
2: from 'Unclassified' to 'Mountain'
3: from 'Unclassified' to 'Field'
4: from 'Unclassified' to 'Urban Land'
5: from 'Unclassified' to 'Road'
6: from 'Unclassified' to 'WoodenLand'
7: from 'Unclassified' to 'Grassland'
8: from 'Unclassified' to 'Barren Land'
9: from 'Water' to 'Unclassified'
10: none
11: from 'Water' to 'Mountain'
12: from 'Water' to 'Field'
13: from 'Water' to 'Urban Land'
14: from 'Water' to 'Road'
15: from 'Water' to 'WoodenLand'
16: from 'Water' to 'Grassland'
17: from 'Water' to 'Barren Land'
18: from 'Mountain' to 'Unclassified'
19: from 'Mountain' to 'Water'
20: none
21: from 'Mountain' to 'Field'
22: from 'Mountain' to 'Urban Land'
23: from 'Mountain' to 'Road'
24: from 'Mountain' to 'WoodenLand'
25: from 'Mountain' to 'Grassland'
26: from 'Mountain' to 'Barren Land'

27: from 'Field' to 'Unclassified'
28: from 'Field' to 'Water'
29: from 'Field' to 'Mountain'
30: none
31: from 'Field' to 'Urban Land'
32: from 'Field' to 'Road'
33: from 'Field' to 'WoodenLand'
34: from 'Field' to 'Grassland'
35: from 'Field' to 'Barren Land'
36: from 'Urban Land' to 'Unclassified'
37: from 'Urban Land' to 'Water'
38: from 'Urban Land' to 'Mountain'
39: from 'Urban Land' to 'Field'
40: none
41: from 'Urban Land' to 'Road'
42: from 'Urban Land' to 'WoodenLand'
43: from 'Urban Land' to 'Grassland'
44: from 'Urban Land' to 'Barren Land'
45: from 'Road' to 'Unclassified'
46: from 'Road' to 'Water'
47: from 'Road' to 'Mountain'
48: from 'Road' to 'Field'
49: from 'Road' to 'Urban Land'
50: none
51: from 'Road' to 'WoodenLand'
52: from 'Road' to 'Grassland'
53: from 'Road' to 'Barren Land'
54: from 'WoodenLand' to 'Unclassified'

55: from 'WoodenLand' to 'Water'
56: from 'WoodenLand' to 'Mountain'
57: from 'WoodenLand' to 'Field'
58: from 'WoodenLand' to 'Urban Land'
59: from 'WoodenLand' to 'Road'
60: none
61: from 'WoodenLand' to 'Grassland'
62: from 'WoodenLand' to 'Barren Land'
63: from 'Grassland' to 'Unclassified'
64: from 'Grassland' to 'Water'
65: from 'Grassland' to 'Mountain'
66: from 'Grassland' to 'Field'
67: from 'Grassland' to 'Urban Land'
68: from 'Grassland' to 'Road'
69: from 'Grassland' to 'WoodenLand'
70: none
71: from 'Grassland' to 'Barren Land'
72: from 'Barren Land' to 'Unclassified'
73: from 'Barren Land' to 'Water'
74: from 'Barren Land' to 'Mountain'
75: from 'Barren Land' to 'Field'
76: from 'Barren Land' to 'Urban Land'
77: from 'Barren Land' to 'Road'
78: from 'Barren Land' to 'WoodenLand'
79: from 'Barren Land' to 'Grassland'
80: none

Fig. 6 Change detection results using thematic change detection approach for image of year April 2017 and final image of May 2019

like open-source or economical ways, then, there is a trade-off as the algorithm will perform differently and may not give the desired result.

Here comes the twist after performing the OBIA, decided to replace the traditional classification process and change it with ML (Fig. 2). So, the challenge was how to replace it and get the whole thing working. Accordingly, the defining of rule-set in the object-based approach for the classification is got finally replaced by ROIs (Table 2) where the training dataset was collected and classified into all the classes of one's interest. Here, the recommendation would be to start with small training data and change accordingly for the classes that do not perform well. It was noted that there may be conflicts in the classes when they are classified by the ML algorithm. In order to overcome this limitation, it was decided to add the classes separately. After all this, once the classification process is over, then do test visually that the algorithm is giving the desired results.

Here, if the algorithm is performing well, then proceed further for the accuracy. But if not then, increase the training data for the classes that one thinks is not performing well. Now, the question is how to select the ML algorithm that is being used in this research. By doing comprehensive [3, 14] exploration, decided to work with support vector machine and maximum likelihood [3, 15] and both performed well (Fig. 5). From the research, it was found/categorized that SVM is giving good results for classes like urban land, fields, water, while maximum likelihood is giving good results for the small classes like road and wooden land. The key feature that really stands out in this approach is the classification part in which the model is doing great with a small amount of training data. Now, referring to the data (i.e., image quality), any medium-quality image [9, 15] to high-quality image will work with this approach. And, if one cannot get hands-on high-resolution [5, 11] data, then go with sentinel-2 [13] data rather than going for landsat-2 [14]. This study was done keeping mainly India in mind as India is a developing nation, so the Chandigarh area was selected for this research. All in all, this approach opens new avenues to explore rather than to only use OBIA.

6 Conclusion

This is a novel approach that is introduced after the object-based approach and pixel-based approach. And, the main motto is to get more accurate results that will benefit the monitoring of change detections and classification. Already object-based approach was giving great results, but it is largely limited to high-resolution data. In order to overcome this limitation, this hybrid approach is developed. Future research has scope to conduct studies by experimenting on different ML algorithms and even apply deep learning algorithms. There is certain deep learning research already underway, but it will take time to achieve the accuracy that this approach and object-based approach provide.

Acknowledgements The support of Planet, Planet Labs, Inc. (formerly Cosmogia, Inc.) is an American private Earth-imaging company based in San Francisco, California, for providing us the high image satellite image data for our research via their Education and Research Program.

References

1. I.L. Castillejo-gonzález, J.M. Pe, F.J. Mesas-carrascosa, F. López-granados, Evaluation of pixel- and object-based approaches for mapping wild oat (Avena sterilis) weed patches in wheat fields using QuickBird imagery for site-specific management. **59**, 57–66(2014). https://doi.org/10.1016/j.eja.2014.05.009
2. G. Chen, Q. Weng, G.J. Hay, Y. He, Emerging trends and future opportunities Accepted us crt. GIScience Remote Sens. 0(0) (2018). https://doi.org/10.1080/15481603.2018.1426092
3. H. Costa, G.M. Foody, D.S. Boyd, Remote sensing of environment using mixed objects in the training of object-based image classifications. Remote Sens. Environ. **190**, 188–197 (2017). https://doi.org/10.1016/j.rse.2016.12.017
4. H.Y. Gu, H.T. Li, L. Yan, X.J. Lu, C. Vi, W.G. Vi, A framework for geographic object-based image analysis (GEOBIA) based on geographic ontology. XL(July), 21–23 (2015). https://doi.org/10.5194/isprsarchives-XL-7-W4-27-2015
5. G.J. Hay, G. Castilla, M.A. Wulder, J.R. Ruiz, An automated object-based approach for the multiscale image segmentation of forest scenes. **7**, 339–359. https://doi.org/10.1016/j.jag.2005.06.005
6. X.U. Jingping, Z. Jianhua, L.I. Fang, W. Lin, S. Derui, W.E.N. Shiyong, W. Fei, Object-based image analysis for mapping geomorphic zones of coral reefs in the Xisha Islands, China. 41201328 (2016). https://doi.org/10.1007/o13131 016 0921 y
7. D. Jovanovi, M. Govedarica, Đ. Ivana, V. Paji,. Object based image analysis in forestry change detection, pp. 231–236 (2010)
8. L. Ma, T. Fu, T. Blaschke, M. Li, D. Tiede, Z. Zhou, Evaluation of feature selection methods for object-based land cover mapping of unmanned aerial vehicle imagery using random forest and support vector machine classifiers (n.d.). https://doi.org/10.3390/ijgi6020051
9. K.V, Mitkari, S. Member, M.K. Arora, R.K. Tiwari, Extraction of Glacial Lakes in Gangotri Glacier Using Object-Based Image Analysis, pp. 1–9 (2017)
10. Planet Team, Planet Application Program Interface: In Space for Life on Earth. (San Francisco, CA, 2017). https://api.planet.com
11. K. Verbeeck, M. Hermy, J. Van Orshoven, An hierarchical object based image analysis approach to extract impervious surfaces within the domestic garden. March 2009, 2009–2012 (2011)
12. T.G. Whiteside, G.S. Boggs, S.W. Maier, International Journal of Applied Earth Observation and Geoinformation Comparing object-based and pixel-based classifications for mapping savannas. Int. J. Appl. Earth Obs. Geoinf. **13**(6), 884–893 (2011). https://doi.org/10.1016/j.jag.2011.06.008
13. A. Whyte, K.P. Ferentinos, G.P. Petropoulos, Environmental modelling & software a new synergistic approach for monitoring wetlands using Sentinels-1 and 2 data with object-based machine learning algorithms. Environ. Model. Softw. **104**, 40–54 (2018). https://doi.org/10.1016/j.envsoft.2018.01.023
14. M. Wieland, Y. Torres, M. Pittore, B. Benito, M. Wieland, Y. Torres, M. Pittore, B. Benito, Object-based urban structure type pattern recognition from Landsat TM with a Support Vector Machine. 1161(July) (2016). https://doi.org/10.1080/01431161.2016.1207261
15. W. Yu, W. Zhou, Y. Qian, J. Yan, Remote sensing of environment a new approach for land cover classification and change analysis: Integrating backdating and an object-based method. Remote Sens. Environ. **177**, 37–47 (2016). https://doi.org/10.1016/j.rse.2016.02.030
16. L. Zhong, L. Hu, H. Zhou, Deep learning based multi-temporal crop classification. Remote Sens. Environ., **221**(March 2018), 430–443 (2019). https://doi.org/10.1016/j.rse.2018.11.032

A Model to Generate Benchmark Network with Community Structure

Shyam Sundar Meena and Vrinda Tokekar

Abstract The studies of social networks focus on the structure and components of networks at different levels. To identify the component in a network, researchers have developed various community detection algorithms. To test the quality of community detection results, networks with well-known community structures are used. But, a very few networks are available for this purpose. Researchers have suggested some models that generate artificial networks with the community. However, most of the proposed models are unable to produce benchmark networks similar to the real-world network. We propose a model that generates benchmark networks for the evaluation of community detection algorithms. The proposed model has been compared with well-known LFR Lancichinetti et al. (Phys Rev 78(4):046110, 2008 [14]) and GLFR Le et al. (2017 26th international conference on computer communication and networks (ICCCN). IEEE, pp 1–9, 2017[15]) models. For performance testing, various structural properties have been analyzed, which are followed by real-world networks. The NMI scores achieved by well-known community detection algorithms were also compared. In experimental analysis, we found that networks generated by our model follow essential properties of real-world networks.

Keywords Community detection · Synthetic networks · Network analysis

1 Introduction

Networks are everywhere! Most of the networks are large, complex and dynamic. Examples of such networks are biological networks [10], social networks:

S. S. Meena (✉)
Department of Computer Science and Engineering, Shri Vaishnav Institute of Information Technology, Shri Vaishnav Vidyapeeth Vishwavidyalaya, Indore, India
e-mail: ssmeena7@gmail.com

V. Tokekar
Department of Information Technology, Institute of Engineering and Technology, Devi Ahilya Vishwavidyalaya, Indore, India

© The Author(s), under exclusive license to Springer Nature Singapore Pte Ltd. 2023   235
Y.-D. Zhang et al. (eds.), *Smart Trends in Computing and Communications*, Lecture Notes in Networks and Systems 396, https://doi.org/10.1007/978-981-16-9967-2_23

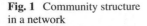

Fig. 1 Community structure in a network

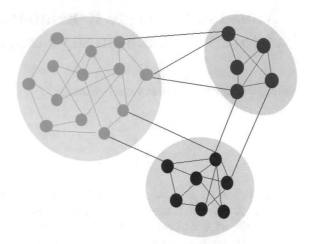

Twitter [12], Facebook [1] and infrastructural networks: electrical power grids [20] and internet routing paths [6].

Some queries related to the complex social networks analysis (SNA) are: How the structure of real-world networks looks? What types of rules the network formation process follows? How nodes make connections to each other in a network? To find answers to these and other related queries, component and node-level studies are required. In recent years, studies have been made to find the meaningful divisions (also known as communities) of networks. The main factors behind the complexity of community detection are the nature and size of networks [7].

Nodes have most of the connections within their community and very few connections to nodes that belong to other communities [8]. In Fig. 1, community structure of a network is shown. In community detection, we identify closely related nodes. It is helpful for epidemic outbreak prevention [11], viral marketing [21], etc. For analysis of the community detection task performance, predicted structure of communities is compared to the original structure of the community. The unavailability of appropriate benchmark networks generated the demand for the development of models to build artificial benchmark networks. We propose a model for building an artificial benchmark network with realistic properties. It works in two steps; first of all, Barabási–Albert preferential attachment model (BA) [2] is used for the generation of network of N nodes. In the second phase, new connections are introduced among nodes, to meet the desired characteristics of real-world networks.

The paper is structured as follows. A review of the popular strategies and models for generating synthetic networks is given in Sect. 2. In Sect. 3, proposed model is described. Experimental setup description is given in Sect. 4. Section 5.1 presents evaluation and discussions on the performance of a proposed model, followed by a conclusion in Sect. 6.

2 Related Work

In this section, a review of the popular strategies and models for the generation of synthetic networks is given. Girvan and Newman [8] presented a benchmark for building a network of 128 nodes. Each node in a group follows the same average degree. But, distribution of degrees of nodes and size of communities is different than real-world networks [14]. Danon et al. [5] proposed changes in Girvan and Newman [8] to generate a network with different sizes of communities. But, generated networks do not follow fat tails in their degree distributions. In the real world, the networks follow power-law degree and community size distribution [2].

LFR benchmark of Lancichinetti et al. [14] can generate networks that follow the power-law distribution of degrees and community sizes. But, it produces artificial networks with low transitivity [18]. According to Newman [17], real-world networks usually have high transitivity. HLFR benchmark [13] can generate networks with hierarchical communities' structure. It is also able to build directed and weighted networks.

In [22], a hierarchical LFR (HLFR) model for the generation of artificial networks with hierarchical community structures was proposed. It follows community structures similar to social networks, biological networks and technical systems. In [15], the authors presented a generalized version of LFR benchmark (GLFR). In place of similar internal and external connections of nodes for each community, the heterogeneous concept was applied.

Though various models for the artificial network have been introduced, it is still hard to achieve satisfactory accuracy and realistic aspects. The serious problem is dissimilarity in structural properties between the generated artificial benchmarks and the available real-world networks.

3 Proposed Benchmark

The possible causes of unrealistic properties observed in the networks created by the LFR [14] benchmark are: (a) initial network generation with the help of configuration model (CM) [16] and (b) assignment of a similar fraction of external links for each node. In the initial step of LFR [14] process, a very flexible model is applied to produce networks of any size, although generated networks do not follow real-world properties such as correlation [19] and transitivity [17]. We propose to adopt a separate generative model with various realistic features. The proposed work, first of all, uses Barabási–Albert preferential attachment model (BA) [2] for initial network generation. In the next step, the fraction of internal connection for each node of communities is selected randomly from a range $[\max(\mu_{min}, \mu - \Delta\mu), \min(\mu_{max}, \mu + \Delta\mu)]$ suggested in [15]. The rest of the steps goes similar to the LFR [14] model. Table 1 lists the input parameters of the proposed model.

Table 1 Parameters used in proposed model

Parameter	Description
N	Number of nodes
$< k >$	Average degree
$k_{\max}$	Maximum degree
μ	Mixing parameter
e_d	Minus exponent for the degree sequence
e_c	Minus exponent for the community size distribution
$C_{\min}$	Minimum size of communities
$C_{\max}$	Maximum size of communities
$\Delta\mu$	Mixing heterogeneity parameter

Algorithm 1 proposed Algorithm

Input: $N, < k >, k_{max}, C_{\min}, C_{\max}, e_d, e_c, \mu$ and $\Delta\mu$
Output: G
1: Generate N nodes and, decide degree sequence $degSeq$ of each node based on the three parameters $< k >, k_{max}, e_d$
2: Assign communities to all N nodes using parameters $C_{\min}, C_{\max}$ and e_c
3: Create a initial graph G using preferential attachment model (BA)
4: **for** each $c \in$ communities **do**
5:      $\mu_c = $ randomChoice[max$(\mu_{\min}, \mu - \Delta\mu)$, min$(\mu_{max}, \mu + \Delta\mu)$]
6:      **for** each $u \in c$ **do**
7:          **while** $degree(u) < (degSeq[u]) * \mu_c$ **do**
8:              $v = $ randomly select a $node$ from c
9:              add an $edge(u, v)$ in G
10:         **end while**
11:         **while** $degree(u) < degSeq[u]$ **do**
12:             $v = $ select a $node$ randomly
13:             **if** (v not in c) and ($degree(v) < degSeq[v]$) **then**
14:                 add an $edge(u, v)$ in G
15:             **end if**
16:         **end while**
17:     **end for**
18: **end for**

The process of benchmark network construction is given in Algorithm 1. In line 1, for the generation of degree sequence of nodes, power-law distribution with exponent e_d is applied. The decision of $k_{\min}$ and $k_{\max}$ is done such that the average degree remains k. In line 2, community assignment for each node is performed such that sum of the sizes of communities equals the number N. The sizes of communities are chosen as per the power-law distribution with exponent e_c. Initially, the assignment of community to nodes is done randomly till the total nodes in a community have not reached the size of the community. The process finally stops when the assignment of the community for each node is done successfully.

In line 3, an initial graph of N nodes and $\alpha * N$ edges is generated using the Barabási–Albert [2] scale-free property, where the value of α is taken 0.5. From 4th to 17th line, rewiring is applied. Here, for each node of each community, new links are attached. The number of nodes that belong to similar community nodes is calculated by μ_c*(degree sequence of node), where μ_c is selected randomly from a range $[max(\mu_{min}, \mu - \Delta\mu), min(\mu_{max}, \mu + \Delta\mu)]$ and remaining $(1 - \mu_c)*$(degree sequence of node) links are connected to the nodes that belongs to other communities.

4 Experimental Setup

The proposed benchmark for the generation of artificial networks is tested for important measures like transitivity, degree distribution and density. Two well-known community detection algorithms are also used, to analyze the behavior and ability of these methods. The selected algorithms are: CNM [4] and Louvain [3]. The implementation of LFR [14] provided by network library [9] is used for performing the all experiments. For the testing of the proposed benchmark, various artificial networks were generated. We used values of various controlling parameters as described below: (1) networks size N= 1000, (2) mixing parameter μ= (0.3, 0.4, 0.5 and 0.6), (3) exponent for the degree sequence $e_d = 3$, (4) value of exponent of community size (e_c) = 1.5, (5) minimum sizes of communities $C_{min} = 20$, (6) average degree $< k > =20$ and (7) mixing heterogeneity parameter $\Delta\mu = 0.2$.

4.1 Structural Measures of Networks

The proposed model was tested on various network properties that most networks seem to share such as density, transitivity and right-skewed degree distributions.

- **Density**: quantifies the fraction of links that belong to the same group from all possible links [7].

$$D(C) = \frac{|\{(i, j) \in E : i \in C, j \in C\}|}{n(n - 1)/2} \tag{1}$$

where n = total nodes in community C and an edge between node i and j is represented by E.

- **Degree distribution**: networks in real-world follow right-skewed degree distribution [2]. It is described by the following equation.

$$P(k) \approx k^{-\gamma} \tag{2}$$

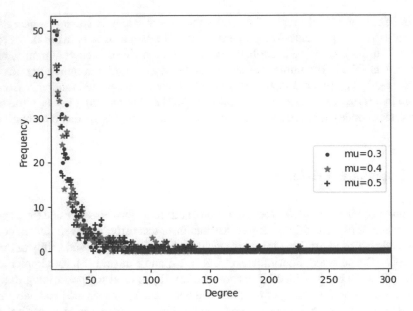

Fig. 2 Degree distribution in network generated by proposed model

where $P(k)$ represents fraction of nodes having degree k and parameter γ generally falls in the range $2 < \gamma < 3$.

- **Transitivity**: Nodes are said in transitive relation if friends of a node are also friends. This effect is quantified by the transitivity C [20], defined by

$$C = 3 * \frac{t_G}{T_G}. \tag{3}$$

where t_G is a count of triangles and T_G represents a number of connected triples of vertices. C values fall in the range [0.3, 0.6] for most of the networks. For a fully connected graph (everyone friend of everyone else), C will be 1.

5 Results and Discussion

5.1 Evaluation of Structural Properties

First, we studied the structure of real-world networks and their essential properties such as degree distribution, transitivity, density.

Fig. 2 shows the degree distribution of networks generated by the proposed model. From the experimental analysis, it is clear that for different values of the mixing parameter (μ), degree follows power-law (see Eq. 1). It illustrates that generated

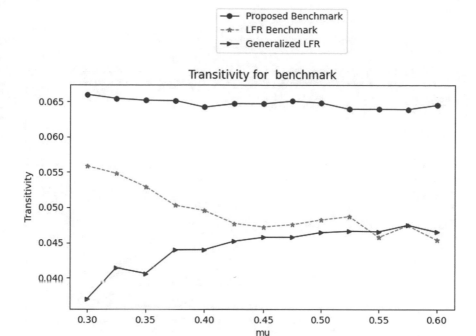

Fig. 3 Transitivity of networks

networks by the proposed model maintain small-world and scale-free characteristics. The consistency of the result ensures that the proposed modification in the LFR [14] benchmark did not affect degree distribution structural characteristics.

Figures 3 and 4 show the impact of μ on transitivity and density. For the LFR [14] network, both decrease as the value of μ increases. The transitivity and density of the GLFR [15] network increase when the value of μ grows. The proposed benchmark shows consistent results for different value of μ. The results state that model proposed in this paper is less sensitive in terms of transitivity and density.

5.2 Community Detection Performances

Normalized mutual information (NMI) [5] values are computed to quantify the similarity between resultant communities—assignment to nodes by an algorithm and known communities of nodes. First, different networks were generated by the proposed model, LFR [14] and GLFR [15], for different values of μ. Communities were detected in these networks by two well-known algorithms. The result of Louvain [3] and CNM [4] is presented respectively in Figs. 5 and 6. The NMI values reveal that the performance of both algorithms decreases for larger values of μ as

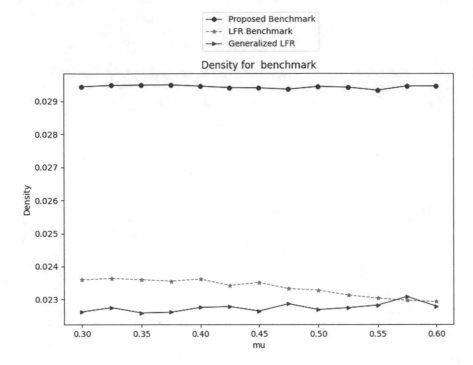

Fig. 4 Density of networks

Fig. 5 Performance of
Louvain algorithm on the
three benchmarks

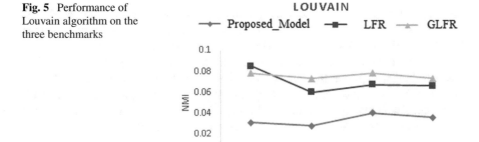

expected. Although, it was noticed that networks generated for various values of
mixing parameter μ by LFR [14] are more susceptible than the proposed model and
GLFR [15] benchmarks. It concludes that the proposed model can maintain similar
structures and properties for the different values of μ.

Fig. 6 Performance of CNM algorithm on the three benchmarks

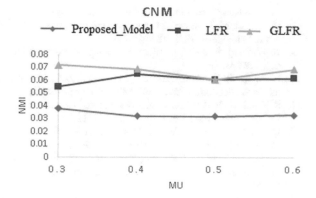

6 Conclusion

In this paper, we studied various properties of real-world networks and proposed a framework for generating artificial networks. We have analyzed various structural properties of networks generated by the proposed model such as transitivity, degree distribution and density. The suggested model is capable of building different sizes of networks akin to real-world networks. Note that our aim is not to prove that the proposed model is better than the LFR [14] and GLFR [15] models. The basic fact we want to highlight is that these models' networks and community structures are topologically different, and algorithms used for community detection should consider these variations when processing benchmark networks.

Acknowledgements I am thankful to my Ph.D. supervisor Dr. Vrinda Tokekar, Professor, Department of Information Technology, Institute of Engineering & Technology, Devi Ahilya Vishwavidyalaya, Indore, India, for her valuable guidance, motivation and support. I am also thankful to each member of my family for their care and support without which this work would have not been feasible for me.

References

1. L. Backstrom, P. Boldi, M. Rosa, J. Ugander, S. Vigna, Four degrees of separation, in *Proceedings of the 4th Annual ACM Web Science Conference* (2012), pp. 33–42
2. A.L. Barabási, R. Albert, Emergence of scaling in random networks. Science **286**(5439), 509–512 (1999)
3. V.D. Blondel, J.L. Guillaume, R. Lambiotte, E. Lefebvre, Fast unfolding of communities in large networks. J. Stat. Mech. Theory Exp. **2008**(10), P10,008 (2008)
4. A. Clauset, M.E. Newman, C. Moore, Finding community structure in very large networks. Phys. Rev. E **70**(6), 066111 (2004)
5. L. Danon, A. Diaz-Guilera, J. Duch, A. Arenas, A, Comparing community structure identification. J. Stat. Mech. Theory Exp. **2005**(09), P09,008 (2005)
6. M. Faloutsos, P. Faloutsos, C. Faloutsos, On power-law relationships of the internet topology. ACM SIGCOMM Comput. Commun. Rev. **29**(4), 251–262 (1999)

7. S. Fortunato, Community detection in graphs. Phys. Rep. **486**(3–5), 75–174 (2010)
8. M. Girvan, M.E. Newman, Community structure in social and biological networks. Proc. Natl. Acad. Sci. **99**(12), 7821–7826 (2002)
9. A. Hagberg, P. Swart, D. Chult, Exploring network structure, dynamics, and function using network (2008)
10. H. Jeong, B. Tombor, R. Albert, Z.N. Oltvai, A.L. Barabási, The large-scale organization of metabolic networks. Nature **407**(6804), 651–654 (2000)
11. S. Kitchovitch, P. Liò, Community structure in social networks: applications for epidemiological modelling. PloS One **6**(7), e22,220 (2011)
12. H. Kwak, C. Lee, H. Park, S. Moon, What is twitter, a social network or a news media? in *Proceedings of the 19th International Conference on World Wide Web* (2010), pp. 591–600
13. A. Lancichinetti, S. Fortunato, Benchmarks for testing community detection algorithms on directed and weighted graphs with overlapping communities. Phys. Rev. E **80**(1), 016118 (2009)
14. A. Lancichinetti, S. Fortunato, F. Radicchi, Benchmark graphs for testing community detection algorithms. Phys. Rev. E **78**(4), 046110 (2008)
15. B.D. Le, H.X. Nguyen, H. Shen, N. Falkner, Glfr: a generalized lfr benchmark for testing community detection algorithms, in *2017 26th International Conference on Computer Communication and Networks (ICCCN)*. (IEEE, 2017), pp. 1–9
16. M. Molloy, B. Reed, A critical point for random graphs with a given degree sequence. Random Struct. Algorithms **6**(2–3), 161–180 (1995)
17. M.E. Newman, The structure and function of complex networks. SIAM Rev. **45**(2), 167–256 (2003)
18. G.K. Orman, V. Labatut, A comparison of community detection algorithms on artificial networks, in *International Conference on Discovery Science*. (Springer, 2009), pp. 242–256
19. M.Á. Serrano, M. Boguñá, Weighted configuration model, in *AIP Conference Proceedings*, vol. 776 (American Institute of Physics, 2005), pp. 101–107
20. D.J. Watts, S.H. Strogatz, Collective dynamics of 'small-world'networks. Nature **393**(6684), 440–442 (1998)
21. L. Weng, F. Menczer, Y.Y. Ahn, Virality prediction and community structure in social networks. Sci. Rep. **3**, 2522 (2013)
22. Z. Yang, J.I. Perotti, C.J. Tessone, Hierarchical benchmark graphs for testing community detection algorithms. Phys. Rev. E **96**(5), 052311 (2017)

Human Activity Recognition Using LSTM with Feature Extraction Through CNN

Rosepreet Kaur Bhogal and **V. Devendran**

Abstract Human activity recognition is important for detecting anomalies from videos. The analysis of auspicious activities using videos is increasingly important for security, surveillance, and personal archiving. This research paper has given a model which can recognize activities in random videos. The architecture has been designed by using BiLSTM layer which helps to learn a system based on time dependencies. To convert every frame into a featured vector, the pre-trained GoogLeNet network has been used. The evaluation has been done by using a public HMDB51 data set. The accuracy achieved by using the model is 93.04% for ten classes and 63.96% for 51 classes from same data set only. Then, this network is compared with other state-of-the-art method, and it proves to be a better approach for the recognition of activities.

Keywords Action recognition · HMDB51 · Neural network · CNN · LSTM · BiLSTM · Video frames

1 Introduction

Computer vision is a multi-stage domain, which essential framework for automatic extraction, analysis, and comprehension from a single image or image sequence. Man-made visualization from videos is one of the real-world challenges of computer vision in various sectors like industrial, educational, security, consumer etc. In the process of recognition, detection, or tracking, there are two broad terms "actions" and "activity", which are generally used in the vision of the survey. And, there is difference

The original version of this chapter was revised: The incorrect last line in the Abstract has been removed. The correction to this chapter can be found at
https://doi.org/10.1007/978-981-16-9967-2_76

R. K. Bhogal (✉)
School of Electronics and Electrical Engineering, Lovely Professional University, Phagwara, India
e-mail: rosepreetkaur12@gmail.com

V. Devendran
School of Computer Science and Engineering, Lovely Professional University, Phagwara, India

Y.-D. Zhang et al. (eds.), *Smart Trends in Computing and Communications*, Lecture Notes in Networks and Systems 396, https://doi.org/10.1007/978-981-16-9967-2_24

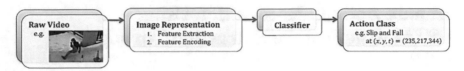

Fig. 1 Steps for activity recognition system [1]

between two words, the action refers to simple movement patterns performed by one person and in a short time such as "raise your hand", "bend", or "swim". They are involved with each other in some way. On the other hand, the activities are in which actions are for long duration, such as when two people shake hands, a soccer team scores a goal, or a group of people hijack a plane, or rob. [1].

An action or action recognition program may include a series of steps. First: insert a video or photo sequence. Second: extraction of feature and action descriptions. Last: interpretation from primitive actions [1] as shown in Fig. 1. Figure 1 shows the different steps to design any activity recognition system. Recognition of human activity remains a challenging research problem because of the following causes. First: there are wide variations of intra-class reason due to the difference in pattern of speed and movement, visual perception, and background clutter. Second: perception of action is closely related to details of complex situations; for example, the characteristics of a scene, the interaction between people and objects. Finally: the diversity and changing nature of the action phase makes it difficult to model the actions (lasting for short time) between successive actions [2].

It is intriguing to know the recent developments in activity recognition which includes its numerous applications. The researchers had heavily relied on clever human and computer-assisted communication programs such as managing a presentation slide or seeing steps to help employees learn and improve their skills. For example, a memory-based attention system can be used to design large public places [3]. There are various other applications which include human activity recognitions like gaming, surveillance, intelligent robots etc. [4]. With that, there is also need to create a smart system for older people that can automatically and quickly detect falls and inform family [5]. There is use of a directionality-base algorithm to track multiple identities and individuals in special care homes [6]. Using deep learning, various frameworks are given by Hasan and Roy-Chowdhury [7] in which the system designed that is calculated hierarchical feature and model which detect actions based on calculated. The researcher is also working on view-oriented activity recognition by pointing to a multi-dimensional hash table followed by a randomized voting system [8]. Niu and Abdel-Mottaleb [9] view invariant recognition is now at a consistent way of seeing human activity using details of movement and structure [3]; it has different sub-systems for behavioural detection and crowd measurement in a single learning mode and did not require camera measurement. Singh et al. [6] used adaptive background–foreground separation and derive directionality-based feature vector for recognition. Maddalena and Petrosino [10] proposed an approach which can handle videos of having moving background, illumination and system execution time also is less for high resolution. In this [11], use the low-level feature vector that contains the action-wide profile prediction depending on the action and simple temporary objects. In [12], the multiview-based method using deep network has been given and designed using LSTM and CNN models. Fan et al. [13] proposed model multiple features expression and improve long short-term memory (LSTM) network performance. The LSTM-based network is the most acceptable network and many HAR system using RNN because this network is working based in past and future time also. Similarly, in this paper also, the network has been designed using LSTM.

This paper presents a learned system using deep learning network layers which can identify the action from videos. The paper is having various sections; Sect. 2 includes explanation regarding Network model for HAR. Section 3 is about the result and discussion. Then, in the last, there is the conclusion based on work performed.

2   Network Model for HAR

Building video recognition system can be divided into two parts as follows:

2.1   Feature Extraction

The feature extraction is the first step for classification. By using pre-trained network, each video has been converted to sequence vector. The Google network is used as pre-trained network in feature extraction process. GoogLeNet is a 22-layer deep convolutional neural network with overall number of layers is 144. This design network has 224-by-224 image input size. GoogLeNet achieves optimal performance by representing the input frame of video in vector while simultaneously storing important location information. With the concept of transfer learning, the global average pooling layer is used for extraction of features for each video. The last four layers are not used here, because of the fact this network is used only for feature extraction not for classification. Each video represents in the sequence has size of [1024 × number of frames]. Where, 1024 are the number of features for each frame from a video. The feature extraction process is shown in Fig. 3.

2.2   BiLSTM Network for Prediction

To design an activity recognition system, the deep learning has been emerging area. The deep learning-based techniques are very famous and proved as to be the effective way for automatic action recognition. Earlier models use backpropagation which increases the time duration of the processing because of insufficient backflow error.

The RNN is the model which works using gradient descent algorithm with less-computational complexity [14]. The long short-term memory (LSTM) is one of the recurrent neural networks (RNN) which is given by Hochreiter [14]. Basically, the LSTM network is designed by using various cells, and it also contains an input gate, output gate as well as a forget gate. The cells are used to remember values over some intervals and gates regulate the pass of an information from one cell to another cell. For understanding of the LSTM network, assume the network contains N blocks and M inputs [15]. It has the various elements as stated below:

Block Input: This is used to combine the current input x^t and output of the LSTM y^{t-1} in the last iteration. The corresponding calculation can be done using below equation:

$$z^t = g\left(W_z x^t + R_z y^{t-1} + b_z\right) \tag{1}$$

where W_z and R_z are called as weights with input and output, respectively, and b_z acts as bias weight vector.

Input gate: This is required to update the input gate that combines the input and output of the last iteration and cell value in the last iteration. Equation used to find the input gate value is given below:

$$i^t = \sigma\left(W_i x^t + R_i y^{t-1} + p_i \Theta c^{t-1} + b_i\right) \tag{2}$$

In Eq. 2, the W_i, R_i and p_i are the weights which are used to update x^t, y^{t-1} and c^{t-1}, respectively, and b_i is called bias vector corresponding to the input. In Eq. 2, Θ is the pointwise multiplication process of vector.

Forget gate: It determines which information should be retained in cell states. For that, activation function of the gate is calculated based on input and output of state of memory cell at previous step. For that calculation, equation is stated below:

$$f^t = \sigma\left(W_f x^t + R_f y^{t-1} + p_f \Theta c^{t-1} + b_f\right) \tag{3}$$

In Eq. 3, the W_f, R_f and p_f are weights which are used to update the x^t, y^{t-1} and c^{t-1}, respectively, and b_f is called bias vector corresponding to the input. In Eq. 2, Θ is the pointwise multiplication process of vector.

Cell: Equation 4 given states how to calculate the value for cell.

$$c^t = (z^t \Theta i^t + f^t \Theta c^{t-1}) \tag{4}$$

Output gate: After combining the input, output, and cell value, the given Eq. 5 used to calculate the output of the gate.

$$o^t = \sigma\left(W_o x^t + R_o y^{t-1} + p_o \Theta c^t + b_o\right) \tag{5}$$

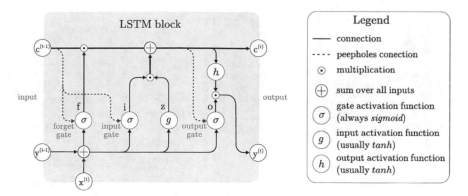

Fig. 2 The LSTM block architecture [15]

Block output: In the last, to calculate the block output using cell value and output gate value, the corresponding Eq. 6 is used:

$$y^t = g(c^t) \Theta o^t \tag{6}$$

where σ, h and g used to pointwise non-linear activation functions. The sigma is logistic sigmoid function, g and h is $\tanh(x)$ functions, respectively [15] (Fig. 2).

Pipelined two LSTM in an opposite direction together in a network to make BiLSTM network.

In the BiLSTM, the output of a single layer of LSTM during the "t" combined, results of the BiLSTM layers are determined not only by traces from previous vectors but also by indications from future vectors. In this way, more details are keeping by network which is continue from the future data. This network is capable of providing a good level of interaction between consecutive data with video signals [16].

In this work, the BiLSTM network has been used, and it has given good accuracy for activity. In other words, using the BiLSTM network in our model, results in long-term bidirectional relationships which have been subtracted by going back and forth several times in the vector sequence embedded in all parts of the video [16].

The extracted features, as given in Fig. 3, have been given to sequence layer of the network which is used for action recognition. The layout of the LSTM network with layers is given in Fig. 3. The network has various layers named as sequence input, BiLSTM, dropout, fully connected, Softmax, and in the last classification layer, of which the output is class labels. Sequence input layers with input size corresponding to the size of the feature vectors. BiLSTM layer with 200 hidden units with a dropout layer after that. For output, only one label for each sequence is used by setting the "output mode" of the BiLSTM layer to the "final", demonstrated in Fig. 3. The BiLSTM layer is used to study the long-term dependence between the time series of each video. Dropout layer is a type of standard method in which the inputs and reconnection of LSTM units are seamlessly extracted from renewals

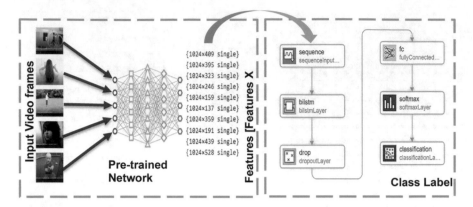

Fig. 3 Network model for HAR

and weight recovery during network training. This layer basically helps to reduce the overfitting and improves the overall model with respect to performance. The Softmax layer used to run neural network for multi-class function used to help us determine multiple classes in the images or videos.

3 Experiments Results

To convert video frames into vectors, pre-trained network configuration, i.e. GoogLeNet is used. In this case, a large database of human movements has been used. The database contains 2 GB of video data with 7000 clips over 51 classes. The CNN network has been used to calculate feature output for each video frame. Each videos in the form of sequence as feature vector has extracted from global average layer of the GoogLeNet. After extraction of the feature, then designing of system which will recognize activities. The network which is train and validate with all videos. For the training and validation of the designed network for HAR, the videos are divided into parts. For the training, took 90% of the videos for the total amount of recognition and validation rest 10% of videos.

The BiLSTM hidden units used are 200 units in number with batch size of 16. The Adam optimizer has been used to train the entire network from end-to-end with the initial reading level of 0.0001. The dropout layer of 0.5 is applied to avoid overfitting. The implementation has been done on i5 processor NVIDIA GEFORCE, 8 GB RAM, and 1 TB SSD.

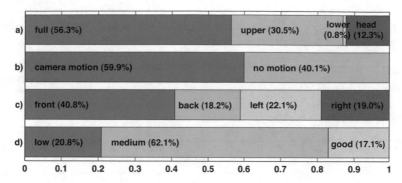

Fig. 4 Distribution of various conditions for the HMDB51, (**a**) visible body part, (**b**) camera motion, (**c**) camera view point, and (**d**) clip quality [17]

3.1 HMDB51 Data set

IIMDB51 (human movement database) contains 51 different categories that are captured with considering the complexity of human actions. The database contains 6849 clips divided into 51 action categories, each containing a minimum of 101 clips. The action categories can be grouped into five categories:

(i)   Ordinary facial expressions: smile, laugh, chew, and speak.
(ii)   Facial actions by deception of something: smoke, eat, and drink.
(iii)   Typical body movements: cartwheel, clapping, climbing, climbing stairs, diving, falling down, backhand flip, handstand, jump, lift up, run, run, sit down, sit down, distract, get up, turn, go, and wave.
(iv)   Physical movement with object contact: hair brushing, gripping, drawing a sword, pulling a golf ball, hitting something, kicking a ball, picking, pouring, pushing something, cycling, horseback riding, shooting a ball, shooting a bow, shooting a gun, swinging a baseball bat, exercise with the sword, and throwing.
(v)   Physical movement of human contact: fencing, hugging, kicking, kissing, punching, shaking hands, and fighting with a sword.

The detailed distribution of database is given as in Fig. 4. Statistically, distribution is divided into various parameters such as visual body, camera movement, camera view point, and clip quality in Fig. 5. This database has been used for training and validation of a system which will predict class.

3.2 Experiment on 10 Classes of HMDB51

The result has been obtained in 10 Classes which are *brush_hair, climb_stairs, cartwheel, catch, chew, clap, climb, dive, draw_sword and driddle.*

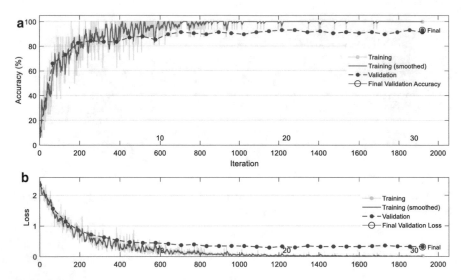

Fig. 5 **a** Training progress using HMDB (10 classes) (blue coloured solid line). **b** Variation of loss (red coloured solid line)

In the training and testing process, total 1150 video have been used. The training progress has been shown in Fig. 5. In the same figure, the first graph is between accuracy and iteration. Initially, there is prompt increase in the accuracy until 100 iterations. After 100 iterations, there is very gradual increase and with maximum value reaching at 93.04%. In Figure 6, the second graph is variation of loss with iterations. The validation loss is very high at 0th iteration, and there is a very sharp reduction in loss from 0 to 200th iteration. After that, it reduces very slowly and end up at 0.3245.

3.3 Experiment on 51 Classes of HMDB51

In Fig. 6, the evaluation has been done on complete data set by taking 51 classes. The observation graph has been given in Fig. 6 same as Fig. 5. There are two graphs; first is about accuracy, and second is about validation loss with respect to iterations. With a greater number of classes, the resultant accuracy is not similar to the previous experiment. Till, 200th iteration, the accuracy become more than 40%, and then afterwards, it increases in small percentage and end up at 63.96%. Similarly, in the validation loss-based observation, there is gradual reduction after 2000th iteration and ends up at approx. 1.56. If we consider all the classes of data set HMDB51, then accuracy is 63.96%. The comparison with other state-of-the-art method has been given in Table 1. Based on comparison, it concluded that BiLSTM-based network can also be used for activity recognition.

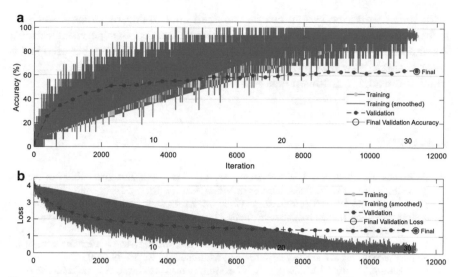

Fig. 6 a Training progress using HMDB (51 classes) (blue coloured solid line). b Variation of loss (red coloured solid line)

Table 1 Comparison with other method for HMDB51

Reference for comparison	Accuracy (%)
Zhu et al. [18]	53.8
Tran et al. [19]	54.9
Simonyan and Zisserman [20]	55.4
Srivastava et al. [21]	44.1
Sun et al. [22]	59.1
Wang et al. [23]	57.2
Bilen et al. [24]	57.2
Pan et al. [25]	49.2
Our method	63.96

4 Conclusion

In the paper, the model based on BiLSTM has been demonstrated. This model has been compared with other methods, given in Table 1. Every system needs the improvement same as this model also. When the number classes taken are 10, this model gives very good recognition rate. When 51 classes are taken, the accuracy ends up at 63.96%. There are various conditions on which accuracy is depending that are how calculated feature vectors, hyperparameters, and number of layers of the network. The most important concept which whole thing is depending upon is the tuning of hyperparameters of the model. It has significant impact on the model which is based on deep learning concepts.

In future work, the system can be designed after tuning hyperparameter with multiple training videos. This model is based on the concept of supervised learning. The model can also be evaluating on unsupervised learning. The data set used in this paper is single view data set. Ideal model can be created by extending this model which can work for multiview data set.

References

1. P. Turaga, R. Chellappa, V.S. Subrahmanian, O. Udrea, Machine recognition of human activities: a survey. IEEE Trans. Circ. Syst. Video Technol. **18**(11), 1473–1488 (2008)
2. A.-A. Liu, N. Xu, W.-Z. Nie, Y.-T. Su, Y. Wong, M. Kankanhalli, Benchmarking a multimodal and multiview and interactive dataset for human action recognition. IEEE Trans. Cybern. **47**(7), 1781–1794 (2017)
3. K.F. MacDorman, H. Nobuta, S. Koizumi, H. Ishiguro, Memory-based attention control for activity recognition at a subway station. IEEE Multimed. **14**(2), 38–49 (2007)
4. M.B. Holte, C. Tran, M.M. Trivedi, T.B. Moeslund, Human pose estimation and activity recognition from multi-view videos: comparative explorations of recent developments. IEEE J. Sel. Top. Signal Process. **6**(5), 538–552 (2012)
5. N. Lu, Y. Wu, L. Feng, J. Song, Deep learning for fall detection: 3D-CNN combined with LSTM on video kinematic data. IEEE J. Biomed. Heal. Inform. **2194**(c), 1–1 (2018)
6. M. Singh, A. Basu, M.K. Mandal, Human activity recognition based on silhouette directionality. IEEE Trans. Circ. Syst. Video Technol. **18**(9), 1280–1292 (2008)
7. M. Hasan, A.K. Roy-Chowdhury, A continuous learning framework for activity recognition using deep hybrid feature models. IEEE Trans. Multimed. **17**(11), 1909–1922 (2015)
8. J. Ben-Arie, Z. Wang, P. Pandit, S. Rajaram, Human activity recognition using multidimensional indexing. IEEE Trans. Pattern Anal. Mach. Intell. **24**(8) 1091–1104 (2002)
9. F. Niu, M. Abdel-Mottaleb, View-invariant human activity recognition based on shape and motion features, in *IEEE Sixth International Symposium on Multimedia Software Engineering* (IEEE, 2004), pp. 546–556
10. L. Maddalena, A. Petrosino, A self-organizing approach to background subtraction for visual surveillance applications. IEEE Trans. Image Process. **17**(7), 1168–1177 (2008)
11. S. Cherla, K. Kulkarni, A. Kale, V. Ramasubramanian, Towards fast, view-invariant human action recognition, in *2008 IEEE Computer Society Conference on Computer Vision and Pattern Recognition Workshops* (IEEE, 2008), pp. 1–8
12. C. Li, Y. Hou, P. Wang, W. Li, Multiview-based 3-D action recognition using deep networks. IEEE Trans. Hum. Mach. Syst. **49**(1), 1–10 (2018)
13. Z. Fan, X. Zhao, T. Lin, H. Su, Attention-based multiview re-observation fusion network for skeletal action recognition. IEEE Trans. Multimed. **21**(2), 363–374 (2019)
14. S. Hochreiter, Long Short-Term Mem. **9**(8), 1735–1780 (1997)
15. G. Van Houdt, C. Mosquera, G. Nápoles, A review on the long short-term memory model. Artif. Intell. Rev. **53**(8), 5929–5955 (2020)
16. X. Wu, Q. Ji, TBRNet: two-stream BiLSTM residual network for video action recognition. Algorithms **13**(7), 1–21 (2020)
17. H. Kuehne, H. Jhuang, E. Garrote, T. Poggio, T. Serre, HMDB: a large video database for human motion recognition. Proc. IEEE Int. Conf. Comput. Vis. 2556–2563 (2011)
18. Y. Zhu, Y. Long, Y. Guan, S. Newsam, L. Shao, Towards universal representation for unseen action recognition. Proc. IEEE Comput. Soc. Conf. Comput. Vis. Pattern Recognit. 9436–9445 (2018)
19. D. Tran, J. Ray, Z. Shou, S.-F. Chang, M. Paluri, ConvNet architecture search for spatiotemporal feature learning, no. section 3 (2017)

20. K. Simonyan, A. Zisserman, Two-stream convolutional networks for action recognition in videos. Adv. Neural Inf. Process. Syst. **1**, 568–576 (2014)
21. N. Srivastava, E. Mansimov, R. Salakhutdinov, Unsupervised learning of video representations using LSTMs, in *32nd International Conference Machine Learning ICML 2015*, vol. 1 (2015), pp. 843–852
22. L. Sun, K. Jia, D.Y. Yeung, B.E. Shi, Human action recognition using factorized spatio-temporal convolutional networks, in *Proceedings of IEEE International Conference Computer Vision*, vol. 2015 Inter (2015), pp. 4597–4605
23. H. Wang, C. Schmid, A. Recognition, T. Iccv, Action recognition with improved trajectories to cite this version : action recognition with improved trajectories (2013)
24. H. Bilen, B. Fernando, E. Gavves, A. Vedaldi, Action recognition with dynamic image networks. IEEE Trans. Pattern Anal. Mach. Intell. **40**(12), 2799–2813 (2018)
25. T. Pan, Y. Song, T. Yang, W. Jiang, W. Liu, VideoMoCo: contrastive video representation learning with temporally adversarial examples (2021)

Sentiment Classification of Higher Education Reviews to Analyze Students' Engagement and Psychology Interventions Using Deep Learning Techniques

K. R. Sowmia, S. Poonkuzhali, and J. Jeyalakshmi

Abstract Globally, higher education institutions are closed due to the COVID-19 pandemic. The sudden shift to online education excites most teachers and students. The professors are researching online learning platforms. They are only involved in face-to-face teaching in traditional teaching platforms. There are many concerns about the quality of online education. This paper proposes a framework for comparing online learning with traditional learning using emotions, learner perception, instructors, student engagement, understanding, effectiveness, learning outcome, peer collaboration, constraints, and comparisons. Deep learning algorithms like LSTM, GRU, and RNN classify the reviews. Students are positive during online learning in higher education according to LSTM, GRU, and RNN experimental analysis. Students are becoming more comfortable with online learning environments for higher education, according to detailed survey results.

Keywords Student engagement · Psychology interventions · Sentiments · Reviews · Online learning · Higher education

1 Introduction

Online education relies heavily on ICT to improve the activities and tools. Sentimental classification is used in higher education to analyze student comments and categorize them as positive, negative, or neutral towards online learning. In higher education, it is common practice to collect student feedback via comments or Google Forms to analyze their attitudes towards various subjects.

To analyze the reviews, machine learning techniques such as multinomial naïve Bayes, stochastic gradient descent, SVM, random forest, and multilayer perceptron classifier can be used. The results can be used to improve online education.

K. R. Sowmia (✉)
Department of IT, Rajalakshmi Engineering College, Chennai, Tamil Nadu, India
e-mail: sowmia.kr@rajalakshmi.edu.in

K. R. Sowmia · S. Poonkuzhali · J. Jeyalakshmi
Department of CSE, Rajalakshmi Engineering College, Chennai, Tamil Nadu, India

© The Author(s), under exclusive license to Springer Nature Singapore Pte Ltd. 2023
Y.-D. Zhang et al. (eds.), *Smart Trends in Computing and Communications*, Lecture Notes in Networks and Systems 396, https://doi.org/10.1007/978-981-16-9967-2_25

The responses from Google Forms help understand student collaboration. Students actively participate in online learning, create activities, discuss and evaluate them, thus improving their overall performance.

Deep learning now plays a major role in improving sentiment analysis results in fields like academic research, healthcare, tourism, and textile industries. The proposed sentiment classification system addresses student psychology and online learning engagement. Sentimental classification can be used to classify student feedback using deep learning neural network algorithms like LSTM, GRU, and RNN. Sentimental classification using deep neural networks will provide reliable data to improve the quality of learning. The overall contribution of this work can be summarized as follows as:

1. The extraction of reviews about online learning in higher education based on parameters such as emotions, learner's perception, instructors, student engagement, student understanding, student effectiveness, learning outcome, peer collaboration, constraints, and comparison with traditional learning.
2. Several kinds of data preprocessing steps have been carried out namely tokenization, stopword removal, and lemmatization
3. In the processed dataset, class imbalance problem has been addressed and rectified by using data augmentation techniques
4. The classification of the positive and negative sentiment using deep learning techniques such as LSTM, GRU, and RNN
5. The visualization of the resultant dataset has been done using word cloud for online education for higher education.

The organization of the paper can be described as follows as Sect. 2 explains about the relevant work, Sect. 3 describes the proposed model for Sentiment Classification, then Sect. 4 demonstrates about the obtained results, and finally, Sect. 5 concludes the proposed work.

2 Related Work

A detailed discussion of the literature review that has been conducted for the sentimental classification of higher education is as follows:

There is no simple definition to reveal the concept of student engagement in higher education. Researchers have come up with various suggestions to identify positive psychology interventions in higher education to improve student engagement in online learning. Research has been done on the positive effects of psychology interventions to enhance student engagement. The parameters that relate to psychology interventions and student engagement are self-perceptions, motivation, happiness, sentimentally connected with the organization, positive feedback in LMS etc [1]. To make students active in an online learning environment, the personal and academic data of students must be related to avoid student disengagement in online learning [2].

Hu et al. [3] conducted research on how and why students learn, as well as factors influencing student engagement. The model proposed for student engagement is based on two goals: future goals and school satisfaction. Kahu [4] frames student engagement in higher education based on four perspectives, namely the behavioral perspective, the psychological perspective, the sociocultural perspective, and the holistic perspective. The behavioral perspective is based on student behavior and institutional practice; the psychological perspective is based on the individual psychosocial process; the sociocultural perspective is based on the socio-political context; and the holistic perspective is based on the broader view of engagement [5].

The academic outcomes of students and their relationship with engagements is an open question for research in higher education [6]. Devito [7] suggested the factors influencing student engagement are grouped into five clusters: Peer collaboration among students and active involvement in learning, interaction between tutors and students, academic challenges faced by students and supporting faculty and family members. Apart from the psychological influence of institutions and tutors on student motivation and achievement, parents have a great influence on their learning styles and activities [8]. Boulton et al. come with a survey on student engagement with the results of positive interaction between student engagement and their happiness [9].

3 Framework of the Proposed Model for Sentiment Classification to Analyze Students' Engagement and Psychology Interventions

Figure 1 shows the architecture diagram about the proposed model for sentiment classification to analyze students' engagement and psychology interventions.

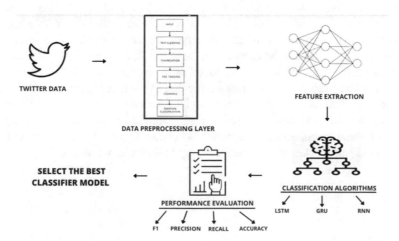

Fig. 1 Architecture diagram for sentiment classification to analyze students' engagement and psychology interventions

In the above depicted sentiment analysis model, the Twitter data collected based on student engagement, emotions, understanding, peer collaboration, psychological interventions pass through various data preprocessing techniques such as removing retweets, uppercase, numerals, special characters, and ascents. Feature extraction is being done and extracted data is classified using different classification algorithms like LSTM, GRU, and RNN. The elaborate process flow in listed below:

3.1 Dataset Generation

The dataset has been collected from Twitter API which consist of reviews based on student engagement and psychology interventions in online learning. The reviews are extracted from Twitter API based on the parameters such as student feedback, student emotions, learners perception, instructor feedback, student engagement, student understanding, student effectiveness, learning outcome, and peer collaboration. Nearly, 3978 tweets are stored as a .csv file for further analysis.

3.2 Data Preprocessing

Data preprocessing is done to make the dataset in a standard format to make learning algorithms more effective. It is done using natural language processing techniques. The preprocessing methods used here are tokenization, stopword removal, and lemmatization [10]. The reviews may be long sentences with a variety of phrases, emoticons, punctuation, and other elements. All non-alphanumeric characters must be omitted from the reviews. All retweets are also removed in this process. The process also includes the deletion of all uppercase letters and converting all of the texts to lowercase letters. Tokenization is applied where we split down long text strings into smaller chunks or tokens. It is also important to remove unnecessary vocabulary. Stopwords are the most common words in any language (for example, articles (a, an, the), prepositions (in, at, on, of, etc.), and these words do not add much detail to the text. Stopword removal is the method of eliminating certain words from a document so that the low-level details can be focused on. The process of lemmatization is also done to transform a term into its simplest form, such as 'learning' to 'learn.' All these processes are done on the tweet dataset to perform classification as positive or negative sentiments experienced by students in online learning.

3.3 Feature Extraction

In this process, the initial raw data is reduced to smaller units for making further processing easier. In this approach, these words are pre-trained, tokenized words are re-joined, and then it is replaced with the existing words.

3.4 Classification

In this step, the classification of the sentiments into two classes such as positive reviews and negative reviews using LSTM, GRU, and RNN. The structure and tuned hyperparameters of these algorithms are described below:

3.4.1 LSTM

The embedded layer, which uses 500 length vectors to construct the LSTM model, comes first. It is followed by the LSTM layer, which has 32 memory units (smart neurons). Finally, as it is a classification problem, the dense output layer with a single neuron and a sigmoid activation function makes a 0 or 1 prediction for two classes, namely positive and negative. Since this is a binary classification problem, log loss is used as the loss function, which is binary cross entropy in keras. The optimization algorithm ADAM is employed. With a batch size of 128 epochs, the model is then suited for 7 epochs. The evaluation metrics such as F1, precision, recall, and accuracy are estimated using the matplotlib library.

3.4.2 GRU

The embedded layer, which uses 100 length vectors to construct the GRU model, comes first. It is followed by the GRU layer, which has 32 memory units, (smart neurons). Finally, as it is a classification problem, the dense output layer with a single neuron and a sigmoid activation function makes a 0 or 1 prediction for two classes, positive and negative. Since this is a binary classification problem, log loss is used as the loss function, which is binary cross entropy in keras. The optimization algorithm ADAM is employed. With a batch size of 128 epochs, the model is then suited for 7 epochs. The evaluation metrics such as F1, precision, recall, and accuracy are estimated using the matplotlib library. The Algorithm 1: GRU for sentiment classification is depicted in (Fig. 2).

Algorithm 1: GRU for Sentiment Classification

Result: h$_t$*iscalculated*

z$_t$*isinitialized*

while *TRUE* **do**

    instructions;

    if *update signal* **then**

        update the gate z$_t$: z$_t$ = $\sigma \left(W^{(z)} x_t + U^{(z)} h_{t-1} \right)$

        reset the memory : r$_t$ = $\sigma \left(W^{(r)} x_t + U^{(r)} h_{t-1} \right)$

        new memory updated using tanh activation

        Final memory is calculated: h$_t$ = $z_t \odot h_{t-1} + (1 - z_t) \odot h_t'$

    else

        Compute: z$_t$ = $\sigma \left(W^{(z)} x_t + U^{(z)} h_{t-1} \right)$

    end

end

Fig. 2 Algorithm 1: GRU for sentiment classification

3.4.3 RNN

The embedded layer, which uses 100 length vectors to construct the RNN model, comes first. It is followed by the RNN layer, which has 32 memory units, (smart neurons). Finally, as it is a classification problem, the dense output layer with a single neuron and a sigmoid activation function makes a 0 or 1 prediction for two classes, positive and negative. Since this is a binary classification problem, log loss is used as the loss function, which is binary cross entropy in keras. The optimization algorithm ADAM is employed. With a batch size of 128 epochs, the model is then suited for 7 epochs. The evaluation metrics such as F1, precision, recall, and accuracy are estimated using the matplotlib library.

4 Results and Discussions

The implementation of these dataset is done on Windows platform with 8 GB size of RAM, x-64-based processor with Intel CORE i5 10th Gen processor. The three deep learning techniques have been implemented in the Google Collaboratory environment.

Model accuracy graph is drawn to represent the number of classifications predicted correctly by a model to the total number of predictions done. Model loss graph is drawn to measure the error in our trained model.

Figure 3a depicts the increase in accuracy after every epoch as the LSTM model is being trained, and Fig. 3b depicts the decrease in loss after every epoch as the LSTM model is being trained.

Figure 4a depicts the increase in accuracy after every epoch as the GRU model is being trained, and Fig. 4b depicts the decrease in loss after every epoch as the GRU model is being trained.

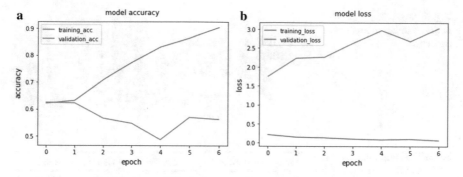

Fig. 3 **a** Accuracy graph for the LSTM model. **b** Loss graph for the LSTM model

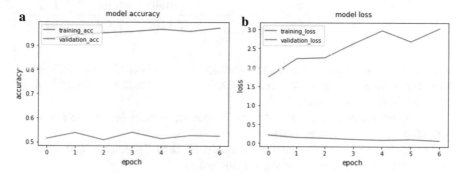

Fig. 4 **a** Accuracy graph for the GRU model. **b** Loss graph for the GRU model

Figure 5a depicts the increase in accuracy after every epoch as the RNN model is being trained, and Fig. 5b depicts the decrease in loss after every epoch as the RNN model is being trained.

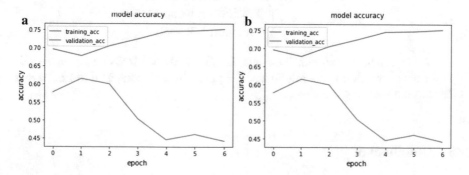

Fig. 5 **a** Accuracy graph for the RNN model. **b** Loss graph for the RNN model

Fig. 6 Word cloud for
online learning in higher
education

Table 1 Comparing LSTM,
GRU, and RNN models with
performance metrics

Performance metrics	LSTM	GRU	RNN
Accuracy (%)	90.38	97.3	75.14
F1 score	0.89	0.9	0.71
Precision	0.89	0.9	0.71
Recall	0.89	0.9	0.71

Figure 6 is a word cloud that displays the most frequently used words in the higher
education reviews. It can be inferred that the words "online," "technical," "career"
etc. are frequently used in the reviews.

The performance metrics for the classification algorithms such as LSTM, GRU,
and RNN are calculated and evaluated.

The performance metrics such as accuracy, F1 Score, precision, and recall are
calculated for the different deep learning algorithms.

$$Accuracy = \frac{Total\ number\ of\ Correct\ Predictions}{Total\ number\ of\ Predictions}$$

$$Recall = \frac{True\ Positive}{Total\ Positive + False\ Negative}$$

$$recision = \frac{True\ Positive}{Total\ Positive + False\ Positive}$$

$$F1\ Score = 2 \times \left(\frac{(Precision \times Recall)}{(Precision + Recall)}\right)$$

Table 1 compares the trained models, LSTM, GRU and RNN based on various
performance metrics. It can infer that GRU performs well compared to LSTM and
RNN as the accuracy is higher.

5 Conclusion

Sentiment analysis is performed using text classification algorithms like LSTM,
GRU, and RNN. The models are trained to evaluate the student engagement and
psychology interventions in online learning using ten parameters such as student

feedback, student emotions, learners perception, instructor feedback, student engagement, student understanding, student effectiveness, learning outcome, and peer collaboration. The results are compared with various performance metrics, and then, the best model is chosen. It can be inferred that the GRU model out performs the LSTM and RNN model in accurate text classification. The work can be further extended by taking online surveys from students based on the above-mentioned parameters. Statistical analysis can be done on the data to work on the analysis of student engagement and psychology interventions in higher education during online learning.

Acknowledgements I present my sincere thanks towards "The Centre for Data Science", Rajalakshmi Engineering College, Chennai, Tamil Nādu, India for extending support to the research work by providing necessary infrastructure and software facilities.

References

1. R. Cerezo, M. Sa´nchez-Santilla´n, M.P. Paule-Ruiz, J.C. Nu´ñez. Students' LMS interaction patterns and their relationship with achievement: a case study in higher education. Comput. Edu. **96**, 42–54 (2016)
2. J. Hammill, T. Nguyen, F. Henderson, Student engagement: the impact of positive psychology interventions on students, Act. Learn. High. Edu. SAGE. (2020)
3. Y-L. Hu, G. Ching, Factors affecting student engagement, in *2012 Conference on Creative Education* (2012)
4. E.R. Kahu, Framing student engagement in higher education. Stud. High. Edu. (2013)
5. G.V. Uma, R. Kumar, S. Poonkuzhali, R. Kishore Kumar, K. Sarukesi, Collapsed concept map game: an innovative online formative knowledge assessment method. Soc. Dev. Teach. Bus. Process. New Net Environ. B H **9**(4), 818–827 (2014)
6. K.S. Na, Z. Tasir, Identifying at-risk students in online learning by analysing learning behaviour: a systematic review, in *2017 IEEE Conference on Big Data and Analytics (ICBDA)* (16–17 Nov 2017)
7. M. DeVito, Factors influencing student engagement. Summer (2016)
8. C.A. Boulton, C. Kent, H.T.P. Williams, Virtual learning environment engagement and learning outcomes ata 'bricks-and-mortar' university. Comput. Educ. **126**, 129–142 (2018)
9. C.A. Boulton, E. Hughes, C. Kent, J.R. Smith, H.T.P. Williams, Student engagement and wellbeing over time at a higher education institution. PLoS ONE **14**(11), e0225770 (2019). https://doi.org/10.1371/journal.pone.0225770
10. K.R. Sowmia, S. Poonkuzhali, Artificial intelligence in the field of education:a systematic study of artificial intelligence impact on safe teaching learning process with digital technology. J. Green Eng. **10**(4), 1566–1583 (2020)

Strategic Network Model for Real-Time Video Streaming and Interactive Applications in IoT-MANET

Atrayee Gupta, Bidisha Banerjee, Sriyanjana Adhikary, Himadri Shekhar Roy, Sarmistha Neogy, Nandini Mukherjee, and Samiran Chattyopadhyay

Abstract In this paper, we discuss how to handle the issue of energy loss, delay during processing and transmission in IoT-MANET for real-time applications such as video in surveillance networks and also interactive applications in smarthome. Here, we propose a hierarchical strategic network model to reduce the cost of computation, latency and energy consumption. The proposed model when compared with random and normal grid topology gives better performance. Results of comparison show that the proposed hierarchical model provides an average of 166.29 Mbps for bandwidth utilization, 63.92 Mbps of throughput, 0.000095 s of jitter and 12.28 % of packet loss for transferring video data in mobile conditions. Also, each supernode receives an average 6.22 frames per second out of 15 transmitted frames.

Keywords Intelligent collaboration · Video streaming · Edge-MANET

1 Introduction

In contemporary researches, mobile edge computing components are integrated with MANET, WSN [6] and IoT to build up smartspaces [1]. IoT-MANET [1] has been built as a small but scalable network to support a variety of applications such as smart-building/homes, precision farming, defence surveillance, augmented reality. For our daily well-being, these technologies support things of various types to interact among themselves. For example, depending on the ambient temperature, the automatic air-conditioning in a smarthome system may adjust its cooling effect. These interactions may not involve exchange of a large amount of data or consuming enough network bandwidth, however, in some cases challenge is low latency and real-time decision

A. Gupta (✉) · B. Banerjee · H. S. Roy · S. Neogy · N. Mukherjee
Department of Computer Science and Engineering, Jadavpur University, Jadavpur, India
e-mail: atrayeegupta@gmail.com

N. Mukherjee
e-mail: nmukherjee@cse.jdvu.ac.in

S. Adhikary · S. Chattyopadhyay
Department of Information Technology, Jadavpur University, Jadavpur, India

© The Author(s), under exclusive license to Springer Nature Singapore Pte Ltd. 2023
Y.-D. Zhang et al. (eds.), *Smart Trends in Computing and Communications*, Lecture Notes in Networks and Systems 396, https://doi.org/10.1007/978-981-16-9967-2_26

making at the edge which involves the use of efficient network design and intelligent collaboration among nodes. Associated edge technologies include fog, cloudlets, edge-based MANET technologies [3]. Also, in these contexts, some processing is best done at the edge due to security reasons. Here, mobile edge computing [4] uses high computing resources at strategic locations to facilitate processing and storage in real time. It aids real-time application by enabling data fusion and intelligent decision making from multiple sensor devices. For example, in a tactile network, processing video surveillance data for target tracking in real-time requires high network bandwidth, low latency and low jitter which are best supported in edge rather than the cloud. Hence, in both surveillance and interactive cases discussed above, the challenge is to design a network that can reduce communication delay (which affects response time) and energy consumption (which directly affects network lifetime). Therefore, the purpose of this research work is to find a network design to reduce the energy consumption (due to complex computation and data transmission) and delay for resource-constrained IoT/edge-based MANET nodes which support different types of applications. Therefore, in this paper, we propose a hierarchical strategic network model and compare the same with random and grid topology. We also check the performance of a video streaming application and an interactive application in these three topologies over various layer-2 and layer-3 routing protocols of MANET. The results of the comparison are obtained in NS3 simulator with selected parameters such as average throughput, bandwidth consumption, packet delay, jitter, packet loss and also video frames per second (FPS) received at the sink.

The paper is organized as follows. Section 2 describes the hierarchical model of the network, Sect. 3 discusses the strategic approach, and Sect. 4 discusses the simulation parameters. Results are discussed in Sect. 5. Paper is concluded in Sect. 6.

2 Hierarchical Model

We consider a heterogeneous network cluster as shown in Fig. 1. Here, twelve ordinary nodes (red), R_{ij}, distributed throughout the network, with limited resources are used primarily for data collection and routing. These are equipped with cameras and stream video feeds to more powerful super nodes. These ordinary nodes may or may not have GPS, from which they ascertain their location. They communicate with other R_{ij} and also with supernodes to transmit video, image, audio, text data and also control information.

The other nodes, in the above scenario, responsible for intelligent processing are called supernodes (SuN) or cluster head (CH) used interchangeably in this paper. In Fig. 1, there are four supernodes (green) equipped with GPS (e.g. laptops). They have higher energy, storage, computation capability and high antenna gain to cover areas larger than R_{ij}'s. Each of them is equipped with a camera. Hence, they can also detect objects under the coverage area they are monitoring. They communicate with the ordinary nodes, collect data, aggregate and process them and send a collective decision to base station (BS). For example, CH can process the image frames of transmitted video from R_{ij}s and decide the number and type of target. It can also find target's possible location and future direction collaborating with other CHs. If no such information is revealed, the supernode may refrain from sending to BS. Moreover, these CH's have no/lower mobility compared to R_{ij}s. The purpose of dividing the area in Fig. 1 is to reduce the number of hops and mitigate interference by restricting the number of control information within a MANET. We call this a hierarchical model for data collection and efficient processing.

3 Strategic Approach

In this approach, we divide the large network coverage area into small grids to reduce the number of hops and induce quicker communication between source to destination. The entire 1000 m × 1000 m area is subdivided into zones to place sixteen nodes as shown in Fig. 1. In a typical zone ($A1$, $B2$, $C3$ and $D4$), we place three mobile R_{ij} that can communicate with CH. We denote CH_A, CH_R, CH_C and CH_D as cluster heads in each zone. For example, R_{11}, R_{12}, R_{13} in zone $A1$ communicate with CH_A within four hops. However, if we used 12 randomly placed nodes, the number of hops and control message exchange could have been higher in case there is route reconfiguration due to mobility of nodes. Therefore, this placement strategy also reduces the number of control messages in the MANET. Here, each R_{ij} moves inside the (250 m × 250 m) region and does not move to other zones. Each CH collects zone information from R_{ij} and communicates among themselves and routes to BS also within four hops. Blue lines in Fig. 1 represent hop count. Hence, with this approach, the number of hops can be reduced from source to sink with less

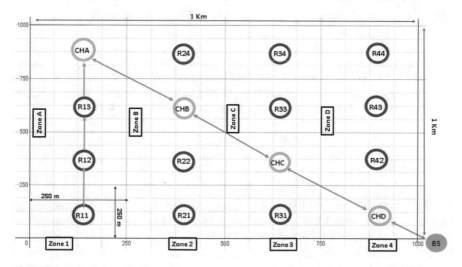

Fig. 1 Strategic placement of nodes

Table 1 Simulation parameters-1

Parameter	Value
General	
Simulation time	300
Total number of nodes	16
Number of source	12
Number of sinks	4
Coverage area	$1000 \times 1000 \text{ m}^2$
Data rate	331,776 kbps(15 FPS)
Packet size	1472 Bytes
Physical layer	
Transmission power	30 dBm
Tx antenna gain	9 dBi
Rx antenna gain	9 dBi
Propagation loss model	FriisPropagationLossModel
PropagationDelayModel	ConstantSpeedPropagationDelayModel
MAC layer	
WiFi standard	WIFI_PHY_STANDARD_80211ac
RemoteStationManager	ns3::ConstantRateWifiManager
DataMode	VhtMcs8
ControlMode	VhtMcs8
MCS	8
ChannelWidth	80
Nodes 0–3 channel number	106 (80 Mhz) 5 Ghz frequency
Nodes 4–7 channel number	122 (80 Mhz) 5 Ghz frequency
Nodes 8–11 channel number	138 (80 Mhz) 5 Ghz frequency
Nodes 12–15 channel number	155 (80 Mhz) 5 Ghz frequency

than five hops reducing latency and data loss during transmission. We also used different frequencies or channel numbers, shown in Table 1, to mitigate inter-cluster interference among nodes during data transmission from R_{ij} to SuN.

4 Simulation

To check how our proposed grid performs compared to random and generic grid topology, we simulated the above scenario in NS3. Here, we perform three test simulations as scenarios-1 (random), 2 (grid) and 3 (proposed grid -p_grid). We have used scenario-1 for random topology as shown in Fig. 2a where all 16 nodes are randomly placed inside 1000 m × 1000 m area. Here, any source node can send

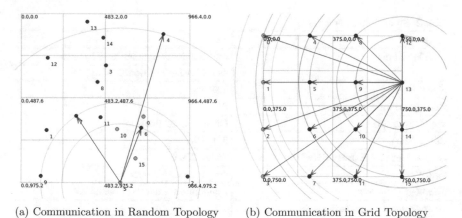

(a) Communication in Random Topology   (b) Communication in Grid Topology

Fig. 2 Communication at different instants in random and grid topologies

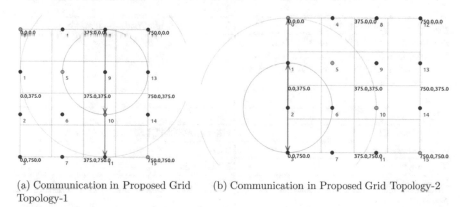

(a) Communication in Proposed Grid Topology-1   (b) Communication in Proposed Grid Topology-2

Fig. 3 Communication at different instants in proposed grid topology

data to any other sink node. In scenario-2, we placed the nodes in a grid ⌊4 × 4⌋ as shown in Fig. 2b. The distance between each node is 250 m. Here, any node can send collected data to other sink nodes. Green nodes represent sink placed in the first column of the grid. Scenario-3 shows the proposed grid in Fig. 3a, where each node is restricted to move around a bounded rectangle of 250 m × 250 m. Nodes under the supervision of a particular SuN cannot move to other zones. When any node hits the boundary, it bounces back and chooses another direction. Simulation parameters are detailed in Tables 1 and 2.

Table 2 Simulation parameters-2

Parameter	Value
Mobility model	ns3::RandomWaypointMobilityModel
	ns3::ConstantPositionMobilityModel
Node speed	1.94 m/s
Node pause time	2 s
Topology (grid)	GridPositionAllocator
MinX, MinY	0.0, 0.0
DeltaX, DeltaY (Distance Between Nodes)	250.0, 250.0
GridWidth	4 nodes per column wise
Topology (Random)	RandomRectanglePositionAllocator
16 nodes placed randomly inside $1000 \times 1000\,m^2$	
X length	1000 m
Y length	1000 m
Topology (Proposed grid)	RandomRectanglePositionAllocator
Single node is randomly placed inside a restricted area of 250×250 m^2 (For each 16 nodes, each rectangle is chosen so that they do not move to other zone)	
X length	250 m
Y length	250 m
Routing protocol	Global, AODV, OLSR, DSDV, HWMP
Packet interval	0.1 s
IP address	10.1.1.0/24
Transport layer	UDP
Application layer	
I-server and I-client	Sends and receives simple text-message
OnOff application	Video streaming
OffTime	ns3::LogNormalRandomVariable
	[Mu=0.4026|Sigma=0.0352]
OnTime	ns3::WeibullRandomVariable
	[Shape=10.2063|Scale=57480.9]
MaxBytes	10,989,173
Server start and stop time	0, 300 s
Client start and stop time	0.1 s after server, 299 s

Table 3 Bandwidth comparison for three scenarios

	Bandwidth (Random)	BandWidth (Grid)	Bandwidth (P_Grid)
Video (Static) (Mbps)	196.34	209.22	217.59
Video (Mobile) (Mbps)	150.86	160.65	166.29
Interactive (Static) (Kbps)	159.67	160.34	162.36
Interactive (Mobile) (Kbps)	158.71	158.12	160.36

Table 4 Comparing packet delay for three scenarios

	Delay (Random) (s)	Delay (Grid) (s)	Delay (P_Grid) (s)
Video (Static)	16.12	13.52	10.47
Video (Mobile)	21.78	22.25	18.99
Interactive (Static)	0.00230	0.00238	0.00237
Interactive (Mobile)	0.00353	0.00272	0.00261

5 Results and Discussion

A. Comparing Bandwidth Utilization

Tabulated results are average of all layer three (L3-AODV, OLSR, DSDV, etc.) and layer two (L2-Hybrid Wireless Mesh Protocol (HWMP)) routing protocols used in this paper. Table 3 provides overall results of the comparison of bandwidth in three scenarios for L3 and L2 protocols. Theoretically, we know a higher bandwidth network allows transferring more data. Also, we can observe in the case of video and interactive applications that random topology has the lowest, but proposed grid has the highest bandwidth utilization on average for all MANET protocols used above. These results show that bandwidth utilization of networks can vary by changing network topology, type of MANET routing protocol and mobility of nodes.

B. Comparing End-to-End Delay

Term delay is used in this paper to refer to packet delay or average end-to-end delay as per [2]. Table 4 shows packet delay for three topologies in static and mobile condition. In Table 4, packet delay in the proposed grid for both video and interactive applications shows minimum delay compared to other topologies.

Table 5 Comparing jitter for three scenarios

	Jitter (Random) (s)	Jitter (Grid) (s)	Jitter (P_Grid) (s)
Video (Static)	0.0001029	0.0000882	0.0000705
Video (Mobile)	0.0001108	0.0001137	0.0000946
Interactive (Static)	0.002133	0.001324	0.001481
Interactive (Mobile)	0.002034	0.001634	0.001435

Table 6 Comparing throughput for three scenarios

	Throughput (Random)	Throughput (Grid)	Throughput (P_Grid)
Video (Static) (Mbps)	74.57	83.78	104.45
Video (Mobile) (Mbps)	60.33	53.29	63.92
Interactive (Static) (Kbps)	81.24	80.72	81.27
Interactive (Mobile) (Kbps)	80.37	79.92	80.85

C. Comparing Jitter

Packets with high jitter can pause the video for few moments, and in the worst case, it may also induce packet loss at the receiver. We calculated mean jitter from paper [2]. In Table 5, mean jitter values of all three topologies are tabulated for video and interactive application where video application in proposed grid shows significantly lower jitter than interactive application.

D. Comparing Throughput

As bandwidth provides an estimate of how much data can travel, while throughput provides actual value of transmitted data. We calculated throughput using formula from paper [5]. Table 6 shows throughput decreases due to mobility in both video and interactive applications. Proposed grid shows a higher throughput in both video and interactive applications.

E. Comparing Packet Loss

Percentage of packet loss is calculated with respect to number of packets sent. Table 7 shows packet loss comparison for three topologies in static and mobile conditions for video and interactive data. We can observe that mobility-induced packet loss is slightly higher in all three topologies. Also, interactive application shows a lower

Table 7 Comparing packet loss for three scenarios

	PacketLoss (Random) (%)	PacketLoss (Grid) (%)	PacketLoss(P_Grid) (%)
Video (Static)	11.39	11.06	7.08
Video (Mobile)	13.92	13.60	12.28
Interactive (Static)	1.74	3.12	0.41
Interactive (Mobile)	4.86	5.87	2.75

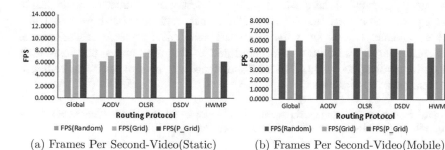

(a) Frames Per Second-Video(Static) (b) Frames Per Second-Video(Mobile)

Fig. 4 Comparing frames per second for video application

Table 8 Comparing FPS for three scenarios

	FPS (Random)	FPS (Grid)	FPS (P_Grid)
Video (Static)	7.28	8.37	10.03
Video (Mobile)	5.28	5.12	6.22

packet loss percentage compared to multimedia data transfer as expected. The proposed grid has a minimum packet loss percentage among three scenarios.

F. Comparing Frames Per Second

We measure the number of frames received at the sink for three topologies for video data transfer at 15 FPS from each source. Figures 4a, b show FPS in static and mobile conditions, respectively. HWMP protocol has 5.5 FPS on average in three topologies in mobile condition. For others, FPS drops in the range of 4.3–7.5 for all three scenarios during mobility. Comparison of FPS is shown in Table 8, which also depicts that the proposed grid has the highest FPS for video application tested here.

6 Conclusion

In this research work, we propose a hierarchical strategic network model to reduce the cost of computation, latency and energy consumption for IoT/edge- based MANET technology for real-time multimedia (surveillance in tactile networks) and interactive (smarthome) applications. We designed a cluster-based distributed network with IoT-MANET over 802.11 ac to mitigate the problem of computation and delay. This network topology considers the reduction of number of hops and number of messages aggregated at the cluster head by restricting the number of nodes inside the zone. Also to mitigate inter-cluster interference, each zone has been assigned a different channel. Results show that proposed hierarchical strategic network model performs better in transmitting both multimedia (streaming video) and text-based interactive applications for IoT-MANET and, hence, can be deployed in surveillance and smarthome applications.

Acknowledgements This research is supported by Centre for Artificial Intelligence and Robotics (CAIR)-the Defence Research and Development Organization (DRDO), India.

References

1. R. Bruzgiene, L. Narbutaite, T. Adomkus, *MANET Network in Internet of Things System* (2017)
2. G. Carneiro, P. Fortuna, M. Ricardo, Flowmonitor: a network monitoring framework for the network simulator 3 (ns-3), in *Proceedings of the Fourth International ICST Conference on Performance Evaluation Methodologies and Tools. VALUETOOLS '09, ICST (Institute for Computer Sciences, Social-Informatics and Telecommunications Engineering)* (Brussels, BEL, 2009). https://doi.org/10.4108/ICST.VALUETOOLS2009.7493
3. H. El-Sayed, S. Sankar, M. Prasad, D. Puthal, A. Gupta, M. Mohanty, C.T. Lin, Edge of things: the big picture on the integration of edge, iot and the cloud in a distributed computing environment. IEEE Access **6**, 1706–1717 (2018)
4. P. Mach, Z. Becvar, Mobile edge computing: a survey on architecture and computation offloading. IEEE Commun. Surv. Tutor. **19**(3), 1628–1656 (2017)
5. R. Patidar, *Validation of wi-fi Network Simulation on ns-3* (University of Washington, Technical report, 2017)
6. R. Rao, G. Kesidis, Purposeful mobility for relaying and surveillance in mobile ad hoc sensor networks. IEEE Trans. Mobile Comput. **3**(3), 225–231 (2004)

Limitation for Single-equation Dependency

Michael Shell, John Doe, and Jane Doe

Abstract Linear regression models are trained with standard errors. We have observed that regression models often give a low accuracy as compared to other techniques. We have evaluated the regression technique and draw a conclusion that these standard errors are the reason for the lower accuracy. The main reason of producing low accuracy is dependency on single-equation denoting relation between the dependent and independent variables. In regression model, on the basis of least square method, the affect of one variable is studied on other variable. However, there comes a time oftenly where the affect of one variable on another drastically changes due to the type of values or range of values present in the data. Such changes in the values might increase the higher accuracy of regression model. So, highly accurate regression model might not work for datasets with such variations. In this paper, we have discussed the issue of single-equation dependency to improve the accuracy. By using the ridge regression technique, we have developed a model that can produce high accuracy for multiple datasets with different variants in the values and the ranges of the values.

Keywords Least square method · Regression model · Ridge regression · Single-equation dependency

1 Introduction

Regression is a statistical procedure that analyzes the relationship between a dependent variable (Y) and one or more independent variables $(X_1, X_2...X_k)$ [1].

M. Shell (✉)
Department of Electrical and Computer Engineering, Georgia Institute of Technology, Atlanta, GA 30332, USA
URL: http://www.michaelshell.org

J. Doe · J. Doe
Anonymous University, Atlanta, GA, USA

If only one independent variable was used, it would be a simple regression analysis, and if more than one independent variable was used, it would be a multiple regression analysis. In order to study the relationship between these variables, it is necessary to establish the functional relationship between them. A first step in determining this possible relationship between the variables is to analyze the graph of observed data. This graph is called a scatter plot and allows you to visually determine whether the variables are related or not. If they are related, the intensity, the sense of the relationship between the variables (direct or inverse) and the type of relationship (linear or nonlinear) existing between them can be intuited.

1.1 Assumptions of the Linear Regression Model

In order to work with a linear regression model, the following assumptions are met [2]:

- The relationship between the variables is linear, and the model is correctly specified.
- The coefficients of the model are constant over time.
- The independent variables are deterministic, and there is no linear relationship between them (absence of multicollinearity).
- The errors are incorrect (there is no autocorrelation), they have constant variance (homocedasticity) and mathematical hope equal to zero, that is, the matrix of variances and covariances of the errors is a scalar matrix where the elements of the main diagonal are the variances of errors.
- The number of observations of the variables must be greater than the number of parameters to be estimated.

1.2 Simple Linear Regression Model

The simple linear regression model is characterized in that to estimate or predict the dependent or endogenous variable, only one independent or exogenous variable is used [3], through the following equation:

$$\gamma_i = \alpha + \beta X_i + \epsilon_i \qquad i = 1...N \tag{1}$$

where N is the number of observations of the variables; the coefficients α and β are the unknown parameters that indicate, respectively, the ordinate in the origin (or estimated value of Y when X = 0) and the slope or coefficient of the regression (or variation the dependent variable before unit variations of the variable independent); ϵ is the random disturbance that includes all those non-observable facts and, therefore, are associated with chance. This disturbance is what gives the model its stochastic character.

The problem of regression consists in estimating values for the unknown parameters α and β from the observations of the variables Y and X in such a way that the equation is completely specified.

Since the observed values of the dependent variable (Y) generally differ from those obtained through the regression line, an error is produced that will be calculated as the difference between the observed and estimated values. These errors are known as waste.

The objective of the regression analysis will be to obtain the estimates of the parameters that make the sum of the squares of the residues minimal. This estimation procedure is generally carried out using the ordinary least squares method (MCO). This term of "least squares" proceeds from the description given by Legendre in 1805 "carrés moindres" and by Gaus in 1809.

Where the estimate of the ordinate at the origin, is the estimate of the regression coefficient, is the sample mean of the independent variable and is the sample mean of the dependent variable.

For example, if we estimate by MCO the relationship between consumption (Y) in thousands of euros and income (X) in thousands of euros received in a given rgion, we could obtain the following results:

$$\hat{Y} = 18.1 + 0.18X \tag{2}$$

$$(3.3) \quad (0.03) \qquad\qquad R^2 = 0.84 \tag{3}$$

where in brackets, we have the standard deviations of each estimated parameter (useful especially for making inference); R^2 is the coefficient of determination; autonomous consumption would be 18,100 euros, and the slope that, in this case, is positive and equal to 0.18 would indicate that if the rent is increased by one thousand euros, consumption would only increase by 180 euros.

1.3 Multiple Linear Regression Model

Multiple linear regression is characterized in that to estimate or predict the dependent variable, (Y), two or more independent variables are used, $(X_1...X_k)$:

$$Y_i = \beta_0 + \beta_1 X_{1i} + ... + \beta_k X_{ki} + \epsilon_i \quad i = 1...N \tag{4}$$

where β_0 is the independent term that each of the coefficients of the independent variables and $\beta_i (i = 1k)$ measures the effect that each of these variables has on the dependent variable.

As in the simple regression, the estimation procedure is ordinary least squares, although it is true that, in this case, increasing the number of variables makes obtaining the estimator the parameters of the model more complicated. The matrix expression of this estimator is as follows:

$$\hat{\beta} = (X^T X)^{-1} X^T Y \tag{5}$$

where $\hat{\beta}$ is the vector of the estimated parameters of the model, X is the matrix of observations of the explanatory variables of the model, X^T its transposed, and Y is the vector of observations of dependent variable.

The estimator properties obtained by ordinary least squares are:

1. It is unbiased, that is, $E(\beta) = \beta$ if it is true that $E(\epsilon) = 0$
2. The variance of the estimator is: $Var(\beta) = \sigma_\epsilon^2 (X^T X)^{-1}$ if it is fulfilled that $Var(\epsilon) = \sigma_\epsilon^2 I$ I, where σ_ϵ^2 is the variance of the term error, e, I is the identity matrix
3. It is consistent
4. The ordinary least squares estimator is the optimal unbiased linear estimator (Gauss Markov theorem).

For example, if we estimate the relationship between consumption (Y), income (X 1) and inflation (X 2) of a giv- en region, we could get the following results:

$$\hat{Y} = 14.3 + 0.15X_1 - 0.07X_2 \tag{6}$$

$$\quad (2.2) \quad (0.02) \quad\quad (0.01) \quad\quad\quad\quad R^2 = 0.94 \tag{7}$$

where we can appreciate that according to the data obtained from each of the variables, autonomous consumption would be 14,300 euros; if there is an increase in income of one thousand euros (keeping the rest of the variables constant), consumption would increase by 150 euros; an increase in inflation by 1% would imply that consumption would decrease by 700 euros (if the rest of the variables remain constant).

1.4 Nonlinear Regression

Linear regression does not always provide adequate results since, sometimes, the relationship between the dependent variable and the independent variables is not linear. The estimation of the parameters of these nonlinear functions is a more complicated process than in the linear functions, since it can even happen that in the nonlinear models, the number of parameters does not match the number of explanatory variables.

However, there are times when the original variables can be transformed to convert the nonlinear function into linear and, thus, apply linear regression techniques. Therefore, if nonlinearity, for example, affects only the explanatory variables but not the coefficients, then new variables (depending on the nonlinear ones) can be defined to obtain the linear model.

An example of a nonlinear model in the parameters would be the one given by the following exponential function:

$$Y = \alpha X^{\beta} \tag{8}$$

where α and β are unknown constants.

However, this nonlinear model can be transformed into a linear one by applying logarithms on both sides of the equation, as follows:

$$\log(Y) = \log(\alpha) + \beta\log(X) \tag{9}$$

In this model, which is called double log regression, the relationship between the $\log(Y)$ and the $\log(X)$ would be estimated and the interpretation of the parameter β is interesting, since it measures the elasticity of Y with respect to X.

However, other times, nonlinear models cannot be transformed into linear ones. In that case, it would be necessary to apply other methods for the estimation of the parameters such as: nonlinear least squares, maximum likelihood, Monte Carlo simulations.

1.5 Applications

The use of linear regression is very widespread and can be applied to virtually any field that is related, among others, to economics, finance, medicine, biology, physics, engineering, etc.; that is, it allows us to relate variables in a rigorous and quantifiable way in different environments.

1.6 Limitations

One of the main limitations of the regression analysis is based on the fact that two variables grow or decay following the same guidelines does not necessarily imply that one causes the other, since it can happen that a spurious relationship occurs between them. Therefore, to establish the relationship between different variables, it is necessary that this relationship be based on a good theory, since this statistical analysis is adequate to quantify a known relationship between variables, but it is not the best instrument to find functional relationships between variables.

2 Literature Review

When addressing any type of machine learning problem, there are many algorithms to choose from, but there is something that we must be clear about and that is that no algorithm is the best for all problems, and each of them has its own pros and cons, which it does not serve as a guide to select the most suitable.

Although remember that the performance of the different machine learning algorithms depends largely on the size and structure of the data, linear regression defines as train the best line through all data points. This algorithm can be simple linear regression, where the prediction is made with a single variable, multiple linear regression, where a model is created for the relationship between multiple independent input variables and the polynomial regression in which the model becomes a nonlinear combination of the characteristic variables.

- Advantages of linear regression includes easy to understand and explain, which can be very valuable for business decisions.
- It is quick to model and is particularly useful when the relationship to be modeled is not extremely complex and does not have much information.
- It is less prone to overfitting. Along with the advantages, there comes the disadvantages too.
- You cannot model complex relationships.
- You cannot capture nonlinear relationships without transforming the input, so you have to work hard to fit nonlinear functions.
- It can suffer with outliers. Such type of regression technique is very useful for:
- Take a first look at a dataset.
- When you have numerical data with many features.
- Perform econometric prediction.
- Modeling marketing responses.

2.1 Related Work

Regression model is a kind of predictive model where the relationship between dependent and independent variable is established. Here, in this model, the predictor is the independent variable and target is the dependent variable. Regression model is used in time series models. Most of the time to present this relationship, a single equation is used, and we will be discussing limitations in the use of single equations in regressions model.

Multiple methods are used in estimating a single equation, and the one very common is ordinary least squares method which is used for linear regression estimation. Ordinary least squares method uses supervised discrete hashing (SDH) for hashing, where SDH solves the optimization in steps like most critical ones, binary hashing optimization in discrete fashion, discrete cyclic coordinate descent (DCC) used multiple iterations for solving SDH, and all this is a time consuming in relationship to the fast supervised discrete hashing (FSDH) described in [4]. FSDH used in [4] is faster than traditional SDH in least square regression in solution to projection matrix.

Thanks to its efficacy and completeness, linear least square regression has fulfilled its purpose as the tool of choice for system optimization. Limitations in shapes that linear models can assume over long ranges, possibly poor extrapolation properties and susceptibility to outliers are the principal drawbacks of linear least squares.

In [5], authors have evaluated the traditional regression methods and prompted the cases where regression analysis failed to conclude with the appropriate results. To counter this, they have evaluated the propensity score analysis which produced much better results.

Addressing the problem of single-equation dependency, authors of [6] have proposed an effective piecewise linear regression with novel linear programming formulation. Their proposed methodology can overcome the issues of single-equation dependency. Their proposed methodology consists of dividing a single variable into multiple segments. After dividing it into pieces, their method fits multivariable linear regression function per segment.

3 Proposed Work

To demonstrate the proposed work, we have developed a problem statement and then demonstrate how our proposed solution works fine to elimate the single-equation dependency problem. In general, the proposed methodology can be explained as: We obtain the problem of optimizing the cost function (empirical risk) (we believe that among the signs, there is a constant, and therefore, a free coefficient is not needed)

$$\mathcal{L}(X, \vec{y}, \vec{w}) = \frac{1}{2n} \sum_{i=1}^{n} (\vec{y}_i - \vec{x}_i^T \vec{w}_i) \tag{10}$$

$$\frac{1}{2n} |\vec{y}_i - X\vec{w}|_2^2 \tag{11}$$

$$\frac{1}{2n} (\vec{y}_i - X\vec{w})^T (\vec{y}_i - X\vec{w}). \tag{12}$$

If we differentiate this functionality by vector wow, equal to zero and solve the equation, we get an explicit formula for the solution:

$$\vec{w} = (X^T X)^{-1} X^T \vec{y} \tag{13}$$

We have tested this proposed work with a Boston Housing Dataset and evaluated the results which are presented in the results section.

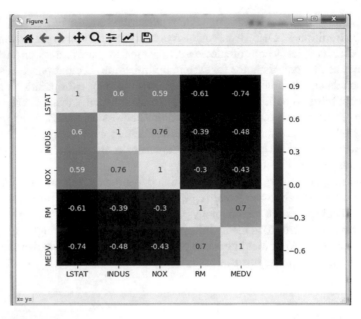

Fig. 1 Heat map

4 Experiments and Results

For experiments and evaluation of the proposed technique, we have used Boston
Housing Dataset consisting of the prices of houses present at different places in the
Boston. Moreover, along with the prices of houses, the dataset also provides a side
information about the Crime, INDUS, AGE and other attributes too. The loaded
dataset is Boston area estate details

MEDV—Mid value
RM—Room per dwellinh)
TAX—Tax rate per 10K $
INDUS—Proportion of non-retail business acres per town
LSTAT—% lower status of the population
NOX—Nitric oxides concentration

The first step of the experiments is to perform the required preprocessing of the
given dataset. Initially as always, we start exploring a new dataset by looking at the
graphs. We visualize the correlation matrix and in the form of a heat map presented
in Fig. 1. It is a heat map that shows relation and dependency of each independent
variable on to dependent variable that the larger the value, the greater the impact of
the variable.

Table 1 Initial regression model constructed

Slope	−0.95
Intercept	34.55

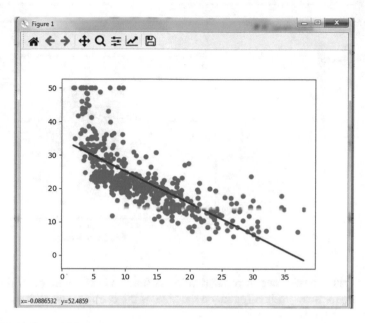

Fig. 2 Regression model

We proceed to the construction of a regression model. Define dependent and independent variables. We have presented the slope and the intercept of the constructed regression model in Table 1.

How our model have performed a regression on the given dataset, we need to evaluate the performance. Before that, we have drawn a graphical representation of the regression model which is represented in Fig. 2.

Possible metrics for checking the quality of the model in sklearn can be found in the documentation , or you can calculate by yourself using the formulas. Since there are several independent variables in our model, we cannot display their dependence on two-dimensional space, but we can plot the relationship between the residual model and the predicted values, which will also help us diagnose the quality of the model. This is called a residuals plot. With it, we can see nonlinearity and outliers, andcheck the randomness of the error distribution. Outputs are represented in Table 2.

Regularization is a way to reduce the complexity of a model in order to prevent retraining or correct an incorrectly posed task. This is usually achieved by adding some a priori information to the condition of the problem (Fig. 3).

Table 2 MSE and R^2 values

MSE train	37.934
MSE test	39.817
R^2 train	0.552
R^2 test	0.522

Fig. 3 Unexpected outcomes

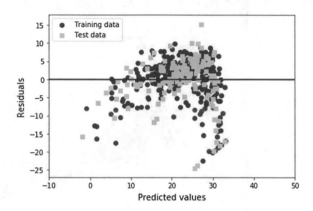

In this case, the essence of regularization is that we create a model with all the predictors, and then artificially reduce the size of the coefficients, adding a certain amount to the error.

Ridge regression or ridge regression (ridge regression)—this is one of the methods of reducing the dimension. Often, it is used to combat data redundancy when independent variables correlate with each other (i.e., multicollinearity occurs). The consequence of this is the poor conditioning of the matrix $XT\ X$ and instability of estimates of regression coefficients. Estimates, for example, may have the wrong sign or meanings that far exceed those that are acceptable for physical or practical reasons.

When we do a linear regression, the loss function looked like this:

$$\mathcal{L}(X, \overrightarrow{y}, \overrightarrow{w}) = \frac{1}{2n} \sum_{i=1}^{n} (\overrightarrow{x}_i^T \overrightarrow{w}_i - \overrightarrow{y}_i)^2 \tag{14}$$

Now in ridge regression, we add to it a parameter that will handle the affect of the independent variable on the dependent variable λ which denotes the size of the fine. If $\lambda = 0$, then this is an ordinary linear regression. Now, the formula looks like this:

$$\mathcal{L}(X, \overrightarrow{y}, \overrightarrow{w}) = \frac{1}{2n} \sum_{i=1}^{n} (\overrightarrow{x}_i^T \overrightarrow{w}_i - \overrightarrow{y}_i)^2 + \lambda \sum_{j=1}^{m} |w_j| \tag{15}$$

Table 3 MSE and R^2 values

MSE train	0.449
MSE test	0.472
R^2 train	0.552
R^2 test	0.522

Now, the output after applying the ridge regression becomes as represented in Table 3 with a ridge coefficient [0.75149352]

5 Conclusion

By using the ridge regression technique, we can easily tackle the single-equation dependency for the model and can improve its performance for multiple type of dataset ranging from low to high values and can obtain increase accuracy.

Acknowledgements The authors would like to thank...

References

1. N. Gogtay, S. Deshpande, U. Thatte, J. Assoc. Phys. India. Principles of regression analysis **65**, 48–52 (2017)
2. H.-F. Wang, R.-C. Tsaur, Fuzzy Sets Syst. Insight of a fuzzy regression model **112**(3), 355–369 (2000)
3. W. Donsbach, W. Donsbach, *The international encyclopedia of communication* (Wiley, 2008)
4. J. Gui, T. Liu, Z. Sun, D. Tao, T. Tan, IEEE Trans. Pattern Anal. Mach. Intell. Fast supervised discrete hashing **40**(2), 490–496 (2017)
5. J. Amoah, E.A. Stuart, S.E. Cosgrove, A.D. Harris, J.H. Han, E. Lautenbach, P.D. Tamma, Comparing propensity score methods versus traditional regression analysis for the evaluation of observational data: a case study evaluating the treatment of gramnegative bloodstream infections. Clin. Infect. Dis. (2020)
6. L. Yang, S. Liu, S. Tsoka, L.G. Papageorgiou, Expert Syst. Appl. Mathematical programming for piecewise linear regression analysis **44**, 156–167 (2016)

Autocorrelation of an Econometric Model

Preeti Singh and Sarvpal Singh

Abstract The treatment of an econometric model requires a clearly defined sequence of tasks. The identification of the model leads us to review the literature, to justify the defined relationship between the dependent variable and the independent variables. Model estimation uses the mathematical apparatus to find the equation of fit. Once the model has been estimated, it must be properly diagnosed using statistical tests. After the diagnosis phase, one can use the model to make predictions. This contribution deals with the identification, estimation, diagnosis, and prediction phases for the treatment of econometric models. Likewise, the diagnosis is deepened by developing the problems of autocorrelation, heteroscedasticity, residual normality, multicollinearity, endogeneity, and others.

Keywords Autocorrelation · Econometric model · Data analysis

1 Introduction

This topic addresses the problem of autocorrelation, a characteristic that the term can present error of an econometric model and that manifests itself more frequently when dealing with data observed over time. This type of data, called time series data, consists of observations of variables collected over time (days, months, quarters, years, etc.) of a certain economic unit (company, consumer, region, country, etc.). These data present a natural ordering [1], since the observations are ordered according to the moment of time in which they have been observed.

It is very important to keep in mind that when explaining the behavior of an observed variable over time, one should keep in mind that observations may be correlated throughout it. For example, a family's consumption [2] in one year is likely to be correlated with consumption from the previous year since, on average,

P. Singh (✉) · S. Singh
MMM University of Technology, Gorakhpur, India
e-mail: singh.preeti294@gmail.com

S. Singh
e-mail: spsingh@mmmut.ac.in

© The Author(s), under exclusive license to Springer Nature Singapore Pte Ltd. 2023   289
Y.-D. Zhang et al. (eds.), *Smart Trends in Computing and Communications*, Lecture Notes in Networks and Systems 396, https://doi.org/10.1007/978-981-16-9967-2_28

ceteris paribus, one would expect a family to not to change much their behavior as far as their consumption is concerned from one year to the next. This would possibly involve introducing into the model, to explain the consumption of a year, for example, the consumption of the previous year [3]. Moreover, one can consider that a change in the level of an explanatory variable, for example the income of this family in one year, has effects on consumption in subsequent years. One way to keep account of this would be to introduce, as an explanatory variable of consumption in a year, not only the income of that year, but the income from previous years. Finally, in addition to observable factors [4] whose influence on consumption is distributed throughout the time, there can be unobservable factors, collected in the disturbance term of the model, influence of which is maintained for several periods, presenting what is called autocorrelation or serial correlation over time [5].

As it can be seen, there are different ways of introducing effects into a model that are distributed throughout it. One can introduce delays of both the endogenous variable, for example $Y t - 1$, and of certain explanatory variables, $X t - j, j = 1 \ldots J$, or to model the error term $u t$ as a function, for example, from its own past $u t - j, j = 1 \ldots p$. That is, what is known as a specification model dynamics. In many cases, it is necessary to reflect on this specification and know to which part to assign, systematic or disturbance, it corresponds to. A dynamic bad specification or simply functional of the systematic part of a model can lead to detect autocorrelation in the disturbance term.

In this topic, this contribution will consider how to detect the presence of serial correlation in the error term, analyze what can be the source of this problem, and if it is not a bad specification problem, how to collect this dynamic characteristic in the error term in the model. We will also study the implications of the existence of autocorrelation in alternative estimation methods and in hypothesis tests on the parameters of interest of the model. In a later topic, we will deal with the dynamics in the systematic part.

1.1 Autocorrelation Tests and Residue Analysis

The disturbances are not observable, so the residuals will be our approximation to these variables, both in the graphic analysis and in the statistical contrasts. As in the subject of heteroscedasticity, one can use as a preliminary [6] to a statistical autocorrelation test, the graph of the time series of the residuals, obtained from estimating by OLS the specification of the model of interest. Here, two contrasts designed to test the null hypothesis of the absence of autocorrelation against various alternatives of autocorrelation in error will be considered.

1.2 Durbin and Watson contrast

This contrast is designed to test the null hypothesis of no autocorrelation

$$H0 : \rho = 0, ut = \varepsilon t \quad ^{iid}(0, \sigma 2\varepsilon) \tag{1}$$

versus the alternative of an AR process (1), where the alternative hypothesis can be

$$H a : u t = \rho u t - 1 + \varepsilon t \quad 0 < \rho < 1 \tag{2}$$

or

$$H a : u t = \rho u t - 1 + \varepsilon t \quad -1 < \rho < 0 \tag{3}$$

The procedure is as follows. The interest model for OLS is estimated, and the residuals û t are calculated for $t - 1 \dots T$. The statistic of Durbin–Watson is calculated as:

$$\text{DW} = \sum T t = 2(\hat{u}t - \hat{u}t - 1)2 \sum T \tag{4}$$

Thus, the value of the DW statistic is between (0.4). If the DW value is close to zero it indicates that the value of $\hat{\rho}$ is close to 1. If $\hat{\rho} \in (0.1)$, then DW $\simeq \in (2.0)$, so DW values of 2 to 0 will indicate evidence for positive order one autocorrelation. If $\hat{\rho} \in (-1.0)$, then DW $\simeq \in (4.2)$, so DW values from 2 to 4 would indicate evidence toward autocorrelation of order one negative.

Formally, to perform the contrast, one needs to know the distribution of the statistic under the null hypothesis of no autocorrelation, but the distribution of this DW statistic is not the usual one and it is shown that it depends on the matrix X. Durbin and Watson have tabulated some heights, $d\ i$ (lower) and $d\ s$ (upper) of critical values, which allow deciding whether to reject the hypothesis null at a chosen level of significance.

Procedure
Given the sample size T and the number of regressors except the term independent $K' = K - 1$, one should obtain in the table of tabulated critical values the lower bounds $d\ I$ and greater than $d\ s$ for the selected level of significance.

- If DW $\in (d\ s, 4 - d\ s)$, then H 0: $\rho = 0$ is not rejected against either of the two alternatives either Ha: $0 < \rho < 1$ or Ha: $-1 < \rho < 0$. Therefore, there is no evidence of autocorrelation of order one.
- If DW $< d\ I$, H0 is rejected: $\rho = 0$ and there is positive one order autocorrelation evidence Ha: $0 < \rho < 1$.
- If DW $> 4 - d\ i$, H0 is rejected: $\rho = 0$ versus the autocorrelation alternative of order one negative Ha: $-1 < \rho < 0$.

- If DW $\in (di, ds)$ or if DW $\in (4 - ds, 4 - di)$, the statistic falls into the indeterminacy zones contrast, so the contrast is inconclusive.

Comments

- This contrast can detect other higher-order AR or MA processes, as long as they give rise to residuals with a significant order one autocorrelation. For the same reason, it can detect problems of bad specification of the systematic part of the model. But contrast does not indicate what may be the case.

1.3   Econometric Methods and Data Analysis

- If the autocorrelation of order one is not significant, this contrast may not detect higher-order AR or MA processes.
- In the indeterminacy zones, the contrast is not conclusive. However, one should have noted that the null hypothesis of no autocorrelation is also not accepted, being cautious with the OLS results, especially in the case of inference using the usual t and F statistics.

This region of indeterminacy is greater the lower the number of degrees of freedom is (lower T and higher K).

- The DW statistic is not reliable in models that include as explanatory variables the lagged dependent variable $Yt - 1$, $Yt - 2$, ... or another type of stochastic regressor.

2   Literature Review

2.1   Background

It can be said that there is autocorrelation when the error term of an econometric model is correlated with itself over time, such that $Cov\,(e\,i,\,e\,j) \neq 0$. This does not mean that the correlation between errors occurs in all periods, but can occur only among some of them. The most common reason is that one or more of the parameters βj are not constant throughout the sample. In the field of economics, it can come from structural changes, special events, etc.

As in the previous case, in the presence of autocorrelation, the OLS estimators continue to be unbiased but do not have a minimum variance, and the generalized minimum square (CGM) estimation method must be used instead. Logically, if the lack of constancy in the value of the parameters is serious, the calculated estimate will be useless.

To address this problem, it is possible to choose to:

- Improve model specification (add, remove, or transform variables) or add terms of type MA to neutralize autocorrelation.

Use consistent estimators for V to $r(\beta)$ Var (β). The existence of autocorrelation in the residues is easily identifiable obtaining the functions of autocorrelation (ACF) and partial autocorrelation (ACP) of the least-quadratic errors obtained in the estimation. If these functions correspond to a white noise, the absence of correlation between the residues will be verified. However, the mere visual examination of the previous functions can be confusing and not very objective, so that in the econometric practice, various contrasts are used for autocorrelation, the most often used being that of Durbin–Watson.

2.2 Related Work

Autocorrelation is the presence of linear relationship between disturbances for different observations:

$$\text{cov } X(u\ s) = EX(u_t u_s) = 0 \quad \forall t, \quad st = s$$

Autocorrelation is expected when working with series data. The disturbance covariance matrix is:

$$V_X(u) = E_X(uu') = \begin{bmatrix} \sigma^2 & \sigma_{12} & \sigma_{13} & \cdots & \sigma_{1T} \\ \sigma_{21} & \sigma^2 & \sigma_{23} & \cdots & \sigma_{2T} \\ \sigma_{31} & \sigma_{32} & \sigma^2 & \cdots & \sigma_{3T} \\ \vdots & \vdots & \vdots & \ddots & \vdots \\ \sigma_{T1} & \sigma_{T2} & \sigma_{T3} & \cdots & \sigma^2 \end{bmatrix} = \Sigma$$

2.3 Consequences of Autocorrelation

- **Properties of the OLS estimators conditioned to X**
  The OLS estimator is linear, unbiased, but it is NOT efficient, in the sense of variance being minimal, because the Gauss–Markov theorem is not fulfilled. There is another estimator better than the OLS, that is, with less variance [7]: the least squares estimator.
- **Inference with the OLS estimator**
  When assumption $S.5$ is not fulfilled and there is autocorrelation, the true matrix of covariances of the OLS estimator is:

Fig. 1 Positive
autocorrelation

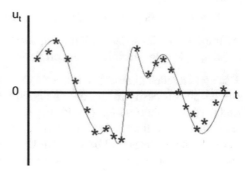

Fig. 2 Negative
autocorrelation

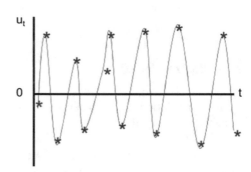

$$V_X(\hat{\beta}) = (X'X)^{-1} X' \Sigma X (X'X)^{-1}$$

Autocorrelation Detection

- Graphic analysis
    The following graph visualizes the temporal evolution of the disturbances when
    there are:
- Positive autocorrelation: long streaks of positive and negative shocks [8].
- Negative autocorrelation: disturbances alternate positive and negative sign (Figs. 1
    and 2).

3 Proposed Work

The hypothesis of non-autocorrelation (one of the basic hypotheses of the RLM)
means that the terms of disturbance are incorrectly interrelated. Moreover, if the
disturbance term follows a normal distribution, the hypothesis of non-autocorrelation
implies independence between the terms of disturbance; that is, the residue of any
observation is not correlated or the residue of another observation. This implies that
the covariance between any pair of disturbance terms is zero.

The above is translated into the fact that the elements outside the main diagonal of the variance and covariance matrices of the disturbance term are all equal to zero. Under the assumption of autocorrelation, there will be at least a couple of disturbance terms [9], for which it will not be true that the covariance between these is zero.

In short, it can be concluded that if the error for one observation is positive, then it will be systematically increased in the next observation in a time-based analysis, and hence, we will have a positive error correlation.

By extracting the error factor within the model for an observation, a variable must be incorporated in the model that will capture the autocorrelation error and will nullify its impact on the observation and hence will decrease the error within the model.

A graph of the autocorrelation of a time series by delay is called a correlogram, or an autocorrelation diagram, or autocorrelation function plot (ACF). When executing the example, a 2D graph is created that shows the delay value along the X-axis (squared waste) and the Y-axis [3] correlation between -1 and 1. These indirect correlations are the ones the partial autocorrelation function seeks to eliminate. Without going into mathematics, this is the intuition for partial autocorrelation.

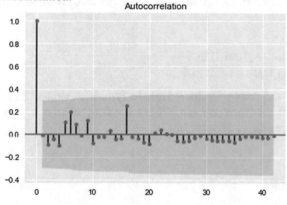

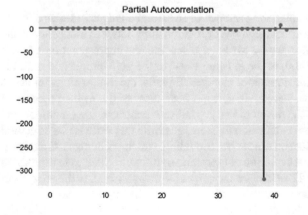

3.1 Ljung–Box Test

The Ljung–Box test refers to a statistical test to determine whether an autocorrelation group taking in a time series is nonzero [1]. Instead of testing randomness at each different delay, this tests generic randomness.

- Null hypothesis: Absent autocorrelation
- Alternative hypothesis: Autocorrelation present.

Given the p-value, we cannot reject the null hypothesis of the absence of autocorrelation.

3.2 Durbin–Watson Statistic

The Python statsmodel incorporates the Durbin–Watson statistic by default. The Durbin–Watson test [10] informs us with a value from 0 to 4, of possible the presence of autocorrelation Ar (1) (first-order autocorrelation, when consecutive residues are correlated), where

- Autocorrelation is null at 2
- Positive autocorrelation ranges from 0 to 2
- Negative autocorrelation ranges from 2 to 4

Values in the range of 1.5–2.5 can be considered relatively normal. Values obtained outside of this range may be a cause for concern. Whereas, values below one or more than 3 need to be addressed on immediate basis.

4 Concluding Remarks

As seen throughout this chapter, the presence of structural changes in series of time-stamped data significantly affects analysis or relationships between independent and dependent variables. In this work, we have discussed the representation of time series with changes structural, of the existing tests for the both simple and multiple changes and to calculate the number of changes. Equally, the implications for series of nonlinear time were considered. Here, it has been discussed how the presence of structural changes affects the development of unit root contrasts, long memory, and prognosis. The predominant conclusion that emerges from this work is that structural changes have a marked effect on time series modeling, which encourages its analysis. A study of the chronological evolution of research on structural changes in time series shows that approximately 83 articles on the subject were published. In the analysis of the literature presented here, a cursory study of some of the important issues in time series analyzes structural changes. However, in this process the wealth and diversity of the subject analyzed and the degree of growth of investigations in the area was acknowledged. While the research on changes structural has had a marked growth, there are still fields of research that require further study, such as the development of formal procedures for specification and model building for time series no linear with structural changes, where both characteristics are considered simultaneously and do not require a strong prior knowledge of the series; developing a set of tests sufficient to allow identification, estimation and representation of series with the characteristics cited; building models oriented to forecast; and the characterization of different types of structural changes.

Acknowledgements Mentorship of Prof. Marcin Paprzycki, Galgotias University, Greater Noida, India, is acknowledged and appreciated.

References

1. N.S. Balke, Detecting level shifts in time series. J. Bus. Econ. Stat. **11**(1), 81–92 (1993)
2. P. Perron, The great crash, the oil price shock, and the unit root hypothesis. Econometrica J. Econometric Soc., 1361–1401 (1989)
3. R.L. Lumsdaine, D.H. Papell, Multiple trend breaks and the unit-root hypothesis. Rev. Econ. Stat. **79**(2), 212–218 (1997)
4. L.C. Nunes, P. Newbold, C.-M. Kuan, Spurious number of breaks. Econ. Lett. **50**(2), 175–178 (1996)
5. G.C. Chow, Tests of equality between sets of coefficients in two linear regressions. Econometrica J. Econometric Soc., 591–605 (1960)
6. R. Quandt, Tests of the hypothesis that a linear regression obeys two separate regimes. J. Am. Stat. Assoc. **55** (1960)
7. D.W. Andrews, Tests for parameter instability and structural change with unknown change point. Econometrica J. Econometric Soc., 821–856 (1993)
8. A. Banerjee, R.L. Lumsdaine, J.H. Stock, Recursive and sequential tests of the unit-root and trend-break hypotheses: theory and international evidence. J. Bus. Econ. Stat. **10**(3), 271–287 (1992)
9. J. Bai, Estimation of a change point in multiple regression models. Rev. Econ. Stat. **79**(4), 551–563 (1997)
10. C.W. Granger, N. Hyung, Occasional structural breaks and long memory with an application to the s&p 500 absolute stock returns. J. Empir. Financ. **11**(3), 399–421 (2004)

Trends of Artificial Intelligence in Revenue Management of Hotels

Smrutirekha⑩, Priti Ranjan Sahoo⑩, and Ashtha Karki⑩

Abstract The authors provide an in-depth summary of artificial intelligence's role in the revenue management of hotels across the globe. Focusing on artificial intelligence, the chapter draws the reader's attention toward its usage and role in the lives of the customers and service providers in the hotel industry, which ultimately leads to the revenue management system. The chapter focuses on the current trends of artificial intelligence to generate revenue for hotels more innovatively. Some of the finest and latent features of artificial intelligence in the revenue management system have been discussed in the chapter by elucidating few examples of few renowned hotels in the world. The revenue management system of the hotel industry has been discussed briefly in the chapter. The essence of the chapter lies in the role and importance of the ever-changing and emerging trends of artificial intelligence in the hotel industry, particularly for revenue generation. Though artificial intelligence cannot replace the emotional intelligence provided by humans, it is the need of the hour to enhance and sustain in the innovative and digital era. AI rules across manufacturing and production sectors, but it also plays a strategic role in the service industry.

Keywords Artificial intelligence · Hotels · Revenue management

1 Introduction

Artificial intelligence (AI) is spreading like wildfire across manufacturing, processing, and services in this digital era. The AI-based technologies are intelligent enough to replace human activities completely. The speed and accuracy of the process also increase while using AI. AI has a positive impact to a greater extent, but at the same time, it has a low value for the human touch, particularly in the service industry. AI technology used in hotels is not a far-fetched imagination anymore.

Smrutirekha (✉) · P. R. Sahoo · A. Karki
KIIT School of Management, KIIT University (Institution of Eminence), Bhubaneswar, India
e-mail: smrutirekha195@gmail.com

P. R. Sahoo
e-mail: prsahoo@ksom.ac.in

Several parts of the world's most well-known businesses are already implementing so [22]. In hotels, artificial intelligence (AI) encompasses everything from robots to intelligent supercomputers. The fundamental reason why a revenue management system is imperative for those within the hotel industry, notwithstanding the measure of their lodging, is since it permits complex calculations to be carried out rapidly and permits for the kind of real-time tracking of advertising information that's outlandish to duplicate physically.

It has instead become imperative to sustain in the world of business. The immediate strategic benefits of incorporating AI in hotels for revenue management come from cost reduction, efficiency in its operations, and increased revenue [22]. The gray areas like people management and social responsibility are also a part of the evolving technology of AI in the hotel industry. The ability for a hotel to make smart estimating decisions in a computerized mold makes the business case for investing in an AI-powered revenue management system persuasive. In terms of boosting productivity, it is a compelling argument. It is also effective in avoiding potential revenue losses that can occur when a hotel falls short of maximizing occupancy or, even worse, experiences a decrease in occupancy. With the research from various literature sources on AI-based technologies of the world, this chapter is based on the discussion of the roles and trends of AI in the hotel industry for revenue management purposes, along with an overview of the revenue management systems and tools.

2 Trends of Artificial Intelligence in the Hotel Industry

A thorough grasp of the technology, its properties, and applications is required to facilitate AI adoption and implementation in the real world. The enormous amount of data that our world offers every moment needs to be catered to enhance the services. Through the acquisition and utilization of data, hospitality firms are rapidly investing in AI-powered platforms, which are becoming vital for providing elevated, authentic, and personalized experiences to customers [2]. The current trends of artificial intelligence in the hotel industry are fascinating enough to draw the attention of both the guests and the manager. Few major and critical trends of AI in the hotel industry are given below.

- **Emotional AI**—It can read both vocal and nonverbal clues to comprehend customer behavior. By researching how people react to different materials, products, and services, one can quickly detect numerous forms of human emotions.
- **Augmented Reality and Virtual Reality**—It is related to experiences that AI can provide to the consumers, which can transform the view of the consumers in terms of personalization and customization [20].
- **AI used in Robots**—AI has made robots more innovative and efficient than ever. Robots can perform most human activities, giving an advantage over human labor [9].

- **AI in the Internet of Things**—Smart and wearable devices helps ease the service provider's process. It is an innovative and efficient way of accessing the required data. Analytical decisions can be made using the data without any human intervention [16, 17].
- **Predictive Analytics**—Provides a better understanding of the customer's needs, wants, and demands. Accordingly, the services can be tailored to meet their expectations [5].
- **Ethical AI**—Ethics and fair play are critical aspects of AI in the hotel industry. Business leaders advocate using ethical AI to maintain transparency and work as per the business requirements [6].
- **AI in Cybersecurity**—Cybersecurity has always been a significant concern, now one of the most enhanced prefaces in the digital world.
- **AI in Computer Vision**—Postures, movements, gestures, facial recognition (of the guests/consumers), etc., are captured and analyzed for understanding human intelligence for improvisation of the services and management.

3 Overview of Revenue Management of Hotels

Revenue management can be defined as increasing the revenue of a business by matching the supply and demand by selling the fitting room to the designated customer at the right time at a perfect price through the best available distribution channel, maintaining the best cost-efficiency. Data analytics is used behind the scenes to assess consumer behavior and forecast demand.

The use of revenue management in the hotel industry has expanded dramatically in recent years. It began in the room sections and has since expanded to include meeting and event facilities and food and beverage shops. At the same time, innovative enhancements empower the improvement of unused merchandise, such as mechanized income administration frameworks. On the one hand, there are opportunities to misuse huge information and move to a science-based income administration framework. Risks, on the other side, include greater susceptibility as a result of hacking or data leaks. In any case, the industry is gradually moving toward computerization and will need to adjust over time.

When a customer submits a booking request, the hotel's RM system records it. The latter comprises four structural elements: data and information, hotel revenue centers, revenue management software and tools, and the revenue management team. The specific booking aspects of the given booking request are the operational results of the RM process—booking status, number of rooms, types, categories of rooms, duration of stay, price, cancelation, and revision terms and conditions. The customer's opinions of the fairness of the hotel's RM system and their intentions for future bookings with the same hotel/hotel chain are influenced by the booking information and the operation of the entire RM system [8]. The RM system is constantly affected by external (macro- and micro-) and internal environmental factors in which the hotel operates (the company's goals, financial situation, legislation, competition, changes

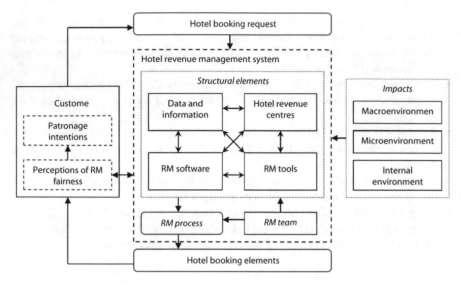

Fig. 1 *Source* Adapted from [12]

in demand, destination's image), and revenue managers' decisions must take all of these factors into account.

As stated by Ivanov and Zhechev [14], the revenue management system of hotels has been given in Fig. 1.

3.1 Key Performance Indicators of Revenue Management System in Hotels

The KPIs help track and maintain an effective revenue management system (RMS). Numerous technology solutions are available for their measurement. The major KPIs include.

- GOPPAR—Gross operating profit per available room—considers total available rooms despite the occupancy status.
- RevPAR—Total revenue per available room—takes into account all revenue from rooms, including room service, meals, and other amenities.
- ARPA—Average revenue per account—average amount of revenue generated per customer account.
- EBITDA—Earnings before interest, taxes, depreciation, and amortization—total revenue minus the expenses (which exclude ITDA).

3.2 Important Revenue Management Tools

- **Property Management System (PMS) Management System (PMS)**—From the front desk (check-in/check-out, room assignment) to finance and revenue management, the PMS oversees all areas of the hotel's operations (managing room rates, billing). Examples of popular PMS—Apaleo, Cloudbeds, and Mindmatrix
- **Channel Manager**—Helps connect and manage online distribution channels for hotels to sell their inventory to OTAs and agents globally. Examples—Siteminder, Lodgable
- **Business Intelligence and Data Analytics**—Business intelligence (BI) platforms are technologies that help collect, integrate, analyze, and present hotel data in the reports. Helps gain specific insights into the current scenario in business. Examples—OTA Insight, Power BI, Looker
- **Upselling tools**—Upselling is one of the most effective strategies to boost revenue and ADR while maintaining the same level of demand. It is one of the most critical pricing strategies used in the revenue management system. Examples—Oaky, Revinate Upsell

4 Major Roles Played by Artificial Intelligence for Revenue Management of Hotels

As studied by Haddad et al. [10], revenue management (RM) is an essential tool for matching supply and demand by segmenting consumers based on their purchase intents and allocating capacity to the various groups in a way that maximizes a company's revenues. In today's digital business industry, AI has a crucial and intricate role to play in carrying out profitable yet sustainable business.

AI-assisted revenue management is all about intelligent estimation. It is all about maximizing room occupancy at the lowest possible cost by combining request figures, competitor rates, and cost sensitivities—all while taking into account a variety of additional factors, including request drivers like regularity, special event dates, and day-of-week differences. In addition to the type of room, the length of stay, and the degree to which a lower cost advancement may weaken income and benefits, in the long run, skillful estimating entails taking into account other factors like the type of room, the length of stay, and the degree to which a lower cost advancement may weaken income and benefits in the long run [24]. The combinatorial difficulties of clever estimating are not to be trifled. The hotel industry is one of its kind where AI is gradually overpowering human activities or legacy systems.

By building up manufactured insights in the hotel industry, inns can make more critical openings to provide excellent guest-friendly administrations extending from coordinating visitor inclinations, recommending books or music, adjacent sports club to complement customers' taste, all the way to consequently alarming inn staff for personalized supper choices, extraordinary benefits, and complimentary administrations, etc. Through interconnected gadgets, sensors, and machine learning, hotel operations framework can associate superbly with the physical world, engaging visitors with a profoundly personalized client involvement. Hotel rooms can use the existing arrangement of technological developments and virtual collaborators to advance and improve the encounter to the following level.

A few of the most critical roles AI play for revenue management in hotels have been stated below.

- Optimization of booking technologies used to book the rooms online. It enhances the preface between the customer and the hotels during the entire process of booking a room or any service online [1]. The software system of the hotel is optimized using artificial intelligence. This, in turn, improves the cost and finance management of the hotels' products and services. Group booking software is a significant development in the hotel reservation system, which has made the task easier. The use of AI technology is also upgrading cancelation policies.
- Social media presence is one of the most crucial effects of using AI-backed technologies in today's world, where a customer is highly dependent on social media to create an impression and image of the service before buying it [4]. The decision-making part heavily depends on the hotels' level of presence and activity on the social media platforms. This helps the customer assess the hotels' quality of services and facilities.
- AI-backed concierges where the guests are offered services of automated check-in and check-out using bots. Room service facilities, mobile keys, and intelligent rooms are essential facilities AI provides through mobile phones [21]. Through AI technologies, the service providers can create consistent guest experiences and provide them with the best of services, eventually leading to an increase in revenue.
- Automated services within the hotel industry have proved to be of advantage in terms of human labor. The chatbots, widgets, and robots used within the premises of the hotels for maintaining a soft touch servicescape, automated food and beverage services with the option for customization, augmented reality, virtual reality tours, etc., have brought about a revolution in the hotel industry [7].
- The after-sales service is gaining momentum in the hotel industry, where the feedback and review of the guests are of utmost importance for revenue management. This feature incorporates the predictive analysis feature of AI-based technologies. The scope for improvement is highlighted by the feedback, rating, and reviewing system of a hotel's software. The artificial neural network comes into play to assess the demands, needs, and expectations of the consumer, which directs toward pricing and revenue management [18].

A list of the segregated AI functions as studied by Ivanov and Webster [13].

- Website automation—XML links to suppliers, white labels, chatbots
- Virtual reality
- Booking engine

- Metasearch aggregator
- Own aggregator platforms
- Online travel aggregators
- Dynamic packaging software
- Digital concierge
- Booking–as-a-service software
- Mobile applications for feedback
- Self-service kiosks
- Augmented reality
- Virtual guide
- QR codes
- Virtual personal assistant software
- Personalized intelligent solutions
- Document issuance software
- Payment transactions applications
- Management of information software
- Integration of all suppliers CRS data through XML links and White-label
- Inventory management software
- Intelligent retargeting and bid management
- User profiling
- Tracking software
- Data management platforms
- Navigation.

5 Discussion and Conclusion

In most cases, a great RMS will utilize the information and claim computations to conduct a real-time assessment of the status of the system and request accordingly to calculate perfect room costs. As a result, the more significant part of fundamental lodging income administration decisions can be made from a single, centralized dashboard.

The futuristic view of the RMS might direct in the discussed direction. RMS, with time, will become much more powerful and strategic. Social media and data analytics will significantly impact and would technologically drive AI [3]. RMS will be more centralized, which would help in gross resource management. Strong communication and organizational behavior will be seen in the future of AI-based RMS. The functional spaces would incorporate revenue management; its measurement would be innovative in nature, and strategies would be more refined and streamlined for better financial decisions [11]. The hotel industry and the customers will grow more aware of RM restoration procedures, making the customer experience an additional strategic factor for the sector [9].

In the hospitality industry, revenue management is a crucial job. Even though AI is still considered an emerging technology trend, several hotels have already reaped the benefits and significantly influenced their bottom line [19].

While present systems rely primarily on previously acquired data (big data) to extract any implications and recommendations to tourists, predictive analytics solutions, aimed at anticipating customer behavior and projecting industry success, will make significant development in the future [23]. Intelligent systems will estimate, forecast, plan, and control activities across channels and suppliers, maximizing resource efficiency, revenue growth, and customer satisfaction [15]. AIs will be the primary instrument for enhancing operations, redesigning, and reengineering business processes with the ultimate goal of putting the customer first while still maximizing revenues and costs.

Robots as self-contained gadgets would not replace intermediaries' communication and coordination but improve hotel industry information management [26].

Traditional legacy system has always prevailed in the revenue management cycle in the hotel industry. But with the evolving landscape of AI technology, the revenue management system has experienced a hurricane. A comparative analysis between the legacy system and the AI-based RMS clarifies the improvisations achieved in Fig. 2.

Mechanical propels, including frameworks integration, will progress the capability to recognize the esteem of each client, expanding the viability and making a difference to estimate requests precisely, boosting development buy reservations. This permits companies to centralize their income divisions, as one director presently can direct more than one property.

Fig. 2 Comparison between legacy system and AI-based RMS

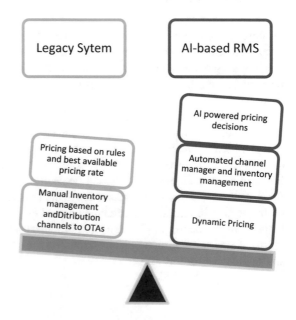

The last decade's technological improvements have had a massive impact on the hospitality business, tourist behavior, and hotel operations. Most lodgings and accommodations have adjusted this current estimating frame after income administration was introduced to the business. While having a revenue management system for hotel rooms is the norm, the industry is increasingly moving to support its implementation in conference and convention venues and food and beverage outlets.

Artificial intelligence-based technologies are rapidly evolving, presenting modern opportunities such as using vast amounts of data. Still, they also provide concerns, such as the worry of current workers losing their jobs. Social networking is a fantastic way for businesses to connect with customers and get information about them. This can be used for short- and long-term planning and a new way to provide rooms. While entire computerization is not yet possible, a beautifully designed revenue management computer software now supports the revenue chief in analyzing and estimating and strategy-oriented intelligent decision-making. The commerce insights gathered from the detailing capabilities, for illustration, can offer assistance progress deals adequacy, create competitive insights, and give good bits of knowledge into inhabitance patterns, visitor socioeconomics, showcase situating, and channel benefit. A promoting office can utilize the estimates as a director for deciding when to extend special spending to a reasonable request. An operations team can know when to develop (or diminish) staffing based on anticipated inhabitance. In brief, the benefits tend to go well past the office known as revenue management, eventually rising above all parts of the organization.

Hotel visitors are ultimately helped from take-off to goal with personalized suggestions that upgrade their travel and after-arrival choices at the lodging (eating, drinks, breakfasts, on-property exercises, etc.). After gathering loads of client information, hotels can utilize machine learning with business-specific calculations to create expectations of what clients are most likely to select. Such expository capacity of AI can keep the inns spry in recognizing the following set of patterns regarding items or benefits offerings. It can be utilized to assess and acknowledge the traveler or client persona to coordinate them with essential administrations [25]. Using the brilliance of proactive analytics and identity branding, hotels can recognize long-term needs. It would be ideal if the clients can be satisfied by the services they have been searching for some time recently and which they indeed inquired for.

Hotel marketing professionals who anticipate guest emotions and motivations ahead of time are more likely to acquire traction in their business growth than those that use a more traditional strategy. To determine the impact of various services on clients, hotel operators must work out a variety of emotions and develop a comprehensive approach.

Hotel marketing leaders will adapt to creative tactics to strengthen their marketing messages as AI improves the means of collecting client information. Even for newcomers to the hospitality industry, the trend of integrating AI-powered predictive analytics in hotels for marketing purposes can be beneficial.

Artificial intelligence has entered the traditional hospitality landscape with the promise of increasing revenue, improving hotel reputation, and elevating the customer experience—all of which legacy systems were incapable of doing efficiently. But still, AI technology's implementation in revenue management system (RMS) with more sophisticated pricing discrimination and real-time demand adaptability is in its early phases.

References

1. N. Antonio, A. de Almeida, L. Nunes, Big data in hotel revenue management: exploring cancellation drivers to gain insights into booking cancellation behavior. Cornell Hospitality Q. **60**(4), 298–319 (2019). https://doi.org/10.1177/1938965519851466
2. M. Bounatirou, A. Lim, A case study on the impact of artificial intelligence on a hospitality company. Adv. Ser. Manag. **24**, 179–187 (2020). https://doi.org/10.1108/S1877-636120200 000024013
3. E. Brynjolfsson, A. McAfee, Will humans go the way of horses. Foreign Aff. **94**, 8 (2015)
4. E. Çeltek, I. Ilhan, *Big Data Artificial Intelligence, and Their Implications in the Tourism Industry* (2020), pp. 115–130. https://doi.org/10.4018/978-1-7998-1989-9.ch006
5. D.J. Connolly, R.G. Moore, Decision making in the context of hotel global distribution systems: a multiple case study. *Hospitality* (1999)
6. T.H. Davenport, *The AI Advantage: How to Put the Artificial Intelligence Revolution to Work.* MIT Press (2018)
7. S.J. DeCanio, Robots and humans–complements or substitutes? J. Macroecon. **49**, 280–291 (2016)
8. N.F. El Gayar, M. Saleh, A. Atiya, H. El-Shishiny, A.A.Y. Fayez Zakhary, H.A.A. Mohammed Habib, An integrated framework for advanced hotel revenue management. Int. J. Contemp. Hosp. Manag. **23**(1), 84–98 (2011). https://doi.org/10.1108/09596111111101689
9. L. González-Serrano, P. Talón-Ballestero, Revenue management and E-tourism: the past present and future. Handb. E-Tourism, 1–28 (2020). https://doi.org/10.1007/978-3-030-05324-6_76-1
10. R. Haddad, A. El, Roper, P. Jones, The impact of revenue management decisions on customers' attitudes and behaviours: a case study of a leading UK budget hotel chain. Angew. Chem. Int. Ed. **6**(11), 951–952, 5–24 (2008)
11. S.H. Ivanov, Tourism beyond humans–robots, pets and Teddy bears, in *Paper to Be Presented at the International Scientific Conference Tourism and Innovations* (2018)
12. S. Ivanov, V. Zhechev, Hotel marketing. Varna Zangador (2011)
13. S. Ivanov, C. Webster, Economic fundamentals of the use of robots, artificial intelligence, and service automation in travel, tourism, and hospitality, in *Robots, Artificial Intelligence, and Service Automation in Travel, Tourism and Hospitality* (Emerald Publishing Limited 2019)
14. S. Ivanov, V. Zhechev, Hotel revenue management–a critical literature review. Tourism Int. Interdisc. J.**60**(2), 175–197 (2012)
15. M. Ivanova, Robots, Artificial intelligence, and service automation in travel agencies and tourist information centers, in *Robots, Artificial Intelligence, and Service Automation in Travel, Tourism and Hospitality* (Emerald Publishing Limited 2019)
16. R.S. Jha, P.R. Sahoo, Internet of Things (Iot)—Enabler for connecting world, in *ICT for Competitive Strategies* (2020), pp. 1–7. https://doi.org/10.1201/9781003052098-1
17. R.S. Jha, P.R. Sahoo, Influence of big data capabilities in knowledge management—MSMEs, in *ICT Systems and Sustainability* (Springer 2021), pp. 513–524
18. J. Kim, S. Wei, H. Ruys, Segmenting the market of West Australian senior tourists using an artificial neural network. Tour. Manage. **24**(1), 25–34 (2003)
19. S.E. Kimes, The future of hotel revenue management. J. Revenue Pricing Manag. **10**(1), 62–72 (2011). https://doi.org/10.1057/rpm.2010.47

20. S.E. Kimes, J. Ho, Revenue management in luxury hotels. J. Revenue Pricing Manag. **17**(4), 291–295 (2018). https://doi.org/10.1057/s41272-017-0113-1

21. V. Liljander, F. Gillberg, J. Gummerus, A. Van Riel, Technology readiness and the evaluation and adoption of self-service technologies. J. Retail. Consum. Serv. **13**(3), 177–191 (2006)

22. K. Nam, C.S. Dutt, P. Chathoth, A. Daghfous, M.S. Khan, The adoption of artificial intelligence and robotics in the hotel industry: prospects and challenges. Electron. Mark. (2020). https://doi.org/10.1007/s12525-020-00442-3

23. A. Parasuraman, C.L. Colby, An updated and streamlined technology readiness index: TRI 2.0. J. Serv. Res.**18**(1), 59–74 (2015)

24. P.R. Sahoo, S.K. Lenka, B.B. Pradhan, Transactional and psychological concerns of all parties involved in opaque hotel room distribution. Int. J. Econ. Res. **14**(8), 1–9 (2017)

25. P.R. Sahoo, B.B. Pradhan, S.K. Lenka, S.K. Patra, Review research on application of information and communication technology in tourism and hospitality industry. Int. J. Appl. Bus. Econ. Res. **15**(11), 311–334 (2017)

26. A.P. Srivastava, R.L. Dhar, Technology leadership and predicting travel agent performance. Tourism Manag. Perspect. **20**, 77–86 (2016)

Sentiment Analysis and Vector Embedding: A Comparative Study

Shila Jawale and S. D. Sawarkar

Abstract Automatic intent/sentiment classification can be done with various machine learning approaches as well as methods. But the success of these techniques majorly depends on the representation of words or documents in vector space. That can be easily consumed by machine toward learning the hidden pattern of text/corpus. To achieve this, various methods have been proposed, and many are commercially accepted as well. In deep learning architecture for intent/sentiment analysis, the vector embedding plays a crucial role. It represents feature extraction. In this paper, various methods for vector embedding are discussed along with their comparison.

Keywords Sentiment analysis · Glove · FastText · BERT · Vector embedding

1 Introduction

The use of the Internet is growing day by day resulting in large amounts of data. The huge data getting occurs in different online applications too. It may be in the form of various types of reviews, opinions generated with multiple public/social media applications such as Amazon and Twitter. [1] To find good insights from this data to make it as information, the processing is a must. To accomplish this process in finding good insights and make use of it for finding the effectiveness of those opinions or intent for further decision-making.

Sentiment analysis as the name reflects to identify the opinion or intent or attitude or consent toward things. The sentiment is generally called opinion mining or review analysis. It is used to identify, study, quantify, obtain, tacit states, and subject-related information [2]. It plays a very important role in varied kinds of decision-making. Sentiment analysis is very well implemented in various kinds of domain such as financial sector [3], e-commerce domain [4], educational domain [5–7], clinical domain [8], disaster management [9], smart city [10], and many more [11–13].

S. Jawale (✉) · S. D. Sawarkar
Datta Meghe College of Engineering, Airoli, Navi Mumbai, Maharashtra, India
e-mail: shila.jawale@dmce.ac.in

© The Author(s), under exclusive license to Springer Nature Singapore Pte Ltd. 2023   311
Y.-D. Zhang et al. (eds.), *Smart Trends in Computing and Communications*, Lecture Notes in Networks and Systems 396, https://doi.org/10.1007/978-981-16-9967-2_30

There are three forms of sentiment analysis: granularity-oriented, task-related, and methodologically oriented. Polarity categorization, level of valence or arousal at specific/beyond polarity, subjectivity/objectivity identification, and feature/aspect-based sentiment analysis are some of the subcategories of task-oriented sentiment analysis. Document, phrase, and word levels, as well as character levels, are all granularity-oriented. Method-based is divided into three categories: supervised learning, unsupervised learning, and semi-supervised learning. Sentiment categorization based on features found in the text. Words, their occurrence, grammatical representation and categorization, the presence of sentiment words, phrases in sentences, language norms, sentiment shifters, and syntactic dependency seem to be instances of these features [14].

To have effective sentiment analysis, many techniques have been applied. Some are based on lexical and some are based on machine learning approaches. Machine learning has its subset as deep learning. Nowadays, it is becoming a popular way to handle massive text data. Due to the larger dimension of text data, deep learning becomes an effective technique to get decision-making accurate with better accuracy.

Deep learning is a multilayer architecture. Each layer is responsible for completing the dedicated task. The different layers such as embedding layer, convolution layer, sequential layer, and bidirectional layer. After defining the model, the very first layer is the embedding layer. This is an important layer in terms of vector representation of the available text. Different approaches till now are available in the literature, and still very few are applicable in the production environment. The performance of deep learning is heavily depending on kinds of word embedding is applied as feature extraction is done by model itself [15].

The paper gets divided into multiple Sect., viz. 2 Embedding method for Sentiment Analysis, Sect. 3. Related Work, Sect. 4. Proposed Model, Sect. 5. Result and discussion followed by Conclusion.

2 Embedding Method for Sentiment Analysis

Various strategies have been used to achieve the automatic sentiment analysis. The computational methods are shown in Fig. 1. The general sentiment analysis process shown in Fig. 2.

There are various methods to handle vector embedding in other words vector conversion to represent each word or entire document with a multiple or single numerical representation. Sometimes, it is just represented in binary or in decimal format. The various vector computational ways in state-of-literature as bag-of-word, TF-IDF, Glove, FastText, EMLO, pre-trained model BERT, and their variations.

Bag-of-words: The bag-of-words (BOW) [16] model is an oldest and very common method used for text representation. The words are represented with a particular number and their occurrences. At best, the BOW is effective for topic-based text classification rather than sentiment analysis. By disrupting word order as well as discarding contextual information, the BOW loses performance (which is

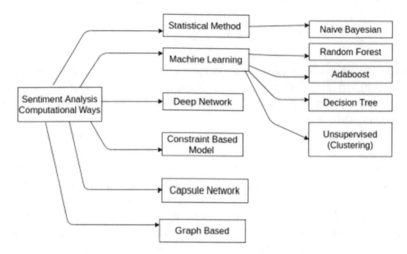

Fig. 1 Computational methods process

Fig. 2 General sentiment analysis process

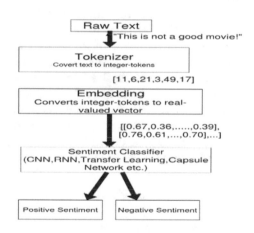

critical for accurate sentiment assessment). The BOW model struggles with polarity shifts like negation.

Term Frequency: TF-IDF is the product of two statistical terms, as shown in Eq. (1). The first is the term frequency (TF), and the second is the inverse document frequency (IDF) [17]

$$W_{t,d} = \text{tf}_{t,d} - \text{idf}_t \tag{1}$$

where $W_{t,d}$ is the weight for feature t (i.e., a word) in document d, and $\text{tf}_{t,d}$, and idf_t are defined in Eqs. (2) and (3), respectively [4]:

$$\text{tf}_{t,d} = \frac{\text{Frequency of term}}{\text{Total number of term in documents}} \tag{2}$$

$$idf = \log \frac{N}{df_t} \qquad (3)$$

where $tf_{t,d}$ is the term frequency, which calculates the number of times term occurs in a document, and $\log(N/df_t)$ is the inverse document frequency, obtained by dividing the total number of documents, N, by the number of documents containing that term, df_t and using the log of that value.

Word2Vec: [18] Due to the lack of keeping the context in TF-IDF model, another model based on the idea of representing the word with pre-trained vocabulary which was computed based on two model such as Skip-gram model and CBOW.

Skip-Gram Model (SG): The idea of SG, in each estimation step you are taking one word as a center word, and you are going to try and predict words in their context out to some window size, and the model is going to define a probability distribution that is the probability of a word appearing in the context given this center word. And we are going to choose vector representation of words so we can try and maximize that probability distribution (we just have one probability distribution of the context word).

Continuous Bag-of-Words Model (CBOW): In this model, the target word (center word) is predicted based on the surrounding words. The context is represented by multiple words, and to predict the missing target word, the sequence of context words must be given, then try to predict the missing word in the context according to the window size. This can be applied to different sizes of windows (10, 15, 20, 100, etc.). According to Nandakumar and Salehi [19], CBOW works several times faster than SG to train and has slightly better accuracy for frequent words, while SG works well with a small amount of training data, and represents well even rare words or phrases. Figure 3 shows the architecture of CBOW and SD algorithms. Figure 3 shows the model [20].

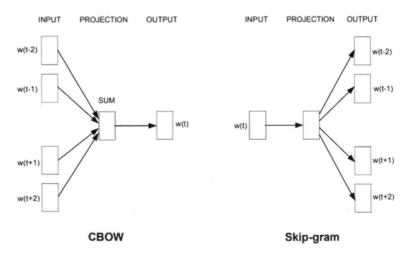

Fig. 3 CBoW and Skip-gram model

Glove: It is a word representation model. Glove, for global vectors, because the global corpus statistics are captured directly by the model [21]. Glove is the most popular count-based method in word embedding models which work upon dimension reduction in the global co-occurrence matrix of words. The global co-occurrence matrix contains statistical amounts of co-occurring events of every word couple in different contexts. This matrix is then decomposed to a much smaller matrix that contains the embedding features for each word. The process is done by matrix factorization and the minimization of a loss function. The continuous bag-of-words (CBoW) model [17] and the Skip-gram model [18] are both included in Word2Vec, a deep neural network that learns word embeddings from text. The CBoW predicts the target word (e.g., "the girl is a television," where " " signifies the target word) given a collection of context words, whereas the Skip-gram model predicts the related terms given the target word [19, 20].

FastText and ELMO: Character-based model [19] is used for creating vector. The author instead of using words made the model convert embedding the character data. Due to this, out of vocabulary words also got treated very well. The model considers sentence structure and arrangement by considering sub-word units and representing words by a sum of their character n-grams. Begin by presenting the general framework to train word vectors, then present the sub-word model and eventually describe dictionary of character n-grams [22].

ELMO: Another character embedding model, the vector for a word is constructed from the character n-grams that compose it. Since character n-grams are shared across words, assuming a closed-world alphabet, these models can generate embeddings for OOV words, as well as words that occur infrequently. EMLO can handle a novel sort of deep contextually relevant word representation that reflects complex aspects of word use (e.g., structure, syntax, and semantics), and how these uses vary across linguistic contexts (i.e., polysemy-words with similar pronunciation). The learnt functions of the internal states of a deep bidirectional language model (Bi-LSTM) that was pre-trained on a large text corpus were represented by the word vectors [23].

Pre-trained word vector Model- BERT: The BERT (bidirectional encoder representations from transformers) model was proposed by author in [21]. This pre-trained BERT model has provided, without any substantial task-specific architecture modifications, provided state-of-the-art performances over various NLP tasks. The BERT pre-trained model [22], which uses Google's vanilla transformer language model [23] as a substructure, has been dubbed the "rediscovery" of the natural language processing (NLP) pipeline because of the higher level of language understanding it provides [24]. "However, the BERT model suffers from fixed input length size limitations, word piece embedding problems, and computational complexities" [25]. Table 1 shows a variety of pre-trained BERT models. The primary BERT models as a function of L (number of encoders) and H (number of hidden units). Because bigger models include a significant number of trainable parameters, smaller models are ideal for environments with limited computational resources: an average-sized model like BERT-Base has roughly 110 million trainable parameters, whereas BERT-Large has over 340 million [26].

Table 1 Various BERT model variations

BERT models	$H = 128$	$H = 256$	$H = 512$	$H = 768$	$H = 1024$
$L = 2$	BERT-tiny	–	–	–	–
$L = 4$	–	BERT-miny	BERT-small	–	–
$L = 8$	–	–	BERT-medium	–	–
$L = 12$	–	–	–	BERT-base	–
$L = 24$	–	–	–	–	BERT-large

3 Related Work

Due to the lack of varied kinds of languages their nature, the context of various words, and different meanings with different contexts, the above-discussed techniques do not give a prominent result in the commercial domain. That is why various other approaches evolved, which might be an improvement in existing methods or adding the domain knowledge and then creating those vector representations of word or combining multiple vectors embedding together etc. Many of the researchers try to combine the different vector models to get the effective word embedding for sentiment analysis and related tasks [27]. For text classification, the author proposes a novel language-independent word encoding method. Using a new method named binary unique number of words "BUNOW," the suggested model translates raw text input to low-level feature dimensions with minimum or no preprocessing steps. Each unique word in a dictionary can have an integer ID that is represented as a k-dimensional vector of its binary equivalent with BUNOW. The output vector of this encoding is fed into a convolutional neural network (CNN) model for classification [28]. The most extensively used approach for choosing features is the TF-IDF algorithm. Even though the TF-IDF method is straightforward to apply, it still suffers from semantic deletion, which means it ignores the semantic information in the text. This study introduces LSA as a solution to the problem. The cosine value was used to check the word similarity. The feature extraction is achieved according to the similarity between the words, which compensates for the TF-IDF insufficiency. Finally, the extracted features are applied into classification algorithms [29]. "Bag-Of-Words model is the most widely used technique for sentiment analysis, it has two major weaknesses: using a manual evaluation for a lexicon in determining the evaluation of words and analyzing sentiments with low accuracy because of neglecting the language grammar effects of the words and ignore semantics of the words. Author propose a new technique with enhancement bag-of-words model for evaluating sentiment polarity and score automatically by using the weight of the words instead of term frequency" [30]. Many researchers' works have focused on refining the concept of embedding. The authors proposed models that learn sentiment-specific word embeddings (SSWE) in [31]. The learned word vectors, as well as the semantic sentiment information, are represented as embeddings. The authors of [32] devised word embedding, which is a combination of sentiment supervision at both document and word levels. The model was trained and learned a sentiment-related embedding. Li and Gong [33] proposes a significant improvement in semantics-oriented word vectors by combining the word embedding model with standard matrix factorization via a projection level. The feature extraction is done by different vector embedding methods such as FastText and Word2Vec given to the RNN module and character-level embedding given to CNN through this hybrid model designed to get the combined feature for sentiment analysis. The author [15] proposes a feature-based fusion adversarial recurrent neural networks (FARNN-Att) composite model with an attention mechanism. Firstly, extracted the long-term dependence of text using the Bi-LSTM network and put forward a novel contextual feature representation way. Then, combined the prediction results of two features vectors in the full connection layer, which can be captured through the feature connection and attention mechanism. To make model more robust and generalized, a regularization method applied can be treated as of adversarial training. In the LE-LSTM model [34], author combined the word sentiment embedding along with its corresponding word embedding. The final layer output of the word sentiment classifier is considered as the sentiment embedding of each word. The word sentiment classifier and word embedding lookup are updated during the main training process. Aiming at the problem of insufficient sentiment word extraction ability in existing text sentiment analysis methods and OOV (out-of-vocabulary) problem of pre-training word vectors, a neural network model combining multi-head self-attention and character-level embedding is proposed by the author in their paper [35]. The author applied the word embedding technique (Word2Vec) along with different types of stemming such as root-based and stem-based, and no stemming for document classification, and found good results as compared to just root model. They also found when using a stem-based algorithm, skip-gram achieves good results with a vector of smaller dimension, while CBOW requires a larger dimension vector to achieve similar performance [36]. The word embedding of word2vec training provides solely semantic information. Before going on to word embedding fusion, the text classification algorithm is employed to learn original text multi-word embedding in morphological, grammatical, and sentiment expressed information. The enhanced convolution neural network [37] is used for sentiment analysis. The author provides an ensemble technique that merges the information from Glove and word2vec embeddings with structured knowledge from the semantic networks ConceptNet and PPDB. For huge text, this reflects a multilingual vocabulary. On several word similarity assessments, the embeddings it generates reach state-of-the-art performance [38]. An embedding method paragraph vector in document ranking given by the author in [39]. They focused on the effects of combining knowledge-based with knowledge-free document embeddings for the text classification task. The author provides a solution to FastText's shortcomings, such as missing words containing contextual information that was not adequately encoded. They initially suggest a shape near word filtering-based approach for retrieving missing nouns. Then, for successful text classification, a self-training word embedding approach based on a knowledge network was combined with pre-training word embeddings to provide the optimum mixed word vector with rich semantics [40]. To overcome the problem in short text news headlines such as sparse data and weak relationship between words, the author proposed a fusion vector. The Word2Vec tool is used to train the word vector in the short text corpus, and the average word vector is incorporated by the additive averaging method. Afterword's gist of the short text corpus is formed by the WTTM model to obtain the topic extended feature vector. Finally, the average word vector and the topic extended feature vector are combined [41]. To address the issues of word embedding such as realize the coexistence of words and phrases in the same vector space, the author proposes an enhanced word embedding (EWE) method. Before completing the word embedding, this method introduces a unique sentence reorganization technology to rewrite all the sentences in the original training corpus. Then, all the original corpus and the reorganized corpus are merged together as the training corpus of the distributed word embedding model, so as to realize the coexistence problem of words and phrases in the same vector space. The glove model is used for this fusion embedding [42, 43].

4 Proposed Model

By looking at the state-of-the-art literature above, we try to address the prominent issue of word embedding along with checking the effectiveness of individual word embedding and a fusion model. This will address the various problems of effective sentiment analysis such as variable vector length, context of sentences, representation of words, and OOV words. The proposed model tries to address these issues by combining the word-level model glove for handling general word embedding, a character-based model such as FastText to resolve the problem of out-of-word vocabulary and to maintain the semantic relationship between the word and phrases the BERT model. The model generates the fused vector further passed different DNN classifiers such as Bi-LSTM, GRU, and CNN model.

Combing different vector representation will create a better embedding as compared to the individual embedding vector. Keeping the vector length constant, we used here padding for BERT representation. It generates variable length vector representation. The modified combined vector then feed to deep neural classifier. The classifier will run on test data to learn the weight of words which determines the polarity of the sentence. The two polarities here we considered as positive and negative of the sentence.

5 Result and Discussion

Here, we considered three-word embedding state-of-the-art methods such as Glove, FastText, and BERT. Along with deep learning architecture such as Bi-LSTM, GRU, CNN, all these DNN architecture are implemented as a shallow network for very few-layer. For the experimental purpose, the dataset used are Amazon Kindle, Yelp, and IMDB all taken from the Kaggle dataset library. The models are trained with

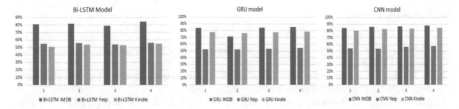

Fig. 4 Performance graph of various embedding such as Glove [21], FastText [28], BERT [22], and custom embedding

supervised approach, and positive and negative polarities have been considered (Fig. 4).

From the various experiments conducted during this study, it has been observed that the performance of deep learning model very much depends on kind of vector embedding applied. Words represented with meaningful numbers (vector). To encode this tacit knowledge for a machine to learn effectively just by applying as it is, available encoding method gives good performance with respect to accuracy; however, due to limitation in this general vocabulary, presence for Amazon kindle dataset does not provide good accuracy, whereas for yelp and IMDB, results are very promising. In future, will going to try other deep learning models to check the performance of vector embedding without presence of domain knowledge.

6 Conclusion

Sentiment analysis is used in broad spectrum of industry. Data is available in abundance for sentiment analysis in various forms or reviews, comments, opinions, and many more ways. To have an effective sentiment analysis encoding, this data is important, and to do these, various techniques are available in literature along with commercial environment. But still there is significant improvement based on the domain and varied nature of available languages. The various vector embedding such as Glove, FastText, and BERT provides a good way to represent word but insignificant in maintaining the context, OOV handling, variable length to represent words. To reap the best from each model, the proposed model creates the combined vector representation. The experimental result shows the significant performance with shallow neural network. In future, we would like to check the interpretability of model by adding attention mechanism look for what model learned.

References

1. Cheng, *Deep Learning for Automated Sentiment Analysis of Social Media* (n.d.). https://doi.org/10.1145/3341161.3344821
2. S. Jawale, *Interpretable Sentiment Analysis Based on Deep Learning: An Overview* (n.d.). https://doi.org/10.1109/PuneCon50868.2020.9362361
3. Esichaikul, *Sentiment Analysis of Thai Financial News* (n.d.). https://doi.org/10.1145/3301761.3301773
4. S. Nimje et al., Prediction on stocks using data mining, in *Proceedings of the 3rd International Conference on Advances in Science & Technology (ICAST)* (2020)
5. Cirqueira, *Improving Relationship Management in Universities with Sentiment Analysis and Topic Modeling of Social Media Channels: Learnings from UFPA* (n.d.). https://doi.org/10.1145/3106426.3117761
6. Whitelaw, *Using Appraisal Groups for Sentiment Analysis* (n.d.). https://doi.org/10.1145/1099554.1099714
7. Wong, *Sentiment Analysis across the Courses of a MOOC Specialization* (n.d.). https://doi.org/10.1145/3287324.3293864
8. Smith. *Cross-Discourse Development of Supervised Sentiment Analysis in the Clinical Domain* (n.d.)
9. Buscaldi, *Sentiment Analysis on Microblogs for Natural Disasters Management: A Study on the 2014 Genoa Floodings* (n.d.). 0.1145/2740908.2741727
10. P. Yenkar, *A Survey on Social Media Analytics for Smart City* (n.d.). https://doi.org/10.1109/I-SMAC.2018.8653707
11. Paul, *Compass: Spatio Temporal Sentiment Analysis of US Election What Twitter Says!* (n.d.) https://doi.org/10.1145/3097983.3098053
12. Yun, *Multi-Categorical Social Media Sentiment Analysis of Corporate Events* (n.d.). https://doi.org/10.1145/3154943.3154957
13. Xu, *Fast Learning for Sentiment Analysis on Bullying* (n.d.). https://doi.org/10.1145/2346676.2346686
14. W. Medhat, A. Hassan, H. Korashy, Sentiment analysis algorithms and applications: a survey. Ain Shams Eng. J. **5**(4), 1093–1113 (2014)
15. M.U. Salur, I. Aydin,A novel hybrid deep learning model for sentiment classification. IEEE Access **8**, 58080–58093 (2020)
16. Y. Zhang, R. Jin, Z.-H. Zhou, Understanding bag-of-words model: a statistical framework. Int. J. Mach. Learn. Cybernet. **1**(1–4), 43–52 (2010)
17. P. Bojanowski et al., Enriching word vectors with subword information. Trans. Assoc. Computat. Linguist. **5**, 135–146 (2017)
18. S. Al-Saqqa, A. Awajan,The use of word2vec model in sentiment analysis: a survey, in *Proceedings of the 2019 International Conference on Artificial Intelligence, Robotics and Control* (2019)
19. N. Nandakumar, B. Salehi, T. Baldwin,A comparative study of embedding models in predicting the compositionality of multiword expressions, in *Proceedings of the Australasian Language Technology Association Workshop* (2018)
20. E. Peters Matthew et al.,Deep contextualized word representations (2018). arXiv preprint arXiv:1802.05365
21. J. Pennington, R. Socher, C.D. Manning,Glove: global vectors for word representation, in *Proceedings of the 2014 Conference on Empirical Methods in Natural Language Processing (EMNLP)* (2014)
22. A. Chiorrini et al.,Emotion and sentiment analysis of tweets using BERT. EDBT/ICDT Workshops (2021)
23. T. Mikolov et al.,Efficient estimation of word representations in vector space (2013). arXiv preprint arXiv:1301.3781
24. R. Al-Rfou et al., Character-level language modeling with deeper self-attention, in *Proceedings of the AAAI Conference on Artificial Intelligence*, vol. 33(01) (2019)

25. J. Devlin et al.,Bert: pre-training of deep bidirectional transformers for language understanding (2018). arXiv preprint arXiv:1810.04805
26. I. Tenney, D. Das, E. Pavlick,BERT rediscovers the classical NLP pipeline (2019). arXiv preprint arXiv:1905.05950
27. J.M.C. Lima, J.E.B. Maia, A topical word embeddings for text classification, in *Anais do XV Encontro Nacional de Inteligência Artificial e Computacional* (SBC, 2018)
28. R. MohammadiBaghmolaei, A. Ahmadi, Word embedding for emotional analysis: an overview, in *2020 28th Iranian Conference on Electrical Engineering (ICEE)* (IEEE, 2020)
29. B.O. Oscar et al.,Sentiment analysis with word embedding, in *2018 IEEE 7th International Conference on Adaptive Science & Technology (ICAST)* (IEEE, 2018)
30. A.A. Helmy, Y.M.K. Omar, R. Hodhod,An innovative word encoding method for text classification using convolutional neural network, in *2018 14th International Computer Engineering Conference (ICENCO)* (IEEE, 2018)
31. Y. Li, B. Shen,Research on sentiment analysis of microblogging based on LSA and TF-IDF, in *2017 3rd IEEE International Conference on Computer and Communications (ICCC)* (IEEE, 2017)
32. D.M. El-Din, Enhancement bag-of-words model for solving the challenges of sentiment analysis. Int. J. Adv. Comput. Sci. Appl. **7**(1) (2016)
33. S. Li, B. Gong,Word embedding and text classification based on deep learning methods, in *MATEC Web of Conferences*, vol. 336 (EDP Sciences, 2021)
34. Y. Ma, H. Fan, C. Zhao, Feature-based fusion adversarial recurrent neural networks for text sentiment classification. IEEE Access **7**, 132542 132551 (2019)
35. X. Fu et al., Lexicon-enhanced LSTM with attention for general sentiment analysis. IEEE Access **6**, 71884–71891 (2018)
36. H. Xia, C. Ding, Y. Liu, Sentiment Analysis model based on self-attention and character-level embedding. IEEE Access **8**, 184614–184620 (2020)
37. H.A. Almuzaini, A.M. Azmi, Impact of stemming and word embedding on deep learning-based Arabic text categorization. IEEE Access **8**, 127913–127928 (2020)
38. X. Sun et al.,Text sentiment polarity classification method based on word embedding, in *Proceedings of the 2018 2nd International Conference on Algorithms, Computing and Systems* (2018)
39. R. Speer, J. Chin,An ensemble method to produce high-quality word embeddings (2016). arXiv preprint arXiv:1604.01692
40. H. Machhour, I. Kassou, Concatenate text embeddings for text classification, in *2017 International Conference on Internet of Things, Embedded Systems and Communications (IINTEC)* (IEEE, 2017)
41. H. Wang, H. Kun Guo, Z. Liu, Mixed word embedding method based on knowledge graph augment for text classification, in *2019 IEEE International Conference on Parallel & Distributed Processing with Applications, Big Data & Cloud Computing, Sustainable Computing & Communications, Social Computing & Networking (ISPA/BDCloud/SocialCom/SustainCom)* (IEEE, 2019)
42. J. Ge, H. Wang, Y. Fang,Short text classification method combining word vector and WTTM, in *2020 IEEE 4th Information Technology, Networking, Electronic and Automation Control Conference (ITNEC)*, Vol. 1 (IEEE, 2020)
43. S. Hu et al.,Enhanced word embedding method in text classification, in *2020 6th International Conference on Big Data and Information Analytics (BigDIA)* (IEEE, 2020)
44. S.P. Shamseera, E.S. Sreekanth, Word vectors in sentiment analysis. Int. J. Curr. Trends Eng. Res. (IJCTER) **2**(5), 594–598 (2016)
45. F.A. Acheampong, F. Adoma, W. Chen, N.-M.Henry, Text-based emotion detection: Advances, challenges, and opportunities. Eng. Rep. **2**(7), e12189 (2020)

3G Cellular Network Fault Prediction Using LSTM-Conv1D Model

N. Geethu and M. Rajesh

Abstract Cellular network plays an important role in daily life by exploring digital world of communication. Cellular network technology continuously evolves in past decades from 1 to 5G and beyond. The evolution results in more network accessibility and data utilization. As the availability of network, mobility, and portability of cellular devices are increasing, the network traffic will also be increasing. Higher the transmission rates, higher will be the fault occurrence possibility. Monitoring network parameters and finding fault in cellular network are key factor in determining consistency of network. Cellular network which is highly dynamic than usual networks needs intelligent way of fault handling as the human over head will be unpredictable and very high. Modeling intelligent network fault identification system can simplify human efforts and improve efficiency with better accuracy. The research is on real-time data of 3G cellular network including various network parameters like uplink threshold and identifies the behavior of data usual or unusual to predict the fault occurrence. The study is on various LSTM techniques such as bidirectional LSTM, vanilla LSTM, and stacked LSTM combined with time distributed Conv1D.

Keywords Fault prediction · Deep learning · Prediction model · Cellular network · Bidirectional LSTM · LSTM-Conv1D

1 Introduction

Emerging cellular network technologies evolve with better connectivity and scalability. The competing cellular business world ensures the quality of service by maintaining stable fault management mechanism. This is one of the key factors to have growth in the market. Mobile applications, online shopping, banking, online

N. Geethu (✉) · M. Rajesh
Department of Computer Science and Engineering, Amrita School of Engineering, Amrita
Vishwa Vidyapeetham, Bengaluru, India
e-mail: geethunellikal@gmail.com

M. Rajesh
e-mail: m_rajesh@blr.amrita.edu

classes, video conferencing, etc., are all part of common life. These are all demanding stable network and consistent performance. The trust factor will be mainly based on effective fault identification and even can be applied for fault prevention as next level.

Increasing network connectivity and traffic demands efficient fault management technique. As the cellular networks increase the mobility and availability, the chances of fault occurrence will also be high. Unhandled faults in connectivity result in performance degradation of the network. There should be effective fault management system with efficient way of managing the network traffic without failure.

The analysis of the cellular network real-time dataset with fixed and variable parameters on normal and fault detected cases is essential to develop model to handle faults in network. The human resource utilization on this consumes more time, cost and prone to errors. Formulate prediction model that can diagnose fault in the presence of certain parameters occurrence using neural networks is highly advanced and has future scope in terms of evolving deep learning techniques. In case of limited data, generative adversarial neural networks can be used to enhance the dataset of the deep learning model.

Existing studies explain the scope of research in this area considering the limitations of confidential real-time cellular network data. Here the objective is to model advanced fault management system for predicting faults for 3G cellular network based on fixed and variable network parameters using various deep learning models and identify the best. The focus is on LSTM techniques such as bidirectional LSTM, vanilla LSTM, and stacked LSTM combined with time distributed Conv1D.

The study identifies advanced deep learning technique for fault prediction in terms of analyzing different types of LSTM models combined with time distributed 1D convolution feature extraction on 3G cellular network data.

2 Literature Review

Tan and Pan [1] explain neural network-based fault prediction model with high efficiency. The logs of network transmission are used as dataset. Network log is preprocessed first for this. It uses time window to get the required samples. This is implemented on two levels. Sample features are extracted using convolutional neural networks. The convolutional networks extract the hidden features from logs. The LSTM model is used for predicting the results. Based on this, a hybrid model is developed for prediction of the network fault. Here it uses CNN-based feature extraction technique. Model has high accuracy in real-time data processing. But the CNN-based feature extraction from logs approach, adds more complexity to system.

Ji et al. [2] explain NLP-CNN-based fault prediction model using network log analysis. The logs from various log files are extracted and processed using NLP. The CNN model has embedded layer to get features from logs. The vectors obtained from this will be the input to next layer. 1D convolution is applied to input. The model is compared with LSTM model. As it is doing convolution for feature extraction of logs on top of embedded layer, the model has more complexity in processing real-time data.

Zheng et al. [3] explain CNN-LSTM prediction model for predicting failure in equipment in industry. The model is basically using industry equipment data for study. The paper involves comparison of LSTM model with hybrid LSTM-CNN model. The dataset contains matrix of values over a period of variables affecting system to predict failure. CNN is used for feature extraction. The convolutional layer extracts the features using filter. The relevant matrix data is extracted and is filtered using kernels. Here it reduces the dimension using pooling. The final fully connected layer reduces features to one dimension. This is fed to LSTM network. The model can be used to predict future fault occurrence for very large dataset effectively. This is effective way to overcome exploding gradient problem compare to other RNN techniques. While modeling cellular network fault prediction, the model needs to be analysed using real-time cellular network data input and related research using advanced LSTM-CNN network models.

Rezaei et al. [4] explain unsupervised machine learning model. For fault diagnosis, the paper applies expectation maximization (EM), an unsupervised method. It uses DBSCAN along with it, which is a method of grouping data based on density, providing random noise. The method is applied in mobile network communication to implement the self-decision-making systems. Here the different algorithms for grouping data are being compared with each other, and the results are being analysed. Each data group is related to a specific condition causing the fault. Also, there will be related guidelines to solve the faults. The machine learning technique of grouping both traffic and signaling is considered as the most relevant sources of fault information. The method is affected by human error during the modeling stage. The efforts will be high and are difficult to manage highly complex system.

The analysis of existing papers on studies conducted in this area mainly gives the statement of developing a most advanced deep learning-based model, considering the evolving highly dynamic cellular network nature. The combined LSTM–Conv1D models should be investigated more with possible varieties, as the advancement in such latest technology can give the most desirable results considering the above analysis.

3 Design and Implementation

3.1 The LSTM-Conv1D Design

Long short-term memory is advanced RNN technic to overcome the short memory. LSTM has three gates, forget gate, input gate, and output gate. The forget gate decides the information to discard from the cell, input gate decides values from the input to update the memory state, and output gate decides what to output based on input and

Fig. 1 Block diagram of
basic fault prediction model
using LSTM

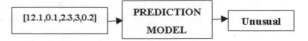

the memory of the cell [5]. This is mainly used for models which consider previous data states for result generation. The LSTM which has memory to store previous state can help on the dependability of output on previous results. Three types of models are:

- Prediction of next input sequence value.
- Prediction of input sequence category.
- Generation of another output sequence.

This paper focuses on the second type of LSTM application which predicts the category, (here the natural/unusual network parameters) of inputs as in Fig. 1. This is commonly used in applications such as

- Detect and predict the anomalies: For input containing various observation of parameters or sensors, predict if the sequence is having anomaly or is normal [5].
- Categories DNA Sequence: For DNA sequence input having A, C, G, and T values, predict whether the sequence is for a coding or non-coding region.
- Analyze Sentiment data: Given a sequence of text such as a review or a tweet, predict whether the sentiment of the text is positive or negative

There are three approaches considered for building the model. The models are all using deep learning techniques for feature extraction and prediction as the traditional ways are found to have less efficiency and less improvement during the analysis on the previous studies itself (even with the regular network data). Time distributed layer is used to wrap convolutional layer to apply to all samples.

a. Vanilla LSTM and Time distributed Conv1D

The approach uses 1D convolution for feature extraction. The 1D convolution has kernel move only in one dimension. Unlike the normal 2D CNN for image data, which has two dimensional kernel movement, it can extract features of text, numeric and audio inputs. One of such examples is to extract the sensors data and time series data [6]. It uses input data of two dimensions, for example, change of network parameter values with respect to time. Here the network parameters can be given as input to the Conv1D layer. The Vanilla LSTM is LSTM model with single hidden layer. Figure 2 shows the fault prediction model using Vanilla LSTM and time distributed Conv1D.

b. Bidirectional LSTM and Time distributed Conv1D

The bidirectional LSTM as name indicates, the input sequence is analysed in forward and backward directions. Then it combines both results. The model uses 1D convolution for feature extraction and later combined with bidirectional LSTM for predicting the fault for given network parameters. Figure 3 shows the fault prediction model using bidirectional LSTM and time distributed Conv1D.

Fig. 2 Fault prediction
model using Vanilla LSTM
and time distributed Conv1D

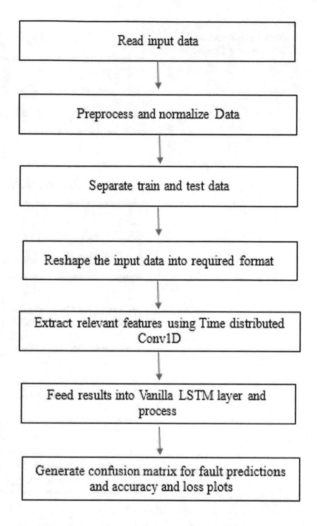

iii.   **Stacked Bidirectional LSTM and Time distributed Conv1D**

The term stacked indicates that multiple layers are combined. The stacked bidirectional model used 1D convolution layer and multiple bidirectional LSTM layers. Figure 4 shows the fault prediction model using stacked bidirectional LSTM and time distributed Conv1D.

3.2 *Implementation*

For all the three model categories, the 12 features are given as input from input layer to time distributed Conv1D with 64 neurons.

Fig. 3 Fault prediction
model using bidirectional
LSTM and time distributed
Conv1D

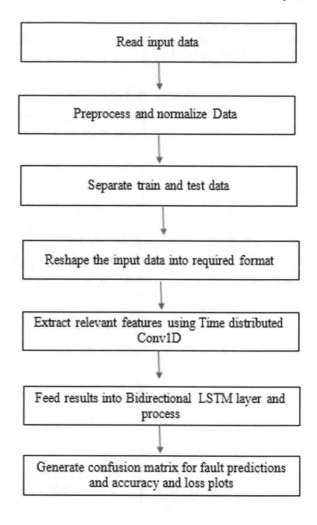

For vanilla LSTM model, it is performing convolution operation in one direction to extract the relevant features. The flatten layer converts the outcome of this into one-dimensional vector. This is fed to LSTM layer. There is no regularization used as the best result from this type of model is without any regularization [7].

For bidirectional LSTM with time distributed Conv1D also, the input layer is fed output to time distributed Conv1D with 64 neurons, to extract the relevant features. The bidirectional LSTM layer processes the input for prediction in two directions and yields best results among this category of models during the implementation.

The stacked bidirectional model has three-layered bidirectional LSTM stack. The extracted features from the convolution are fed to the first layer of the LSTM stack. The model is three-layered stack.

Fig. 4 Fault prediction
model using stacked
bidirectional LSTM and time

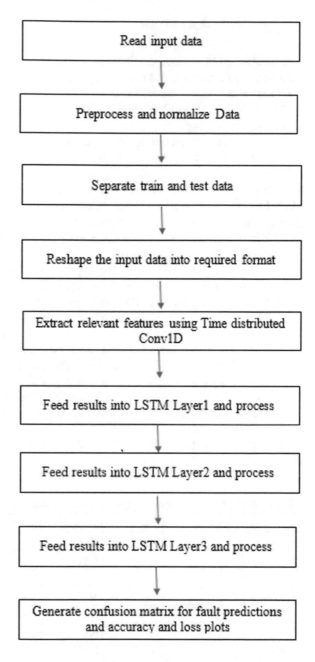

4 Results and Analysis

The Vanilla LSTM and time distributed Conv1D model are implemented with various parameters such as number of neurons, epochs, test train splits, loss functions, output activation function, optimizer, and the regularization.

The dropout is the regularization mechanism which will remove unwanted weight and reduce overfitting.

This L2 regularization also reduces overfitting by reducing the unimportant weight values. Table 1 indicates the notable models with various parameter values.

Table 2 indicates the accuracy and loss plots of Vanilla LSTM model. It shows that the accuracy is maximum with second model which is using Adam optimizer, MSE loss function, and RELU activation function. The SGD optimizer is giving poor results compared to Adam optimizer. The accuracy and loss graphs in Table 0.2 show there is no bias or variance for this model as the test and validation loss curves are almost syncing with each other.

The bidirectional LSTM-Conv1D models with various parameters and the accuracy and loss plot table are shown in Tables 3 and 4, respectively.

The best accuracy result is without regularization. The model3 without dropout and L2 regularization are showing the best result without any bias or variance. The loss curves and accuracy curves are showing a good sync in this model.

The stacked bidirectional LSTM-Conv1D models with parameter changes are shown in Table 5. The high accuracy is with model2 without any regularization, but it is showing high variance in Table 6 explaining the accuracy and loss plots. Regularization model has less accuracy than the bidirectional LSTM model.

The best model for the fault prediction is using bidirectional LSTM. The confusion matrix is given in Table 7.

The precision value is 0.97, recall is 0.83, and the F1 score is 0.8. This indicates the better performance of the model.

Table 1 Vanilla LSTM models with various parameter values

	Model 1	Model 2	Model 3	Model 4
Accuracy	72.5	83.18	82.1	81.01
Validation loss	0.19	0.12	0.13	0.13
Regularization	–	–	Dropout (0.2)	Dropout and L2
Epochs	50	100	100	100
Optimizer	SGD	Adam	Adam	Adam
Test split	0.1	0.1	0.1	0.1
Loss function	MSE	MSE	MSE	MSE
Output activation function	RELU	RELU	RELU	RELU

Table 2 Accuracy and loss plots of Vanilla LSTM models

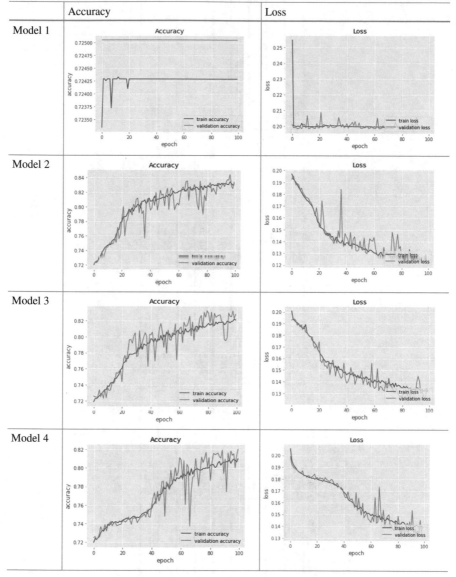

5 Conclusion and Future Scope

The cellular network fault prediction model provides better results using bidirectional LSTM and time distributed Conv1D architecture. As the cellular network data is highly dynamic and large in amount, the deep learning model will provide an efficient fault detection mechanism compared to traditional mechanisms.

Table 3 Bidirectional LSTM models with various parameter values

	Model 1	Model 2	Model 3	Model 4
Accuracy	83.5	82.6	85.18	81.66
Validation loss	0.12	0.12	0.11	0.13
Regularization	–	Dropout	–	Dropout and L2
Epochs	100	100	200	100
Optimizer	Adam	Adam	Adam	Adam
Test split	0.1	0.1	0.1	0.1
Loss function	MSE	MSE	MSE	MSE
Output activation function	RELU	RELU	RELU	RELU

Table 4 Accuracy and loss plots of bidirectional LSTM models

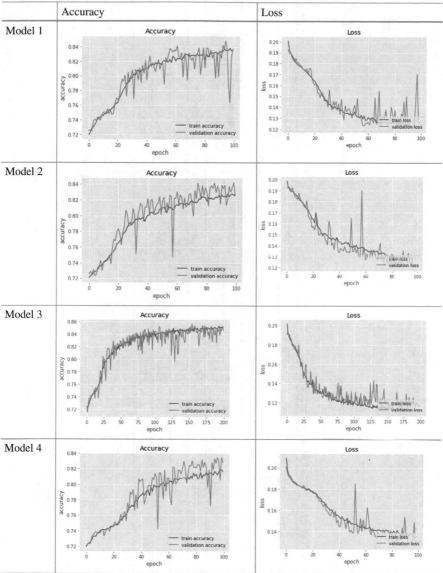

Cellular network fault management using deep learning techniques is a research area of wide scope as it is less under investigation compared to other areas. The limited and incomplete real-time confidential data is a major hurdle. There can be used generative adversarial neural networks (GAN) in further steps to generate data based on available one. The model can be evaluated using other cellular network data from 4G or 5G networks as well.

Table 5 Stacked bidirectional LSTM models with various parameter values

	Model 1	Model 2	Model 3	Model 4
Accuracy	74.6	87.5	81.4	76.04
Validation loss	0.11	0.18	0.13	0.16
Regularization	–	–	Dropout	Dropout and L2
Epochs	50	100	100	100
Optimizer	SGD	Adam	Adam	Adam
Test split	0.1	0.1	0.1	0.1
Loss function	MSE	MSE	MSE	MSE
Output activation function	RELU	RELU	RELU	RELU

Table 6 Accuracy and loss plots of stacked bidirectional LSTM models

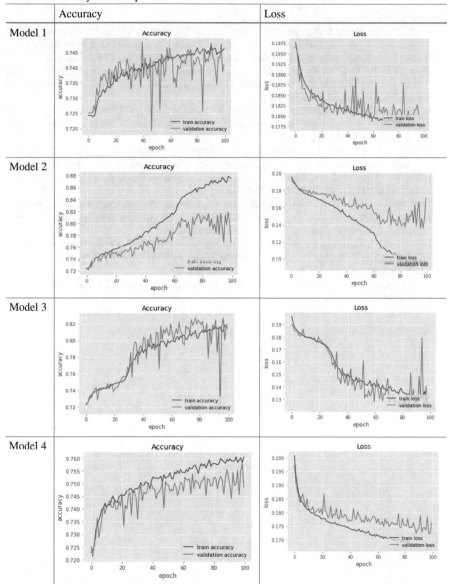

Table 7 Confusion matrix of prediction model

	Actual positive (1)	Actual negative (0)
Predicted positive (1)	2572	75
Predicted negative (0)	515	520

References

1. Z. Tan, P. Pan, Network fault prediction based on CNN-LSTM hybrid neural network, in *2019 International Conference on Communications, Information System and Computer Engineering (CISCE)* (Haikou, China, 2019), pp. 486–490, https://doi.org/10.1109/CISCE.2019.00113
2. W. Ji, S. Duan, R. Chen, S. Wang, Q. Ling, A CNN-based network failure prediction method with logs, in *2018 Chinese Control Decision Conference (CCDC)* (2018), pp. 4087–4090 https://doi.org/10.1109/CCDC.2018.8407833
3. L. Zheng et al., A fault prediction of equipment based On CNN-LSTM network, in *2019 IEEE International Conference on Energy Internet (ICEI)* (Nanjing, China, 2019), pp. 537–541. https://doi.org/10.1109/ICEI.2019.00101
4. S. Rezaei, H. Radmanesh, P. Alavizadeh, H. Nikoofar, F. Lahouti, Automatic fault detection and diagnosis in cellular networks using operations support systems data, in *NOMS 2016—2016 IEEE/IFIP Network Operations and Management Symposium* (Istanbul, 2016), pp. 468–473. https://doi.org/10.1109/NOMS.2016.7502845
5. https://github.com/Ahmed-Hussein2009/Jason-Brownlee-books/blob/main/8.%20Long%20Short-Term%20Memory%20Networks%20With%20Python%20Develop%20Sequence%20Prediction%20Models%20With%20Deep%20Learning.pdf
6. https://towardsdatascience.com/understanding-1d-and-3d-convolution-neural-network-keras-9d8f76e29610%20WRITTEN%20BY%20Z%C2%B2%20Little%20Data%20Science%20Diary%20Follow%2010%2010
7. https://machinelearningmastery.com/how-to-reduce-overfitting-in-deep-learning-with-weight-regularization/

Deep Learning for Part of Speech (PoS) Tagging: Konkani

Annie Rajan⊙, Ambuja Salgaonkar, and Arshad Shaikh⊙

Abstract This is the first time that an experiment using deep learning has been attempted with a Konkani language data set. For this study, over 100,000 PoS tagged Konkani sentences were used. The f-scores for deep learning are 90.73% for training data and 71.43% for test data. These results are better than the ones previously reported for Konkani. We have provided a list of references of PoS tagging for Indian languages specifically using deep learning to place our research in perspective.

Keywords Part of speech tagging · Konkani · Deep learning · Part of speech tagger · Computing methodologies · Machine learning

1 Introduction

Tagging can be performed by employing supervised learning if a manually tagged data set is available; otherwise, the learning is unsupervised [1]. Among the various machine learning approaches, the most widely used for PoS tagging are rule-based and stochastic. The rules in a rule-based tagger are initially handcrafted by manual analysis of the data [2, 3]. PoS tagging methods include hidden Markov model (HMM) [4], support vector machine (SVM) [5], decision tree [6], maximum entropy [7], conditional random field (CRF) [8], and deep learning [9–12].

A. Rajan (✉) · A. Salgaonkar
Department of Computer Science, University of Mumbai, Mumbai, India
e-mail: ann_raj_2000@yahoo.com

A. Salgaonkar
e-mail: ambujas@udcs.mu.ac.in

A. Rajan
DCT's Dhempe College, Panaji, India

A. Shaikh
Learning and Personalization, BYJU'S, Bangalore, India

© The Author(s), under exclusive license to Springer Nature Singapore Pte Ltd. 2023
Y.-D. Zhang et al. (eds.), *Smart Trends in Computing and Communications*, Lecture Notes in Networks and Systems 396, https://doi.org/10.1007/978-981-16-9967-2_32

The research results presented in this paper are: With deep learning, the f-score obtained is 90.20% after training and 71.43%, when employed on a set of 108,546 sentences of a Konkani database developed by the Indian Languages Corpora Initiative (ILCI).[1]

As far as we can ascertain, this is the first time that deep learning methods have been implemented for PoS tagging of Konkani language on a data set of such a large size.

2   Literature Survey

Comparable work in various Indian languages is summarized for an appraisal of our PoS tagging experiments in Konkani: Table 1 describes data sets and a performance results when deep learning is employed.

Konkani, the official language of the small but important Indian State of Goa, is making rapid progress in putting NLP applications to use. The Bureau of Indian Standards (BIS) has published annotation standards for Konkani [20]. The tag set has eleven parts of speech: verb, adverb, noun, pronoun, demonstrative, adjective, conjunction, postposition, quantifiers, particles, and residuals.

3   Brief History of Konkani Language and Its Speakers

The central portion of the narrow coastal strip between the Arabian Sea and Western Ghats in peninsular India is referred to as Konkan. The region is about 800 km long and about 45–75 km wide. The Konkani language is spoken along a total of about 600 km of the coast, along the central and southern parts of Konkan, and further south. As such, the geographical extent of the main Konkani speaking area is relatively small. Furthermore, it is not the only language spoken in this region; the major Indian languages like Marathi and Kannada are also spoken alongside. It is spoken by a population of 2.2 million people, spread across neighboring states of Goa, Karnataka, Maharashtra, and Kerala. Konkani attained its full potential around the eleventh century, and by the fifteenth century, it had evolved into a full-fledged literary language [21].

Small migrant communities of Konkani speakers[2] are found in Uganda, Kenya, Pakistan, Portugal, and the Persian Gulf [21, 22]. As a result, the language has been written in a be-wildering multiplicity of scripts: Devanagari, Roman, Kannada, Malayalam, Perso-Arabic. This led to the outward splitting up of one and the same language as it is spoken [23]. For historical as well as technical reasons, resources

---

[1] http://tdil.meity.gov.in/

[2] https://en.wikipedia.org/wiki/Konkani_language.

Table 1 PoS tagging in Indian languages using deep learning

S. No.	Language	Size of data set (words)		Performance measure (%)
		Training data	Test data	
1	Malayalam [13]	9000 tweets	915 tweets	92.54 (f-score)
2	Kannada [14]	10,500 sentences	1600 sentences	92.00 (f-score)
3	Sanskrit [15]	34,270 words	4321 words	97.86 (accuracy)
4	Bengali [16]	102,933 words	–	93.28 (f-score)
5	Hindi [17]	23,565 words	23,281 words	92.19 (accuracy)
6	Nepali [18]	100,720 words	–	92.00 (f-score)
7	Telugu [19]	67,861 sentences	–	81.73 (accuracy)

for Konkani have remained paltry, and their development, which although began several years ago, is slow. Sunaprant, a Konkani newspaper in Devanagari script, was operational during the year 1987–2015.[3] Bhaangar Bhuin, another Konkani newspaper in Devanagari script, was started in the year 2016 and made its presence in digital form in the year 2017.[4] Over the past two centuries, many newspapers were published in Devanagari, Kannada, and Roman script, but none has survived up to the present day.

4 Data Set

A manually annotated corpus from ILCI in the domains of tourism, health, entertainment, agriculture, and history with a total of 108,546 sentences is used for the experiment. Initially, the PoS tagger is trained on 86,837 sentences (1,006,177), and the test data has 21,709 sentences (125,389 words). The number of tags for the two data sets is 39, while the BIS tag set for Konkani has only 29 tags [20].

5 Experiments

The steps taken for PoS tagging using deep learning are shown in the block diagram of Fig. 1, and each step is explained in sufficient detail.

Embedding: A sequence of words to be tagged is the input to a hidden layer that converts each word of the sequence into a dense vector. It has three parameters, as follows: The input dimension is the size of the vocabulary of the test data. The words are encoded by the integers 0, 1, 2,..., because what is relevant is not the

---

[3] https://en.wikipedia.org/wiki/Sunaparant.

[4] https://en.wikipedia.org/wiki/Bhaangar_Bhuin.

Fig. 1 Block diagram for the biLSTM experiment

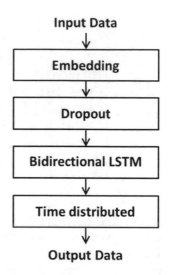

exact string of letters making up each word, but the various contexts in which it occurs. The input length is the length of the sequence of words that is input to the layer. The sequence length is a constant; if an actual sequence is of smaller length, it is to be padded with dummy integers. F-scores were calculated for word sequences of length of 7, 15, 23, 30, 40, and 50 obtained from the ILCL data set. It was found that sequence length of 23 gave the highest score among all, see Table 2. The output dimension is the size of the vector space, i.e., the number of components of the vector, into which the word will be embedded. The model is trained for 10 epochs, where an epoch is when the complete data set is passed through the model. It was found that the accuracy did not improve after 10 epochs, so the model was trained for 10 epochs.

Dropout: To reduce the overfitting of the dense vectors to the training data, i.e., to regularize the overfitted data, some components of the dense vector are randomly reset to 0 and the weightage of the remaining components is increased correspondingly.

Bidirectional: In bidirectional LSTM, the training data is supplied to two LSTM net- works (Fig. 2), one in the forward direction and another in the reverse direction. This technique is employed to learn the prefix and suffix of a given word.

The overall structure of bidirectional LSTM is same as LSTM. In biLSTM, the sequence is passed in both directions, two set of hidden states output form the biLSTM, and these states are combined to get a rich hidden state.

Consider the sentence "səptəpuriʧẽ d̪əɾʃən gʰeʈʄjaɾ mokʃə meʈʈa" (सप्तपुरीचें दर्शन घेतल्यार मोक्ष मेळटा), as an example to predict part of speech for the word "gʰeʈʄjaɾ" (घेतल्यार) in the sentence mentioned above. In bidirectional LSTM, it will be able to see the past information and also the information further down the road, i.e., forward LSTM: "səptəpuriʧẽ d̪əɾʃən gʰeʈʄjaɾ" and backward LSTM: "gʰeʈʄjaɾ

Table 2 F-scores using biLSTM method for different sequence lengths of input data

S. No.	Sequence length	Recall (%)	Precision (%)	F-score (%)
1	7	53	58	55
2	15	56	66	61
3	23	60	72	66[a]
4	30	60	71	65
5	40	61	68	64
6	50	50	57	54

[a]Sequence length of 23 gives the best results

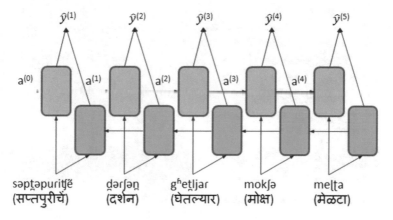

Fig. 2 Example of biLSTM-based analysis

mokʃə meḷṭa". The computed hidden state of both forward and backward LSTM is passed to the next layer. The equations used for this experiment are as per the reference of the paper [24].

Time-distributed dense layer: The last layer of the model is used to assign an appropriate PoS tag to each word. The dense layer generates a sequence of one-hot vectors, each of which contains probabilities of tags to be assigned to a single word. Since the dense layer needs to run on each word of the sequence, we used a time-distributed modifier. The predicted one-hot vectors are converted back to tagged sentences, i.e., sequences of words with PoS tags, after prediction. For training and testing, we used Keras with TensorFlow. (TensorFlow is an open-source software library for machine learning applications like neural networks. Keras is a framework for designing and implementing neural networks, with libraries such as TensorFlow, Theano, or the cognitive toolkit (CNTK).

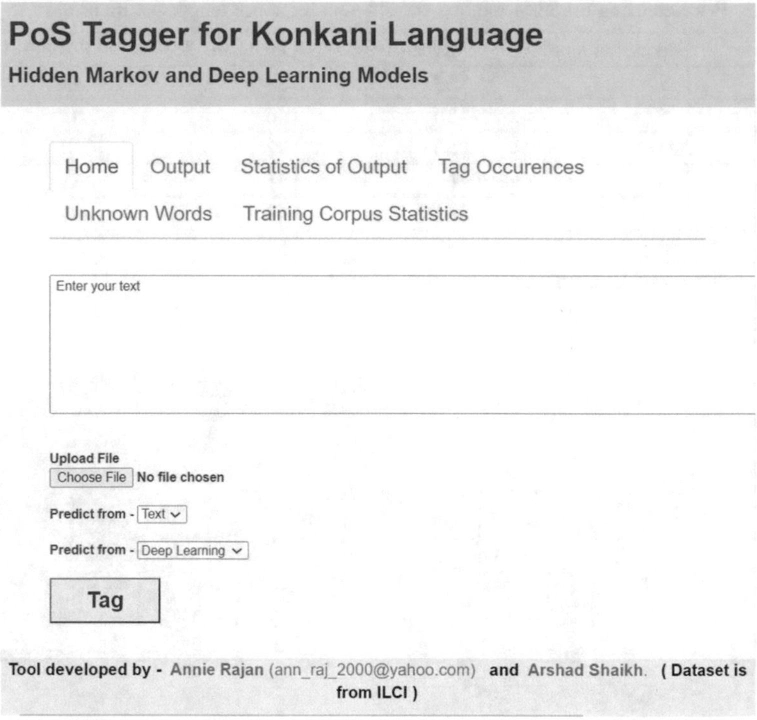

PoS Tagger for Konkani Language
Hidden Markov and Deep Learning Models

Home   Output   Statistics of Output   Tag Occurences

Unknown Words   Training Corpus Statistics

Enter your text

Upload File
Choose File No file chosen

Predict from - Text ˅

Predict from - Deep Learning ˅

Tag

Tool developed by - Annie Rajan (ann_raj_2000@yahoo.com) and Arshad Shaikh. (Dataset is from ILCI)

Fig. 3 PoS tagger for Konkani language

6 POS Tagger

A PoS tagger for Konkani language was developed using the Django framework. The tagger[5] allows the user to enter a text in Devanagari script, or upload a TXT file. The text can be tagged using a choice of deep learning. The tagged text is displayed in the output tab or can be downloaded in TXT format. Other tabs display the following information: the statistics of the output, the tags that occurred in the given input text, and the words that could not be tagged by deep learning, and finally, statistics about the training corpus (Fig. 3).

---

[5] http://103.86.177.205:5000/

Table 3 Tenfold validation output of training data for deep learning

S. No.	Cross-validation	Recall (%)	Precision (%)	F-score (%)	Accuracy (%)
1	Run 1	90.25	90.70	90.38	99.84
2	Run 2	90.06	90.43	90.16	99.84
3	Run 3	90.40	90.74	90.48	99.85
4	Run 4	90.22	90.59	90.33	99.84
5	Run 5	90.38	90.54	90.37	99.84
6	Run 6	90.55	90.63	90.51	99.84
7	Run 7	90.41	90.65	90.45	99.84
8	Run 8	90.19	90.45	90.23	99.84
9	Run 9	90.33	90.55	90.35	99.84
10	Run 10	90.28	90.61	90.35	99.84

Table 4 Fivefold validation output of test data for deep learning

S. No.	Cross-validation	Recall (%)	Precision (%)	F-score (%)	Accuracy (%)
1	Run 1	72.91	71.41	71.83	98.88
2	Run 2	73.18	71.43	71.94	98.87
3	Run 3	73.31	71.56	72.09	98.87
4	Run 4	73.95	71.74	72.23	98.88
5	Run 5	72.98	71.70	72.00	98.87

7 Results

A tenfold cross-validation has been was conducted by randomly selecting 43,418 sentences each time from the training data set for deep learning (Table 3). A fivefold cross-validation has been carried out by randomly choosing 10,854 sentences each time from the test data set for deep learning (Table 4).

8 Discussion

The number of misclassified words for the deep learning algorithm is 44,731 or 4.56% of the training data set of 1,006,088. Table 5 contains the list of tags and the misclassified word count.

From Table 5, we observe that the five tags with the highest errors are N_NN, RB, JJ, V_VM_VF, V_VM_VNF, which resulted in maximum errors. Further analysis of these words will help us to improve the f-score and accuracy of the tagger.

Table 6 contains examples of incorrectly tagged words by biLSTM, respectively.

Table 5 Word count of misclassified words using biLSTM

S. No.	Tag	Incorrectly tagged	Incorrectly tagged (%)
1	N-NN	7932	6.72
2	RB	4973	6.21
3	JJ	4157	3.52
4	V_VM_VF	3282	2.78
5	V_VM_VNF	2912	2.47
6	N_NNP	2289	1.94
7	DM_DMD	2020	1.71
8	PSP	1762	1.49
9	PR_PRP	1434	1.21
10	N_NST	1431	1.21
11	QT_QTF	1398	1.18
12	CC_CCS	1365	1.16
13	DM_DMR	1064	0.90
14	PR_PRL	790	0.67
15	QT_QTC	693	0.59
16	V_VAUX_VF	675	0.57
17	V_VM_VNG	626	0.53
18	DM_DMQ	583	0.49
19	RD_UNK	564	0.48
20	V_VM_VINF	557	0.47
21	CC_CCD	555	0.47
22	CC_CCS_UT	472	0.40
23	RP_INTF	457	0.39
24	QT_QTO	394	0.33
25	PR_PRQ	380	0.32
26	PR_PRI	375	0.32
27	PR_PRF	301	0.25
28	DM_DMI	275	0.23
29	PR_RPD	214	0.18
30	V_VAUX_VNF	213	0.18
31	RP_NEG	160	0.14
32	RP_INJ	123	0.10
33	V_VNG	95	0.08
34	PR_PRC	75	0.06
35	V_VM	46	0.04
36	RD_ECH	38	0.03
37	RD_RDF	23	0.02
38	RP_CL	19	0.02
39	RD_SYM	9	0.01

Table 6 Examples of words tagged incorrectly by biLSTM

S. No.	Word	Tagged in golden standard	Incorrectly tagged by biLSTM
1	təʃi (तळी)	RB	DM_DMD
2	səgljãṭ (सगलययीत)	QT_QTF	N_NN
3	gəʊrəw (गौरळ)	JJ	N_NN
4	mədʰjə (मध्यप)	N_NNP	N_NST
5	patəlilli (पयतळललरॢ)	V_VM_VNF	V_VM_VF

9 Conclusion

In this paper, we have presented deep learning strategies for POS tagging of Konkani language. This is the first attempt to create tagger that uses deep learning. F-scores of 90.73% were obtained using deep learning, while the error rates were 4.55%. A detailed analysis of this situation will aid in enhancing the f-score of the PoS tagger.

References

1. F.M. Hasan, N. UzZaman, M. Khan, *Comparison of different POS Tagging Techniques (N-Gram, HMM and Brill's tagger) for Bangla* (2007)
2. B.B. Greene, G.M. Rubin, *Automatic Grammatical Tagging of English* Department of Linguistics (Brown University, 1971)
3. S. Klein, R.F. Simmons, A computational approach to grammatical coding of English words. J. ACM (JACM) **10**, 334–347 (1963)
4. L. Rabiner, B. Juang, An introduction to hidden Markov models. IEEE ASSP Mag. **3**, 4–16 (1986)
5. C. Cortes, V. Vapnik, Support-vector networks. Mach. Learn. **20**, 273–297 (1995)
6. S.R. Safavian, D. Landgrebe, A survey of decision tree classifier methodology. IEEE Trans. Syst. Man Cybern. **21**, 660–674 (1991)
7. J. Skilling, *Classical MaxEnt Data Analysis, Maximum Entropy and Bayesian Methods* (Kluwer Academic Publishers, 1989)
8. J. Lafferty, A. McCallum, F.C. Pereira, Conditional random fields: probabilistic models for segmenting and labeling sequence data (2001)
9. S. Hochreiter, J. Schmidhuber, Long short-term memory. Neural Comput. **9**, 1735–1780 (1997)
10. M.I. Jordan, Attractor dynamics and parallelism in a connectionist sequential machine (1990)
11. K. O'Shea, R. Nash, An introduction to convolutional neural networks, arXiv preprint arXiv:1511.08458 (2015)
12. A. Sherstinsky, Fundamentals of recurrent neural network (RNN) and long short-term memory (LSTM) network. Phys. D Nonlinear Phenom. **404**, 132306 (2020)
13. S. Kumar, M.A. Kumar, K.P. Soman, Deep learning based part-of-speech tagging for Malayalam Twitter data (Special issue: deep learning techniques for natural language processing). J. Intell. Syst. **28**, 423–435 (2019)

14. K.K. Todi, P. Mishra, D.M. Sharma, Building a kannada pos tagger using machine learning and neural network models, arXiv preprint arXiv:1808.03175 (2018)
15. B. Premjith, K.P. Soman, P. Poornachandran, A deep learning based Part-of-Speech (POS) tagger for Sanskrit language by embedding character level features., in *FIRE*, pp. 56–60
16. M.F. Kabir, K. Abdullah-Al-Mamun, M.N. Huda, Deep learning based parts of speech tagger for Bengali, in *2016 5th International Conference on Informatics, Electronics and Vision (ICIEV)*, pp. 26–29
17. Parikh, Part-of-speech tagging using neural network. Proc. ICON (2009)
18. G. Prabha, P.V. Jyothsna, K.K. Shahina, B. Premjith, K.P. Soman, A deep learning approach for part-of-speech tagging in Nepali language, in *2018 International Conference on Advances in Computing, Communications and Informatics (ICACCI)*, pp. 1132–1136
19. M. Jagadeesh, M.A. Kumar, K.P. Soman, Deep belief network based part-of-speech tagger for Telugu language, in *Proceedings of the Second International Conference on Computer and Communication Technologies*, pp. 75–84
20. M. Sardesai, J. Pawar, S. Walawalikar, E. Vaz, BIS Annotation standards with reference to Konkani Language, in *Proceedings of the 3rd Workshop on South and Southeast Asian Natural Language Processing*, pp. 145–152
21. S. Borkar, *Let us Learn Konkani* (Abhiman Publications, Panaji, 1984)
22. M. Almeida, *A Description of Konkani* (2003)
23. O.J. Gomes, Old Konkani language and literature: the Portuguese role, Konkani Sorospot Prakashan, Goa (1999)
24. H. Sak, A.W. Senior, F. Beaufays, Long short-term memory recurrent neural network architectures for large scale acoustic modeling (2014)

DAMBNFT: Document Authentication Model through Blockchain and Non-fungible Tokens

Uday Khokhariya, Kaushal Shah, Nidhay Pancholi, and Shambhavi Kumar

Abstract Hard copies of documents can easily be forged, resulting in the decrease of credibility of issuing institutions and unfair use of forged documents by certain individuals as well. The model that we propose aims to authenticate documents through the use of blockchain technology, non-fungible tokens, and interplanetary file system. When a document is stored on the blockchain, a non-fungible token is created, which contains the unique address of the issuing institution and the hash of the document itself. The ownership of this token is then transferred to the document holder by the corresponding issuing authority. In this way, when someone wants to verify the authenticity of the document, they can use the address mentioned in the token to trace back the creator. If the document's hash differs from the one stored in the token, we know that the document has been altered. Even when unauthorized users are successful in adding the forged documents to the blockchain, they will not have the same unique signature as that of the authorized institution. The proposed model allows autonomous authentication of documents using public blockchain technology.

Keywords Non-fungible token (NFT) · Blockchain technology · Interplanetary file system · Smart contracts · Document forgery

1 Introduction

The issue of hardcopy document forgery is still prevalent. In South Africa, a series of cases have been reported where academic documents are forged [1]. There have been multiple cases of document forgery in India too, where documents related to land and academics have been forged [2]. The digitized records can be preserved for a longer time [3]. Hence, it is essential to combine the digitization property along with the validation criteria to create an efficient document authentication system. The system we propose works using the concepts of blockchain technology, non-fungible tokens (NFTs), and smart contracts. Blockchain is a shared ledger that has the property of

U. Khokhariya · K. Shah (✉) · N. Pancholi · S. Kumar
Pandit Deendayal Energy University (PDEU), Gandhinagar, India
e-mail: shah.kaushal.a@gmail.com

© The Author(s), under exclusive license to Springer Nature Singapore Pte Ltd. 2023   347
Y.-D. Zhang et al. (eds.), *Smart Trends in Computing and Communications*, Lecture Notes in Networks and Systems 396, https://doi.org/10.1007/978-981-16-9967-2_33

being averse to malicious changes. All the participants of the network have access to the records, but no participant has the ability to alter the contents of the blockchain. Therefore, once a document is added to the blockchain, it is infeasible to alter [4]. Smart contracts are pieces of code, or in simpler words, a certain set of rules which are executed whenever a document is added to the blockchain.

The system also uses non-fungible tokens (NFTs), which is an irreplaceable and immutable token stored on the blockchain [5]. These tokens are used to represent ownership and also for unique identification of an entity. The blockchain is a distributed system, and all the contents of the blockchain are stored on every node of the network. Therefore, storing the complete documents on the blockchain becomes impractical as a lot of storage space will be needed. As all the documents have to be stored on each node of the network, the method also causes a lot of redundancy. To solve that problem, our proposed system utilizes an interplanetary file system (IPFS) as the off-chain storage network [6]. In this system, documents are stored at certain locations in the network, and a content identifier (CID) which can be used to locate the document, is stored on the blockchain.

2 Background

The development of blockchain technology has paved the way for managing documents in a decentralized manner. If we consider an example, certificates are being stored on ledgers, and these ledgers constitute a blockchain [7]. This illustrates that the blockchain technology can be effectively used for managing documents. The evolution of technology would create another challenge of protecting the documents which are present in the ledgers.

Blockchain was first proposed in 1991. After 2014, another domain of blockchain was explored, and "blockchain 2.0" was born [8]. Eventually, several other implementations of blockchain such as the Ethereum blockchain have developed scope for various implementations [9]. Blockchain is a technology that is presently undergoing extensive research [10–13]. A blockchain is made up of several blocks or ledgers that are linked together. These ledgers are strict, and each one is linked by a cryptographic hash key which ensures that the transaction's validity is preserved [12].

The non-fungible token (NFT) provides a unique authority over a digital asset which is made on the concept of Ethereum blockchain. These are present on public ledgers and are unalterable [13]. NFT has been in discussion since 2018. The NFT market gained its popularity with the development of cryptoart. It is considered to be the most secured and easy method being used in transfer of digital assets [14]. The security of copyright has become a major challenge. Therefore, it is important to secure the rights of the owner. This is where NFTs come in picture because they ensure the ownership authority and are unalterable [15]. The digital assets such as land titles and college degrees can be converted to digital assets, stored in ledgers and are secured using NFTs [2, 3, 7].

Smart contracts are basically predefined code which is stored in blockchain. They get executed when the prior specified conditions are fulfilled [16]. The concept of smart contract was first implemented for storing bond agreements. These agreements are directly signed between the two parties which are involved. Smart contracts result in a successful transaction, providing access to encrypted content, changing of names on business titles, etc. [17]. This reduces the need of trusted third party. These scripted codes are stored on ledgers of blockchain. The script is assigned an address to store it on the ledger of blockchain. To execute the scripted code, one has to call for the assigned address [18]. As described in [19]; smart contracts provide a fast, secured, and transparent transfer of digital assets.

3 Related Work

Blockchain has created a lot of opportunities and has provided us with a scope of development of decentralized and secured applications [20, 21]. The ledgers are minted with NFTs, and each NFT has one defined unique address. The NFTs are unalterable as discussed in [22]. As mentioned in [23], one such breakthrough is evident as Blockcerts. Blockcerts are blockchain-based credentials with a decentralized and trustless verification mechanism. Users have direct access to permanent, transparent, and sustainable evaluation and management tools for learner credentials. Personal encrypted credentials allow users to customize lifetime learning paths and tailor education to their own beliefs and requirements. The original answer to Internet credentialing was badges. Mozilla's open digital badges have established a de facto worldwide standard, and the specifications are still available for free. They are accessible through e-portfolios and social media. Even if badges are genuine, they become worthless if issuers stop hosting them [23]. TrustChain is a three-layered trust management architecture that uses a consortium blockchain to log supply chain contracts and dynamically award trust and reputation scores based on these interactions. TrustChain is the first worldwide jewelry industry partnership to use blockchain technology to trace diamonds and precious metals from mine to completed jewelry and finally into the hands of the consumer [24]. The TrustChain architecture is based on the concept of agents conducting transactions with one another. The exchange of data, the purchase or sale of products, and the movement of funds are all examples of transactions in the real world [25].

4 The Proposed Model

Digital documents need an authentication system to track the creation timestamp, content changes, ownership, publishers, and validity [26]. Blockchain helps in tracing all these steps in the form of an immutable ledger. It prevents tampering of documents and helps verify the document issuer using digital signature methodology

[27]. Our model is divided into two spaces to minimize blockchain nodes' storage and computational requirements, namely on-chain and off-chain storage.

4.1 Participants Overview

- **Blockchain Network (BN)**: It is the leading network of the system where all NFTs will be stored. Smart contracts deployed on BN will be carrying out all the processes in the form of transactions. Further, the transaction log would also be stored on the BN, which gives credibility to every transaction happening on the network (Fig. 1).
- **Interplanetary File System (IPFS)**: It is a decentralized peer-to-peer framework which stores files in the form of hashes [6]. It works on the principle of content-based addressing and would act as off-chain storage in our model. IPFS network provides availability, reliability, and security of stored data [28]. To ensure synchronization of generated hashes, both blockchain and IPFS network would be required to use the same hashing algorithm to identify any document uniquely in the system.
- **Document Authority (DA):** The concept of DA is derived from [29]. DAs are responsible for creating, maintaining, and updating the required documents on BN and IPFS. They sign the documents and carry out the issuing process. Since all the documents are stored on a public blockchain network, it allows any institution or an individual to be a part of the system and register itself as DA.
- **Document Holder (DH)**: Individuals having ownership of particular documents are regarded as DHs. DH needs to hold a secured copy of the document with

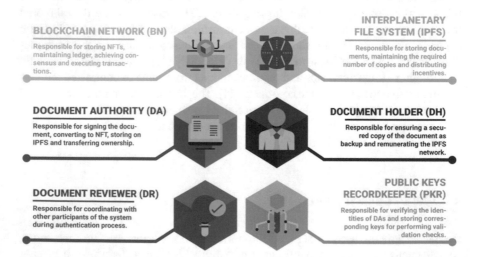

Fig. 1 Outline of the participants of the proposed system

him/her to safeguard the document from getting lost in case the IPFS network gets offline or most participants log out from the network. A DA can also act as DH when the institution itself requires secured storage of documents.

- **Document Reviewer (DR)**: When a DH claims the ownership of his/her owned document, DR verifies the claim by checking the authenticity of the claim. Since all the documents lie as NFTs in the public domain, DR does not need permission from the network to perform validation checks.
- **Public Keys Record-keeper (PKR)**: Whenever an institution or individual requests to become a DA, its corresponding address and public key are maintained by PKR. PKR is also responsible for verifying the identity of registered DAs to keep the network safe from identity imitation threats. It would act as a central ledger for all DAs that are a part of the network to ensure transparency and accountability of their signed documents.

4.2 Design Structure

- **Incentive Providence**: Documents issued by DAs occupy storage space on the IPFS network. The storage contributors to this network need to be incentivized to encourage active participation and keep the required number of copies available in the system. DA provides incentives on a short-term basis as agreed upon by the DH. After the expiration time of the agreement, DH needs to remunerate the IPFS network for as long as documents are required to be a part of the system. This process allows more participants to be a part of the system and allows IPFS participants to discard inactive documents after they no longer need the storage requirements. The distribution of incentives will be managed by the IPFS network, where they will be proportionally allocated according to the active time and storage space provided by individual participants (Fig. 2).
- **Membership Registration**: Whenever an entity wants to issue documents on the blockchain, they would need to register themselves with PKR. Users would need to confirm the claimed identity in order to join the network. PKR would be responsible for verifying the user's identity to secure the network from identity impersonation threats. PKR will generate a unique address for each requesting partaker. Public key and address are stored on the central ledger of PKR, whereas private key remains secured with the user. After joining the network, the user is known as DA.
- **NFT Creation**: DA carries out the process of converting required documents to NFT. Hash is used as data for creating the NFT while the original document would be replicated and distributed within the IPFS nodes. Generated hash of each document remains the same on IPFS and BN. When NFT gets successfully created on the blockchain and stored on the IPFS network, its ownership is transferred to its corresponding DH. NFT creation and ownership transfer take place through smart contracts which ensure all transactions get stored on the blockchain for traceability.

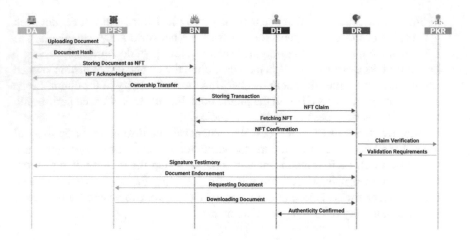

Fig. 2 Sequence diagram of the proposed document authentication system

- **Claim Validation**: When DH claims an NFT on the blockchain, DR carries out the validation process and checks the legitimacy of the claim by verifying the digital signature. Authentication requirements are fetched from PKR. Once the signature has been verified, it can also be sent to the DA for testimony. When a claim gets authenticated, the required document is fetched from the IPFS network using the hash stored on the NFT.

5 Discussion and Future Work

As mentioned in [30], the documents related to land entitlements are not properly maintained across India. The documents are not verified and result in forgery. Similarly, making of duplicate certificates is another major issue. With the advent of COVID-19, a lot of data had to be collected and analyzed. Once the vaccination of people started, India had shifted to digital storage of records. These digital assets were present in fragments. Another major challenge faced was to remove this data fragmentation. The government of Gujarat developed "Anveshan," a connector database to overcome data fragmentation, track data alteration, and to track dynamic data [31].

The document authentication model discussed here aims to prevent forgery of documents and identify the institution that issued the document through the use of blockchain and non-fungible tokens (NFTs). As the document issuer is being recognized through the NFT, forged documents can be easily identified based on the authenticity of the digital signature. The model also provides an added advantage in the case of documents that expire after a specific period. The initial transaction fees required for converting the document to NFT and storing it on the interplanetary file system (IPFS) are paid by DA on a short-term basis. However, if DH wants to renew the document, he/she needs to reward the IPFS nodes; otherwise, the document would be removed from the IPFS network. This characteristic of document removal from the IPFS allows the storage space to be freed up for the new documents to be added.

6 Conclusions

It becomes challenging to identify the authenticity of a physical copy of any document. Many methods have been implemented which focus on finding anomalies in the document, but none is always successful. The model we propose uses blockchain and non-fungible tokens to identify each document uniquely on the network. Whenever a document is issued by its corresponding authority, it is added on the blockchain in the form of a token which contains the hash of that particular document. Interplanetary file system acts as the off-chain storage where the document can be retrieved using its specific hash. The digital signature of the stored token ensures the trust, reliability, and authenticity of the issuing institution. Any person can join the network as an issuer and add documents to the blockchain, but the digital signature of the verified issuing institution cannot be replicated, and hence, the document could be easily identified as being forged. The model achieves its purpose of eliminating document forgery along with being transparent, decentralized, and immutable.

References

1. N. Dlamini, S. Mthethwa, G. Barbour, Mitigating the challenge of hardcopy document forgery, in *2018 International Conference on Advances in Big Data, Computing and Data Communication Systems (icABCD)* (IEEE, 2018), pp. 1–6
2. D.K.S. Puri, Study of a copied forgery. J. Secur. Adm. **3**(2), 79–87 (1980)
3. M. Baldi, F. Chiaraluce, MigelanKodra, L. Spalazzi, Security analysis of a blockchain-based protocol for the certification of academic credentials. arXiv:1910.04622 (2019)
4. P. Snow, B. Deery, J. Lu, D. Johnston, P. Kirby, Factom white paper. Retrieved from Factom: https://www.factom.com/devs/docs/guide/factom-white-paper-1-0 (2014)
5. U.W. Chohan, Non-fungible tokens: blockchains, scarcity, and value. In *Critical Blockchain Research Initiative (CBRI) Working Papers* (2021)
6. M. Naz, M, A. F.A. Al-zahrani, R. Khalid, N. Javaid, A.M. Qamar, M.K. Afzal, M. Shafiq, A secure data sharing platform using blockchain and interplanetary file system. Sustainability **11**(24), 7054 (2019)
7. O.S. Saleh, O. Ghazali, M.E. Rana, Blockchain based framework for educational certificates verification, in *Studies, Planning and Follow-up Directorate. Ministry of Higher Education and Scientific Research, Baghdad, Iraq. School* (2020)
8. M. Ulieru, Blockchain 2.0 and beyond: adhocracies, in *Banking Beyond Banks and Money* (Springer, Cham, 2016), pp. 297–303
9. A. Bogner, M. Chanson, A. Meeuw, A decentralised sharing app running a smart contract on the ethereum blockchain, in *Proceedings of the 6th International Conference on the Internet of Things* (2016), pp. 177–178

10. C. Lepore, M. Ceria, A. Visconti, U.P. Rao, K.A. Shah, L. Zanolini, A survey on blockchain consensus with a performance comparison of PoW, PoS and pure PoS. Mathematics**8**(10), 1782 (2020)
11. V. Patel, K.S. FenilKhatiwala, Y. Choksi. A review on blockchain technology: components, issues and challenges, in *ICDSMLA 2019* (Springer, Singapore, 2020), pp. 1257–1262
12. D. Shrier, W. Wu, A. Pentland, Blockchain & infrastructure (identity, data security). Massachusetts Inst. Technol.-Connect. Sci. **1**(3), 1–19 (2016)
13. L. Kugler, Non-fungible tokens and the future of art. Commun. ACM **64**(9), 19–20 (2021)
14. A. Guadamuz, The treachery of Images: non-fungible tokens and copyright. Available at SSRN 3905452 (2021)
15. R. Raman, B. Edwin Raj, The world of NFTs (non-fungible tokens): the future of blockchain and asset ownership, in *Enabling Blockchain Technology for Secure Networking and Communications* (IGI Global, 2021), pp. 89–108
16. Z. Zheng, H.-N.D. ShaoanXie, W. Chen, X. Chen, J. Weng, M. Imran, An overview on smart contracts: Challenges, advances and platforms. Future Gener. Compu. Syst. **105**, 475–491 (2020)
17. M. Bartoletti, L. Pompianu, An empirical analysis of smart contracts: platforms, applications, and design patterns, in *International Conference on Financial Cryptography and Data Security* (Springer, Cham, 2017), pp. 494–509
18. K. Christidis, M. Devetsikiotis, Blockchains and smart contracts for the internet of things. IEEE Access **4**, 2292–2303 (2016)
19. Z. Allam, On smart contracts and organisational performance: a review of smart contracts through the blockchain technology. Rev. Econ. Bus. Stud. **11**(2), 137–156 (2018)
20. G. Rathee, A. Sharma, R. Kumar, R. Iqbal, A secure communicating things network framework for industrial IoT using blockchain technology. Ad Hoc Netw. **94**, 101933 (2019)
21. P. Pant, R. Bathla, S.K. Khatri., A model to implement and secure online documentation using blockchain, in *2019 4th International Conference on Information Systems and Computer Networks (ISCON)* (IEEE, 2019), pp. 174–178
22. J. Santos, K.H. Duffy, A decentralized approach to blockcerts credential revocation, in *A White Paper from Rebooting the Web of Trust V* (2018)
23. M. Jirgensons, J. Kapenieks, Blockchain and the future of digital learning credential assessment and management. J. Teacher Educ. Sustain. **20**(1), 145–156 (2018)
24. L.E. Cartier, S.H. Ali, M.S. Krzemnicki, Blockchain, chain of custody and trace elements: an overview of tracking and traceability opportunities in the gem industry. J. Gemmol. **36**(3) (2018)
25. P. Otte, M. de Vos, J.A. Pouwelse, TrustChain: a sybil-resistant scalable blockchain. Future Gener. Comput. Syst **107** (2017). https://doi.org/10.1016/j.future.2017.08.048
26. A. Ahmad, S.B. Maynard, G. Shanks, A case analysis of information systems and security incident responses. Int. J. Inf. Manage. **35**(6), 717–723 (2015)
27. H. Stančić, Model for preservation of trustworthiness of the digitally signed, timestamped and/or sealed digital records (TRUSTER preservation model), *InterPARES Trust Project* (2018), pp. 1–42
28. S. Krishnan, V.E. Balas, E. Golden Julie, H.R. Yesudhas, S. Balaji, R. Kumar (eds.), *Handbook of Research on Blockchain Technology* (Academic Press, 2020)
29. S. Yao, J. Chen, K. He, R. Du, T. Zhu, X. Chen, PBCert: privacy-preserving blockchain-based certificate status validation toward mass storage management. IEEE Access**7**, 6117–6128 (2018)
30. N. Gupta, M.L. Das, S. Nandi, LandLedger: blockchain-powered Land property administration system, in *2019 IEEE International Conference on Advanced Networks and Telecommunications Systems (ANTS)* (IEEE, 2019), pp. 1–6
31. J.C. Liu, S. Banerjee, A. Bhagi, A. Sarkar, S.D. BhriguKapuria, V. Sethuraman, S. Patil, S. Banerjee, Blockchain technology for immunisation documentation in India: findings from a simulation pilot. Lancet Glob. Health **9**, S22 (2021)

A Novel Hybrid Translator for Gujarati to Interlingual English MTS for Personage Idioms

Jatin C. Modh⑩ and Jatinderkumar R. Saini⑩

Abstract Gujarat is one of the states in the western part of India, and the Gujarati language is the official language of Gujarat. The Gujarati language is more than 700 years old and is spoken by more than 55 million people around the world. A machine translation system (MTS) is needed for the communication between people knowing different languages. Idioms are used in almost all the languages. An idiom is a phrase or an expression whose meaning does not necessarily relate to the literal meaning of its individual words. The idiom generally means something different than what is directly conveyed by its individual words. Translation of idioms for any language, as with Gujarati idioms, into any other language is a challenging task. All existing MTS including Google Translate and Microsoft Bing fail to translate Gujarati idioms suitably. In the current paper, a particular category of Gujarati idioms, called the personage idioms, has been treated for detection from the input text and translation into the English language. We have deployed a dictionary-based algorithm and context-based search, respectively, for idioms having one and multiple meanings. This is a first of its kind work in the world. From a broad technical view point, this is an application of Gujarati to interlingual translation for the MTS sub-domain of natural language processing (NLP). The interlingual language is considered to be English for the present research work.

Keywords Gujarati · Interlingual · Machine translation system (MTS) · Natural language processing (NLP) · Personage idioms

J. C. Modh
Gujarat Technological University, Ahmedabad, India

J. R. Saini (✉)
Symbiosis Institute of Computer Studies and Research, Symbiosis International (Deemed University), Pune, India
e-mail: saini_expert@yahoo.com

© The Author(s), under exclusive license to Springer Nature Singapore Pte Ltd. 2023   355
Y.-D. Zhang et al. (eds.), *Smart Trends in Computing and Communications*, Lecture Notes in Networks and Systems 396, https://doi.org/10.1007/978-981-16-9967-2_34

1   Introduction

The Gujarati language is the official language in the state of Gujarat, India, and it is also the official language in the union territory of Dadra and Nagar Haveli, Daman, and Diu. The Gujarati language is more than 700 years old, and it is the sixth most widely spoken language in India as per 2011 data. Gujarati is the 26th most widely spoken language in the world based on the number of local speakers. Gujarati is the fastest growing Indian language in Canada, America, and many other countries [1].

In the Gujarati language, idioms play a very important role. An idiom is a term whose meaning is different from the literal meaning of the words it is made of. An idiom is a phrase whose collective meaning is different than its literal meaning. Idioms are part of Gujarati culture. Idioms are used frequently by the Gujarati people to express feelings. Elders use idioms in day-to-day interaction, and this way idioms are passed to the new generation.

In the twenty-first century, people are migrating from one country to other countries for study, business, and earnings. Gujarati people are business-oriented, and they are ready to migrate to foreign countries. For communication between different language people, a machine translation system is required. English is the universal language and widely accepted language in the region of Gujarat for the purpose of migration. Many machine translation systems are available in the market. But, all are facing issues related to the translation of idioms.

1.1   Gujarati Personage Idioms

Idioms are found in the Gujarati language like any other language. Idioms are terms that are not to be taken exactly. Personage idiom represents a dramatic, fictional, or historical character in a play or other work. It represents a person or human individual [2]. For example, બાબરોભૂત 'babaro bhut' is a Gujarati bigram personage idiom, and its actual meaning in Gujarati is દેખાવમાં બિહામણી વ્યક્તિ 'dekhavamam bihamani vyakti'; its meaning in English is derived as 'scary guy in appearance'. Another example પોપાબાઈ 'popambai' is Gujarati unigram personage idiom, and its actual meaning in Gujarati is શિથિલ રાજ્યકર્તા 'shithila rajyakarta'; its meaning in English is derived as 'lazy ruler'. Gujarati personage idioms are misunderstood by other language readers as they do not know the culture, history, and popular characters of Gujarat and Gujarati language [3]. The focus of this paper is on such Gujarati personage idioms.

The rest of the paper is organized as follows: Sect. 2 represents the literature review related to translation of Gujarati text and Gujarati idioms translation; Sect. 3 covers the methodology including data collection and the steps of proposed model; Results and analysis are discussed in Sect. 4; finally, conclusion, limitations, and future work are described in Sect. 5.

2 Related Literature Review

The scope of this paper is the translation of Gujarati text containing Gujarati personage idioms into English or any other language. Technically, using the parlance of the domain of natural language processing (NLP), this type of translation can be called translation from Gujarati to interlingual language.

Google Translate [4] is a multilingual machine translation system developed by Google Company. It supports 109 different languages for machine translation. Google Translate uses the Google Neural Machine Translation system. Google Translate also translates the Gujarati language into English or any other language [5]. It is a very popular translator but not able to translate Gujarati idioms correctly.

Microsoft Translator [6] is a multilingual machine translation system developed by Microsoft. It supports 90 different languages for machine translation. Microsoft Translator uses Microsoft cognitive services integrated with other Microsoft products. Microsoft Translator also translates Gujarati language text into English or any other language [7]. It also fails to deal with the translation of Gujarati idioms.

Goyal and Sharma [8] described a neural machine translation system for the Gujarati–English news. They tested multilingual neural machine translation models. They tested and advocated that for low-resource Gujarati–English language pairs; multilingual neural MT models are more effective than individually trained MT models.

Saini and Modth [9] implemented a dictionary-based machine translation system GIdTra for translating only Gujarati bigram idioms into English. They classified bigram idioms on the basis of their possible meanings in the English language.

Modh and Saini [10] discussed various machine translation approaches for the Gujarati language. The researchers have implemented a context-based machine translation system (MTS) for translating Gujarati bigram and trigram idioms into the English language [11]. They used IndoWordNet for finding context and improving machine translation of Gujarati idioms [12]. They detected all possible forms of Gujarati idioms from the input text using diacritic and suffix-based rules and by dynamic phrase generation technique [13].

Sen et al. [14] depicted a constrained news translation system for Gujarati–English language pairs. They used original parallel corpus acquired from back-translation of monolingual data and trained transformer-based neural machine translation system. They did not use any external data and advocated to improve the BLEU score.

Agrawal et al. [15] introduced a multilingual parallel idiom dataset of 2208 idioms for English and other seven Indian languages including the Gujarati language. They illustrated sentiment analysis and machine translation and advocated significant improvement in handling idioms without using the idiom dataset. They concluded that phrase-based SMT is better than NMT to handle idioms. They used English idiomatic sentences as a source and converted them into Indian languages.

Based on this literature review and study on machine translation of Gujarati language, all researchers are facing problems in dealing with the translation of idioms.

All researchers had issues in handling idiomatic text of any language. No researchers have analyzed specifically Gujarati personage idioms in detail.

3   Methodology

In the Gujarati language, 3200 distinct idioms are available [13]. Personage idioms are other exceptional idioms used for the character in a literary work or a history, a play, a novel, etc. [3]. Personage idioms are used for important or strange or any distinguished person.

3.1   Data Collection

Overall, 88 Gujarati personage idioms were collected from different sources [16, 17]. Personage Gujarati idioms are stored in the idiom database. Tables 1 and 2 show the two different snapshots of idiom database containing personage Gujarati idioms. Idiom database contains fields like id, idiom, gmeaning (Gujarati meaning), emeaning (English meaning), gcwords (Gujarati context words), and popularity field. Id denotes sequence number; idiom column stores personage idiom; gmeaning stores the actual Gujarati meaning of particular personage idiom; emeaning stores the derived English meaning of particular personage idiom. Field gcwords stores the Gujarati context words to identify the correct gmeaning from the database. Field popularity stores the integer number. Both the fields Gcwords and popularity are needed only when the Gujarati personage idiom has more than one meaning [12]. Table 1 shows the snapshot of database containing single-meaning Gujarati personage idioms, whereas Table 2 shows the snapshot of database containing Gujarati personage idioms having more than one meaning. In Table 1, gcwords and popularity fields are not shown as they are having NULL and 0 values, respectively, but they are used for the actual execution of proposed model.

3.2   Proposed Model

MySQL was used to create and maintain the idiom database for storing Gujarati personage idioms, while PHP was used as a scripting language. XAMP cross-platform was used for the local Webserver to implement the proposed model, while Python 3.7 was used for executing the Python code. Notepad + + was used as an editor for writing Python code and PHP code. Googletrans free and unlimited Python library were used for the implementation of Google Translate API [18, 19]. The algorithmic representation of the steps followed toward systematic methodology is presented below:

Table 1 Snapshot of database containing single-meaning Gujarati personage idioms

Id	Idiom	Transliteration of (2)	Gmeaning	Transliteration of (4)	Emeaning
(1)	(2)	(3)	(4)	(5)	(6)
1	ઊંટવૈદ	untavaida	લાયકાત વગરનો વૈદ	layakat vagarano vaida	an unqualified doctor
2	કડકો	kadako	ખિસ્સે તદ્દન ખાલી	khisse taddana khali	having no money
3	કુબ્જા	kubja	બેડોળ સ્ત્રી	bedol stri	mis-shapen woman
4	કોડાફાડ	kodaphad	આખાબોલી વ્યક્તિ	akhaboli vyakti	outspoken person
5	ગુંદરિયો	gundariyo	ચિટકી રહેનાર વ્યક્તિ	chitaki rahenar vyakti	clingy person
6	ઝ્રેડ	jhoda	વળગાડ જેવી ત્રાસજનક વ્યક્તિ	valagad jevi trasajanak vyakti	annoying person
7	ઢ	dha	મંદબુદ્ધિ	mandabuddhi	incompetent person
8	પ્રજ્ઞાચક્ષુ	prajnachakshu	અંધજન	andhajan	blind person
9	લલ્લુ	lallu	મૂર્ખ વ્યક્તિ	murkha vyakti	stupid person
10	લુખ્ખો	lukhkho	ખોટી ગુંડાગીરી કરનાર	khoti gundagiri karanar	grabing money by bossiness
11	હંસલો	hansalo	આત્મા	atma	soul
12	રાધા	radha	પ્રેમિકા	premika	sweetheart
13	અક્કલનો બારદાન	akkalano baradan	મૂર્ખ માણસ	murkha manas	foolish man
14	ધોબીનો કૂતરો	dhobino kutaro	ઠામઠેકાણા વગરની વ્યક્તિ	thamathekana vagarani vyakti	person with no whereabouts
15	બારમો ચંદ્રમા	baramo chandrama	દુશ્મનાવટ	dusmanavata	animosity
16	બીકણ બિલાડી	bikana biladi	ખૂબ બીકણ વ્યક્તિ	khub bikan vyakti	very timid person
17	મંથરા	manthara	કાવતરાખોર અને કાન ભંભેરણી કરનાર સ્ત્રી	kavatarakhor ane kanabhambhcrani karanar stri	conspiratorial woman
18	તાડકા	tadaka	અતિશય ઊંચી સ્ત્રી	atishaya unchi stri	extremely tall woman
19	છાપેલું કાટલું	chapelum katalum	નામચીન	namachin	notorious person
20	ચમચો	chamacho	ખુશામત કરનાર વ્યક્તિ	khushamat karanar vyakti	flatterer
21	કુંભકર્ણ	kumbhakarna	ઊંઘણશી વ્યક્તિ	unghanashi vyakti	drowsy person
22	નારદ	narad	બંને પક્ષોને લડાવી મારનાર	banne pakshone ladavi maranar	one who makes persons quarrel
23	શકુની	shakuni	મુત્સદી અને કપટી વ્યક્તિ	mutsadi ane kapati vyakti	diplomat and insidious person
24	દિલ્હીનો શાહુકાર	dilhino shahukar	મોટો ઠગ	moto thaga	rogue person

(continued)

Table 1 (continued)

25	ભૂખડી બારસ	bhukhadi baras	કંગાળ અને ખાઉંખાઉની દાનતવાળું	kangal ane khaumkhauni danat valum	poor and voracious
26	બાબરો ભૂત	babaro bhut	દેખાવમાં બિહામણી વ્યક્તિ	dekhavamam bihamani vyakti	scary guy in appearance
27	બલિનો બકરો	balino bakaro	કોઈના ફૂડા કરતૂતનો ભોગ બનતી વ્યક્તિ	koina kuda karatutano bhoga banati vyakti	scapegoat
28	અંગૂઠા છાપ	angutha chhap	અભણ કે નિરક્ષર	abhana ke nirakshar	illiterate person
29	જૂનો જોગી	juno jogi	અનુભવી	anubhavi	experienced person
30	સહદેવ જોશી	sahadev joshi	પૂછ્યા વિના ઉત્તર ન આપે તેવી વ્યક્તિ	puchhya vina uttara na ape tevi vyakti	a person who does not answer without asking
31	દશમો ગ્રહ	dasamo grah	કનડગત કરનાર જમાઈ	kanadagat karanar jamai	harassment by son-in-law
32	પોપાંબાઈ	popambai	શિથિલ રાજ્યકર્તા	shithil rajyakarta	lazy ruler
33	ચારસો વીસ	charaso vis	ઠગ	thaga	rogue
34	પાંચમી કતારિયો	panchami katariyo	દેશદ્રોહી વ્યક્તિ	deshadrohi vyakti	traitor person

- Step1: Overall, 88 Gujarati personage idioms were collected.
- Step2: As a preprocessing activity, extraneous spaces and characters were removed from the collected idioms.
- Step3: Idiom database for Gujarati personage idioms was created with the fields shown in Table 2.
- Step4: Accept input Gujarati text from the user containing Gujarati personage idiom(s).
- Step5: The proposed algorithm searches for personage idioms from the input text by comparing them with the idiom field of the database.
- Step6: Proposed algorithm replaces the Gujarati personage idiom (idiom) with the actual Gujarati meaning text (gmeaning) using the database. Also, note that if a personage idiom has more than one meaning, then Gujarati context words (gcwords field) and popularity field are used to decide the suitable meaning of a specific personage idiom [12]. Other input Gujarati text remains the same. This intermediate output contains literal Gujarati text only.
- Step7: The proposed algorithm translates intermediate Gujarati literal text into English language using Google Translate API.
- Step8: Display input Gujarati text, intermediate Gujarati text output, and final English translation.

Table 2 Snapshot of database containing multiple-meaning Gujarati personage idioms

Id (1)	Idiom (2)	Transliteration of (2) (3)	Gmeaning (4)	Transliteration of (4) (5)	Emeaning (6)	Gcwords (7)	Popularity (8)
1	નંગ	nang	ભાગ અથવા ટુકડો	bhag athava tukado	piece	અંશ અદદ અનુરાગ અભિપ્રાય અરથ અર્ણવ અર્થ અવકૃય આધ આશાય ઇશક ઇશ્ક ઉચિત ઉદ્દેશ એટલે કંદર કિંમત કિલો કીમત કેટલા કોતર ખંડ ખરીદ ખો ગહ્વર ગુફા ગુહા ગોપુર ગ્રાહક ચિત્તવૃત્તિ જવાહિર જાયઝ ટુકડો ડઝન તાત્પર્ય તોલ તોલમાપ થાય દમોડ દર દરવાજો દરિ દામ દ્વાર દ્વાર ધોરણ નંગ નગીન નગીનો નેડો પણ પૃથ્વીગૃહ પૈસા પ્યાર પ્રણય પ્રમાણ પ્રમાણે પ્રાઇસ પ્રીતિ પ્રેમ બખોલ બદલો બાકું બારણું બિલ ભાગ ભાવ ભાવતાલ ભાવાર્થ ભોણ મણ મણિ મત મતલબ મનોભાવ મનોભાવના મનોવિકાર મનોવૃત્તિ મનોવેગ મફત માનસિક ભાવ માને માપ મુજબ મુનાસિબ મૂલ મૂલ્ય મોંઘુ મોલ મોલભાવ યથાર્થ યોગ્ય રતન રત્ન રાગ રૂપિયા રેટ લગન વજન વળતર વાજબી વિવર વેચાણ વેપારી વ્યાજબી શાક શાકભાજી સંગત સંબદ્ધ સનેડો રાત્નું રાર રોદો સ્નેહ હેત હેતુ	2
2	નંગ	nang	મૂર્ખ કે ખાંધી વ્યક્તિ	murkha ke khandhi vyakti	a word indicating stupidity	અક્કલ અઘાદક અધમ ઘનિર્મલ અનિષ્ટ અનુગ અનુચર અનુચિત અપભ્રંશ અપભ્રંશિત અપવિત્ર અપુનિત અપ્રશંસનીય અપ્રિય અબદ્ધ અભિચર અભિસર અયોગ અયોગ્ય અવદ અવિનયી અવિમલ અવિશુધ્ધ અવિહિત અશિષ્ટ અશુચિ અશુદ્ધ અશુધ્ધ અશ્લાધ્ય અશ્લીલ અસંસ્કારી અસભ્ય અસાધુ અસ્વચ્છ આચારભ્રષ્ટ આજ્ઞાંકિત આદમ આદમી આધાર આધીન ઇન્સાન ઉદ્ધત કપટી કેવો ખરાબ ખરાબ રીતે ખલ ગંદ ગંદું ગંધારુ ગાંડા ગ્રામ્ય ચાકર છે જંગલી જન જેવું જેવો ટેકો તદ્દન તાબેદાર દાસ દુષ્ટ દૂષિત દોષ્યુક્ત ધૂર્ત નઠારું નર નરસું નિંદનીય નિંદાત્મક નિંદ નીચ નોકર પરાધીન પરિચારક પાપી પામર બગડેલું બદ બદતર બિલકુલ બીભત્સ બુદ્ધિ બૂઠું બેઅદબ બેકાર ભરોસો ભૂંડ ભ્રષ્ટ મનુષ્ય મલિન માણસ માનવ માનવી માનુષ મૂર્ખ ગેલુ મ્લાન લક્ષ વાહિયાત વિકારી વિકૃત વિચારશક્તિ વ્યક્તિ શખ્સ શઠ સાવ સેવક હલકટ	1
3	નંગ	nang	રત્ન	ratna	a crystallized jewel	અંગૂરી અંગુલિ અંગુષ્ટ અગ્ર અનુભાવ અપરાધ અપલક્ષણ અરુણ અર્ણવ અવગુણ અસર આંગલી આંગણું આગસ આમદ આમદાની આમિષ આય આવક આશ્રવ ઇચ્છા ઇનકમ ઉંગલ ઉંગલી ઉપલબ્ધિ ઉલઝન એબ કમાણી કલ્યાણ કામયાબ ક્રૂટેવ ફુલક્ષણ કૃતકૃત્ય કૃતાર્થ કેતુ ખરાબી ગુત્થી ગુનો ગુરુ ગ્રહ ચંદ્ર ચાંદી છાપ જય જવાહિર જાતક જીત જીર્ણવજ્ર જુર્મ જ્યોતિષ તાસીર દુર્ગુણ દોષ ધારણ નંગ નગીન નગીનો નક્શી નીલમ નુકસાન પંખરાગ પક્ષરાગ પક્ષરાગ મણિ પક્ષરાજ પન્ના પહેરવું પાપ પોખરાજ પ્રતાપ પ્રભાવ પ્રશ્ન પ્રાપ્તિ પ્લેનેટ ફતેહ ફાયદો બગાસ બરકત બુધ મંગલ મંગળ મણિ મલાતર મસલો માણિક માણેક મોતી યશ ચાફ્ત રંગ રક્તોપલ રતન રત્ન રાતો મણી રાશિ રાહુ લબ્ધિ લાભ લાલ મણી વિજય વિજયશ્રી વીંટી વૃદ્ધિ વૈકાંતમણિ શનિ શુક્ર	3

(continued)

Table 2 (continued)

4	બલા	bala	આ ફત	aphat	trouble	શોણરત્ન સંસર્ગ સફળ સફળતા સમસ્યા સમૃદ્ધિ સમૃદ્ધિજીત સાર્થક સિદ્ધ સિદ્ધિ સુખ સૂર્થમણિ સોનું સ્પર્શ હિત હીરો હેતુસિદ્ધિ અંતરાય અક અધ અડયણ અયોગ અરિષ્ટ અવગ્રહ અવરોધ અવસન્નતા અવસન્નત્વ અશર્મ અસુખ અસુવિધા આડખીલી આદીનવ આપત્તિ આપદ આપદા આફત આભીલ આરામ ઉદ્વેગ કઠોર કષ્ટ કાયો ખરડો કુદરતી સ્થિતિ ક્લેશ ખરબચઈ ખાડાટેકરાવાળું ખોડખામીવાળું ગભરામણ ગર્દિશ ચેન ડોલિયું તકલીફ તસ્દી તોફાની થાક ઉતારવો દ દુઃખ દોચન નડતર નિરાંત પરેશાની પીડા બાધા ભીડ મવાલી મુશ્કેલી મુસીબત રાહત લાગણીવિહોણું વિક્ષેપ વિક્ષેપણ વિઘ્ન વિપત્તિ વિપદ વિપદા વિલોપ વિશ્રાન્તિ વિશ્રામ વૃજિન વ્યથા વ્યાકુલતા વ્યાકુલપણું વ્યાકુળતા વ્યાઘાત શામત સંકટ સંતાપ સુક્ષન સુખ હરકત હેરાનગત	1
5	બલા	bala	વણ આ મં ત્રિ તમ હેમા ન	vana amantrit mahem an	uninvited guest	અચાનક અતિથિ અન્ન અન્નગ્રહણ અભ્યાગત અશન આગંતુક આગત આમંત્રણ આવકાર આવાહન આહાર આહ્વાન ઈંજન કંકોતરી કંકોત્રી કહેણ કુટુંબ કુટુંબકબીલો ફળ ખટલો ખબર ખાવું ખોરાક ઘર ઘરખટલો જમણ જમવાનું જમવું જાણ જાહેરાત તલબ તેડું નિમંત્રણ નિમંત્રણપત્ર નોતરું નોતરૂં પરિવાર પરોણો ફેમિલી ભક્ષ્ય પદાર્થ ભોજન ભોજનકાર્ય મહેમાન મિજમાન રસોઈ રહેઠાણ રાંધણું. વગર વસ્તાર શિકાર સંદેશો સમાચાર સુવિધા સૂચના	2

4 Results and Analysis

For the output authentication, three linguistics were consulted and asked to perform human intelligence task (HIT). As part of HIT, they were clearly explained the motivation behind the research as well as the requirement of precision of their translation. All the experts were having a qualification of at least post-graduation in Gujarati. Two experts were males, while one was Female. For example, input text is રામુ પાંચમી કતારિયો છે 'Ramu panchami katariyo chhe'; the intermediate output is is રામુ દેશટ્રોહી વ્યક્તિ છે 'Ramu desadrohi vyakti chhe', and the final English translation by the proposed algorithm is 'Ramu is a traitor person'. Table 3 shows the translation of Gujarati language text containing Gujarati personage idioms into the English language by the proposed model. The instances presented in Table 3 make use of imaginary and fictitious names Raju and Ramu to exemplify the input text used for testing the proposed model. We submit that like any other NLP test data used by other researchers, these names are used by us just for improving the comprehension of the presentation. In Table 3, input text and intermediate output both are in Gujarati language, but transliteration is shown for understanding only.

The same Gujarati text is given to Google Translate and Microsoft Bing Translator for the comparison of the proposed model results. If Gujarati input text રામુ પાંચમી કતારિયો છે 'ramu panchami katariyo chhe' is given to Google Translate and Microsoft Bing Translator, Google Translate translates it into English as 'Ramu is fifth rower' incorrectly; Microsoft Bing Translator also translates it into English as 'Ramu is the fifth Qatari' incorrectly; Google Translate and Microsoft

Table 3 Translation of Gujarati personage idioms into English language by proposed model

Sr. No	Input text	Intermediate output	Final English translation output
1	રાજુ ઊંટવૈદ છે. raju untavaida che	રાજુ લાયકાત વગરનો વૈદ છે. raju layakat vagarano vaida chhe	Raju is an unqualified doctor.
2	રામુ ચારસો વીસ છે. ramu charaso vis chhe	રામુ ઠગ છે. ramu thag chhe	Ramu is a thug.
3	રાજુ લુખ્ખો છે. raju lukhkho chhe	રાજુ ખોટી ગુંડાગીરી કરનાર છે. raju khoti gundagiri karanar chhe	Raju is a false bully.
4	રામુ પાંચમી કતારિયો છે. ramu panchami katariyo chhe	રામુ દેશદ્રોહી વ્યક્તિ છે. ramu deshdrohi vyakti chhe	Ramu is a traitor person.
5	મનુષ્યના મરણ પછી હંસલો જતો રહે છે. manushyana maran pachhi hansalo jato rahe chhe	મનુષ્યના મરણ પછી આત્મા જતો રહે છે. manushyana maran pachhi atma jato rahe chhe	The soul is going after the death of a human being.
6	રામુ બીકણ બિલાડી છે. ramu bikan biladi chhe	રામુ ખૂબ બીકણ વ્યક્તિ છે. ramu khub bikan vyakti chhe	Ramu is a very timid person.
7	રામુ તો રાજુનો ચમચો છે. ramu to rajuno chamacho chhe	રામુ તો રાજુનો ખુશામત કરનાર વ્યક્તિ છે. ramu to rajuno khushamat karanar vyakti chhe	Ramu is a man who flatters Raju.
8	રામુ અભ્યાસમાં સાવ ઢ છે. ramu abhyasamam sav dha chhe	રામુ અભ્યાસમાં સાવ મંદબુદ્ધિ છે. ramu abhyasamam sav mandabuddhi chhe	Ramu is very dull in his studies.
9	રાજુ તો સાવ નંગ છે, તેના પર ભરોસો ના રાખી શકાય. raju to sav nang chhe, tena par bharoso na rakhi shakay	રાજુ તો સાવ મૂર્ખ કે ખંધી વ્યક્તિ છે, તેના પર ભરોસો ના રાખી શકાય. raju to sav murkh ke khandhi vyakti chhe, tena par bharoso na rakhi shakaya	Raju is a very stupid person, he cannot be trusted.

Bing could not recognize the Gujarati personage idioms, so both could not able to translate it correctly, whereas the proposed model first detects Gujarati personage idiom પાંચમી કતારિયો 'panchami katariyo' from the input and replaces it with Gujarati actual meaning દેશદ્રોહી વ્યક્તિ 'desadrohi vyakti', so intermediate output is રામુ દેશદ્રોહી વ્યક્તિ છે 'Ramu desadrohi vyakti chhe' and then using Google Translate API to translate the intermediate Gujarati result into the English language correctly as 'Ramu is a traitor person'. Table 4 shows the comparison of the English translation of Gujarati text containing Gujarati personage idiom(s) by Google Translate, Microsoft Bing Translator, and the proposed model.

Some interpretations, results, and ambiguities faced during the experimentation.

1. Proposed algorithm provides most nearby English translation of Gujarati personage idioms; whereas Google Translate and Microsoft Bing Translator do not provide correct translation of Gujarati personage idioms. For example, for Gujarati sentence રાજુ ઊંટવૈદ છે 'raju untavaida chhe', Google Translate translates in English as 'Raju is a camel doctor' that is absolutely wrong one;

Table 4 Snapshot of comparisons of proposed model with translation by Google Translate and Microsoft Bing

Sr. No	Input text	English translation		English Translation (Proposed Algorithm)	English Translation by HIT
		By Google Translate	By Microsoft Bing Translator		
1	રાજુ ઊંટવૈદ છે. raju untavaida che	Raju is a camel doctor.	Raju is a camel vaid.	Raju is an unqualified doctor.	Raju is an unqualified doctor.
2	રામુ ચારસો વીસ છે. ramu charaso vis chhe	Ramu is four hundred and twenty.	Ramu is four hundred and twenty.	Ramu is a thug.	Ramu is a thug.
3	રાજુ લુખ્ખો છે. raju lukhkho chhe	Raju is lurking.	Raju is a lukhkho.	Raju is a bully.	Raju is a bully.
4	રામુ પાંચમી કતારિયો છે. ramu panchami katariyo chhe	Ramu is the fifth rower.	Ramu is the fifth Qatari.	Ramu is a traitor person.	Ramu is a traitor person.
5	મનુષ્યના મરણ પછી હંસલો જતો રહે છે. manushyana maran pachhi hansalo jato rahe chhe	Laughter is gone after the death of a human being.	After the death of a man, the goosegoes go away.	The soul is going after the death of a human being.	The soul is leaving after the death of a human being.
6	રામુ બીકણ બિલાડી છે. ramu bikan biladi chhe	Ramu is a timid cat.	Ramu is a beaded cat.	Ramu is a very timid person.	Ramu is a very timid person.
7	રામુ તો રાજુનો ચમચો છે. ramu to rajuno chamacho chhe	Ramu is Raju's spoon.	Ramu is the spoon of Ramu.	Ramu is a man who flatters Raju.	Ramu is a man who flatters Raju.
8	રામુ અભ્યાસમાં સાવ ઢ છે. ramu abhyasamam sav dha chhe	Ramu is in the study.	Ramu is very good at studies.	Ramu is very dull in his studies.	Ramu is very dull in his studies.
9	રાજુ તો સાવ નંગ છે, તેના પર ભરોસો ના રાખી શકાય. raju to sav nang chhe, tena par bharoso na rakhi sakay	Raju is completely naked, he cannot be trusted.	Raju is very naked, he cannot be trusted.	Raju is a very stupid person, he cannot be trusted.	Raju is a very stupid person, he cannot be trusted.

Microsoft Bing Translator translates it in English as 'Raju is a camel vaid' that is also wrong one, whereas proposed algorithm translates as 'Raju is an unqualified doctor' that is very similar to correct translation 'Raju is an unqualified doctor'.

2. Common Gujarati idioms are n-grams where $n = 2$ to 9 [11], whereas our research shows that Gujarati personage idioms are either 1-g or 2-g only. The number of unigram or 1-g Gujarati personage idioms is 58, and the number of bigram or 2-g personage idioms is 30. Hence, the total number of Gujarati personage idioms is 88.

3. Most of the Gujarati personage idioms are found to have single meaning except નંગ 'nang' and બલા 'bala'; નંગ idiom has three different literal meanings depending on the surrounding context words as (1) મૂર્ખ કે ખાંધી વ્યક્તિ 'murkh

ke khandhi vyakti' a stupid person (2) ભાગ અથવા ટુકડો 'bhag athava tukado' piece (3) રત્ન 'ratna' a crystallized jewel, whereas personage idiom બળી has two different literal meanings depending on the surrounding context words as (1) વણઆમંત્રિત મહેમાન 'vana amantrit maheman' uninvited guest (2) આફત 'aphat' trouble.

4. Literal meaning of Gujarati personage idioms are shown in Gujarati and English both the languages in the idiom database. It provides useful personality insight about the particular personage. For example, for Gujarati personage idiom શકુની 'shakuni', actual meaning in Gujarati is મુત્સદી અને કપટી વ્યક્તિ 'mutsadi ane kapati vyakti', and its English meaning is derived as 'diplomat and insidious person', so proposed model helps other language users about the correct meaning as well as popular character meaning.

5. An idiom database was created with 88 distinct personage idioms and tested with different Gujarati sentences containing personage idiom(s). The proposed algorithm can able to find out personage idiom within Gujarati text by searching it in the idiom database if it exists in the idiom database.

5 Conclusion, Limitations, and Future Work

All the existing MTSs, including Google Translate and Microsoft Bing Translator, face the problem of accuracy in translating the Gujarati idioms into English or any other language. The proposed algorithm detects all the Gujarati personage idioms present in the input Gujarati text and translates them, as an intermediate step, into Gujarati itself but with semantically correct meaning. This Gujarati intermediate output contains no Gujarati personage idioms. Finally, intermediate Gujarati output is translated into the interlingual language. We have considered English as an interlingual language for the present work by using Google Translate API.

We deployed a hybrid approach comprising of a dictionary-based algorithm and context-based search for the translation of personage Gujarati idioms. These two approaches have been used, respectively, for idioms with literal meanings and the idioms with multiple idioms. To the best of our study of related literature, this is a first of its kind implementation for accurate translation of personage idioms of any language in world. Based on the results obtained from the proposed algorithm, it is advocated that the proposed algorithm is promising and can be extended for all the Gujarati idioms available in the Gujarati language. The proposed algorithm is specifically useful for translating Gujarati personage idioms into the English language accurately. Furthermore, all Gujarati personage idioms can be translated into any other language by implementing the proposed model. The limitation of the proposed model is in differentiating between the Gujarati personage idiom and the same word used as a noun. For the future work, we have already started to work on a special case where the same idiom is used as a noun as well as in the idiomatic sense simultaneously.

The proposed model opens the pathway for translating Gujarati idioms into any other language. We believe that the proposed model is free from the geographic boundaries on the translator, and the proposed work will definitely drive other researchers in the area of MTS in specific and NLP in general.

References

1. Wikipedia, Gujarati language. https://en.wikipedia.org/wiki/Gujarati_language . Accessed 6 Aug 2021
2. Merriam Webster Dictionary, Definition of personage. https://www.merriam-webster.com/dictionary/personage. Accessed 11 Aug 2021
3. Your Dictionary, Personage meaning. Available online https://www.yourdictionary.com/personage. Accessed 11 Aug 2021
4. Google Translate, Google translate. Google Corporation Ltd. Available online https://translate.google.co.in/. Accessed 11 Aug 2021
5. Wikipedia, Google translate. Available online https://en.wikipedia.org/wiki/Google_Translate. Accessed 11 Aug 2021
6. Bing Microsoft Translator, Microsoft bing. Microsoft Corporation Ltd. Available online https://www.bing.com/translator. Accessed 11 Aug 2021
7. Wikipedia, Microsoft translator. Available online https://en.wikipedia.org/wiki/Microsoft_Translator. Accessed 11 Aug 2021
8. V. Goyal, D. Sharma, The IIIT-H Gujarati-English machine translation system for WMT19, in *Association for Computational Linguistics; Proceedings of the Fourth Conference on Machine Translation (WMT), vol 2: Shared Task Papers (Day 1)* (2019), pp. 191–195
9. J.R. Saini, J.C. Modh, GIdTra: a dictionary-based MTS for translating Gujarati Bigram Idioms to English, in *Fourth International Conference on Parallel, Distributed and Grid Computing (PDGC) 22–24 December 2016* (2016), pp. 192–196, Available online https://ieeexplore.ieee.org/document/7913143
10. J.C. Modh, J.R. Saini, A study of machine translation approaches for Gujarati language. Int. J. Adv. Res. Comput. Sci. **9**(1), 285–288 (2018). Available online: https://ijarcs.info/index.php/Ijarcs/article/download/5266/4497
11. J.C. Modh, J.R. Saini, Context based MTS for translating Gujarati Trigram and Bigram Idioms to English, in *2020 International Conference for Emerging Technology (INCET), Belgaum, India* (2020), pp. 1–6. https://doi.org/10.1109/INCET49848.2020.9154112. Available online https://ieeexplore.ieee.org/document/9154112/
12. J.C. Modh, J.R. Saini, Using IndoWordNet for contextually improved machinetranslation of Gujarati Idioms, Int. J. Adv. Comput. Sci. App. (IJACSA) **12**(1) (2021). https://doi.org/10.14569/IJACSA.2021.0120128
13. J.C. Modh, J.R. Saini, Dynamic phrase generation for detection of Idioms of Gujarati language using diacritics and suffix-based rules. Int. J. Adv. Comput. Sci. Appl. (IJACSA) **12**(7) (2021). https://doi.org/10.14569/IJACSA.2021.0120728
14. S. Sen, K.K. Gupta, A. Ekbal, P. Bhattacharya, IITP-MT system for Gujarati-English news translation task at WMT 2019, in *Association for Computational Linguistics; Proceedings of the Fourth Conference on Machine Translation (WMT), Volume 2: Shared Task Papers (Day 1)* (2019), pp. 407–411
15. R. Agrawal, V.C. Kumar, V. Muralidharan, D. Sharma, No more beating about the bush: a step towards Idiom handling for Indian language NLP, in *LREC 2018, Eleventh International Conference on Language Resources and Evaluation*. Available online http://www.lrec-conf.org/proceedings/lrec2018/pdf/1052.pdf. Accessed 12 Aug 2021
16. GujaratiLexicon, Gujaratilexicon.com, Available online http://www.letslearngujarati.com/about-us. Accessed 11 Aug 2021

17. Rudhiprayog Ane Kahevat Sangrah, Published by Director of languages, Gujarat State, Gandhinagar (2010)
18. Googletrans 3.0.0, Free google translate API for Python. Translates totally free of charge. Available online https://pypi.org/project/googletrans/. Accessed 10 August, 2021
19. Googletrans 3.0.0 Documentation, Googletrans: free and unlimited google translate API for Python. Available online https://py-googletrans.readthedocs.io/en/latest/. Accessed 10 Aug 2021

Alleviation of Voltage Quality-Related Issues: A Case Study of Bahir Dar Textile Share Distribution System

Dessalegn Bitew Aeggegn, Yalew Werkie Gebru, Takele Ferede Agajie, Ayodeji Olalekan Salau⊙, and Adedeji Tomide Akindadelo

Abstract This paper presents a unified power quality conditioner (UPQC)-based compensation approach for voltage quality improvement in Bahir Dar textile share company distribution system. The existing system was modeled and simulated to evaluate its performance. The base case waveform shows that power quality is deteriorating in the factory. The voltage sag and swell at the base case are 65% and 140%, respectively, with voltage unbalance and harmonic content for a duration of 0.1 s. After using the proposed fuzzy logic controller (FLC)-based UPQC, both voltage sag and swell are alleviated by 100% and the waveform is maintained and kept at 1.0 p.u.

Keywords Fuzzy logic controller · Voltage sag · Voltage quality · Harmonics

1 Introduction

Power quality (PQ) and reliability of the supply are vital for the proper running of industrial processes that have critical, sensitive, and nonlinear loads. The major concerns of the industries are short duration voltage variations such as voltage sag, voltage swell, and short interruptions. Due to the advancement of semiconductor technology, these factories use a wide range of sensitive and nonlinear equipment and loads during production phases. If this short duration power quality (PQ) issues lasts for a long period, they may cause production loss, power outage, and other costs [1–5].

D. B. Aeggegn (✉) · Y. W. Gebru · T. F. Agajie
Department of Electrical and Computer Engineering, Debre Markos University, Debre Markos, Ethiopia
e-mail: dessalegnbitew29@gmail.com

A. O. Salau
Department of Electrical/Electronics and Computer Engineering, Afe Babalola University, Ado-Ekiti, Nigeria

A. T. Akindadelo
Department of Basic Sciences, Babcock University, Ilishan Remo, Nigeria

The root causes of voltage quality-related issues are majorly uneven distribution of single-phase loads among the three-phase supply in distribution grids, technical and non-technical faults, large loads and motor drives, and the utility grid [6–10].

The rest of this paper is structured as follows. Section 2 presents the proposed method. The experimental results and discussion are presented in Sect. 3, and Sect. 4 concludes the paper.

2 Methodology

2.1 System Design

In this study, Bahir Dar textile share company distribution system was considered as the case study. To alleviate power quality problems such as voltage sag, voltage swell, voltage unbalance, and harmonic distortion, a fuzzy logic-based UPQC is proposed. System modeling and simulations were performed using MATLAB/Simulink. The system was simulated in scenarios of without and with UPQC at steady-state conditions. The overall diagram is consisting of all existing features, and UPQC components (like series transformers, controllers, VSI converters, filters, and others complimentary features) are shown in Fig. 1, while all the system parameters used are presented in Table 1.

The load variety was made using a load survey during the assessment. Finding the load parameters was not required, due to the available clear load data such as power factor, real and reactive power values from the company. These values are presented in Table 2.

2.1.1 Sizing of the System

The system was sized based on the load survey and assessment. The sizing of both series and shunt APFs is sized in parallel for maximum demand. This supports the reactive power and partial compensation of active power. The system has a supply grid/feeder of 15 kV and MDB of 400/230 V and the total load of 4546.10 kW with a power factor of 0.65. Table 2 presents the summary of the load assessment.

Figure 2 shows the series APF subsystem block diagram that is used to improve voltage quality in Bahir Dar textile share company distribution system. Figure 3 shows the shunt APF which regulates the voltage by improving the voltage unbalance.

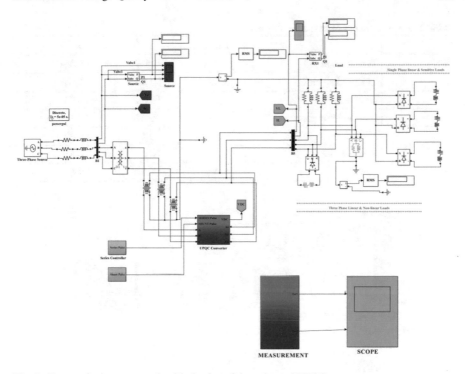

Fig. 1 Proposed system network with the use of the proposed UPQC

Table 1 System parameters

S/No	System model	Parameter description
1	Source	Three phase, 15 kV rms (L-L), 15 kV base, $R = 1.5\ \Omega$, $L = 2.4\ \mu H$
2	VSI converter	IGBT based, 3 arm, 6 pulse, 4 kHz
3	PLL	Three phase, discrete, $K_p = 20$, $K_i = 50$, Sampling time = 50 μs
4	LC filter	$L = 3\text{mH}$, $C = 1\ \mu\text{F}$
5	DC link	$C = 400\ \mu\text{F}$, $V_{\text{DCref}} = 700$ V
6	PI controller	$K_i = 12$, $K_p = 0.5$
7	Transformer	One three phase: 7.5 MVA, 15/0.4 kV, 50 Hz;
		Three single-phase: 2.5 MVA, 0.4/0.4 kV, 50 Hz
8	Load	Linear: three phase, 400 V, 50 Hz, 2.49 MW, 1.34 MVAr
		Single phase, 230 V, 50 Hz, 0.276 MW, 0.14 MVAr
		Nonlinear: Three phase, 4000 V, 50 Hz, 0.592 MW, 0.32 MVAr
		Single phase, 230 V, 50 Hz, 0.592 MW, 0.32 MVAr

Table 2 Load summary from load assessment

		Capacity		
S/No.	Type of load	MVA	MW	MVAr
1	Linear three phase	3.83	2.49	1.34
2	Nonlinear three phase	0.91	0.59	0.32
3	Linear single phase	0.42	0.276	0.148
4	Nonlinear single phase	0.91	0.59	0.32
5	Reserved for future	1	0.6	0.35
6	Total	6.421	4.546	2.478

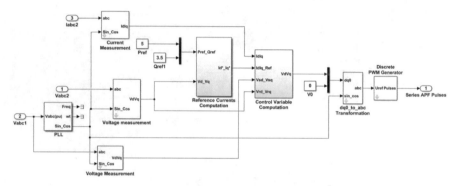

Fig. 2 Simulink model of the series APF

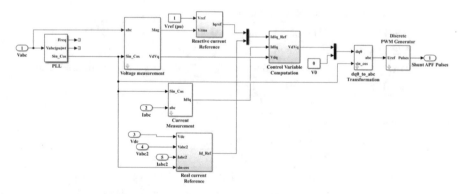

Fig. 3 Simulink model of the shunt APF

3 Results and Discussion

3.1 Steady-State Simulation Result of Existing System Without UPQC

The power flow was simulated using the latest version of PSAT 2.1.10 toolbox added to MATLAB toolbox directory. The existing system PSAT single line diagram is shown in Fig. 4. Power flow simulation was performed using PSAT to determine the load voltage profile, power loss, load real, and reactive power flow. The results are presented in Table 3.

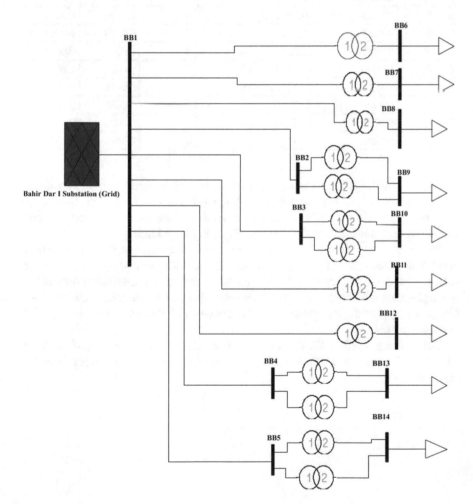

Fig. 4 PSAT single line diagram of the existing system

Table 3 Network components and simulation results

Buses	10	Number of iterations	4
Transformers	13	Maximum P mismatch (p.u.)	− 0.002
Generators/Grid	1	Maximum Q mismatch (p.u.)	0.008
Loads	9	Power rate [SMVA]	10

Table 4 Power flow results (all values in p.u.)

Bus	Voltage	Angle	P_{gen}	Q_{gen}	P_{load}	Q_{load}
Bus_1	1.0 (if...)	0	0.403	0.224	0	
Bus_2	0.97179	−0.07061	0	0	0.0208	0.0068
Bus_3	0.96554	−0.03511	0	0	0.0208	0.0068
Bus_4	0.94554	−0.03511	0	0	0.0208	0.0068
Bus_5	0.93854	−0.03511	0	0	0.0208	0.0068
Bus_6	0.9243	−0.01746	0	0	0.0622	0.0017
Bus_7	0.8596	−0.11437	0	0	0.1418	0.1075
Bus_8	0.8929	−0.07061	0	0	0.0208	0.0068
Bus_9	0.9256	−0.03499	0	0	0.0622	0.0017
Bus_10	0.9156	−0.11437	0	0	0.1418	0.1075

As shown in Table 4, only buses 7 and 10 are below the required voltage according to the IEEE standard, because the p.u voltage value is less than 0.95. Hence, this indicates that the PQ issue is caused by the sensitive, nonlinear loads and improper grounding and non-uniform distribution of single-phase loads.

Power loss is experienced in some sections, even when the overall voltage status is good. This is because of the adaptive under voltage, the overloading of transformers (i.e., not using of parallel operation of transformers) and non-uniform distribution of single-phase loads in the company's distribution system. Therefore, the line flows (P_{flow} and Q_{flow}) can be negative from the source and load bus and vice versa. This is shown in Table 5.

The existing system PSAT power flow result is almost the same with the MATLAB/Simulink scope display reading of voltage, real, reactive power parameters discussed in Sect. 3.2.

Table 5 Line flow (all values are in p.u.)

From bus	To bus	Line	P_{flow}	Q_{flow}	P_{Loss}	Q_{Loss}
Bus_8	Bus_2	1	−0.02208	−0.068	8e−04	0.0020
Bus_9	Bus_2	2	−0.055	−0.002	1e−04	0.0019
Bus_10	Bus_2	3	−0.14018	−0.1075	0.0083	0.0165
Bus_2	Bus_2	4	−0.02208	−0.068	8e−04	0.0157
Bus_7	Bus_2	5	−0.1418	−0.1075	0.00083	0.0165
Bus_3	Bus_2	6	−0.014	−0.0034	2e−04	0.00039
Bus_3	Bus_2	7	−0.014	−0.0034	2e−04	0.00039
Bus_4	Bus_2	8	−0.014	−0.034	2e−04	0.00039
Bus_4	Bus_2	9	−0.014	−0.034	2e−04	0.00039
Bus_5	Bus_2	10	−0.014	−0.034	2e−04	0.00039
Bus_5	Bus_2	11	−0.014	−0.034	2e−04	0.00039
Bus_6	Bus_2	12	−0.028	−8e−03	0	5e−04
Bus_6	Bus_2	13	−0.028	−8e−03	0	5e−04

3.2 MATLAB/Simulink Simulation Result of the Existing System

3.2.1 Existing System Steady-State Load Side Simulation Result

The existing system load side steady-state simulation result is shown in Fig. 5, and the results show the extent to which both the load voltage and current has been distorted.

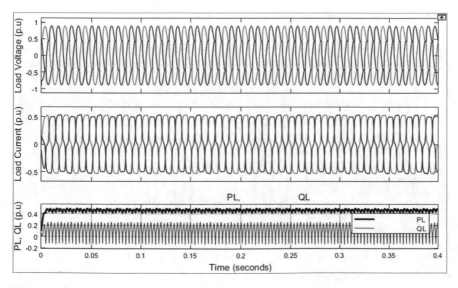

Fig. 5 Load side (base case) system simulation result

This is generally due to PQ issues in the system. Figure 5 shows both the voltage and current waveforms. It is clear that harmonics are detected in the system with real power of about 4.5 MW and the reactive power of about 2.3 MVAR. This shows that the PSAT simulation result is similar to the MATLAB/Simulink result.

A. Simulation of Voltage Sag and Voltage Swell

A fault was applied to the existing system, and simulations were performed. A result of 0.65 p.u sag was achieved as shown in Fig. 6.

Voltage swell was also simulated using a step signal and breaker for tripping OFF heavy loads for a duration of 0.1 s. The resultant waveform is shown in Fig. 7. The instantaneous tripping OFF loads of the system help to examine the capability of the

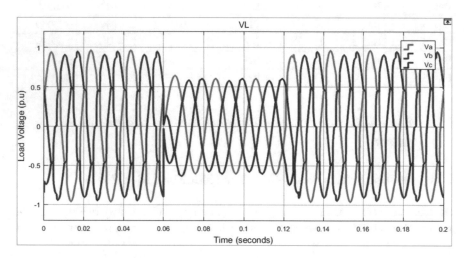

Fig. 6 Voltage sag simulation result without using UPQC

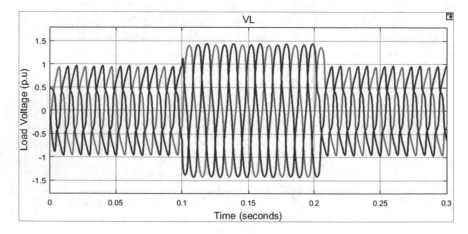

Fig. 7 Voltage swell simulation result without using UPQC

base case system to withstand voltage swell. As seen in Fig. 7, the base case systems voltage swell reaches up to 1.2 p.u for a duration of 0.1 s. The application of UPQC is used to reduce this effect.

3.3 Dynamic Analysis and Simulation Result with UPQC

The waveforms of the system after the insertion of UPQC are shown in Fig. 8. The results show that both the grid and load voltages are perfect at 1.0 p.u, and the load current magnitude is purely sinusoidal at 0.4 p.u. This therefore shows that UPQC has improved the systems PQ.

Voltage unbalance alleviation was achieved with the use of UPQC. The waveform in Fig. 9a shows the extent the system was affected by non-uniform distribution of single-phase loads among the three-phase supply. The results show that all the three-phase voltages are under the standard limit and unbalanced. But after the insertion of UPQC as shown in Fig. 9b, the load voltage becomes balanced at 1.0 p.u.

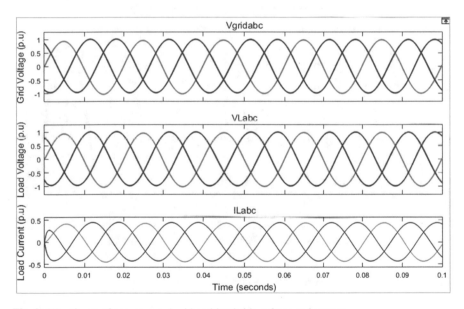

Fig. 8 Waveforms of compensated grid and load side voltage and current

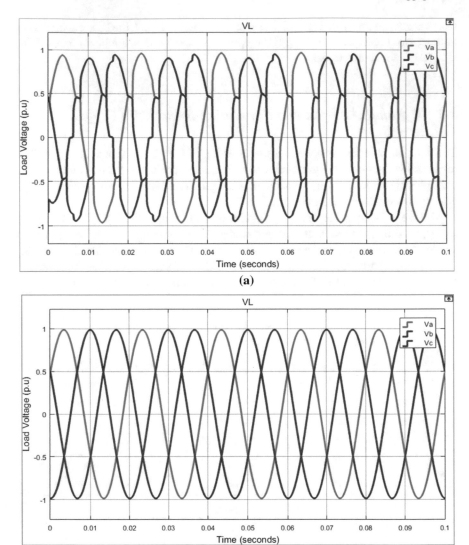

Fig. 9 Load voltage unbalance **a** without and **b** with

4   Conclusion

This paper presented a fuzzy logic controller (FLC)-based UPQC for the alleviation or regulation of voltage sag/swell and voltage unbalance. The proposed UPQC model mitigated all voltage quality-related issues in the investigated distribution system. The voltage sag and swell at the base case were 65% and 140%, respectively, with voltage unbalance, harmonic content, and other issues for a duration of 0.1 s. After the introduction of the FLC-based UPQC, both voltage sag and swell are mitigated to 100% and the variations are maintained and kept at 1.0 p.u. The overall results show that the UPQC is effective in regulating the voltage quality-related issues with the ability to inject partial real power aimed at minimizing power losses.

References

1. T. Kang, S. Choi, A.S. Morsy, P.N. Enjeti, Series voltage regulator for a distribution transformer to compensate voltage sag/swell. IEEE Trans. Ind. Electron. **64**(6), 4501–4510 (2017)
2. D.B. Aeggegn, A.O. Salau, Y. Gebru, Load flow and contingency analysis for transmission line outage. Arch. Electr. Eng. **69**(3), 581–594 (2020). https://doi.org/10.24425/aee.2020.133919
3. S.S. Rao, P.S.R. Krishna, S. Babu, Mitigation of voltage sag, swell and THD using dynamic voltage restorer with photovoltaic system, in *International Conference on Algorithms, Methodology, Models and Applications in Emerging Technologies (ICAMMAET)* (2017), pp. 1–7
4. M. Addisu, A.O. Salau, H. Takele, Fuzzy logic based optimal placement of voltage regulators and capacitors for distribution systems efficiency improvement. Heliyon **7**(8), e07848 (2021), https://doi.org/10.1016/j.heliyon.2021.e07848
5. R.K. Patjoshi, V.R. Kolluru, K. Mahapatra, Power quality enhancement using fuzzy sliding mode based pulse width modulation control strategy for unified power quality conditioner. Int. J. Electr. Power Energy Syst. **84**, 153–167 (2017)
6. M. Prasad, A.K. Akella, S. Das, S. Kuma, Voltage Sag swell and harmonics mitigation by solar photovoltaic fed ZSI based UPQC. Int. J. Renew. Energy Res. **7**(4), 1646–1655 (2017)
7. T.F. Agajie, A.O. Salau, E.A. Hailu, Y.A. Awoke, Power loss mitigation and voltage profile improvement with distributed generation using grid-based multi-objective harmony search algorithm. J. Electr. Electron. Eng. **13**(2), 5–10 (2020)
8. A.O. Salau, J. Nweke, U. Ogbuefi, Effective implementation of mitigation measures against voltage collapse in distribution power systems, in *Przegląd Elektrotechniczny* (2021), pp. 65–68. https://doi.org/10.15199/48.2021.10.13
9. A.O. Salau, Y.W. Gebru, D. Bitew, Optimal network reconfiguration for power loss minimization and voltage profile enhancement in distribution systems. Heliyon **6**(6), e04233. https://doi.org/10.1016/j.heliyon.2020.e04233
10. J.N. Nweke, A.O. Salau, U.C. Ogbuefi, "Bus voltage sensitivity index based approach against voltage collapse in distribution systems," in *International Conference on Decision Aid Sciences and Application (DASA)* (2021), pp. 1062–1066. https://doi.org/10.1109/DASA53625.2021.9682359.

Customer Perception, Expectation, and Experience Toward Services Provided by Public Sector Mutual Funds

R. Malavika and **A. Suresh**

Abstract In recent years, mutual funds have gained popularity as a means of ensuring the financial security. Mutual funds have benefited families in capitalizing on India's wealth while also contributing to the country's economic record. As knowledge and understanding of mutual funds grow, most people are reaping the benefits of investing in them. This study is focused on the various customer perceptions, expectations, and experiences toward services provided by public sector mutual funds. This study mainly analyzed the main features and major factors affecting the public sector mutual fund's customers' experience, expectation, and their perception. A quantitative study was undertaken through a structured questionnaire to analyze major factors that attracted customers toward public sector mutual funds, and based on the analysis, it was found that customers are satisfied because they have got better responsiveness from public sector mutual funds to customer complaints. As part of the study, many customers' opinions and reviews have been collected. This project will help public sector mutual funds to develop new strategies to create a more customer-friendly approach for increasing the sales and improving the customer satisfaction and their experience with public sector mutual funds.

Keywords Customer expectation · Customer perception · Customer experience · Mutual funds

1 Introduction

A mutual fund is a trust that pools the reserve funds of several investors with a common budgetary goal according to the website. The money is subsequently invested in various sorts of protections by the reserve director, based on the plan's goal. These could range from debentures to foreign exchange trading products. The

R. Malavika (✉) · A. Suresh
Department of Management, Amrita Vishwa Vidyapeetham, Amritapuri, Kollam, Kerala, India
e-mail: ramachandranmalavika@gmail.com

A. Suresh
e-mail: suresh@am.amrita.edu

most cost-effective investment for the average individual is a mutual fund, which allows them to invest in a better, carefully controlled portfolio with minimum effort. In a mutual fund, the asset director, who is otherwise called the portfolio administrator, exchanges the asset's fundamental protections, acknowledging capital increases or misfortunes, and gathers the profit or interest pay. The venture proceeds are then given to the individual financial backers. The worth of a portion of the common asset, known as the net resource esteem per share, net asset value (NAV) is determined everyday dependent on the all-out worth of the asset isolated by the quantity of offers at present given and remarkable. A trust that accumulates the reserve funds of numerous financial backers with a shared monetary purpose is called a common asset. This sum of money is being put into a project with the goal of reaching a specified goal. The joint reserve obligation is common, and the asset, for example, has a place among all financial backers. The money is subsequently invested in capital market assets such as offers, debentures, and various protections. A mutual fund is a form of financial instrument that pools money from several different participants in order to achieve a shared objective. The funds are then invested in a variety of assets.

A mutual fund is a type of investment vehicle that gives small investors access to a diverse selection of assets, securities, and other forms of protection. Diversification decreases risk because all stocks cannot move in the same direction in the same proportion around the same time. Investors are given units determined by the amount of money they have placed in mutual funds. Unit holders are people who invest in mutual funds. When an investor buys units in a mutual fund, he becomes a part owner of the fund's assets in the same proportion as the sum he contributes to the corpus (the total amount of the fund). A mutual fund investor is also known as a unit holder or a mutual fund shareholder. The growth of mutual funds is of three stages: The Unit Trust of India, that had an ultimate resource of Rs. 6700 crores at the end of 1988, was the lone player throughout the first stage, which took place from 1964 and 1987. The next stage occurred from 1987 and 1993, during which time eight funds were established (6 by banks and one each by LIC and GIC). By the end of 1994, the resources available under management had grown to 61,028 crores, with 167 programs in place. The third stage started with the passage of private and unfamiliar areas in the mutual fund industry in 1993. Kothari Pioneer Mutual Fund was the principal fund to be set up by the private sector in a relationship with an unfamiliar fund. As toward the finish of the monetary year 2000 (31st walk), 32 funds were working with Rs. 1,13,005 crores as complete resources under administration. As of the end of August 2000, there were 33 funds under management, with 391 plans and capital totaling Rs 1,02,849 crores. In 1993, the Securities and Exchange Board of India (SEBI) issued a comprehensive guideline that defined the architecture of Mutual Fund and Asset Management Companies. In 1993 and 1994, a few private segments mutual funds were distributed. From that point on, the number of private participants has rapidly increased. In India, there are 34 mutual fund companies that manage a total of 1,02,000 crores. An investor can create many folios.

1.1   Objectives

- To identify the customer perception and experience about the services provided by public sector mutual funds.
- To know about the preferences of investors toward mutual funds with special reference to public sector mutual funds as well as in other mutual fund companies.
- To evaluate the factors affecting the preference of various schemes.
- To determine the level of investor satisfaction with public sector mutual funds.

1.2   Scope of the Study

The study has got a very significant scope to add the advancement of public sector mutual funds by determining the customers' interests and expectations in terms of portfolio, mode of investment, choice, and rate of return. The research can be used to help the organization make better decisions and prepare its marketing campaign. The analysis will be useful in determining the respondents' general knowledge of mutual funds.

2   Literature Review

Following studies done by researchers on mutual fund have been referred for this study.

For most of the households, mutual funds are a good source of returns. It is especially beneficial to those who have reached retirement age. Ordinary investors are still limited to traditional investments such as gold and fixed deposits. This is due to a shortage of knowledge about how mutual funds operate. In reality, many people who invest in mutual funds do not understand how they function or how to administer them. As a result, organizations that offer mutual funds must provide prospective investors with comprehensive information about mutual funds [1].

India's mutual fund business has a big untapped market. More consumers are relying on low-cost, low-risk professional management of their assets. As more funds enter the business, these companies' strategic marketing strategies are becoming increasingly important for their survival. The success of a mutual fund is contingent on a thorough knowledge of the psychology of small investors. Saving products with varying risk–return combinations compete with MF businesses. Investors have become more vigilant and selective [2].

A mutual fund is the best investment for the ordinary individual since it allows them to invest in a diversified, professionally managed range of assets at a cheap cost. Investors buy units in a mutual fund scheme with a specific investment objective and strategy in mind. The income generated by these assets, as well as any capital gains realized by the plan, is distributed to the scheme's units in proportion to their ownership [3].

Customers of mutual funds are willing to invest in mutual funds if they can guarantee a link to the equity market, given that investing in mutual funds yields higher returns than investing in gold, according to the findings of the study. The primary motivation for developing a mutual fund is that it is tied to the stock market and can guarantee a particular rate of return. Consumers of mutual funds regard them as a superior investment alternative than other investment options available, and they choose to invest in them over gold. Mutual fund investment has been reported to be treated as a channel for investors and is capable of attracting clients [4].

When it comes to making investments, most investors relied on historical mutual fund performance. Prospectuses/newsletters have been the most popular source of information for investors. The total return on mutual fund schemes has been used by the majority of investors to measure their performance. Investors in public sector mutual funds expressed dissatisfaction with the services given, while private sector mutual fund investors cited a lack of awareness as a key issue [5].

The purpose of this study is to determine the level of mutual fund knowledge among investors. Investor perceptions and opinions on topics such as mutual fund types, satisfaction levels, and the function of financial advisors and brokers have all been investigated. Investors' views on elements that entice them to invest in mutual funds, sources of information, shortcomings in mutual fund managers' services, issues facing the Indian mutual fund business, and so on [6].

The benefits of mutual fund investment may be divided into many categories. This includes the safety of money invested in mutual funds, the fund's/favorable scheme's credit rating from recognized credit organizations, full disclosure of all relevant information, and frequent updates. The financial advantages offered by funds/schemes in the form of wealth development, liquidity, return on investment (ROI), early bird rewards, fringe perks, and charge discounts are the next category of financial benefits [7].

3 Research Methodology

3.1 Methodology Adopted

Sampling procedure: The sampling procedure used in this study is judgmental sampling, also called purposive or authoritative sampling.

Sample Size: The sample size is 250 which consists of public sector mutual funds investors in Kerala.

Sampling Unit: These comprise customers of public sector mutual funds in Kerala.

Sampling Area: The area of the research was concentrated mainly in Kerala, India.

3.2 Data Collection

Data collection was done using questionnaires. The questionnaire was framed in an easily understandable way, so the respondents may not have any difficulty in answering them. A total of 18 questions make up the questionnaire. There is also 1 suggestion question to get the opinion and suggestions of the people regarding public sector mutual funds.

4 Analysis and Interpretation

4.1 Customer Perception

- *Time period that you would like to stay invested*

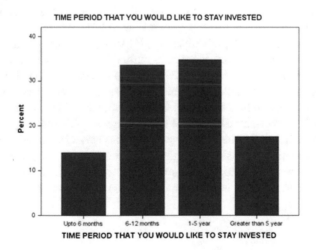

Interpretation

The above chart shows that 14% of the customers like to stay invested for time period up to 6 months, 33.8% of the customers like to stay invested for time period 6–12 months, 34.8% of the customers like to stay invested for time period from 1 to 5 years, 17.6% of the customers like to stay invested for time period greater than 5 years.

- *Features that attracted customers to public sector mutual funds*

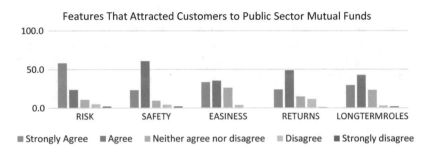

Interpretation

The goal of the study was to discover the characteristics that drew customers to public sector mutual funds and kept them pleased with their investment. Among the 250 respondents, 58.8% strongly agree, 23.6% agree, 10.8% have neutral opinion, 5.2% disagree, 2% strongly disagree that the low risk is a feature which is better in mutual funds. Among the 250 respondents, 23.2% strongly agree, 60.8% agree, 9.6% have neutral opinion, 4.4% disagree, 2% strongly disagree that the safety of their investment is a feature which is better in mutual funds. Among the 250 respondents, 3.6% strongly agree, 35.6% agree, 29.4% have neutral opinion, 4.0% disagree, 4% strongly disagree that the ease of their investment procedure is a feature which is better in mutual funds. Among the 250 respondents, 24% strongly agree, 48.8% agree, 14.8% have neutral opinion, 11.0% disagree, 0.8% strongly disagree that the returns from their investment as a feature which is better in mutual funds. Among the 250 respondents, 29.6% strongly agree, 42.8% agree, 23.2% have neutral opinion, 2.8% disagree, 1.6% strongly disagree that they can achieve their long-term goals with their investment procedure as a feature which is better in public sector mutual funds.

- *Type of mutual fund that is preferred*

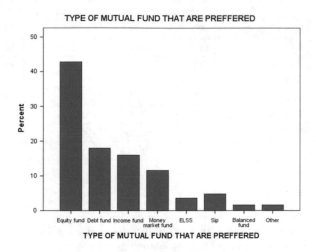

TYPE OF MUTUAL FUND THAT ARE PREFFERED

Interpretation

From the above analysis, 42.8% customers have interested in equity funds, 18% invest in debt fund, 16% invest in income fund,11.6% of customers invest in money market fund, 3.6 invest in ELSS, 4.8 invest in SIP, 1.6% invest in balanced fund and 1.6% in other schemes.

Customer Expectations

- *Improvements are needed in the service provided by public sector mutual funds*

		Frequency	Percent	Valid percent	Cumulative percent
Valid	Yes	91	36.4	36.4	36.4
	No	159	63.6	63.6	100.0
	Total	250	100.0	100.0	
Total		250	100.0		

Interpretation

From the above chart, 36% of the customers suggest that they need some improvements and 64% of the customers do not prefer any improvements or suggestions for the services of public sector mutual funds. They are satisfied with the current services.

- *Responsiveness of public sector mutual funds toward mutual fund*

		Frequency	Percent	Valid percent	Cumulative percent
Valid	Strongly agree	96	38.4	38.4	38.4
	Agree	111	44.4	44.4	82.8
	Neither agree nor disagree	37	14.8	14.8	97.6
	Disagree	3	1.2	1.2	98.8
	Strongly disagree	3	1.2	1.2	100.0
	Total	250	100.0	100.0	
Total		250	100.0		

Interpretation

The above graph shows that 38% strongly agree that the public sector mutual funds are highly responsive toward the complaints of the customers, 44.4% agree that the public sector mutual funds are highly responsive, 14.2% neither agree neither agree nor disagree they have a neutral statement toward the responsiveness of the public sector mutual funds and 1.2% disagree and strongly disagree that is the public sector mutual funds is highly responsive toward the complaints of the customers.

4.2   Customer Experience

- *Investment time period of customers*

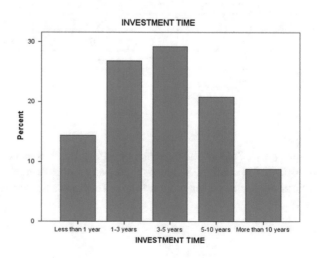

Interpretation
According to the graph above, 14.1% of consumers are investing for less than one year, 26.8% for one to three years, 29.2% for three to five years, 20.8% for five to ten years, and 8.8% for more than ten years.

- *Investment amount of respondents*

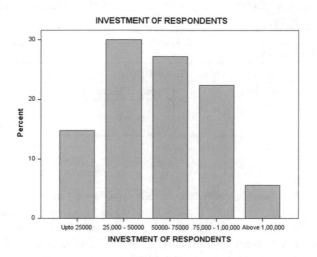

Interpretation
The above chart shows that 14.8% of the investors invest up to 25,0000 rupees, 30.0% of the investors invest from 25,0000–50,000 rupees, 27.2% of the investors invest from 50,000–75,0000 rupees, 22.4% of the investors invest from 75,0000–1,00,000 rupees, 5.6% of the investors invest above 1,00,000 rupees.

- *Level of satisfaction with the services provided*

		Frequency	Percent	Valid percent	Cumulative percent
Valid	Very satisfied	111	44.4	44.4	44.4
	Satisfied	97	38.8	38.8	83.2
	Neutral	28	11.2	11.2	94.4
	Dissatisfied	10	4.0	4.0	98.4
	Very dissatisfied	4	1.6	1.6	100.0
	Total	250	100.0	100.0	
Total		250	100.0		

Interpretation
The above chart shows that 44.4% of customers are highly satisfied with the services provided by public sector mutual funds, 38.8% of customers are satisfied with the

services provided by public sector mutual funds, 11.2% are neutral with their opinion, 4.2% are dissatisfied, and 1.6% of customers are very dissatisfied with the services provided by public sector mutual funds.

- *Factors prompted you to invest in public sector mutual funds*

For factors that influence characteristics that attracted people to public sector mutual funds, a principal component analysis with varimax rotation yielded the following findings.

KMO and Bartlett's test		
Kaiser–Meyer–Olkin measure of sampling adequacy		0.540
Bartlett's test of sphericity	Approx. Chi-square	9.198
	Df	6
	Sig	0.013

Interpretation

Out of the 250 samples collected, the Kaiser–Meyer–Olkin measure of sampling adequacy is 0.5 so the sample is adequate and has a p value less than level of significance so that the value is significant. There are two components: complaints and services obtained from the factor analysis. 63% of customers agree that they consider low risk as a factor that prompted them to stay invested in public sector mutual funds, 54% customers agree that they consider better service as a factor, 43% agree for solving complaints, and 52% customers agree that they consider financial discipline as a factor that prompted them to stay invested in public sector mutual funds.

5 Findings

- Most of the male customers from age group 4–55 and above 55 invest an amount of 25,000–50,000 and female customers of age group 45–55, and there are less customers at age group below 25 invest above 1,00,000.
- Most customers (31%) become mainly aware about public sector mutual funds through newspapers, journals, articles. 22% were influenced by television, 17.6 were influenced by public sector mutual funds employees and 14.8% were influenced by Internet, friends, or relative toward public sector mutual funds.
- Government employees have more investment from the amount of 25,000–50,000, and homemakers have the least investment of up to 25,000 in public sector mutual funds.
- Investment time period is maximum for 5–10 years with an amount of 50,000–75,000, and 25,000–50,000 is the minimum for a time period of less than 1 year.
- The main feature that attracts customers to public sector mutual funds is reduction in risk and cost from investment time period of 5–10 lakhs, safety principle from

investment time period of 5–10 lakhs, easiness to access from investment time period of 5–10 lakhs, returns on investment from investment period from 1 to 5 lakhs, long-term goals from investment period 5–10 lakhs.

- The customers mostly prefer to stay invested (12%) from 3 to 1 year for an amount of 25–50 thousand. And the least preference of investment (5.2%) is for an amount of 75,0000–1,00,000 from a time period of 3–5 years and 25,000–50,000 for less than 1 year.
- Government employees have mainly attracted by the features of safety principle, easiness to access, achieving long-term goals, reduction in risk and transaction cost as a customer of public sector mutual funds
- Safety principle as the main feature attracted by male and female customers to public sector mutual funds and ease of access as the least attractive feature by male and female customers to public sector mutual funds.
- Low risk was recognized as a major factor that influenced consumers to invest in public sector mutual funds among respondents aged 45–55, while addressing complaints was indicated as a minor factor that influenced customers aged 55 and above to participate in public sector mutual funds.
- The majority of male investors cite low risk as a key element in getting consumers to buy in public sector mutual funds, while female investors cite low risk and financial discipline as key reasons in getting people to engage in public sector mutual funds.
- Customers become aware about public sector mutual funds through newspaper, journal, articles, and public sector mutual funds employee as the mode of information by female respondents mainly in equity and income fund and male customers become aware about public sector mutual funds through public sector mutual funds employees as the mode of information through equity fund by female respondents.
- The greatest investment time period was from 3 to 5 years of investment amount 50,000–75,000 and need to stay updated to 6–12 months. The lowest investment time period was more than 10 years for investment amounts above 1,00,000 and needed to stay updated to 1 to 5 years.
- From the analysis, it is identified that customers are satisfied because they have got better responsiveness from public sector mutual funds to customer complaints.

6   Conclusion

The primary objectives were to determine consumer perceptions and experiences with public sector mutual funds as well as to analyze customer experiences with public sector mutual funds. The secondary goal is to learn about investor preferences for mutual funds, with a focus on public sector mutual funds and other mutual fund companies, as well as to investigate the factors that influence investor preferences for various schemes, as well as to learn about the extent of investor satisfaction with public sector mutual funds and to scrutinize investor satisfaction with service quality provided by public sector mutual funds. Most of the customer's perception and expectation about the services provided by public sector mutual funds have been met with their experience from the services provided by public sector mutual funds, and also customers were highly satisfied with the services provided through this study and to find out the customer perception and experience about the services provided by public sector mutual funds and the experience obtained from the public sector mutual funds by the customers are satisfied with the services provided. The customers have satisfaction; hence, they have better responsiveness from public sector mutual funds to customer's complaints. Most customers prefer public sector mutual funds because they are satisfied that their perception and expectation have been met with their experience. Customers were extremely satisfied with the services, and they are highly interested to invest in other schemes. So it can be concluded that when there is proper awareness and knowledge about mutual funds then the customer will be able to understand all aspects and also make them understand the benefit. The customers have been influenced by the features that prompted public sector mutual funds and investment factors that prompted throughout their investment and also from service provided by the employees have made a great empire to the customers.

References

1. A. Naik, S.G. Pramod, A study on investors' perception towards mutual funds with due reference to 'SBI Mutual Funds'
2. D.K.P. Shah, A study of consumer activities towards SBI mutual funds company with reference to Nadiad City (2020)
3. S. Saha, M. Dey, Analysis of factors affecting investors' perception of mutual fund investment. IUP J. Manage. Res. **10**(2) (2011)
4. R. Sharma, Investment perception and selection behaviour towards mutual fund. Int. J. Techno-Manage. Res. 1 (2013)
5. N.J. Bhutada, D. Pingale, An exploratory study on perception of customers and consumer behavior towards mutual fund
6. C. Madhavi, Performance and evaluation of public sector mutual funds in India
7. K. Rajesh, N. Goel, An empirical study on investor's perception towards mutual funds. Int. J. Res. Manage. Bus. Stud. 1 (2014)

DigiDrive: Making Driving School Management Effective

Manasi Khanvilkar, Atreya Rastradhipati, and Wricha Mishra

Abstract The Indian driving school system consists of schools ranging from small-scale setups to huge franchises with branches spread nationwide. Irrespective of the scale of the school, it needs to have a management system to handle its student and instructors data to ensure seamless communication. In today's world, technology has transformed how industries handle their data management processes. However, from this study, it was observed that the Indian driving schools follow conventional and manual methods for managing their data which generate possibilities for errors. Furthermore, it also leads to several communication problems among the users. This study explores the data management methods, and the problems faced in driving schools through contextual inquiry, semi-structured interviews, and competitive analysis. Based on the problems identified, design intervention is proposed consisting of an in-vehicle infotainment system linked to a mobile and Web application to provide a seamless management experience.

Keywords Data management methods · IVI systems · Driving schools

1 Introduction

Driving is a critical maneuver that provides an individual with a sense of responsibility, freedom, and confidence. Although, according to World Road Statistics 2018, India ranks 1st in the number of road accident deaths across the 199 countries followed by China and US. As per the WHO Global Report on Road Safety 2018, India accounts for almost 11% of the accident-related deaths in the world. Vehicles driven by untrained and unqualified drivers are a serious traffic hazard and can cause accidents, death, and injuries. Driving without valid license/learner's license accounts for about 15% of accidents [1]. One safety measure to prevent driving accidents is effective driving education of the drivers. Effective training of the drivers, related to the increasing tolerance for roadside stimuli and other road environment-related

---

M. Khanvilkar (✉) · A. Rastradhipati · W. Mishra
MIT Institute of Design, MIT-ADT University, Pune, MH 412201, India
e-mail: manasikhanvilkar0906@gmail.com

© The Author(s), under exclusive license to Springer Nature Singapore Pte Ltd. 2023   393
Y.-D. Zhang et al. (eds.), *Smart Trends in Computing and Communications*, Lecture Notes in Networks and Systems 396, https://doi.org/10.1007/978-981-16-9967-2_37

aspects, helps in creating a positive attitude and safe response among the drivers, e.g., for reacting safely toward the roadside stimuli which in turn reduces the chances of aggressive driving and increases road safety [2]. Driving schools provide driving education to prepare a new driver to obtain a learner's permit or driver's license. Due to increasing awareness around driving education, there has been an increase in the number of driving schools and driving learners. As a result, the driving schools need a management system to manage all the students' and instructors' data. A management system is a set of policies, processes, and procedures used by an organization to ensure that it can fulfill the tasks required to achieve its objectives [3]. The Indian driving school system consists of small-scale setups to huge franchises with branches spread nationwide. There have been several studies concerning the domain of driving schools, e.g., use and effectiveness of driving simulators [4–6]; driving license testing procedures [7]; formation of driving skills [8], etc. However, the previous work deals with the learning methods in driving schools, and there is a lack of research on the data management methods in the domain of driving schools in India. This paper aims to understand different data management problems faced by the driving schools and propose a design intervention to minimize them.

2 Methods

Since the study targeted the driving schools domain as a whole and not a specific user segment, all three stakeholders - driving school managers, driving school instructors, and students were considered. A driving school manager is responsible for the administrative and organizational tasks in a driving school. Driving school instructors are responsible for teaching the students. The students are the users who pursue driving education from the driving schools. Figure 1 shows the methods followed during this study.

A contextual inquiry was performed across 12 driving schools in Pune. Contextual inquiry is a widely used process, to gather field data from users with the aim of understanding who the users are and how they work in their day-to-day basis [9]. The visited driving schools included small-scale setups as well as large franchises. Selection of the driving schools was done through convenience sampling. During the contextual inquiry, 20–30-min-long semi-structured interviews were conducted with 12 driving school managers and 10 driving instructors from those driving schools. The

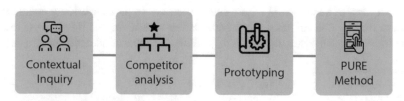

Fig. 1 Methods used in the study

interview questions were based on attendance management, tracking student progress and fee payments, license procedure, use of technology in teaching, selection of road routes while learning, and their session scheduling process. To understand the system from the students' perspective, 15 students who had attended a driving school were interviewed. The interview questions were based on satisfaction with the teaching methods, communication with the driving school, etc.

In order to design an intervention that aimed to solve the identified problems, a stakeholder's analysis was conducted to identify the role and influence of each stakeholder group in the driving school. Stakeholder analysis aims to evaluate and understand stakeholders from the perspective of an organization, or to determine their relevance to a project or policy [10].

After identifying the problems, a competitor's analysis was conducted to understand the existing solutions in the market. Competitor analysis is a method for identifying the strengths and weaknesses of competing products or services before starting work on prototypes [11].

Since usability testing holds great importance in a product development life cycle, the developed designs were tested with 5 users, who were in the process of learning to drive from a driving school, or had gone through the process in the near past. In this case, pragmatic usability rating by experts (PURE) method was used. PURE is a usability-evaluation method in which usability experts assign one or more quantitative ratings to a design based on a set of criteria and then combine all these ratings into a final score and easy-to-understand visual representation [12].

3   Results

3.1   Contextual Inquiry

A flow diagram (Fig. 2) was created based on the insights from the contextual inquiry to understand the interactions that take place in a driving school. It appears that the manager is an important stakeholder in the system as they carry out the critical tasks of communicating between the students, instructors, and any external entities in the system.

Fig. 2 Flow diagram for driving school managers

3.1.1 Driving School Managers and Instructors

Attendance Recording and Student Data Management Methods
Most schools used attendance cards (75% schools), register books (20% schools), excel sheets (5% schools), and other conventional methods to record student attendance (Fig. 3). The handling of student data (documents, forms, etc.) was done manually by maintaining physical records. On enquiring about the issues faced due to these methods, problems like losing data, difficulty in accessing records, space consumption due to physical records, etc., were mentioned.

Teaching Methods in Driving Schools
Figure 4 shows the teaching methods provided by the driving schools. All the schools offer physical training (driving a real car), which is the most crucial part of the learning process. Whereas, methods like online/offline study material (books, videos on driving, traffic rules, etc.), car physical study (explaining internal functioning of a car), and driving simulators are offered in selective driving schools.

Overview of User Groups in Driving Schools
There were some institutes with a large number of cars, students, and limited managers and instructors. This makes managing the data and sessions of all the students, maintenance of cars, appointment of instructors to students, providing sufficient student details to the instructors highly inconvenient for the managers. It also leads to several miscommunication problems. In case of emergencies, the managers face difficulties to track their vehicles and the respective instructors.

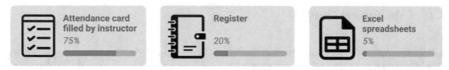

Fig. 3 Attendance recording and data management methods

Fig. 4 Teaching methods in driving schools

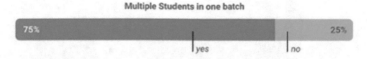

Fig. 5 Multiple students in one batch

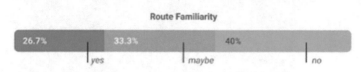

Fig. 6 Route familiarity

3.1.2 Driving School Students

Physical Training in Batches
From the student interviews, it was observed that 75% of students were provided physical training in batches along with other students, and these batches consisted of 2 to 4 students (Fig. 5).

Locating Driving Routes
Most driving schools teach driving on main roads. Since the traffic, pedestrians, and road conditions cannot be estimated beforehand, the students get stuck on the roads which lead to a reduction of the overall session time, thereby affecting student satisfaction. It was observed that around 40% of students practiced on routes that were not familiar to them (Fig. 6).

3.2 Stakeholder's Analysis

These are the insights from the stakeholder' analysis (Fig. 7).

3.2.1 Driving School Managers

The managers have the highest amount of responsibility among all the stakeholders, since they oversee the administrative and organizational tasks of the entire driving school. Hence, they become critical stakeholders in the intervention.

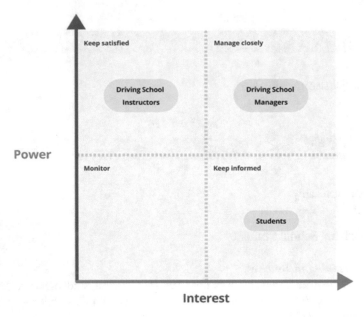

Fig. 7 Stakeholder's analysis

3.2.2 Driving School Instructors

Even though the instructors do not have responsibilities as critical as the manager, they are important stakeholders in a driving school. They have the primary responsibility of a driving school, imparting driving education. Hence, the intervention should equip the instructors to carry out their primary responsibility with ease.

3.2.3 Students

The students are at the receiving end of the driving school services. As a result, their interaction with the intervention would only include information relevant to them such as progress, performance, fees, etc. Their learning experience with the driving school will influence the overall performance and reputation of the driving schools. Hence, the intervention should keep the students well-informed.

3.3 Competitor's Analysis

After understanding the issues, a competitor's map and a competitor's analysis were conducted to understand the existing solutions in the market. From the competitor's map (Fig. 8), several available solutions for tasks like attendance management,

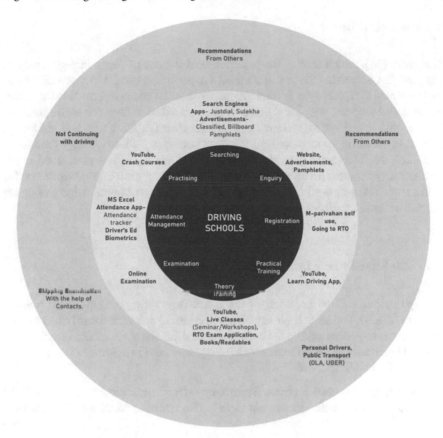

Fig. 8 Competitor's map

theory training, license registration, etc., were discovered. The competitor's analysis compared the existing popular driving school management platforms. Table 1 shows the key highlights of the competitor's analysis.

Table 1 Competitor's analysis key highlights

Features	Drive Scout	TeachWorks	DSManager	Microsoft Excel
Creating a custom learning plan	Yes	No	No	No
Class scheduling	Yes	Yes	Yes	No
Personal student account	No	Yes	No	No
Manage car schedule	Yes	No	Yes	No
Access for instructors	No	Yes	No	No

4 Intervention

The paper [13], which studies the persistence of the use of paper in e-governance, suggests that, in making the transition from a paper-based system to a new electronic system aimed at replacing or minimizing paper, internal actors responsible for implementing and working with for the new system are asked to abandon well-rehearsed routines and longstanding systems, and transfer their faith to a new and unfamiliar system, establishing new routines or integrate new systems into existing routines.

Hence, the intervention we are proposing to tackle the issues mentioned in the results is a system that involves a user interface suited for each user group. Since the data are updated across all three interfaces, it reduces miscommunication among the users. The three interfaces are as follows:

4.1 In-Vehicle Infotainment (IVI) System

The IVI system will be accessed by the instructors during their physical training sessions (Fig. 9). It aims to limit any interaction before and after the driving sessions to reduce the cognitive load on the users during the driving session. The interface was designed with adherence to the Android Auto guidelines [14]. Some of the important features of this interface are

- **Daily Schedule and Session Summary**—The instructor's schedule will be displayed along with the student details like name, contact number, learning

Fig. 9 IVI system screens

module, etc., after logging in. Once a physical training session is completed, the summary of the session is displayed and is updated on the mobile and Web applications.

- **Real-Time Location Sharing**—The real-time location of the vehicle is shared with the manager on the Web application along with the details of the instructor in the vehicle. Similarly, if another vehicle from the same driving school is detected in the vicinity, it will be displayed on the IVI system.
- **Learning Prompts**—This feature intends to make the learning experience more engaging. It provides real-life driving scenarios on the screen based on the route. The instructor can read out and teach the right approach to deal with such situations. E.g., if the map detects a school nearby, it will show the prompt "what should you do when you see a traffic sign of a school nearby?" The instructor can choose to get voice-based prompts to avoid looking at the screen.
- **Calling Feature**—The interface provides the instructor an option to call the students. There is also an emergency calling feature, which connects the instructor with the driving school contact.

4.2 Web-Based Application

The managers interact with the system using the Web application. They have access to all the students' data, instructors' data, schedules, appointments, etc. Saving the documents online will mitigate the need to maintain physical records. They can process payments, track vehicles, send reminders to the students, track student progress, auto-schedule sessions, etc. The Web app will provide a one-stop solution to the manager for managing all the tasks in the school, thereby reducing the need for multiple applications for managing each aspect. The manager will also have the ability to track the real-time location of all the vehicles through this platform.

4.3 Mobile Application

The students will be accessing this system through the mobile application. The application will provide them with all the details regarding their learning sessions, license appointments, instructor contacts, etc. They will also get access to theory learning content like videos, articles, etc. They can contact the instructor and the driving school through the app. As their driving sessions are completed, the summary of those sessions is updated on the app. This way, the students can observe their own progress. The ubiquitous nature of mobile phones makes this an ideal interface for students since the students come from varying age groups, socio-economic backgrounds, etc.

Fig. 10 PURE method
results

5 User Testing and Results

The mobile application was tested with 5 users, who were in the process of learning to drive, or had gone through the process in the near past. They were asked to use the designed application and were observed while performing the tasks. Observations were converted into a numeric scale and put in a graphical form where it could be easily visualized. User's verbal feedback was also noted at each step, thus creating a balance between qualitative and quantitative study. Figure 10 shows the results from the PURE method testing.

After the user testing, it was found that users were having difficulty in using the application when they were provided with more complicated visualizations like calendar or reports, where as they thought visibility of different features was good and made learnability of the application fairly easy. Some users pointed out concerns regarding the use of various gradients across the application, which would hinder the visibility of the content. This gives us an insight that crucial screens of the application where PURE analysis suggests users are finding difficulty needs to be reworked and further tested. Figure 11 shows the mobile application screens after the iterations based on the feedback.

6 Discussion

Despite the abundance of research in the domain of driving schools, there is a lack of research that deals with the data management methods in driving schools. This paper explored the different data management methods and problems faced by the driving schools and proposed a design intervention to minimize them. Through contextual inquiry and semi-structured interviews, it was revealed that the driving schools conform to conventional data management methods such as attendance cards,

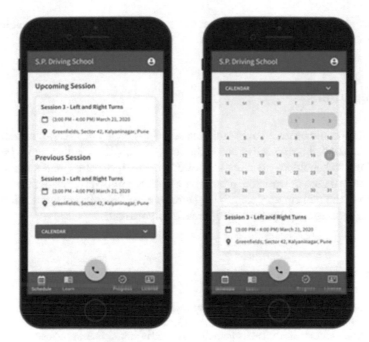

Fig. 11 Mobile application screens

registers, excel sheets, and storing physical documents. This leads to several data management problems like losing data, difficulty in accessing records quickly, space consumption due to physical documents, miscommunication among the stakeholders, etc. The competitor's analysis reflected insights about existing solutions for these problems and the scope for an intervention.

The proposed design intervention consists of in-vehicle infotainment (IVI) system linked to a mobile and Web application. The intervention was built in consideration to the observation from the paper [13] that designers should aim to make the transition between e-governance systems easy for both internal actors and end-users, and recognize that new systems will co-exist with the existing ecosystem long past the planned transition period. As a result, the intervention builds on the existing data management methods of the driving schools, by transitioning them to a digital platform. The intervention is designed with regards to the needs of all the stakeholders - driving school managers, driving instructors, and students. The IVI system focuses on minimizing the interactions and cognitive load during the driving sessions. The Web application and mobile application facilitate seamless communication between the students and the driving school.

The future scope of this study would encompass testing the proposed intervention across varying school sizes in India based on their infrastructure, the number of users, financial capacity, etc., and iterating based on the feedback. The intervention can expand its interaction beyond the existing interfaces like IVI systems, mobile, and Web applications, incorporating more advanced technologies like voice user interfaces, augmented reality, etc.

7 Limitations

The solution could be tested only with one set of stakeholders, i.e., students. In order to achieve a holistic perspective about the intervention, the solution needs be tested with all the stakeholders—driving school managers, instructors, and students. Due to the remote nature of work, the prototype was tested remotely. Testing the interface solutions in the real context, i.e., IVI system in a car, Web-based application in the driving school, etc., will provide deeper usability insights.

References

1. Government of India, Ministry of Road Transport and Highways, Transport Research Wing, New Delhi. *Road Accidents in India—2019*. (New Delhi, 2020)
2. A.S. Neelima Chakrabarty, Driver training: an effective tool for improving road safety in India. J. Eng. Technol. **2**(2) (2012)
3. Management System, Retrieved from Wikipedia: https://en.wikipedia.org/wiki/Management_system (n.d.)
4. C.F. Joost, S.M. de Winter, Advancing simulation-based driver training: lessons learned and future perspectives, in *MobileHCI '08: Proceedings of the 10th International Conference on Human Computer Interaction with Mobile Devices and Services* (2008), pp. 459–464
5. F.B. Pierro Hirsch, Transfer of skills learned on a driving simulator to on-road driving behavior. Transport. Res. Rec.: J. Transport. Res. Board **2660**, 1–6 (2017)
6. P.A. Gunhild Birgitte Sætren, Simulator training in driver education—potential gains and challenges, in *Safety and Reliability–Safe Societies in a Changing World*, (2018), pp. 2045–2051
7. I.M. Akshay Uttama Nambi, ALT: towards automating driver license testing using smartphones, in *SenSys '19: Proceedings of the 17th Conference on Embedded Networked Sensor Systems*, (2019), pp. 29–42)
8. V. Dmitry, K.E. Gribanov, Formation of driving skills in driving schools. **15**, 123–126 (2019)
9. K.H. Hugh Beyer, *Contextual Design: Designing Customer-Centered Systems* (Elsevier Science & Technology, 1997)
10. R.F. Brugha, Z. V., Stakeholder analysis: a review. Health Policy Plan. **15**(3), 239–246 (2000)
11. *Competitor Analysis*. (n.d.). Retrieved October 2021, from Usability Body of Knowledge: http://www.usabilitybok.org/competitor-analysis
12. C. Rohrer, *Quantifying and Comparing Ease of Use Without Breaking the Bank*. Retrieved from Nielsen Norman Group: https://www.nngroup.com/articles/pure-method/ (2017)
13. P.C. Megh Marathe, Officers never type: examining the persistence of paper in e-governance, in *CHI '20: CHI Conference on Human Factors in Computing Systems* (2020), pp. 1–13
14. *Android Auto*. (n.d.). Retrieved from Google Design for Driving: https://developers.google.com/cars/design/android-auto

Revamping an E-Application for User Experience: A Case Study of eSanjeevaniOPD App

Remya Vivek Menon and G. Rejikumar

Abstract E-governance initiatives are likely to succeed only if the applications created for those purposes offer an excellent user experience (UX). For UX to improve, many components associated with the application should meet expectations and make user interactions easier, intuitive, and relaxing. This study aimed to verify the user experience developed by the 'eSanjeevaniOPD' app and suggest ways to improve it. The sequential incident technique (SIT) revealed user concerns in every stage of the user journey and, therefore, improvement opportunities. Accordingly, a few attributes were chosen to improve UX by providing them in the best possible manner. The Taguchi experiment with ten selected attributes by capturing user perceptions identified an optimum combination of these attributes for maximum UX.

Keywords eSanjeevaniOPD · User experience · Taguchi experiment · App design

1 Introduction

E-governance involves the application of information technology for efficient, effective, transparent, flexible, and stakeholder-friendly implementation of various public policy initiatives for positive outcomes [1, 2]. Technology contributes to achieving several goals related to E-governance such as enhancing the quality of service delivery to citizens, improving interactions with government and stakeholders, accelerating information access to citizens, and eradicating social inequalities [3]. For example, E-governance initiatives in health care are vital to upholding health equity by removing barriers to healthcare access and ensuring safe management of individual health records, society health data, and healthcare information systems [4].

R. V. Menon (✉)
Department of Management, Amrita Vishwa Vidyapeetham,
Amritapuri, Kollam, Kerala 690525, India
e-mail: remyavmenon@am.amrita.edu

G. Rejikumar
Department of Management, Amrita Vishwa Vidyapeetham, Kochi 682046, Kerala, India
e-mail: g_rejikumar@asb.kochi.amrita.edu

UN Sustainable Development Goal 3 emphasizes the need for everybody to live a healthy life and highlights the well-being of everyone on the planet [5]. To ensure well-being as specified in SDG3, eliminating barriers to healthcare access [6] and achieving health equity [7, 8] are an essential prerequisite. The integration of information and communication technologies (ICT) into health care provides better governance of the healthcare systems in the country and helps to (1) improve the overall quality and efficiency; (2) reduce healthcare expenses; (3) improve healthcare access to a wider population; (4) minimize health disparities [9].

1.1 'eSanjeevaniOPD' for E-governance in Indian Health Care

To vitalize the digital health ecosystem and minimize healthcare access disparities and achieve the National Digital Health Mission, the government of India introduced the 'eSanjeevaniOPD', National Telemedicine Service, to citizens for outpatient health services, in November 2019. The 'eSanjeevaniOPD' is a doctor-to-patient telemedicine system to provide health advice and prescriptions through digital support without visiting the hospital. Telemedicine provides medical services remotely using telecommunications networks [10]. It helps to address the unequal distribution and shortages in the healthcare system by connecting rural patients to medical specialists and doctors in metropolitan areas. The 'eSanjeevaniOPD' platform has completed 120 lakh consultations as of September 2021 and now serves around 90,000 patients daily across the country, indicating broad acceptance from patients as well as doctors. The facility offered immense support during the COVID-19 pandemic when the public could not physically visit hospitals due to travel restrictions. Despite this, patients have criticisms regarding the quality of the user experience from the application. These include the shortage of doctors, especially specialists, accessibility challenges, language limitations, long waiting lines, inadequate two-way communication possibilities, prior booking options, to list a few [11].

1.2 Motivation Behind the Study

ISO defines user experience (UX) as 'the perceptions and reactions people have when using or anticipating to use a product, system, or service' [12]. UX includes both utilitarian and hedonic aspects attached to each stage of user interaction with the technology or application [13]. Optimizing UX involves building an attractive and efficient user interface (UI) that facilitates interaction with an application or website [14]. An effective UI minimizes the user's effort to achieve the maximum desired results from the services [15]. Literature suggests that approaches like 'user-centered design' [16], 'evidence-based design [17], etc. are for enhancing UX. In user-centered design, the designers give extensive attention to user requirements in each service design stage. Similarly, in evidence-based design, designers use reliable, quantitative, and qualitative inputs from users for finalizing designs and revamping designs for better UX. Hence, exploring and understanding user experiences while interacting with web applications is useful for improving UX. Given the above observations, we attempt to qualitatively understand user experience with the 'eSanjeevaniOPD' using the sequential incident technique (SIT) in this study. Additionally, we attempt to identify certain design modifications and test their impact on UX using the Taguchi approach of robust design. The current study has many practical implications because much research about user perceptions and their experience with telemedicine services offered under 'eSanjeevaniOPD' is not available. We expect that these insights will help technologists identify significant issues in the web design and modify it for better UX and thus helpful in popularizing the E-governance initiative aimed at improving healthcare access in India.

2 Methods and Materials

The first stage in this research involves explorations to understand the general perceptions of users about the 'eSanjeevaniOPD' application. We have used a sequential incident technique (SIT) at this stage to achieve this objective.

2.1 Sequential Incident Technique (SIT)

An SIT uses an exploratory, qualitative interview of identified stakeholders to recollect and report critical incidents that caused pleasant or painful experiences during the service journey [18]. Then, using a visual representation of the customer path diagram, SIT identifies incidents affecting UX in service processes. Figure 1 presents the patient journey map prepared for eliciting patient responses with the 'eSanjeevaniOPD' app.

A modified version of SIT, sequence-oriented problem identification (SOPI), that captures only adverse incidents [19], is more useful when the objective is to identify areas in need of intervention. Therefore, the participants are motivated to recollect the experiences across different episodes in the customer service journey. The participants for SIT were ten patients who have used 'eSanjeevaniOPD' multiple times for various health consultations. The selection of participants was purely judgmental. Since this study aimed at identifying a few modifications contributing to better UX, we asked the participants to recall the negative incidents experienced in each stage of their service journey described in Fig. 1.

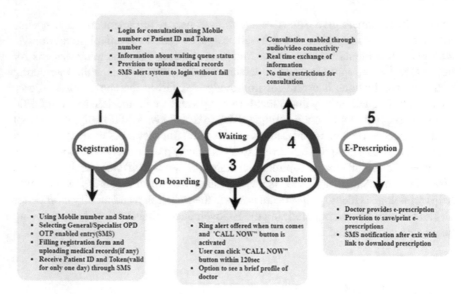

Fig. 1 Patient service journey in eSanjeevaniOPD app

2.2 Results of SIT

The SIT results indicated that, in general, participants seemed to prefer the application and were unanimously satisfied with its concept and how it helped them to get medical consultations during the COVID-19 period. Additionally, they pointed out that despite several glitches regarding connectivity, availability of doctors, and general service quality issues, the application still can improve healthcare access considerably. However, the participants are informed about many negative incidents that have affected their UX in each stage of the patient journey.

In the second phase, we have attempted to develop multiple ideas to improve UX based on inputs from the SIP, in consultation with a few experts. For many modification ideas related to service coverage, on-boarding procedure, prior appointment provisions, etc, experts viewed that available alternative options have merits and demerits. For example, an expanded array of services is possible if service coverage is national, though language barriers in consultations can affect effectiveness. Similarly, the option to choose a doctor can increase user freedom but can create waiting-line management issues. There were similar uncertainties about the ideal level concerning many important attributes contributing to UX, and hence, a multi-criteria decision-making experiment is ideal. We have used the Taguchi experiment to decide the relative importance of the above attributes and their most preferred levels that optimize UX. Table 1 provides the attributes and their chosen levels for the Taguchi experiment shortlisted based on expert consultations.

Table 1 The attributes and their levels considered for experimenting UX

Attributes	Options (levels)
Service coverage (SCO)	State, national
Prior appointment booking (PAB)	Yes, no
On-boarding (ONB)	Own password, only OTP, token-based
Option to choose doctor (OCD)	Yes, no
Login (LOG)	Specifically to a selected doctor, without doctor specificity
Doctor profile (PRO)	Name and qualification, name, age, qualification, and experience; name, age, qualification, experience, and contact details
Gender-based appointment (GEN)	Yes, no
Repeat consultation (REP)	Same doctor, different doctor, option to choose
Facility to rate doctor (RAT)	Yes, no
Prescription (PRE)	Only generic name, only brand name, both the brand and generic name

2.3 Taguchi Experiment

A Taguchi experiment [20] attempts to identify the best levels of design attributes that minimize quality variations due to noise factors (uncontrollable) associated with any product/service production. In 'eSanjeevaniOPD', noises emerge due to connectivity issues, heterogeneity in doctor counseling quality, patient characteristics, etc. Hence, to minimize the adverse effect on UX due to noises, the design features of the application should be user-centric. Traditionally, mean-based calculations without estimating variations evaluated the quality of performance of a product or service, and therefore, assessment of optimum performance suffers [21]. The measure named 'signal to noise ratio' (S/N) was obtained by conducting a Taguchi experiment that considers both mean and variance to evaluate the quality. Taguchi proposed three ways to calculate S/N ratios depending on the target performance objective [22]. They are 'Larger the better', 'Smaller the better', and 'Nominal the better'. We have used the 'Larger the better' criterion since the performance objective was to maximize UX as per participants' perceptions in the experiment.

The Taguchi experiment offers insights to improve the system design, parameter design, and tolerance design [23] and helps create a robust design capable of handling noises. The system's design informs the relative importance of each attribute in the design, whereas the parameter design helps identify the best working level of the attribute that makes the system robust, i.e., maximum resistance to noises. Lastly, the tolerance design offers insights into the extent of variations attached to attributes to decide tolerance limits. The Taguchi experiment provides the main effects of each attribute in the design on UX perceptions calculated from participant responses. The Taguchi experiment uses an orthogonal array (OA), a representative sample from the total possible combinations of all attributes and their levels, for estimating metrics helpful to decide the optimum design [24]. The different stages in a Taguchi experiment are selecting attributes and their levels, the decision of orthogonal array, conducting the experiment and calculating response variable, evaluations using S/N ratios, and deciding optimal settings.

This experiment involved six factors in two levels and four factors in three levels. Hence, the number of combinations is $2^6 * 3^4 = 5184$. An orthogonal array (OA) is a subset of all these combinations [25] and is therefore used as a sample of design combinations for the experiment to choose the best combination. The 'Design of Experiments' (DOE) tab of the tool Minitab v17 includes the Taguchi experiment and has a provision for 'Create Taguchi Design' for generating the OA. Since this study used a multiple-level design of ten factors, of which six are in two levels, and four are with three levels, the tool suggested an L36 OA with 36 combinations (runs) for the experiment. The response variable for each run was calculated based on the importance ranking given by selected evaluators to attributes and levels ($n = 12$). The evaluators were users of 'eSanjeevaniOPD' application. In finalizing the evaluators, we conducted an initial round of discussion, explained the purpose of the experiment to them, and assessed their expertise in offering rankings/ratings. The methodology adopted for calculating the response variable was as follows.

First, the evaluators ranked all attributes (factors) from rank 1 to 10 based on their importance perceptions of these attributes in developing UX. Second, they ranked the levels attached to each attribute based on a preference from rank1 to rank 2 or 3 depending on the number of levels. The weightage scheme for factors and levels was; rank 1 factor = 10, rank 2 factor = 9, so on to rank 10 factor = 1; rank 1 level = 3; rank 2 level = 2, rank 3 level = 1 for three-level factors; and rank 1 level = 2 and rank 2 level = 1 for two-level factors. Thus, the weighted factor level score is factor rank weightage * level rank weightage. Additionally, the response score of the run for each evaluator will be the sum of the weighted factor level score of each attribute level attached to a run in the experiment. The response scores calculated above for each evaluator for the 36 runs formed the response matrix ($36 * 12$) used in the Taguchi experiment. In this study, we adopted the 'Larger the better' criterion for calculating the S/N ratio, since higher values of response score explain better UX. The Taguchi experiment using Minitab v17 yields the OA and S/N ratios.

The residual plots obtained as part of the output appeared as a straight line to imply that residuals are normally distributed, and hence, results are valid. Based on the Minitab output, 'R-sq (adj)' reported was 52.6, confirming that the model sufficiently explains the variation in the response.

3   Results

The Taguchi experiment identified the relative importance of the factors and their most preferred level. The 'Delta' measure representing the difference between the highest and lowest average performance score (S/N, or Standard deviation) of a

Table 2 Response table for factor and their levels

Level	SCO	PAB	LOG	OCD	GEN	RAT	ONB	PRO	REP	PRE
1	36.24	37.33*	36.76	36.77	36.52	37.09*	36.99*	37	37.18	37.1*
2	37.51*	36.42	37*	36.99*	37.24*	36.66	36.92	36.57	35.87	36.92
3							36.72	37.06*	37.58*	36.61
Delta (S/N)	1.27	0.91	0.24	0.22	0.72	0.43	0.27	0.48	1.71	0.49
Rank	2	3	9	10	4	7	8	6	1	5
Delta (std. dev)	1.77	2.27	0.59	1.64	2.43	0.48	1.76	5.03	2.8	4.8
Rank (std. dev)	6	5	9	8	4	10	7	1	3	2

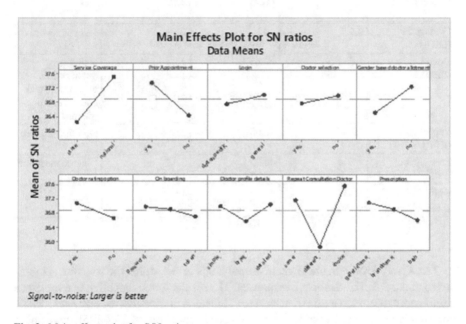

Fig. 2 Main effects plot for S/N ratio

factor indicates its importance in creating UX. The higher the delta values indicates the greater importance of the factor. The S/N ratio is more accurate than the mean for identifying the most important factor, and its preferred level is less affected by noises. Table 2 provides the details of the importance of factors based on the S/N ratio and standard deviation. Figure 2 illustrates the main effects plot of the S/N ratio to understand the effect of each factor on UX. The main effect exists when different levels of a factor affect UX differently.

Table 3 Contribution of factors and optimum levels for UX

Source	Seq SS	p-value	Contribution (%)	Best level for UX
Service coverage (SCO)	14.489	<0.05	23.88	National
Prior appointment (PAB)	7.506	<0.05	12.37	Yes
Login (LOG)	0.512	<0.1	0.84	Without doctor specificity (general)
Doctor selection (OCD)	0.423	<0.1	0.70	No
Gender-based (GEN)	4.630	<0.05	7.63	No
Doctor rating (RAT)	1.696	<0.05	2.80	Yes
On-boarding (ONB)	0.479	<0.1	0.79	Own password
Doctor profile (PRO)	1.686	<0.05	2.78	Name, age, qualification, experience, and contact details (detailed)
Repeat visit (REP)	19.221	<0.05	31.68	Option to choose
Prescription (PRE)	1.504	<0.05	2.48	Generic medicine
Residual error	8.521	–	14.05	
Total	60.667			

The ANOVA table in the Taguchi output helps to determine the statistical significance of the effect of factors on responses. The results found that all the factors are a significant association with UX at least to 0.1 level. Additionally, the error percentage contribution was only 14.05, less than the permissible limit of 15% [25], confirming role of all factors in creating the UX. The maximum contribution to UX was due to the selection of the doctor for repeat consultation (31.68%), followed by service coverage (23.88%) and prior appointment feature (12.37%). In this experiment, the optimal combination was not available in the sample of 36 runs. The optimal combination offering maximum UX is identified from the last column of Table 3. We have validated the experiment by predicting the S/N ratio of the optimal combination using the 'Predict Taguchi Results' option in Minitab. The results showed an S/N ratio of 42.16 much higher than any of the 36 runs to confirm the experiment's validity.

4 Conclusions

E-governance initiatives will be successful only if the applications created for such purposes develop enriching UX. In improving UX, many attributes associated with the application should perform according to user expectations and make user inter-actions more simple, intuitive, and relaxing. This study observed that the 'eSanjee-vaniOPD' app developed for better governance of healthcare access in India has many drawbacks affecting UX. Every stage in the user journey contains areas of concern and thus the potential for improvement. We found that technological chal-lenges, doctor quality, and service quality of doctors during the consultation are the major noises affecting UX. As part of this study, we have shortlisted a few attributes identified to contribute to UX and conducted a Taguchi experiment to understand their relative importance and choose the optimum way to incorporate them in the app design. Similarly, password-protected on-boarding to the app is most preferred for UX. We observed that users prefer to avail telemedicine services from doctors nationally without the limitations of the state of domicile. A prior appointment facil-ity can significantly enhance UX. The users do not prefer logging into doctor-specific consultation rooms, patient gender-based doctor allotment, or facility to choose the doctor in the first instance. Users prefer to know a detailed profile of the doctor with contact details, the ability to offer ratings to doctors, and a facility to choose a doctor for repeat consultation. In addition, users wish to see generic names of the medicines in the prescriptions.

Based on the study's findings, we suggest that a five-dimensional approach is ideal for improving UX and improving E-governance in health care. These dimensions are (1) technology innovation, (2) service quality of medical professionals, (3) IT infras-tructure, (4) ethics, and (5) policy support. Technological innovations focused on UX should provide a more participatory, creative, and real-time experience to users. Also, a favorable change in the attitude of medical professionals is essential for populariz-ing telemedicine. A serious avoidance from senior doctors is visible. There should be sufficient training programs and technology updates required for empowering doc-tors to be more patient-friendly during online consultations. IT infrastructure-related challenges adversely affect UX and penetration of telemedicine. More private par-ticipation may be invited for infrastructural augmentations. Concerns of privacy and security among users prevent them from adopting telemedicine. A clear legal frame-work offering trust to users about handling issues related to privacy breaches, security, etc. can significantly improve UX. Successful implementation of E-governance in India would require the government, medical institutions, corporate ICT players, and non-profit organizations to collaborate, remove major roadblocks, and remove major bottlenecks. Hence, a clear policy framework is an overarching requirement for the fast diffusion of telemedicine initiatives.

There are a few limitations for this study. First, the majority of the attributes considered were doctor-specific. Second, the calculation of the response variable was based on rankings and ratings offered by evaluators, and a more quantifiable outcome would have reduced the subjectivity element in the experiment. Third, even though the SIT revealed that users prefer to get many post-consultation services such as online pharmacy, health plans, tie-up for health check-ups, hospital referrals, such aspects were not included in the experiment. Future studies involving more attributes and experiments focusing on each stage of the service journey can help in revamping the telemedicine services for better UX.

References

1. M. Jain, N. Abidi, A. Bandyopadhayay, Int. J. Technol. Manage. Sustain. Dev. E-procurement espousal and assessment framework: A case-based study of Indian automobile companies **17**(1), 87–109 (2018)
2. J.S. Ojo, J. Public Admin. Policy Res. E-governance: An imperative for sustainable grass root development in Nigeria **6**(4), 77–89 (2014)
3. Y.M. Asi, C. Williams, Equality through innovation: Promoting women in the workplace in low–and middle–income countries with health information technology. Journal of Social Issues **76**(3), 721–743 (2020)
4. I.C. Señor, J.L.F. Alemán, A. Toval, Computer. Personal health records: new means to safely handle health data? **45**(11), 27–33 (2012)
5. L.M. Fonseca, J.P. Domingues, A.M. Dima, Sustainability. Mapping the sustainable development goals relationships **12**(8), 3359 (2020)
6. J.E. Carrillo, V.A. Carrillo, H.R. Perez, D. Salas-Lopez, A. Natale-Pereira, A.T. Byron, J. Health Care Poor Underserv. Defining and targeting health care access barriers **22**(2), 562–575 (2011)
7. J. Sundewall, B.C. Forsberg, Lancet. Understanding health spending for SDG 3 **396**(10252), 650–651 (2020)
8. M. Kludacz-Alessandri, R. Walczak, L. Hawrysz, P. Korneta, J. Clin. Med. The quality of medical care in the conditions of the COVID-19 pandemic, with particular emphasis on the access to primary healthcare and the effectiveness of treatment in Poland **10**(16), 3502 (2021)
9. I. Gole, T. Sharma, S.B. Misra, J. Manage. Public Policy. Role of ICT in healthcare sector: an empirical study of Pune city **8**(2), 23–32 (2017)
10. S.K. Mishra, L. Kapoor, I.P. Singh, Telemed. e-Health. Telemedicine in India: current scenario and the future **15**(6), 568–575 (2009)
11. N. Bajpai, M. Wadhwa, *National Teleconsultation Service in India.* eSanjeevani OPD (2021)
12. T. Lindgren, M. Bergquist, S. Pink, M. Berg, V. Fors, Experiencing expectations: extending the concept of UX anticipation, in *Scandinavian Conference on Information Systems* (Springer, 2018), pp. 1–13
13. J. Häkkilä Designing for smart clothes and wearables user experience design perspective, in *Smart Textiles* (Springer 2017), pp. 259–278
14. A. Pitale, A. Bhumgara, A.: Human computer interaction strategies-designing the user interface, in *2019 International Conference on Smart Systems and Inventive Technology (ICSSIT)* (IEEE, 2019), pp. 752–758
15. R. Oppermann, User-interface design, in *Handbook on Information Technologies for Education and Training* (Springer, Heidelberg, 2002), pp. 233–248
16. R.W. Veryzer, B. Borja de Mozota, The impact of user-oriented design on new product development: an examination of fundamental relationships. J. Product Innov. Manage. **22**(2), 128–143 (2005)
17. D.K. Hamilton, Healthcare Des. The four levels of evidence-based practice **3**(4), 18–26 (2003)
18. B. Stauss, B. Weinlich, Eur. J. Market. Process-oriented measurement of service quality: applying the sequential incident technique **31**(5), 33–55 (1997)

19. G. Botschen, L. Bstieler, A.G. Woodside, J. Euromark. Sequence-oriented problem identification within service encounters **5**(2), 19–52 (1996)
20. G. Taguchi, *System of Experimental Design*, vols. 1 and 2. UNIPUB/Krauss International, White Plains, New York (1987)
21. G. Rejikumar, A.A. Ajitha, M.S. Nair, TQM J. Healthcare service quality: a methodology for servicescape redesign using Taguchi approach **31**(4), 600–619 (2019)
22. G. Taguchi, S. Chowdhury, Y. Wu, S. Taguchi, H. Yano, *Taguchi's Quality Engineering Handbook* (Wiley, 2005)
23. S.K. Karna, R. Sahai, Int. J. Eng. Math. Sci. An overview on Taguchi method **1**(1), 1–7 (2012)
24. G. Taguchi, S. Konishi, *Orthogonal Arrays and Linear Graphs: Tools for Quality Engineering* (ASI Press, Michigan, 1987)
25. M. Tanco, E. Viles, L. Ilzarbe, M.J. Alvarez, Appl. Stochast. Models Bus. Ind. Implementation of design of experiments projects in industry **25**(4), 478–505 (2009)

Secured Data Transmission in Low Power WSN with LoRA

Arabinda Rath, S. Q. Baig, Bisakha Biswal, and Gayatri Devi

Abstract Power consumption factor is very important in WSN and other IOT setup. As the nodes remain in distant locations they need to run powered by battery. When we try to send private data through it we can encrypt it. But a single key for a long period may be less secure. Hence, we need to change the key. But in a low power low processing power setup the conventional key generation algorithm may not be quite feasible as it may need a lot of processing and power. Therefore, we study low power alternative for dynamic key generation and implement it in a LoRa-based WSN.

Keywords Encryption · Decryption · AES · Steganography · Dynamic key · Algorithm · WSN · LoRa · Arduino

1 Introduction

Wireless sensor network (WSN) is a group of specialized electronic devices, which is capable of sensing, computation and wireless communication for monitoring and recording the physical conditions of the environment. The collected informations are organized at the central location. Cryptography is important because it provides for secure communication techniques that allow only the sender and the intended recipient of a message to view its contents even if in the presence of malicious third parties. Businesses, government and many individual use it to protect their personal information.

In this paper while using IOT and wireless sensor network the energy and processing constraint has to be considered. Keeping that into mind we have created a novel algorithm that can generate new keys for AES encryption system making it more secure.

A. Rath (✉) · S. Q. Baig · B. Biswal · G. Devi
Department of Computer Science and Engineering, ABIT, Cuttack, Odisha, India
e-mail: arabindarath73@gmail.com

2 Advanced Encryption Standard (AES)

The symmetric AES encryption algorithm is widely used because AES is much faster than triple DES [1, 2]. The features of AES are as follows

- Symmetric key symmetric block cipher
- 128-bit data, 128/192/256-bit keys
- Provide full specification and design details
- Stronger and faster than Triple DES
- Software implementable in C, C#, Java, etc.

AES is an iterative process based on 'substitution–permutation network'. It consists of a sequence of linked operations, where some involve replacing inputs by specific outputs using substitution and others involve shuffling bits around by using permutations.

All the computations of AES is based on bytes instead of bits. So, AES considers the 128 bits of a plaintext block as 16 bytes. These 16 bytes are arranged in 4×4 matrix.

2.1 AES Security Analysis

In AES, the length of the plain text of size 128-bit, 192-bit and 256-bit supporting three different key lengths of 128-bit, 192-bit and 256-bit, respectively, are used in this paper. The size of the input key must be same as size of the plain text.

The length of the key varies which creates the difference. The strongest level of encryption is provided by the 256-bit key which is the longest. This would ensure that a hacker would have to try 2256 different combinations for 256-bit key to make sure the right one is included. So we require 128- or 256-bit encryption for sensitive data.

AES consists of multiple rounds for processing different key bits such as 10 round to process 128-bit keys, 12 rounds to process 192-bit keys and 14 rounds to process 256-bit keys. This shows that 256-bit key is the most secure form of AES implementation which concludes that the more the number of the round is, the more complex is the encryption. Large key size and more rounds lead to more consumption of energy and space.

3 Key Management

Key management means to manage the cryptographic keys in a cryptosystem. It deals with the generation, exchange, storage, use, crypto-shredding (destruction) and replacement of keys. It includes cryptographic protocol design, key servers, user procedures and other relevant protocols.

3.1 Key Management Cost

There are various algorithms to establish a key between two nodes. But they can be very costly in terms of processing and energy cost. For low power IOT devices [3] it can be a problem. Hence, we have developed a key establishment technique which relies on having a secret formula in both the devices and passing some random characters and generate the key using the formula. This can be an alternative to the conventional methods for low energy devices that work on battery.

4 Method

We transmit sensitive data in a secured way through Low Power LoRa [4, 5]. For this we use AES algorithm, we alter the keys whenever we need just by pressing a push button. We have a Sender Node and a Receiver Node. The Sender node Encrypts sensitive information and the receiver node decrypts it. There is a default key in the program which starts working as soon as we power up the device. The receiver node has a push button connected to it. Which if pressed initiated the key changing process. We program both nodes accordingly. We have used Arduino microcontroller for this.

4.1 Sender Node

We need to initialize software serial so that we can use it for transmission and receiving process while the default serial can be used for displaying the information. We also define baud rate.

Then we include AES Library so that we can perform AES encryption and decryption process. Then we initialize the AES key with a default value. It is a byte array of size 16. Each byte has 8 bits hence 2^8 different numbers can be stored in it. Which are 0 to 255. We can also store hexadecimal values as shown in the program.

Then we define a function that prints the keys in the default serial, so we can view it in serial monitor. This can be useful when we want to view the new key after we have altered it. The Arduino program starts from void setup() we begin the default serial and software serial. And wait until serial is available. Then we display booting to mark the beginning of the program. Then we define and initialize various constants which we will use in our program. We free the stack for avoiding the error. To demonstrate transmission of secured data Every 5 s we send a generate a string and we encrypt and transmit it to the receiver node.

In Arduino void loop() keeps executing continuously. We need to check if there is any key change request continuously hence we need to put that within void loop(). We read the serial character by character until we come across new line. In that case we conclude that the message is complete. The string is sent by the receiver node which is key change request. We use our custom formula to generate the new key using the string received. Our custom formula performs some different arithmetic operations for each byte and then adds some random value of our choice. In the end we find the remainder of it after division with 256 because it can have value between 0 and 255.

4.2   Receiver Node

The receiver node receives the sensitive information which is encrypted and it decrypts the message and finally gets the message in plaintext which is the original message or the sensitive information.

Like the Sender node we initialize software serial so that we can use it for transmission and receiving process while the default serial can be used for displaying the information. We also define baud rate. Then we include AES Library so that we can perform AES encryption and decryption process. Then we initialize the AES key with a default value which should be the same value as sender node. Then we define a function that prints the keys in the default serial, so we can view it in serial monitor.

After this we define the decrypt function it takes the encrypted message and key as argument and returns the plain text original message.

Similar to sender node we initialize serial free the stack and declare and initialize some variables, although we have different variables in receiver node because the operation is different.

In the loop section of the receiver node we check if the push button is pressed or not. We delay for a little time while the button is pressed so that a single press does trigger the program to run several times. When the button is pressed we generate random numbers and save it in a string we send that string to the receiver node as a key changing request. Then we calculate the key value using the same formulae used in sender node and hence getting the same result. Finally, it decrypts the encrypted message received. If the button has never been pressed the decryption takes place with the default key.

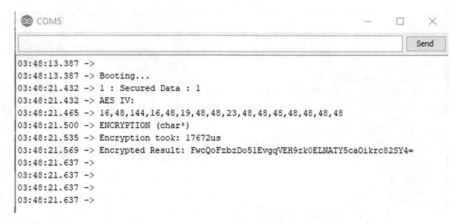

Fig. 1 Sender sends first secured data

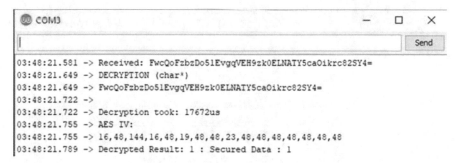

Fig. 2 Receiver receives the data

5 Simulation Result

5.1 Testing the Implementation

When we press the push button the key changes and encryption and decryption take place with the new key (Figs. 1, 2, 3, 4, 5 and 6).

6 Conclusion

While using IOT and wireless sensor network the energy and processing constraint has to be considered. Keeping that into mind we have created a novel algorithm that can generate new keys for AES encryption system making it more secure.

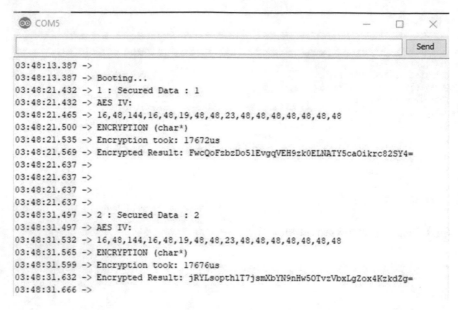

Fig. 3 Second data

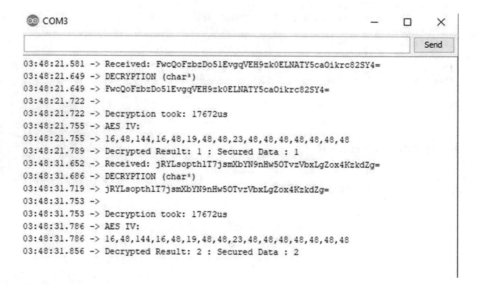

Fig. 4 Receiver decrypts the data

Fig. 5 Changing the key sender's side

424   A. Rath et al.

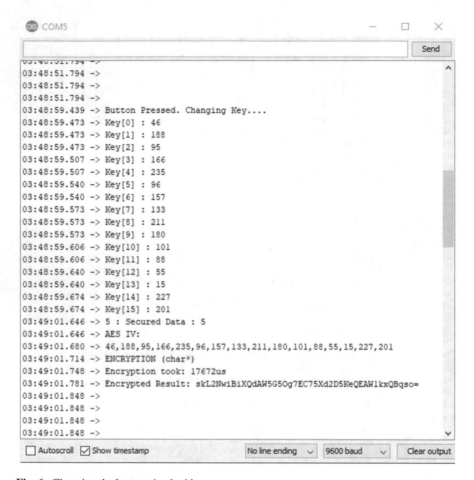

Fig. 6 Changing the key receiver's side

References

1. H.S. Deshpande, K.J. Karande, A.O. Mulani,. Efficient implementation of AES algorithm on FPGA, in *2014 IEEE International Conference Communications and Signal Processing (ICCSP)* (2014)
2. F.J. D'souza, Advanced encryption standard (AES) security enhancement using hybrid approach, in *2017 International Conference on Computing Communication and Automation (ICCCA)*, pp. 5–6 (2017)
3. A. Ishfaq, A. Muhammad, Applying security patterns for authorization of users in IoT based applications, in *2018 International Conference on Engineering and Emerging Technologies (ICEET)*, pp. 22–23 (2018)
4. J. Petajajarvi, K. Mikhaylov, A. Roivainen, T. Hanninen, M. Pettissalo, On the coverage of LPWANs: range evaluation and channel attenuation model for LoRa technology, in *14th International Conference on ITS Telecommunications (ITST)*, pp. 55–59 (2015)
5. M.O.A. Kalaa, W. Balid, N. Bitar, H.H. Refai, Evaluating bluetooth low energy in realistic wireless environments, pp. 1–6 (2016)

Safeguarding Cloud Services Sustainability by Dynamic Virtual Machine Migration with Re-allocation Oriented Algorithmic Approach

Saumitra Vatsal and Shalini Agarwal

Abstract Data centres are networking platforms which exhibit virtual machine workload execution in a dynamic manner. As the users' requests are of enormous magnitude, it manifests as overloaded physical machines resulting in quality of service degradation and SLA violations. This challenge can be negotiated by exercising a better virtual machine allocation by dint of re-allocating a subset of active virtual machines at a suitable destined server by virtual machine migration. It is exhibited as improved resource utilization with enhanced energy efficiency along with addressing the challenge of impending server overloading resulting in downgraded services. The aforesaid twin factors of enhanced energy consumption and enhanced resource utilization can be suitably addressed by combining them together as a single objective function by utilizing cost function based best-fit decreasing heuristic. It enhances the potentials for aggressively migrating large capacity applications like image processing, speech recognition, and decision support systems. It facilitates a seamless and transparent live virtual machine migration from one physical server to another along with taking care of cloud environment resources. The identification of most appropriate migration target host is executed by applying modified version of best-fit decreasing algorithm with respect to virtual machine dynamic migration scheduling model. By executing the selection algorithm, the hotspot hosts in cloud platform are segregated. Subsequently, virtual machine-related resource loads are identified in descending order with respect to hotspots. The resource loads pertaining to non-hotspot hosts are identified in ascending order. Next, the traversing manoeuvring in non-hotspot hosts queue is exercised for identification of the most appropriate host to be reckoned as migration target host.

Keywords Data centre · Load balance · Virtual machine migration · Cloud computing

S. Vatsal (✉) · S. Agarwal
Shri Ramswaroop Memorial University, Barabanki, Uttar Pradesh, India
e-mail: s.vatsall@gmail.com

S. Agarwal
e-mail: shalini.cs@srmu.ac.in

Y.-D. Zhang et al. (eds.), *Smart Trends in Computing and Communications*, Lecture Notes in Networks and Systems 396, https://doi.org/10.1007/978-981-16-9967-2_40

1   Introduction

The robustness of the data centre is indispensable as it has to address business processing peak demands at certain times which require high-resource availability. But it culminates into waste of resources in off-peak periods. This challenging problem can be alleviated by scalable and dynamic resource allocation [1]. It is achieved by dynamically expanding or contracting server hosts' related allocation of resources according to fluctuating scenario of users' demand load. It serves to conserve effectively waste of resources along with addressing server host-related problem of being infested with highly loaded or too lightly loaded situation. Resource allocation is adjudged by infrastructure-related physical resource allocation along with virtual resource allocation at task or application level [2, 3]. The resource allocation with respect to first start-up time of the system is addressed by virtual machine placement by selecting and tapping the physical resources of most appropriate physical machine. Subsequently, it is addressed by exercising re-allocation of virtual machine from one to another physical machine by using migration techniques as and when required. Live migration permits virtual machine migration within cloud with a transparency in its running state. It addresses server virtualization with an added perk of judicious load balancing along with avoidance of overloaded "hotspot" physical machines, in running state, thus avoiding system-related down-time because virtual machines are not turned off during maintenance. The virtual machines on lightly loaded hosts can be clubbed together for them to be placed on fewer physical machines, thus securing avoidance of hotspot physical machines and resource requirement considerations. As a consequence freed up physical machines can be safely turned off for securing power saving. It results into new virtual machines-related high-resource availability [1, 4]. In addition, a new strategy can be added further by clubbing migration-worthy virtual machines into groups on the basis of their physical machines and magnitude of memory shared between them, leading to live gang migration [5]. The traditional static resource scheduling algorithms like destination hashing scheduling, round-robin scheduling, source-hashing scheduling, and weighted round-robin scheduling do not consider user requests' related system load changes or dynamic scheduling of resources and hence results into unnecessary virtual machine migrations [6–9]. It can be better addressed by utilizing energy-aware virtual machine migration algorithms which are capable of virtual machine migration to suitable target hosts for de-activation of idle physical hosts resulting in minimization of energy consumption. Since virtual machine dynamic migration scheduling represents as NP-hard combinatorial optimization bin-packing problem [10, 11], it is negotiated by optimization methods which enhance the efficacy of bin-packing algorithms and associated heuristics and meta-heuristics.

The paper is organized is as follows:

- Section 2 incorporates related work.
- Sections 3 and 4 are pertinent to the problem statement and system model, respectively.

- Section 5 describes a relevant algorithmic approach for dynamic virtual machine migration.
- Section 6 puts forward the performance analysis and the results.
- Section 7 concludes the paper.

2 Related Work

Dynamic virtual machine allocation challenge basically originates from virtual machine placement strategy. It is suitably addressed by static approaches represented by round-robin scheduling and destination hashing scheduling along with dynamic approaches incorporating bin-packing approach through first-fit (FF) algorithm, best-fit (BF) algorithm, first-fit-decreasing (FFD) algorithm, and best-fit decreasing (BFD) algorithm [12]. By using live migration-related clustering technique live gang migration, the optimization of memory and network bandwidth usage can be secured along with minimization of transfer of identical shared data between co-located virtual machines [5]. By addressing a clustering co-location mechanism through the minimization of multiple co-located virtual machine migrations, the system-related load balanced state can be preserved along with continuing the migration process by virtue of utilizing system standard deviation [13]. Virtual machine re-allocation problem subsequent to migration can be reckoned as integer programming quadratic constraint for minimization of data centre-related aggregation communication costs [14]. The servers' related cooling cost within a data centre can be minimized by server placement heuristic approach [15]. The minimization of power consumption of a data centre can be secured by resource aware virtual machine placement algorithm which addresses the resource usage factor [16]. The challenge in virtual machine placement with respect to available bandwidth can be suitably addressed by the use of whale optimization algorithm [17]. To address the energy efficiency issue, the approach has to be profile-based dynamic virtual machine placement which secures minimization of total energy consumption of all physical machines during dynamic virtual machine placement [18]. In order to maximize resource utilization along with minimization of energy consumption, the multi-objective functions are coalesced as one objective function in order to secure an optimal point existing between resource utilization and energy consumption by taking Euclidean distance into consideration [19]. On the basis of multi-objective genetic algorithm and Bernoulli simulation, the minimization of resource wastage pertaining to each individual host along with number of active hosts can be obtained simultaneously [20]. For securing best virtual machine migration-related re-allocation manoeuvring, a multi-objective weighted sum function as mono-objective addresses energy cost reduction along with minimized energy cost and SLA penalty cost [21]. A cross entropy algorithm addressing virtual machine migration-related re-allocation can address the issue of data centre-related total thermal cost minimization by dint of reduced energy consumption [22]. By addressing twin factors of energy consumption and migration cost, modified particle swarm optimization can facilitate re-allocation of migrated virtual machines on the basis of fitness function [23]. By working on relaxed convex optimization framework, a multi-level join of migration and re-allocation algorithms can facilitate optimal solution for addressing migration costs, load distribution, and cross traffic-costs by clubbing them into mono-objective weighted sum function [24].

3 Explication of Problem Statement

Cloud data centre-related prime functionality ensures creation, migration, and cancellation of host-related virtual machines. Further, it substantiates deployment of virtual machine on to a suitable host by re-allocation in order to address the processing of users application-related service requests. It involves certain scheduling strategies with respect to virtual machine dynamic migration for executing load monitoring programme by virtue of retrieving load data of CPU, bandwidth, memory along with other resources related with each virtual machine and physical host. The cloud data centre retrieves resource usage information by executing load monitoring programme. The virtual machine migration scheduling is executed in accordance with platforms load balance strategy whenever host-related load is too heavy or too light. The system processing performance requirements are safeguarded by dynamically migrating virtual machines from one host to another suitable host in order to give the quality of service fully compatible with users' demands. The process of virtual machine migration scheduling is confronted with challenging factors which include right assessment of time for determining most appropriate time instant for virtual machine migration and assessment of most suitable inherent migration scheduling strategy. It also demands a judicious judgement to ascertain instant of time when to commence the migration operation and to confirm if whether the server-related platform is competent enough to sustain load balanced state.

4 System Model

The virtual machine dynamic migration scheduling model as in Fig. 1 encompasses the ability to process users-related service requests by the scheduling main controller which evokes the execution of host-related task. Each host is provisioned with its own load monitoring programme. The associated load monitoring program addresses load state of host and virtual machines with respect to real time. Thus, it provides load data for identifying in order to select hotspot hosts which are listed into a queue. The selected virtual machines undergo migration from this hotspot hosts queue to selected target hosts in order to achieve a load balanced state in data centre platform. This whole process thus involves hotspot host selection algorithm programme along with virtual machine scheduling strategy for ensuring dynamic migration.

The core strategy of cloud computing dynamic resources scheduling inherits the concept of migrating virtual machines amongst data centres' hosts in such a way that

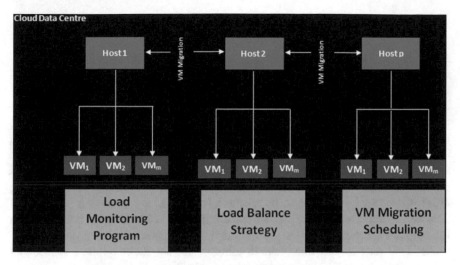

Fig. 1 Dynamic migration scheduling mechanism of virtual machines

each host remains enabled to run at optimal load state. The SLA violation results if host is overloaded resulting in degradation of quality of service (QoS) due to degraded task execution. On contrary, if host is loaded too lightly, it culminates into underutilized host. It exhibits as a useless energy waste which is consumed by an underutilized host. This challenge is addressed by load monitoring programme which monitors resource utilization with respect to each and every host. It identifies the hotspot host on the basis of identifying a host if it exceeds the maximum threshold or the minimum threshold and designating it as a hotspot host.

Specific criteria are expressed as follows in equation set 1 where $A_{CPU}(b_j)$, $A_{RAM}(b_j)$, and $A_{Band}(b_j)$ represent CPU-related load state, memory, and network bandwidth in relation to physical host b_j whilst predetermined maximum load threshold with respect to CPU, memory, and network is represented as Max_{CPU}, Max_{RAM}, and Max_{Band}. Likewise predetermined load threshold related with CPU, memory, and network bandwidth is represented as Min_{CPU}, Min_{RAM}, and Min_{Band}, respectively.

$$\left.\begin{array}{l} A_{CPU}(b_j) > Max_{CPU} \text{ or } A_{CPU}(b_j) < Min_{CPU} \\ A_{RAM}(b_j) > Max_{RAM} \text{ or } A_{RAM}(b_j) < Min_{RAM} \\ A_{Band}(b_j) > Max_{Band} \text{ or } A_{Band}(b_j) < Min_{Band} \end{array}\right\} \tag{1}$$

The fulfilment of any aforesaid condition confirms the host to be a hotspot host.

5 Proposed Algorithmic Approach

By applying hotspot host selection algorithm virtual machine re-allocation algorithm (VMRA), the identification of hotspot hosts is achieved for their arrangement in a queue. But the real challenge is to achieve load balancing amongst hosts whilst using virtual machine dynamic migration scheduling strategy. The algorithm identifies hotspot host as an item and the target host as a packing bin. Subsequently, hotspot hosts are sorted in descending order according to size of items along with due consideration of load state of CPU, network bandwidth, and memory of virtual machines in each hotspot host which forms the basis of load state assessment of all virtual machines by cumulative analysis. The sorting of hotspot hosts in descending order is exercised according to its resource load state. By traversing non-hotspot host queue, the most appropriate target for migration is identified as a packing bin for migratory virtual machines. It infers that the difference between present host-related load state and the maximum threshold value should be minimal subsequent to re-allocating virtual machine on to the target host.

A hotspot host queue is drawn in accordance with following equation subsequent to combining host load state data and hotspot host selection algorithm.

$$\left. \begin{array}{l} HS_b = \{b_1, b_2, b_3 \ldots, b_m\}; \quad m \in N^* \\ A_{b_j} \geq \text{Max or } A_{b_j} \leq \text{Min} \forall j \in [1, m] \end{array} \right\} \tag{2}$$

In Eq. 2, the physical host is represented as HS_b.

Subsequent to ascertaining the hotspot host queue, the sorting in descending order of virtual machines in hotspot hosts is exercised taking their resource load state in consideration. The virtual machine resource load status is calculated according to following equation.

$$C_{VM} = x_1 c_{CPU} + x_2 c_{RAM} + x_3 c_{Band} \ \& \ \sum_{d=1}^{3} x_d = 1 \tag{3}$$

In Eq. 3, the virtual machine-related CPU, memory, and network bandwidth are represented as c_{CPU}, c_{RAM}, and c_{Band}, respectively. Different weights are represented by x_1, x_2, and x_3.

The sorting of virtual machines in hotspot hosts zone in descending order is secured with respect to resource load state by following equation.

$$\left. \begin{array}{l} Y_{VM} = \{vm_1, vm_2, \ldots vm_r\}; \quad VM \in HS_b, r \in N^* \\ C_l \leq C_k \quad \forall l, k \in [1, r] \end{array} \right\} \tag{4}$$

The virtual machine queue is represented as Y_{VM}. The virtual machine resource load is represented by C_l.

The sorting of all the non-hotspot hosts in ascending order is exercised considering their resource load state according to following:

$$
\left.\begin{array}{l}
W_b = \{b_1, b_2, b_3, \ldots, b_t\}; \quad t \in N^* \\
A_{b_j} \geq A_{b_l} \quad \forall j, l \in [1, t] \\
\text{Min} < A_{b_j} < \text{Max}
\end{array}\right\} \tag{5}
$$

The target host sorted in the queue is represented by W_b.

The selection of most appropriate target whilst migrating virtual machines is procured by traversing manoeuvring of target host queue for identifying the most suitable destination. It is achieved by applying equation:

$$
\left.\begin{array}{l}
FW_b = b_j \\
\text{Min}\{\text{Max} - (A_{b_j} + c_l)\} \\
C_l \in Y_{VM}, b_j \in W_b
\end{array}\right\} \tag{6}
$$

The load balancing degree is calculated for measuring the load balance magnitude of old data centre subsequent to exercising virtual machine dynamic migration scheduling. It is expressed by following equation:

$$
(\text{Balance})^2 = \frac{1}{g}\left[\left(A_1 - \overline{A}\right)^2 + \left(A_2 - \overline{A}\right)^2 + \cdots + \left(A_g - \overline{A}\right)^2\right] \tag{7}
$$

The average value of load resource state is represented by $\overline{A}$.

The virtual machine dynamic migration scheduling algorithm is as under:

Algorithm: VMRA

Input: Vir_Mchn_List, Host_List

Output: Dest_Host

```
1.  Vir_Mchn_List.sortDecreasingCPUUtilization();
2.     for Vir_Mchn in Vir_Mchn_List do
3.        maximum_memb←0;
4.        Dest_Host←NULL;
5.        for Host in Host_List do
6.           if Host has sufficient resource for Vir_Mchn then
7.              memb←estm_Power_Utilization(Host, Vir_Mchn);
8.              if memb≥maximum_memb then
9.                 maximum_memb←memb;
10.                Dest_Host←Host;
11.          if Dest_Host≠NULL then
12.             assign Vir_Mchn to Dest_Host;
13. return Dest_Host;
```

The following steps are involved in virtual machine dynamic migration scheduling process:

1. Identification of hotspot hosts: It is load monitor programme which determines each physical host-related resource load state in order to ascertain the virtual machine to accept the migration on to a host.

2. Selection of target host: Sorting of non-hotspot hosts in ascending order to determine the most suitable host fulfilling all pre-conditions necessary for a migration.
3. Virtual machine dynamic migration: The network connection information, memory, and load state of CPU of the virtual machine identified for migration on to a target host are taken in to consideration.
4. The success of migration is adjudged by confirmation message from hotspot hosts to the target confirming that migration process is finished. This message infers that target is enabled with a replica image of virtual machine which was destined to undergo migration from hotspot host. This concludes a successful migration.

6   Evaluation of Performance

6.1   Simulation Setup

The evaluation of virtual machine management and resource provisioning is based on CloudSim simulator [25, 26]. This simulation is exercised considering five physical hosts with identical configuration providing 50 identical virtual machines simulating 50 tasks in cloud environment. The load status of each host is monitored by assuming that one cycle is made up of ten times of experimental execution. The host load-related maximum threshold is considered akin to 80% of system-related resource load whilst the minimum threshold is considered to be 20% of system resource load. The dynamic migration of virtual machines is executed in accordance with hosts-related current resource load state.

6.2   Results

According to Table 1, it is evident that VMRA exhibits SLA violation percentage to be lesser as compared to its competitors along with minimization of virtual

Table 1 Different approach-related energy consumption

Approaches	Energy consumption (in kWh)	SLA violation (in %)	No. of migrations
VMRA	1.39	9.97	297
CA1	1.69	11.29	329
CA2	1.60	10.73	323
NPA	4.57	–	–
DVFS	2.39	–	–

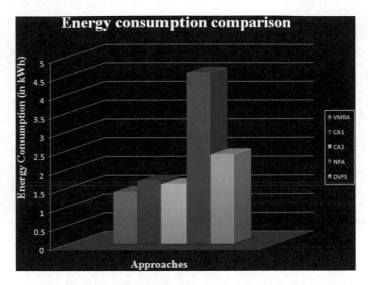

Fig. 2 Different approaches related comparative illustration of energy consumption

machine migrations. It is evident that strategy model for VMRA incurs enhanced energy savings with respect to NPA, DVFS, CA1, and CA2 in terms of less energy consumption, thus exhibiting its superiority over its competitors.

The comparative experimental analysis as shown in Fig. 2 is conducted by incorporating dynamic voltage and frequency scaling (DVFS) [27], non-power aware (NPA) [28, 29], and simple threshold (ST)-based competitive algorithm 1 (CA1) and competitive algorithm 2 (CA2) for depiction of benefits pertaining to minimized energy consumption, reduced number of migrations, and minimized service level agreement (SLA) violations.

The comparison of load balance degree as shown in Fig. 3 with respect to virtual machine dynamic scheduling clearly indicates that load balance degree of VMRA achieves superior load balancing with due consideration of system overhead parameters which include migration number, energy consumption, and SLA violation percentage.

7 Conclusion

For the sake of conserving the data centre-related networking viability, the functioning of virtual machines in cloud environment has to remain uninterrupted by virtue of countering all challenging threat factors which may jeopardize its functioning. The dynamic virtual machine migration and re-allocation on the basis of identifying its load status of target host along with the resource reservoir are the factors of great importance. This approach revamps the concept of green cloud computing. A

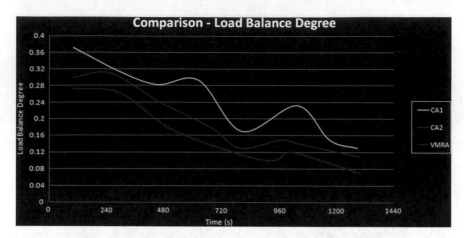

Fig. 3 Comparative performance analysis of load balance degree

minimized number of virtual machine migrations are ensured by turning-off the idle hosts subsequent to a judicious dynamic virtual machine migration and re-allocation on a suitable target host, thus conserving the energy consumption. The assigning of a virtual machine to a target host is ensured in accordance with present and future resource requirements. The algorithmic approach addressed here exhibits its effectiveness encompassing minimized number of migrations, energy consumption along with less SLA violations.

Acknowledgements I am thankful to all concerned with this endeavour.

References

1. M. Mishra, A. Das, Dynamic resource management using virtual machine migrations. IEEE Commun. Mag. **50**(9), 34–40 (2012)
2. W. Hashem, H. Nashaat, R. Rizk, Honey bee based load balancing in cloud computing. KSII Trans. Internet Inf. Syst. (TIIS) **11**(12), 5694–5711 (2017)
3. M. Gamal, R. Rizk, H. Mahdi, Bio-inspired load balancing algorithm in cloud computing, in *Proceedings of International Conference on Advanced Intelligent Systems and Informatics (AISI)*, Cairo, Egypt, pp. 579–589 (2017)
4. A. Strunk, Costs of virtual machine live migration: a survey, in *Proceedings of 8th IEEE World Congress on Services (SERVICES)*, Honolulu, HI, USA, pp. 323–329 (2012)
5. U. Deshp, X. Wang, K. Gopalan, Live gang migration of virtual machines, in *Proceedings of 20th International Symposium on High Performance Distributed Computing*, San Joes, CA, USA, pp. 135–146 (2011)
6. R.V. Rasmussen, M.A. Trick, Round Robin scheduling—a survey. Eur. J. Oper. Res. **188**(3), 617–636 (2008)
7. Weighted round robin, Available online at: http://en.wikipedia.org/wiki/Weighted_round_robin

8. B.X. Chen, X.F. Fu, X.Y. Zhang, L. Su, D. Wu, Design and implementation of intranet security audit system based on load balancing, in *Proceedings of IEEE International Conference on Granular Computing*, pp. 588–591 (2007)

9. K.S.J. Hielscher, R. German, A low-cost infrastructure for high precision high volume performance measurements of web clusters, in *Computer Performance Evaluation. Modelling, Techniques and Tools*. Lecture Notes in Computer Science, vol. 2794, pp. 11–28 (2003)

10. C. Chekuri, S. Khanna, On multi-dimensional packing problems, in *Proceedings of 10th Annual ACM-SIAM Symposium on Discrete Algorithms, Society for Industrial and Applied Mathematics*, pp. 185–194 (1999)

11. H. Youssef, S.M. Sait, Iterative computer algorithms with applications in engineering, Chapter 2

12. A. Beloglazov, R. Buyya, Optimal online deterministic algorithms and adaptive heuristics for energy and performance efficient dynamic consolidation of virtual machines in cloud data centers. J. Concurr. Comput. Pract. Exp. **24**(13), 1397–1420 (2012)

13. N. Ashry, H. Nashaat, R. Rizk, AMS: adaptive migration scheme in cloud computing, in *Proceedings of 3rd International Conference on Intelligent Systems and Informatics (AISI2018)*, Cairo, Egypt, vol. 845 (Springer, 2018), pp. 357–369

14. D. Zeng, S. Guo, H. Huang, S. Yu, V.C. Leung, Optimal VM placement in data centres with architectural and resource constraints. Int. J. Auton. Adapt. Commun. Syst. **8**(4), 392–406 (2015)

15. H. Sun, P. Stolf, J.M. Pierson, G. Da Costa, Energy-efficient and thermal-aware resource management for heterogeneous datacenters. Sustain. Comput. Inf. Syst. **1**(1), 292–306 (2014)

16. M.K. Gupta, T. Amgoth, Resource-aware virtual machine placement algorithm for IaaS cloud. J. Supercomput. **74**(1), 122–140 (2018)

17. M. Abdel Basset, L. Abdle Fatah, A.K. Sangaiah, An improved Lévy based whale optimization algorithm for bandwidth-efficient virtual machine placement in cloud computing environment. Cluster Comput., **22**, 1–16 (2018)

18. F. Alharbi, Y.C. Tian, M. Tang, W.Z. Zhang, C. Peng, M. Fei, An ant colony system for energy efficient dynamic virtual machine placement in data centers. Expert Syst. Appl. **120**, 228–238 (2019)

19. N. Sharma, R.M. Guddeti, Multi-objective energy efficient virtual machines allocation at the cloud data center. IEEE Trans. Serv. Comput. **12**, 158–171 (2016)

20. M. Riahi, S. Krichen, A multi-objective decision support framework for virtual machine placement in cloud data centers: a real case study. J. Supercomput. **74**(7), 2984–3015 (2018)

21. A.F. Antonescu, P. Robinson, T. Braun, Dynamic SLA management with forecasting using multi-objective optimization, in *Proceedings of IFIP/IEEE International Symposium on Integrated Network Management*, pp. 457–463 (2013)

22. X. Chen, Y. Chen, A.Y. Zomaya, R. Ranjan, S. Hu, CEVP: cross entropy based virtual machine placement for energy optimization in clouds. J. Supercomput. **72**(8), 3194–3209 (2016)

23. S.E. Dashti, A.M. Rahmani, Dynamic VMS placement for energy efficiency by PSO in cloud computing. J. Exp. Theor. Artif. Intell. **28**(1–2), 97–112 (2016)

24. T.H. Duong Ba, T. Nguyen, B. Bose, T.T. Tran, A dynamic virtual machine placement and migration scheme for data centers. IEEE Trans. Serv. Comput. (2018)

25. R.N. Calheiros, R. Ranjan, A. Beloglazov, C.A. De Rose, R. Buyya, Cloudsim: a toolkit for modelling and simulation of cloud computing environments and evaluation of resource provisioning algorithms. Software: Pract. Exp. **41**(1), 23–50 (2011)

26. CloudSim: A framework for modeling and simulation of cloud computing infrastructures and services. Available online at: http://www.cloudbus.org/cloudsim/

27. K.H. Kim, A. Beloglazov, R. Buyya, Power-aware provisioning of cloud resources for real-time services, in *Proceedings of 7th International Workshop on Middleware for Grids, Clouds and e-Science*, pp. 1–6 (2009)

28. P. Bohrer, E.N. Elnozahy, The case for power management in web servers, in *Power Aware Computing* (Kluwer Academic Publishers, US, 2002), pp. 261–289

29. A. Wierman, L. Andrew, A. Tang, Power-aware speed scaling in processor sharing systems: optimality and robustness. Perform. Eval. **69**(12), 601–622 (2012)

Survey on Cloud Auditing by Using Integrity Checking Algorithm and Key Validation Mechanism

M. Mageshwari and R. Naresh

Abstract Cloud computing can be used to access and storing data and delivery of different services over the internet. Using cloud storage, users can remotely store their data. Cloud service provider (CSP) provides data owners to store and access their valuable data in the cloud server and offers them to make use of on-demand data access without maintaining a local copy of their data. Even though this service avoids the data owners from making use of their third-party auditor, it certainly possesses serious security threats in maintaining the data owners cloud data. In addition, integrity is also an important issue in maintaining the data owners data stored in the cloud server. This survey presents an overview of integrity check and continuous auditing. The review work based on creating secure clouds by continuous auditing and cloud certification system (CCS) is used for high level security. This survey helps to provide the security by continuous auditing and overcome the integrity issues and to avoid the attacks by integrity checking algorithm and key validation mechanism. In this survey paper, various researchers' ideas based on integrity checking and key schemes have been analyzed as literature review.

Keywords Cloud computing · Third-party auditor · Integrity checking algorithm · User ID · Key validation mechanism

1 Introduction

Cloud storage can be used to save the files remotely. Different types of services are offered by cloud such as Software as a service, Platform as a service, and Infrastructure as a service. When the organizations increased, the data can be stored by multiple ways. But when we are using cloud services which provide data security, trust is good when compared to other storage device [1]. Cloud auditing can be defined as the on-site verification activity. Cloud security audits must check the security. Auditing is taken out by a third-party auditor in cloud computing. The TPA is mainly used for

M. Mageshwari · R. Naresh (✉)
Department of Computer Science and Engineering, SRM Institute of Science and Technology, Kattankulathur, Chengalpattu, Chennai, Tamil Nadu 603203, India
e-mail: nareshr@srmist.edu.in

© The Author(s), under exclusive license to Springer Nature Singapore Pte Ltd. 2023   437
Y.-D. Zhang et al. (eds.), *Smart Trends in Computing and Communications*, Lecture Notes in Networks and Systems 396, https://doi.org/10.1007/978-981-16-9967-2_41

IT auditors. The IT auditor does not access cloud resources directly. The auditor must understand what can be audited and tested. To obtain information, all reports can be collected via TPA which is the main role of TPA. Based on that, integrity checking algorithm hash value can be created for each file from cloud servers, then key validation mechanism provides the separate key for all the data users to avoid the attacks [2]. However, cloud services are important to changing environment, and the organizations are used because of the safety, anonymity, and dependability of cloud services. Cloud service can be used to establishing the trust and increase the transparency. Certification services and companies that sell cloud-related products, such as Amazon, give cloud computing a lot of publicity and exposure. Amazon Web Services and Google Microsoft and VMware are two companies that work together [3]. Various security management methods have been proposed in business auditing in the past literatures for ensuring the reliability and privacy. Continuous auditing can maintain the original and accurate details in business sides. When using index hashing algorithm, data tag replacement algorithm, proof of retrieval algorithm computation cost is high. When using data tags, it required more space.

1.1 These Review Contributions Are Summarized as Follows

- The establishment of a private key is done to reduce the danger of data owner theft.
- An integrity checking algorithm is designed to generate the hash key value before sending to the cloud. The algorithm does not need the high storage space, therefore, reduces the cost of the computation.
- Integrity and security can be maintained by continuous auditing.
- Security and privacy of the data are analyzed and proved.
- To reduce different attacks when using multiple data users.
- When using key validation mechanism, correct user can be able to access the data.

1.2 Cloud Auditing

Cloud auditing must check the data integrity, and it also improves the security and efficiency. The user needs to generate the key. Third-party auditor gives the key to all the data users, and it verifies if the key is sent to correct user or not [4].

1.3 Privacy Protection

While users access the data in cloud easily, cloud storage providers access their privacy information. Privacy security mechanisms are used to guarantee that these data can be secret under malicious clients. It should allow the multiple data users without worrying attackers [5].

1.4 Continuous Auditing

As a result, continuous auditing (CA) of accreditation standards is necessary to ensure that the process is simple and reliable. Continuous auditing is maintaining accurate and original details that will be stored in cloud server [6].

1.5 Integrity Checking Algorithm

For protecting client data in a cloud environment, remote data integrity must be verified. Clients used it to submit and upload critical documents to the cloud. The uploaded file is divided into small blocks using the dynamic block age algorithm. If an intruder taints data in a multi-cloud environment, the continuous auditing process forces the verifier to perform file-level check. When data is stored in a remote storage location, the cloud user reduces control of the data; at this point, consumers may be unaware of security policies, vulnerabilities, and malware information. An intruder might, for example, use a back channel attack and a Denial of Service attack to gain access to cloud data [7].

1.6 Key Generation Mechanism

Users' public keys are generated using their unique identities, but their private keys are generated by either a private key production center or a private key production center (PKG). Once hacked, PKG may easily impersonate any user in order to forge tags while remaining undetected. Cryptography, on the other hand, is a secure alternative because the user's private key contains a concealed value as well as a partial key. The partial key is generated via the key generation procedure (KGM) [8]. Because the user's public key is provided by, certificateless cryptography further avoids the problem of certificate administration. The bulk of cloud auditing solutions in certificateless environments are primarily concerned with personal data. When people wish to share their data, other obstacles arise [9]. When users want to share data with other members of their community, other obstacles arise. Routine data audits, such as address tracking, may result in major assaults for the time being. As a result, it is critical to keep the names of community members disguised during data audits. Confidentiality, on the other hand, protects group members' identities while still revealing them [10].

2 Outline of the Review Work

Integrity checking gives the assurance for safe data in the cloud. The data table describes various algorithms methodologies advantages and disadvantages (Table 1).

3 Prevention of Integrity Security Attacks and Computation Time

This table describes to reduce and prevent various attacks like forgery attack, malicious attacks, and the computation time (Table 2).

4 Various Computation Cost of Auditing

See Figs. 1 and 2.

5 Conclusion

The paper based on a various surveys of different techniques is used to provide data integrity and verify the keys for each data user. It also provides the various techniques used, such as security reliability and privacy. Main thing of this survey is to achieve data integrity and verification of computation cost. As future work, it is in the way of achieving integrity and reducing attacks and computational cost.

Table 1 Outline of review work

S. No.	Author	Methodologies	Algorithm	Advantages	Limitations
1	Lin [11]	Embedded audit module and interceptor approach	Homomorphic algorithm	Automatic data monitoring with improved transparency and efficiency	EAM does not fulfill CA's technical requirement
2	Yang [12]	High-level design of the SAaaS architecture and business flow language can be implemented	Tag generation algorithm, prove and verification algorithm	Shared resources, multi-tenancy, infinite scalability and resources	Denial of service attack, cloud-wide incident detection

(continued)

Table 1 (continued)

S. No.	Author	Methodologies	Algorithm	Advantages	Limitations
3	Zhu [13]	Dynamic audit service for outsourced storage and integrity verification	Periodic sampling algorithm, index hash table	Low cost, scalable. To improve the performance of TPAs	Internal audits damage the integrity
4	Liu et al. [14]	Support formal analysis and fine-grained update requests	Ranked Merkel hash tree, bilinear mapping algorithm	Efficient data security, flexibility	Quality of service is inefficient when using low-level metrics
5	He et al. [15]	Bilinear pairing	Ternary hash tree algorithm, ranked Merkel hash tree algorithm	Verification shows the provable security in the cloud	High computation cost
6	Shen et al. [16]	Auditing is used for group user along with bilinear map, and Hellman is used to reduce the computational cost	Bilinear map algorithm	Decryption time reduced for the user	Computational cost is high
7	Ren et al. [17]	To reduce corrupted data	Privacy preserving algorithm	To recover the data corruption	It can increase data integration efficiency
8	Stephanoward and Fallenbeck [18] [R3]	Bottom up approach and low-level metrics can be used to provide the security	Index hash table	Auditing is trust. Continuous auditing provides the security	Low-level metrics formed complex metrics
9	Shen et al. [19]	Public auditing protocol and blockless verification	MD5 algorithm	Communication cost is less	The proof and verification cost is high
10	Sridhar and Smys [20]	AES encryption and hybrid multilevel authentication algorithm	AES algorithm	Against internal and external attacks by the AES encryption method	The computation and correctness could be checked

(continued)

Table 1 (continued)

S. No.	Author	Methodologies	Algorithm	Advantages	Limitations
11	Yang and Jia [21]	Well-organized storage protocol can be used to provide security Privacy	Proof of retrieve	Privacy is better compared other techniques	Risks can be identified difficult
12	Praveena and Smys [22]	To secure the cloud data. The owner of the private key performs the decryption	Encryption and decryption algorithm	Possible for high security	To be evaluated based on different attacks
13	Shen et al. [23]	To check the data integrity by using algebraic signature	Ternary hash tree algorithm	High efficiency. Computation cost is less	More privacy
14	Yu et al. [24]	Identity-based key homomorphic cryptographic primitives to develop a system for remote integrity checking	Remote integrity, cryptography algorithm	The computation cost is low	Computation cost needs to be reduced
15	Shen et al. [19]	To prevent key exposure based on strong key exposure resilient	AES algorithm	High efficiency. More privacy	High storage overloads. More download time to be needed
16	Aujla et al. [25]	Cloud, safe storage verification, and auditing are used	Merkel hash tree algorithm	Storage and performance should be considerable	High communication and low effectiveness
17	Sebastian [26]	Create a conceptual architecture identify key components	SHA algorithm	Increasing openness and building confidence in the cloud sector to shorten the auditing process	It should not clarify manage certification violations

(continued)

Table 1 (continued)

S. No.	Author	Methodologies	Algorithm	Advantages	Limitations
18	Wu [27]	POS system and POR protocol to prove the security and authentication	Multiple precision algorithms	It provides security and privacy. Evaluation time and cost are low	To achieve security, privacy preserving does not support validation of time stamp
19	Yang [28]	Using the block chain technology to create a compelling opportunity for data visitors	Index hashing algorithm	The TPA and consumers are stateless. They do not need to keep track of data index information	Malicious attack occurred when group of user access the data 0at same time. Malicious attack affects the internal and external audit
20	Li [29]	Lattice-based privacy preserving implemented stateless auditing scheme used	Privacy preserving algorithm	To avoid quantum attack. To achieve provable security	If any hardware device failure, malicious hacks and software bugs raised

Table 2 Security attacks in integrity checking

Author	Security against attack	Computation time
Cao et al. [30]	Forgery attack	Less computation time
Apolinario et al. 31]	Spoofing attack	Minimizes storage overhead
Mahalakshmi and Suseendran [32]	Malicious attack	High storage overhead
Liu et al. [33]	Denial attack, malicious attack	Lower computational cost
Tang et al. 34]	Prevents from replay attack	Minimize computation and storage overhead
Hong et al. [35]	Prevents from forgery attack	Lower computational cost
Lu et al. [36]	Prevents from Dos attacks	Lower computation cost
Ramanan and Vivekanandan [37]	Reduced malicious attack	Lower storage overhead
Sasikala et al. [38]	Prevent quantum computer attacks	Lower computational overhead
El Ghoubach [39]	Prevents from forgery attack	Computational and communication overhead
Rady [40]	Prevents from DDOS attack	Computational overhead is more

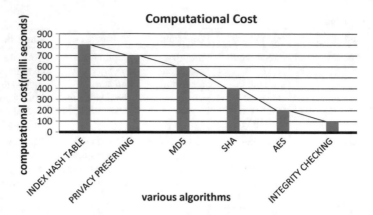

Fig. 1 Computation cost

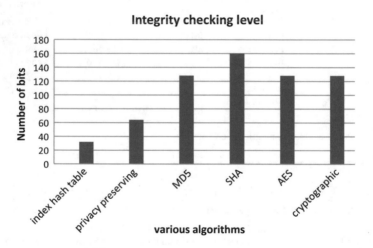

Fig. 2 Level of integrity checking

By reviewing previous studies using TPA, key validation mechanism and integrity checking algorithm are the best solution for cloud auditing.

References

1. A. Saranya, R. Naresh, Cloud based efficient authentication for mobile payments using key distribution method. J. Ambient Intell. Human. Comput. (2021). https://doi.org/10.1007/s12 652-020-02765-7
2. R. Naresh, P. Vijayakumar, L. Jegatha Deborah, R. Sivakumar, A novel trust model for secure group communication in distributed computing. Special Issue for Security and Privacy in Cloud Computing, J. Organizational End User Comput. **32**(3), 1–14 (2020). https://doi.org/10.4018/

JOEUC.2020070101

3. A. Saranya, R. Naresh, Efficient mobile security for E-health care application in cloud for secure payment using key distribution. Neural Process Lett (2021). https://doi.org/10.1007/s11 063-021-10482-1

4. K. Yang, X. Jia, An efficient and secure dynamic auditing protocol for data storage in cloud computing. IEEE Trans. Parallel Distrib. Syst. **24**(9), 1717–1726 (2013)

5. R. Naresh, M. Sayeekumar, G.M. Karthick, P. Supraja, Attribute-based hierarchical file encryption for efficient retrieval of files by DV index tree from cloud using crossover genetic algorithm. Soft Comput. **23**(8), 2561–2574 (2019). https://doi.org/10.1007/s00500-019-03790-1

6. Y. Yu, M.H. Au, G. Ateniese, X. Huang, W. Susilo, Y. Dai, G. Min, Identity-based remote data integrity checking with perfect data privacy preserving for cloud storage. IEEE Trans. Inf. Forensic Secur. **12**, 767–778 (2017)

7. P. Vijayakumar, R. Naresh, L. Jegatha Deborah, S.K. Hafizul Islam, An efficient group key agreement protocol for secure P2P communication. Secur. Commun. Netw. **9**(17), 3952–3965 (2016). https://doi.org/10.1002/sec.1578/abstract

8. P. Vijayakumar, R. Naresh, S.K. Hafizul Islam, L. Jegatha Deborah, An effective key distribution for secure internet Pay-TV using access key hierarchies. Secur. Commun. Netw. **9**(18), 5085–5097 (2016). https://doi.org/10.1002/sec.1578/full

9. R. Naresh, A. Gupta, Sanghamitra, Malicious URL detection system using combined SVM and logistic regression model. Int. J. Adv. Res. Eng. Technol. (IJARET) **10**(4), 63–73 (2020)

10. M. Meenakshi, R. Naresh, S. Pradeep, Smart home: security and acuteness in automation of IOT Sensors. Int. J. Innov. Technol. Explor. Eng. (IJITEE) **9**(1), 3271–3274 (2019)

11. C.-C. Lin, F. Lin, D. Liang, An analysis of using state of threat technologies to implement real-time continuous assurance. in *Proc. 6th World Congress Serv.*, Miami, FL, USA, pp. 415–422 (2010)

12. N.K. Yang, X. Jia, Data storage auditing service in cloud computing: challenges, methods and opportunities. World Wide Web **15**(4), 409–428 (2012)

13. Y. Zhu, G.-J. Ahn, H. Hu, S. Yau, H. An, C.-J. Hu, Dynamic audit services for outsourced storages in clouds. IEEE Trans. Serv. Comput. **6**(2), 227–238 (2013)

14. C. Liu, J. Chen, L. Yang, X. Zhang, C. Yang, R. Ranjan, K. Ramamohanarao, Authorized public auditing of dynamic big data storage on cloud with efficient verifiable fine-grained updates. IEEE Trans. Parallel Distributed. Syst. **25**(9), 2234–2244 (2014)

15. K. He, C. Huang, J. Shi, J. Wang, Public integrity auditing for dynamic regenerating code based cloud storage, in || *IEEE Symposium on Computers and Communication (ISCC)*, (2014)

16. W. Shen, J. Yu, H. Xia, H. Zhang, X. Lu, R. Hao, Lightweight and privacy-preserving secure cloud auditing scheme for group users via the third party medium. J Netw. Comput. Appl. **82**, 56–64 (2017)

17. Z. Ren, L. Wang, Q. Wang, M. Xu, Dynamic proofs of retrievability for coded cloud storage systems. IEEE T Serv. Comput. **11**(4), 685–698 (2015)

18. P. Stephanow, N. Fallenbeck, Towards continuous certification of Infrastructure-as-a-service using low-level metrics, in *Proceedings of the 12th IEEE International Conference on Advanced Trusted Computing*, Beijing, China, pp. 1–8 (2015)

19. J. Shen, D. Liu, D. He, X. Huang, Y. Xiang, Algebraic signatures-based data integrity auditing for efficient data dynamics in cloud computing, in IEEE Trans. Sustain. Comput. (2017)

20. S. Sridhar, S. Smys, A hybrid multilevel authentication scheme for private cloud environment, in *Proceedings of IEEE 10th International Conference on Intelligent Systems and Control (ISCO)*, pp. 1–5 (2016)

21. K. Yang, X.H. Jia, An efficient and secure dynamic auditing protocol for data storage in cloud computing, (in English), IEEE Trans. Parallel Distrib. Syst. **24**(9), 1717–1726 (2017)

22. A. Praveena, S. Smys, Ensuring data security in cloud based social networks, in *Proceedings of IEEE International conference of Electronics, Communication and Aerospace Technology (ICECA)*, vol. 2, pp. 289–295 (2017)

23. W. Shen, J. Qin, J. Yu, R. Hao, J. Hu, Enabling identity-based integrity auditing and data sharing with sensitive information hiding for secure cloud storage. IEEE Trans Inf Forensic Secure **14**(2), 331–346 (2016). http://dx.doi.org/10.1109/TIFS.2018.2850312

24. J. Yu, H. Wang, Strong key-exposure resilient auditing for secure cloud storage. IEEE Trans. Inf. Forensic Secure **12**(8), 1931–1940 (2017). http://dx.doi.org/10.1109/TIFS.2017.2695449
25. G.S. Aujla, R. Chaudhary, N. Kumar, A.K. Das, J.J. Rodrigues, SecSVA: secure storage, verification, and auditing of big data in the cloud environment. IEEE Commun. Mag. **56**, 78–85 (2018)
26. W. Sebastian, J. Yu, H. Xia et al., Light-weight and privacy-preserving secure cloud auditing scheme for group users via the third party medium. J. Netw. Comput. Appl. **82**, 56–64 (2018)
27. H. Yu, Y. Cai, R.O. Sinnott, Z. Yang, ID-based dynamic replicated data auditing for the cloud. Concurrency Comput. **31**(11), e5051 (2019)
28. Y. Yang, J. Yu, R. Hao, C. Wang, K. Ren, Enabling efficient user revocation in identity-based cloud storage auditing for shared big data. IEEE Trans. Dependable Secure Comput. **17**(3), 608–619 (2020)
29. J. Li, H. Yan, Y. Zhang, Efficient identity-based provable multi-copy data possession in multi-cloud storage. IEEE Trans. Cloud Comput. (2020)
30. L. Cao, W. He, Y. Liu, X. Guo, T. Feng, An integrity verification scheme of completeness and zero-knowledge for multi-cloud storage. Int. J. Commun. Syst. 1–10 (2017)
31. F. Apolinario, M.L. Pardal, M. Correia, S-audit: efficient data integrity verification for cloud storage, in *IEEE International Conference on Trust, Security and Privacy in Computing and Communications*, pp. 465–474 (2018)
32. B. Mahalakshmi, G. Suseendran, An analysis of cloud computing issues on data integrity, privacy and its current solutions. Research Article, pp. 467–482 (2019)
33. A. Liu, H. Fu, Y. Hong, J. Liu, Y. Li, LiveForen: ensuring live forensic integrity in the cloud. IEEE Trans. Inf. Forensics Secur. **14**(10), 2749–2764 (2019)
34. X. Tang, Y. Huang, C.-C. Chang, L. Zhou, Efficient real-time integrity auditing with privacy-preserving arbitration for images in cloud storage system. IEEE J. **7**, 33009–33023 (2019)
35. H. Jiang, M. Xie, B. Kang, C. Li, L. Si, ID-based public auditing protocol for cloud storage data integrity checking with strengthened authentication and security. Wuhan Univ. J. Nat. Sci. **23**(4), 362–388 (2018)
36. X. Lu, Z. Pan, H. Xian, An integrity verification scheme of cloud storage for internet-of-things mobile terminal devices. J. Comput. Secur. **92**, 1–17 (2019)
37. M. Ramanan, P. Vivekanandan, Efficient data integrity and data replication in cloud using stochastic diffusion method. Cluster Comput. 14999–15006 (2018)
38. C. Sasikala, C. Shoba Bindu, Certificateless remote data integrity checking using lattices in cloud storage. Neural Comput. Appl. 1513–1519 (2018)
39. I.E. Ghoubach, R.B. Abbou, F. Mrabti, A secure and efficient remote data auditing scheme for cloud storage. J. King Saud Univ. Comp. Inf. Sci. 1–7 (2019)
40. M. Rady, T. Abdelkader, R. Ismail, Integrity and confidentiality in cloud outsourced data. Ain Shams Eng. J. Sci. Dir. 275–285 (2019)

Privacy-Preserving Min–Max Value Retrieval Using Data Obfuscation Algorithm in Distributed Environment

Bhople Yogesh Jagannath

Abstract A comprehensive and efficient p-based algorithm is being proposed to retrieve minimum and maximum of particular data field of database tuple distributed as a horizontal partitioning of single database table among the multiple servers without losing the privacy of individual server. Data obfuscation technique is used in the proposed algorithm to preserve the privacy of individual server. This algorithm is useful in many practical scenarios where privacy-preserving data computing is desirable.

Keywords Privacy-preservation · Distributed environment · Data obfuscation · Horizontal partitioning · Min–max value retrieval · Distributed servers

1 Introduction

Innovation has brought the drastic fall down in semiconductor prices so that system planner over the globe is more inclined toward the distributed system design connected through network link instead of single stand-alone system. Distributed system has multiple merits like scalability, fault-tolerant, reliability, load balancing, and improved response time. The necessities like multiuser configuration, resource sharing, and some form of communication between the workstations have created a new set of problems with respect to privacy, security, and protection of the system as well as the user and data [1].

B. Y. Jagannath (✉)
Department of Information Technology, Government Polytechnic Washim, Washim, India
e-mail: ybhople90@gmail.com

© The Author(s), under exclusive license to Springer Nature Singapore Pte Ltd. 2023   447
Y.-D. Zhang et al. (eds.), *Smart Trends in Computing and Communications*, Lecture Notes in Networks and Systems 396, https://doi.org/10.1007/978-981-16-9967-2_42

2 Proposed Data Obfuscation Algorithm

2.1 Simulation Example

Consider a class with 10 students, each student is having his/her unique rank. Teacher wants to retrieve minimum and maximum rank in his class. No student wants to share his/her rank with anyone, even not with the class teacher. Students are ready to offer any help until and unless their privacy is preserved. Using the following data obfuscation algorithm, teacher retrieves the minimum and maximum rank values with due respect to each student's privacy.

Steps of the proposed data obfuscation algorithm [3] for simulated example.

1. Generate a random number 'X' and pass it to first student. Ask him/her to add his/her rank with 'X' and pass the sum $X + R_1$ to 2nd Student.
2. Second student will add his/her own rank with sum he/she got and pass the sum $X + R_1 + R_2$ to 3rd student, 3rd will pass to 4th, 4th will perform similar action and so on till the 10th students. 10th student will pass the whole sum, i.e., $X + \sum_{i=1}^{10} R_i$), to teacher.
3. Teacher will simply subtract X from the sum and will get sum as $\sum_{i=1}^{10} R_i$.
   Calculate average of all rank as

$$\text{Sum} = \sum_{i=1}^{10} R_i$$

$$\text{Avg} = \text{Sum}/10$$

Initialize the counter variable CountMin $= 0$, CountMax $= 0$, SumMin $=$ Sum, SumMax $=$ Sum, AvgMin $=$ Avg, AvgMax $=$ Avg and pass these value students along with following instruction.

(i) if($R_i \geq$ AvgMin) {
        SumMin=SumMin-R$_i$;
    }
    else
    CountMin++;
(ii) if($R_i \leq$ AvgMax) {
        SumMax=SumMax-R$_i$;
    }
    else
    CountMax++;

For steps 1, 2 and 3, refer Fig. 1.
4. Calculate new average values as AvgMin $=$ SumMin/CountMin
    AvgMax=SumMax/CountMax.

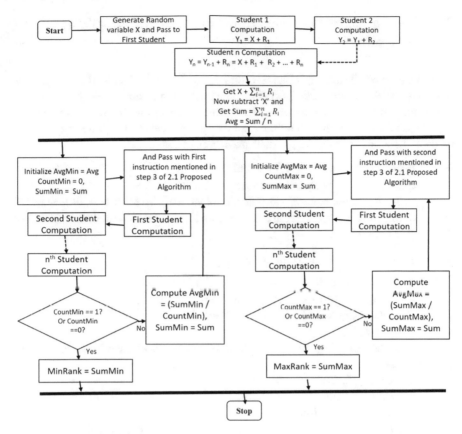

Fig. 1 Flowchart of minimum–maximum rank retrieval

Pass AvgMin, AvgMax. CountMin=0, CountMax=0, SumMin=Sum, SumMax=Sum to students along with above mentioned two instruction to do.

5. Repeat Step 4 along with two instructions in step 2 until we get CountMin = 1 or 0, CountMax = 1 or 0.

6. Value in SumMin indicates Minimum Rank and value in SumMax indicates Maximum Rank.

Minimum–maximum rank retrieval of steps 4, 5, and 6 are given under the synchronization bar in Fig. 1.

2.2   Illustration of Proposed Algorithm with Working Data Set Example

Consider a set of 10 students with their Ranks 1, 2, 3, 4, 5, 6, 7, 8, 9, 10, respectively. Applying the above-mentioned algorithm stepwise on this data set. Purpose is to find minimum and maximum rank of student without losing the privacy of students as their rank is secret for them, and they are ready to reveal it to anyone, but they are ready to offer any help as long as their privacy is preserved.

1–2.   Pass a random no. X to 1st student and get $(X + \sum_{i=1}^{10} i)$. Subtract X

3.   Get Sum $= \sum_{i=1}^{10} R_i$ Calculate Avg $=$ Sum/10 $= 5.5$

    Initialize the counter variable CountMin=0, CountMax=0, SumMin=55, SumMax=55 . And pass this to students along with Instruction in Step 3 of above algorithm.

4.   Get CountMin $= 5$, SumMin $= (55 - \sum_{i=6}^{10} i) = 15$,

    CountMax $= 5$, SumMax $= \left(55 - \sum_{i=1}^{5} i\right) = 40$,

    Calculate AvgMin=SumMin/CountMin=15/5=3,

    AvgMax=SumMax/CountMax=40/5=8.

    Next Iteration-

    Pass the following values along with two instructions mentioned in step 3 of Sect. 2.1.

    CountMin=0, CountMax=0,

    SumMax $=$ sum, SumMin $=$ sum,

    AvgMin $= 3$, AvgMax $= 8$.

    Get CountMin $= 2$, SumMin $= \left(55 - \sum_{i=6}^{10} i\right) = 3$, CountMax $= 2$, SumMax

$= (55 - \left(55 - \sum_{i=1}^{8} i\right)) = 19$, AvgMin $=$ SumMin/CountMin $= 3/2 = 1.5$, AvgMax $=$ SumMax/CountMax $= 19/2 = 9.5$

    Next Iteration-

    Pass the following values along with two instructions mentioned in step 3 of Sect. 2.1

    CountMin $= 0$, CountMax $= 0$,

    SumMin $=$ Sum, SumMax $=$ Sum,

    AvgMin $= 1.5$, AvgMax $= 9$.

5.   Get CountMin $= 1$, SumMin $= \left(55 - \sum_{i=2}^{9} i\right)$, CountMax $= 1$, SumMax $= \left(55 - \sum_{i=1}^{9} i\right)$.

6.   CountMin $= 1$, so value in SumMin indicates minimum Rank.

    CountMax $= 1$, so value in SumMax indicates maximum Rank.

EmpId	Name	Dept	Salary	Rank	DoJ
123	John	Manu.	20000	100	2010
234	Alice	Prod.	30000	200	2009
456	Bob	HR	40000	300	2001
231	Coop	Acc	20000	301	2002
233	Sanjay	Invent.	30000	201	2010
465	Yogesh	Markt.	40000	302	2009
654	Om	Manu.	20000	400	2001
789	Vikas	Prod.	20000	501	2002
987	Bob	HR	30000	700	2011
125	Sanjay	Manu.	40000	800	2012

Partition 1 (rows 123–456), Partition 2 (rows 231–465), Partition 1 (rows 654–125)

Fig. 2 Horizontal partitioning of data sets

2.3 Analogy with Distributed Servers

In the above-simulated example, consider that instead of students, 10 different distributed servers are having a copy of horizontally partitioned data set, of which particular data field (e.g., rank in the above-simulated example) is private for that server and does not want to disclose with any client or server. Aim is to retrieve the minimum and maximum value of that particular data field without compromising the privacy of any individual server. The above-mentioned algorithm when executed in iterative manner produces minimum and maximum values without compromising the privacy of any server in very efficient manner as these servers publish obfuscated data, not the real data. [4]

Figure 2 explains the concept of horizontal partitioning of data sets, where first partition consists of 3 tuples, second partition consists of 3 tuples, and third partition consists of 4 tuples.

3 Complexity Analysis of Proposed Data Obfuscation Algorithm

3.1 Time Complexity Analysis

Assuming that there are n servers, and rank values are uniformly distributed between minimum and maximum, $\sqrt{n}$ iterations are carried out, and each iteration has to go through n servers, so average-case and worst-case time complexity of the algorithm will be $O(n\sqrt{n})$. [5]

In the best case, where all servers having same value of rank, in just 2 iteration, the algorithm produces minimum and maximum rank $O(2 * n) \equiv O(n)$.

Assuming that there are 8 to 10 finite number of servers, average-/worst-case time complexity will be $O(\text{Constant} * \sqrt{n}) \equiv O(\sqrt{n})$, and best case complexity will be $O(\text{constant}) \equiv O(1)$.

3.2 Space Complexity Analysis

Each server requires constant space (space to store 4 integer variables). So, n servers will require $4 * n$ integer space. So time complexity will be $O(4 * n) \equiv O(n)$. Assuming that there are fixed no. of 8 to 10 distributed servers, space complexity will be $O(4 * \text{Constant}) \equiv O(1)$.

4 Conclusion

This p-based data obfuscation algorithm (worst/average time complexity $O(\sqrt{n})$) can have different kinds of utility in distributed environment where a particular relational schema is horizontally partitioned and distributed among n number of servers. Privacy-preserving min-max value retrieval is one of those utilities. In the future, if some distributed system design needs this utility, it can be easily implemented. It can be a most suitable algorithm for distributed key-value stores where privacy-preserving min-max value retrieval is desirable. As its time complexity is $O(\sqrt{n})$, it is computationally efficient, algorithmic latency will be less (ignoring network delays/latency).

References

1. M.R. Ogiela, L. Ogiela, U. Ogiela, Security and privacy in distributed information management, in *2014 International Conference on Intelligent Networking and Collaborative Systems*, INSPEC Accession Number: 14985068. https://doi.org/10.1109/INCoS.2014.108
2. S. Rao, Distributed system: an algorithmic approach. IEEE Distrib. Syst. **9**(11) (2008)
3. D.E. Bakken, R. Parameswaran, D.M. Blough, A.A. Franz, T.J. Palmer, Data obfuscation: anonymity and desensitization of usable data sets. IEEE Secur. Privacy Mag. **2**, 34–41 (2004)
4. J. Hamm, A.C. Champion, G. Chen, M. Belkin, D. Xuan, Crowd-ml: a privacy-preserving learning framework for a crowd of smart devices, in *IEEE International Conference on Distributed Computing Systems* (2015)
5. T.H. Cormen, C.E. Leiserson, R.L. Rivest, C. Stein, *Introduction to Algorithms*, 2nd edn. (The MIT Press, 2001)

Intelligent Virtual Research Environment for Natural Language Processing (IvrE-NLP)

**Prashant Chaudhary, Pavan Kurariya, Shashi Pal Singh,
Jahnavi Bodhankar, Lenali Singh, and Ajai Kumar**

Abstract In the 21st century, natural language processing (NLP) has obtained much prominence for human–machine interaction (HMI). With this interest in natural language processing (NLP) has grown significantly, numerous NLP tools (e.g., morphology, the tagger, and a parser, etc.) have been developed all over the world. Despite having huge importance and requirements, we have noticed gaps for having a comprehensive single framework or platform, which encompass all NLP-related tools and technologies for promoting the research in NLP and sharing the knowledge and resources among NLP researchers required for understanding and building the solution for HMI. Our objective is to apply Software engineering in natural language processing with the concept of an object-oriented model by using a collection of reusable objects by defining the communication protocol, consisting of a set of rules that must be applied to exchange data between two NLP modules. We proposed state of art ivrE—A virtual environment for creating, modifying, executing, and analyzing various NLP solutions and technology. The proposed idea is broadly based on to define own ivrE-NLP object framework model that permits the developer to create, modify, and execute the application and analyze their outcomes by operations on visual representations of the modules. A variety of NLP-based applications (tools, modules, and plugins) already exist, they can publish into store available with environment so it can be used by research community at large. To develop complete NLP

P. Chaudhary (✉) · P. Kurariya · S. P. Singh · J. Bodhankar · L. Singh · A. Kumar
Centre for Development of Advanced Computing, Pune, India
e-mail: cprashant@cdac.in

P. Kurariya
e-mail: pavank@cdac.in

S. P. Singh
e-mail: shashis@cdac.in

J. Bodhankar
e-mail: jahnavib@cdac.in

L. Singh
e-mail: lenali@cdac.in

A. Kumar
e-mail: ajai@cdac.in

© The Author(s), under exclusive license to Springer Nature Singapore Pte Ltd. 2023   453
Y.-D. Zhang et al. (eds.), *Smart Trends in Computing and Communications*, Lecture Notes
in Networks and Systems 396, https://doi.org/10.1007/978-981-16-9967-2_43

framework or platform, we require much more than just assembling or collecting these tools or modules at one place, no matter how good any tool or module is working individually. It requires not only the standards and a set of protocols, but also requires a compliant composition than a pre-defined algorithms and their implementation. In brief, we require a comprehensive open framework to bundle, manage, and integrate set of NLP tools, modules, components, applications, algorithms, and define their associated rules, comprehensive data structures, and knowledge.

Keywords Natural language processing (NLP) · Human machine interaction (HMI) · Intelligent virtual research environment for natural language processing (ivrE-NLP) · Object-oriented model (OOM)

1   Introduction

The original motivation for Intelligent Visual Research Environment for natural language processing (ivrE-NLP) is to support reusing of the existing software and to encourage NLP research groups to explore, analyze, modify, enhance, and develop NLP tools/applications/modules, etc. by combining NLP components already developed. The principal objective of a virtual environment is to accelerate the research in NLP field for developing fast, robust, reusable, and authentic NLP tools/modules/applications which in turn help in the development of the application/system required for HMI.

The objective is to apply Software engineering in NLP with concept of object-oriented model by using collection of reusable objects. Basic idea is to provide such a compliant framework that one can design an NLP system in the best possible way for any given purpose. Furthermore, the ability to insert module into an existing application enables a system to grow in a flexible and modular fashion.

Proposed ivrE-NLP framework/platform will be helpful in developing, executing, and evaluating the outcome of complex and multi-functional NLP systems furthermore provided Visual environment to analyze the components belongs to the built system. Its efforts to meet the following objectives:

- Define information interchange protocol between components/modules of a built NLP system providing lose coupling without changing original logic or theoretical approach. It will allow modules to define their own custom information share protocol mutually accepted by components of system
- Define the integration protocol to incorporate modules in the NLP system written in any source language includes source code, binary code, or executable format
- Support comprehensible graphical interface provide facilities for managing NLP resources during building a system/application.

2   Literature Survey

Dedicated development environments for NLP have been considered or developed for various purposes for the NLP research in the past. Some development environments are conceptualized to aim at the syntactic aspect of natural language and are targeted toward language experts as audience rather than software developers and less work has been done to capture the entire paradigm of NLP. One of the similar works INTEX platform [1] is designed for the language experts to use natural language in general, while LexGram [2] intends on categorical grammars. Some of the IDEs that are most related to development are GATE [3], VisualText4, and NL-OOPS [4]. However, GATE focuses on a uniform platform for NLP-related tasks and Visual Text is specialized in the development of user-defined text analysis applications. However, lips [5] use them in the background thereby considering integrating natural language as a major language for the design flow of software and systems. NL-OOPS is the method nearer to lips, however, it does not insist on the integration with other modeling and programming languages in a common Integrated Development Environment [IDE].

The idea of intelligent information extraction for natural language specifications has been intensively presented in past research work. Saeki et al. [6, 7] studied an approach like [8, 9] that associates lexicon in English text to software concepts found in the object-oriented programming paradigm. Much similar research approaches were considered for pre-programmed methods and restricted the input language to achieve this goal. Some approaches make use of an algorithm [10] or determine recurring designs in property specifications [11].

3   Research, Development, and Integration Environment for NLP

Introducing a Virtual research environment in the field of natural language processing in which researcher allows developing their own application using various NLP components, which are either pre-existing within setup or their own plugin or in the form of services. Thus, ivrE-NLP works as a platform for combining software programming with theoretical knowledge. ivrE-NLP provide initial building blocks and comprises three principal elements:

- Configuration Manager: defines communication protocol based on the object-oriented data model approach, information about various applications is stored in configuration file in form of XML file, which contains annotations for each application, i.e., unique id, a type and dependency file (required resource to run the application).
- Developer Kit: Provide access to various applications and collection of reusable components or modules in the form of object, these objects communicate with each other based on protocol defined by the ivrE-NLP.

- Graphical User Interface: Provide an integrated environment of all available NLP-based tools/modules/applications for design development and execution. It is an easy-to-use visual interface with a 'plug-and-play' design.

Most NLP modules are developed by Iterative and incremental development software models, which require regular analysis and improvement repetitively. Every NLP module must define its dependency in the configuration file along with input and output format so another module can use that information to communicate with it without knowing the internal algorithm.

The researcher can assemble a system with the help of a visual representation of NLP Modules in form of object. Each object can be executed individually or in an integrated pipeline design by the developer. ivrE-NLP provides the facility to analyze results systematically for building, running, and investigating various NLP modules.

We designed ivrE-NLP as a virtual development and integration research environment to support the visual representation of NLP modules for developing analysis and executing entire systems such as machine translation, speech, information extraction, and retrieval.

The visual environment offers executable data objects in form of a tree data structure connected by a leaf or communication protocols, data dependency declarations. Each module contains a configuration file for linking with other modules and generates a log file for viewing and comparing modules results by relating data produced by modules, thus an integrated visual development environment leads to the rapid development of NLP-related research (Fig. 1).

ivrE-NLP allows researcher to develop new applications with the help of pre-existing modules comes with in environment by 'drag and drop' objects and connecting the dots by flowcharts diagram. ivrE-IDE offers various features such as:

- A graphical user interface (GUI) that allows users to organize the NLP tools
- A complete environment with pre-existing NLP modules that work together seamlessly
- Project explore to create/ update application with version control
- Flexible architecture for combining multiple modules
- Review or evaluation facility within development environment (Fig. 2).

ivrE-NLP IDE enables contributors other than the creators of the application to use, enhance, and analyze the NLP modules. The benefits of this facility include:

- Stimulated research in NLP field with reusable components
- Organized and managed engineered application with version control
- Evaluation and review for improvising performance of individual module
- Building profession and maintainable applications for real world
- Tight integration of modules, which improve overall productivity.

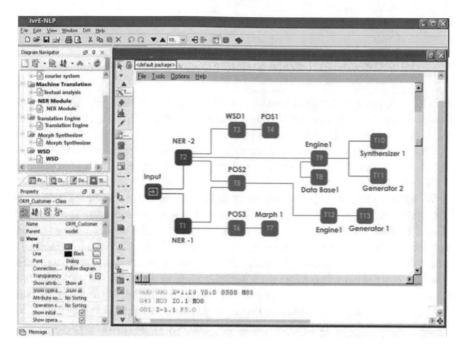

Fig. 1 ivrE-NLP IDE package

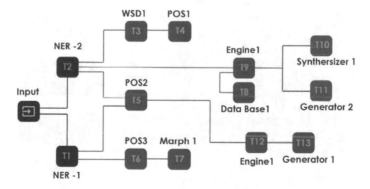

Fig. 2 ivrE-NLP IDE component integration

4 System Architecture

4.1 An Overview of Workflow Model

ivrE-NLP consists of Language resources and Engineering Solutions as an object. Both data and Grammar are types of Language Resource (LRs); all LRs have a data structure linked with them that stores information in form of attribute/value pairs.

Each Document = content + annotations + features where Annotations* where annotations have a start symbol and an end symbol, an ID, a type, and a Map (Fig. 3).

*Annotations (start Node, end Node, ID, type, Map).

Workflow model is the distinct object model which has been used for the ivrE-NLP. This model has been extensively used in natural language processing for a variety of applications. In the above model, modules are typically represented as visual diagrams and connecting lines between the diagrams. Each NLP module performs specific task ranging from preprocessing of text to parsing of sentences and generation of whole texts. The user equipped various facilities to assemble the entire system by using visual diagrams and even execute the system by simply run a command and analyzing results by examining log files generated by the NLP module, so an integrated visual environment helps users for building, running, and analyzing NLP modules.

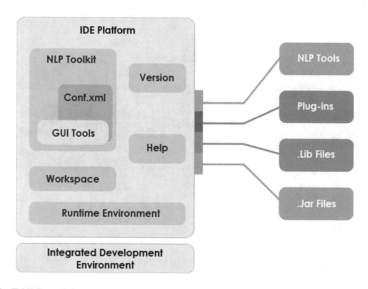

Fig. 3 ivrE-NLP workflow

4.2 ivrE-NLP Paradigm

An ivrE-NLP is a visual environment into which various NLP modules reside. These NLP modules are loosely coupled with each other, they communicate entirely by means of the standard protocol defined by the framework. Inter-modules communication is handled by the workflow model described above along with NLP tools, language resources, and events.

The basic concept that needs to be understood by the researcher before using ivrE-NLP environment is the NLP data model which defines all the resources including input and output for all modules. It uses a standard model (Document = content + template + features) where template (start Node, end Node, ID, type, Map) into the sources document which is characterized by a type and a set of information represented in Map as attribute-value pairs. The Map stores information datasets. The data format is not dependent or restricted to any NLP formalism, to enable the use of modules based on different indigenous algorithms. This generalization helps to facilitate the representation of a wide variety of linguistic information, ranging from very simple tokenizer to very complex parser (Fig 4)

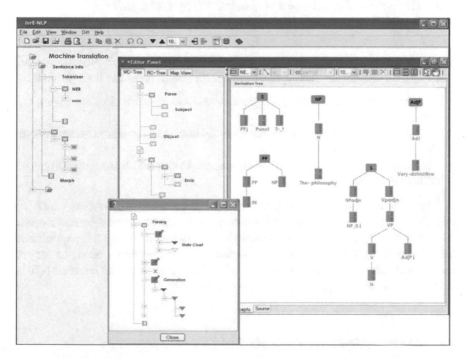

Fig. 4 A ivrE-NLP paradigm

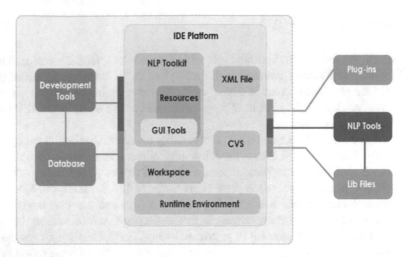

Fig. 5 A ivrE-NLP framework detailing

4.3   ivrE-NLP Framework: (Model, View, Controller Approach)

To provide simple and convenient Graphical Environment in ivrE-NLP framework an integrated architecture having three major sub-components has been provided:

- Language Resources (Model)—represent entities such as lexicons, corpora, and grammar
- Visual Resources (View)—represent visualization and user interfaces components that participate in GUIs.
- Programming Resources (Controller)—represent entities that are primarily algorithms/logic (Fig. 5).

The set of resources provided with ivrE-NLP is known as NLP objects which is a collection of reusable components for natural language processing-based research. All the resources are in the form executable.lib along with the configuration xml file. While using IDE to develop any application, the developer needs to use the toolbar that comes with the framework that involves Graphical User interface (GUI), business logic, and language resources altogether (Fig. 6).

4.4   Communication Protocols and Data Dependency

An ivrE-NLP model consists of NLP tools and technology that could interact with each other in two ways. First, they could express their dependencies using the required statement in their plugin.xml. If plugin A requires plugin B, plugin A can see all the

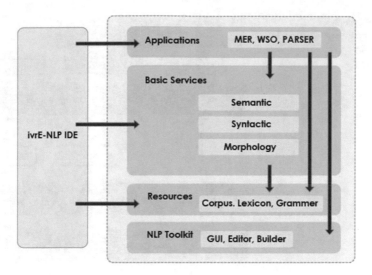

Fig. 6 A ivrE-NLP IDE framework detailing

NLP classes and resources from B, LP class visibility conventions. Each module had a namespace, and they could also specify the versions of their dependencies.

In the NLP data model, the pipeline to be executed is not specified explicitly. Instead, only the requirement of the individual module along with pre-conditions and post-conditions are specified in the configuration file. These dependencies are analyzed, and then a module execution sequence is arranged.

Module dependencies require a commutation protocol to be established for interacting between modules. The configuration file provided along with each module includes a set of annotations or attributes which required being set before execution.

There is a provision to set dependencies via parameters in configuration file, i.e., config.xml.

- Each parameter has one or more attributes, where each attribute has a set of the possible input value
- Each parameter has one or more attributes, where each attribute has a set of the possible output value
- One or more parameter included with set of attributes where each attribute may have a specified value.

An ivrE-NLP model consists of NLP tools and technology that could interact with each other in two ways. First, they could express their dependencies using the required statement in their plugin.xml. If plugin A requires plugin B, plugin A can see all the NLP classes and resources from B dependencies (Fig. 7).

Basically, declarations of pre-and post-conditions are required for each module. A concrete example of these declarations, as specified in an XML configuration files in ivrE-NLP. The pre-conditions indicate that parameter with attributes of object Tagger must be available before it will run. The post-conditions indicate that the attribute

Fig. 7 A ivrE-NLP
communication protocol and
dependency

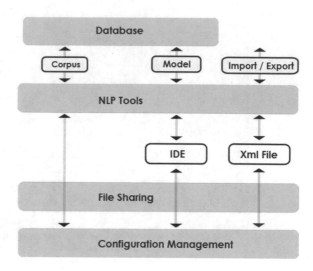

that the parser has run will be set and that parameter of object Parse Derivation Tree
with attributes of type Derived Tree (produced by Generator) will be produced.

4.5   Developing New NLP System in ivrE-NLP

The Researcher creates a new project by selecting various options available in ivrE-
NLP IDE. If necessary, the developer needs to configure the config.xml file to the
registered dependency of the system. Once the project has been constructed developer
can add various.lib or jar file with the project which he can select from NLP lib section.
Developers can write specific wrappers to invoke the open-source tools available with
the framework. There is a configuration file associated with each module in which
the module's data dependencies must be defined, as well as the specifications for
obtaining its result specified. Each project has a configuration manager interface and
the developer need to specify database connectivity and port before execution of the
system. Once the configuration file has been completed, the developer selects "Run
Command" from the NLP IDE top-level menu to compile and execute the application.
Developer can "save" the existing project in workspace provided by IDE for future
use.

4.6   Assemble Customized System via ivrE-NLP

Allow researchers to build customized NLP systems easily from a widely available
language processing components/modules is a key feature of an ivrE-NLP platform.

In any research or development environment the set of compatible modules are obtainable and only few of them are relevant for other modules to develop system. ivrE-NLP provide an option for researchers where it views the set of available modules, their data dependency and can select the application specific module could be integrated straightforwardly. The visual approach taken to implement following functionality:

- Super-graph visualizes instantly available modules with their linked data dependencies.
- A sub-graph visualizes the sub modules details can be easily achieved by simple mouse operations on sub-graph symbol in IDE and yields a system that can be run directly.
- A new system module can be incorporated visually into the existing graph hierarchy of modules.

4.7 Executing NLP System in ivrE-NLP

This is a helpful feature for NLP component developers, where developer possible to obtain instant comment on changes made to the underlying code or resource having dependency exist with system components. It provides a facility to execute the module using IDE, obtain result, evaluate observation, rest the module and replay execution on the upgraded requirement. It encompasses system graph thought of as a detailed assembled graphical process of executable system and useful to evaluate the workflow of system. The system graph uses the color symbols to indicate the current execution state of modules such as green modules indicate ready to execute, red modules tell modules have already been executed, and cannot be executed further. To execute module in ready mode, user should select the ready option with one mouse click and perform execution on second click. In the processing, module symbol becomes red, and the dependent downstream modules automatically turn green.

4.8 Viewing and Analyzing Results in ivrE-NLP

ivrE-NLP provide framework cover one of the important aspects where the researcher can visualize the result of system modules help to analyze the integrated NLP system. NLP systems are complex includes components, interaction protocols, and their development efforts consists of iterative refinement on test data. It provides a provision to define separate results viewers that can be reusable by various NLP modules. It includes the comparison interface to evaluate the outcome of two modules based on the input data, useful to improve the modules of the system. ivrE-NLP also gives the plugin option to integrate the standard tool encourage reusability to compare results among sets of output from a module improve efficiency on outcome of system.

4.9 Statistics and Report Generation in ivrE-NLP IDE

The Report generation feature is designed to provide a user-friendly interface for managing system evaluation reports within ivrE-NLP. Moreover, the Report generation is also useful for comparing various versions of applications with the help of graphs and charts, so the researcher is able evaluate the performance of individual modules in terms of compilation time and memory. In addition, the researcher also allows assembling system with open-source available modules and investigating which module is working well in pair with their NLP system.

5 Conclusions

In this paper, we presented ivrE-NLP a virtual environment for NLP-based solutions and technologies to support software reuse and knowledge sharing within the NLP research community to facilitate NLP researchers for exploring, analyzing, and developing NLP applications. We have also described the recent studies and efforts for building a library to develop NLP application, but our motivation is to design a common unified framework by applying software engineering with NLP for accelerating research and development.

We have also introduced an integrated development environment for NLP components, algorithms, and their associated technology to build a complete NLP solution. We also defined a standard communication protocol where various modules can exchange their information with each other without knowing their internal algorithms; furthermore, developer can execute the NLP solutions and analyze the results of associated modules in isolation and publish their solutions make it available for NLP research community.

References

1. M. Silberztein, INTEX: a corpus processing system, in *Int'l Conf. on Computational Linguistics*, pp. 579–583 (1994)
2. E. König, LexGram-a practical categorial grammar formalism. CoRR, vol. cmp-lg/9504014 (1995)
3. H. Cunningham, D. Maynard, K. Bontcheva, V. Tablan, N. Aswani, I. Roberts, G. Gorrell, A. Funk, A. Roberts, D. Damljanovic, T. Heitz, M.A. Greenwood, H. Saggion, J. Petrak, Y. Li, W. Peters, Text Processing with GATE (2011)
4. L. Mich, Nl-oops: from natural language to object oriented requirements using the natural language processing system lolita. Nat. Lang. Eng. **2**, 161–187 (1996)
5. O. Keszocze, M. Soeken, E. Kuksa, R. Drechsler, Lips: an IDE for model driven engineering based on natural language processing, in *2013 1st International Workshop on Natural Language Analysis in Software Engineering (NaturaLiSE)*, San Francisco, CA, USA (2013)

6. M. Saeki, H. Horai, H. Enomoto, Software development process from natural language specification, in *Proceedings of the 11th International Conference on Software Engineering*, ser. ICSE '89 (ACM, New York, NY, USA, 1989)
7. I.S. Bajwa, B. Bordbar, M.G. Lee, OCL constraints generation from natural language specification, in *International Conference on Enterprise Distributed Object Computing*, pp. 204–213 (2010)
8. M. Soeken, R. Wille, R. Drechsler, Assisted behavior driven development using natural language processing, in *International Conference on Objects, Models, Components, Patterns*, pp. 269–287 (2012)
9. M. Soeken, R. Wille, E. Kuksa, R. Drechsler, *Supporting the Formalization of Requirements Using Techniques from Natural Language Processing*, in preparation (2013)
10. M.B. Dwyer, G.S. Avrunin, J.C. Corbett, Patterns in Property Specifications for Finite-State Verification, pp. 411–420 (1999)
11. M.E.C. Hull, K. Jackson, J. Dick, Requirements Engineering, 2nd edn. (Springer, 2005)

Analyzing the Performance of Object Detection and Tracking Techniques

Suchita A. Chavan, Nandini M. Chaudhari, and Rakesh J. Ramteke

Abstract Objects are detected in computer perspective widely in many real-world applications. In case of video processing, detection and tracking of objects should be very proper and effective. Objects are detected and tracked by traditional methods such as background subtraction, optical flow, and frame differencing method. Convolution neural network, which is a deep learning-based approach, is recently adopted by many developers to identify the object. In this paper, methods of object detection are implemented, analyzed, compared, and discussed. Out of which a robust method has been suggested which satisfies the parameters of precision, recall, and accuracy, also the visualized parameters like object localization, classification, and forms a bounding box to the object are observed and analyzed. It is observed that convolution neural networks detect all relevant objects more accurately than traditional methods. CNN locates the identified object in a video frame using a bounding box that extracts the feature and trains the image for classification. Here, CNN is considered the most promising method for object detection and tracking and can be used in further study where complex work to be handled based on object detection like video inpainting or video restoration.

Keywords Object detection · Foreground mask · Convolution neural network

1 Introduction

Object detection belongs to the techniques of computer vision. It traces the object in a digital image, stored video, or CCTV footage. Object detection plays a significant role in video surveillance, defense and border security, medical image processing, astronomy [1], traffic monitoring system, counting object, image annotation system, extraction of object, and face detection system [2]. In video processing, object

S. A. Chavan (✉) · R. J. Ramteke
KBC North Maharashtra University, Jalgaon, Maharashtra, India
e-mail: suchita19.c@gmail.com

N. M. Chaudhari
KSET, DRS Kiran and Pallavi Patel Global University, Vadodara, Gujarat, India

tracking is a crucial task because detected objects at the first frame should be tracked properly throughout the consecutive frames of the video stream with its motion. Objects can be a person, car, bus, or any moving thing. In video inpainting systems, object detection and tracking are the prior process to perform on a sequence of frames [3]. A video is a collection of consecutive frames. This frame is divided into two parts. One is the foreground object that changes from frame to frame and can be static or moving. Another is the background object that appears in the background of an image.

Background subtraction for object detection can be performed by median filtering, Kalman filtering, and a mixture of Gaussian distributions [4], but a mixture of Gaussian background subtraction techniques is widely used and can easily apply on dynamic background scenes [5]. Optical flow method detects the object by Lucas-Kanade method [6] as well as dense optical flow method. The Lucas-Kanade method computes only sparse feature sets, where dense optical flow computes all points of frames [7]. In this paper, conventional and deep learning-based object detection methods are implemented, analyzed, and compared.

2 Methods of Object Detection and Tracking

Object detection can be possible with methods background subtraction, frame differencing, and optical flow method [8] as well as convolution neural networks.

Background Subtraction Method: In this method, video or image is segmented into foreground object and background object. Each frame is processed to subtract background and update the background, known as background modeling. The general flow of background subtraction is shown in Fig. 1. Foreground mask is generated by subtracting the background image. Foreground mask is a binary image of moving object with learning rate 0 and 1. Background subtraction is formulated with formula.

$$F_m = \begin{cases} 1, & \text{if } I_c - B_m > T \\ 0 & \text{Otherwise} \end{cases} \tag{1}$$

Fig. 1 Generic flow of background subtraction method

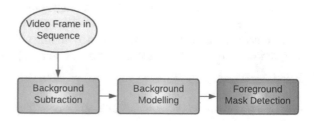

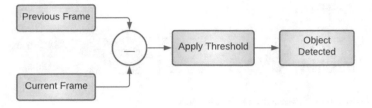

Fig. 2 Working flow of frame differencing method

where F_m is foreground mask with respective to time t, I_c is current image/frame at time t, B_m is background model at time t, and foreground mask is computed by a Gaussian mixture-based model where every pixel is a fusion of K-Gaussians [5].

Frame Differencing: In this method, pixel-wise difference between successive frames is calculated from the sequence of video to detect the objects [9]. Each colored (red, green, and blue) frame is converted into a gray image. Gaussian filters are applied to reduce blurring of the image. The previous frame is converted into gray image. Gray image and the current frame are compared. The measured vail-ation between both images is then checked with threshold. If the value of pixcl becomes bigger, then it is allotted as white, else it is assigned as black [10, 11]. Figure 2 describes the working flow of the frame differencing method.

Optical Flow: Optical flow is the movement of object in between two successive frames caused by displacement of object or camera. If first image is $I(x, y, t)$ and movement of its pixel at t, then new image will obtain $I(x + dx, y + dy, t + dt)$. Dense optical flow estimates the two-dimensional vectors for all pixels in the frame of video based on Gunner Farneback's algorithm [7]. The magnitude and direction of optical flow from a 2D channel exhibit of flow vectors that are processed for the optical flow issue. Hue and value of HSV (hue, saturation, and value) color code are used for visualization of angle of flow, i.e., direction and distance of flow, i.e., magnitude, respectively. The strength of HSV is constantly set to a limit of 255 for ideal perceivability [12].

All above methods are a traditional approach of object detection and tracking. For efficient object detection, classification and localization of objects are very important. Deep learning-based algorithms for detecting objects in all circumstances of video capturing come under recent trends of research.

Convolution Neural Network: CNN is a deep learning algorithm able to detect the objects of image/video. Image is reduced by using ConvNet, so that it becomes easy to process an image without losing features and to achieve good prediction. In a convolution neural network, input image is processed through the convolution layer, then its output passed to the pooling layer, dimension of output feature flatten to one dimensional, and fully connected network shows output as shown in Fig. 3, then fully connected layer with softmax function identifies the classes of objects with probabilistic value [13].

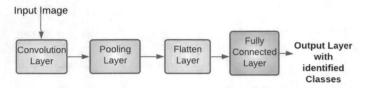

Fig. 3 Architecture of convolution neural network

Convolution layer extracts the feature of input image with volume size $H * W * C$ (height, width, and channel) with filter size $F_h * F_w * C$. Feature map is passed to the pooling layer, and here, the parameters count is reduced by preserving main information about convolved feature. Max pooling is used to reduce the dimension of the image and also suppress the noise. Image is flattened into a single column vector. In fully connected networks, these feature vectors combine together on which softmax function is performed to classify the objects like person, truck, bus, etc.

Object detection using convolution neural network trains the model for the presence of a specific class object. The feature of the object gets localized and classified in output layer with the class information and bounding boxes. For object detection using CNN, basic parameters need to be considered, that is, anchor size, number of class with its name, stride, padding, and filter size.

Object localization finds the value of a bounding box having probability of object detection in terms of 0 and 1, x-coordinate, y-coordinate of the object, height and width of the bounding box.

3 Quantitative Analysis

Implementation of object detection methods explained in the previous section has been performed on Python 3.7 version. After implementing the methods of object detection, its evaluation has been done, and results are retrieved in the form of precision, recall, and accuracy. As per Table 1, confusion matrix is used with the values of true positive, false positive, false negative, and true negative.

Here, the meaning of TP, FP, FN, and TN related to object detection is described below.

True Positive: An object is available in the original frame; also, it is able to detect with the method applied on frame.

False Positive: An object exists in the original frame; however, it is not able to detect or partially detected with the method applied on frame.

Table 1 Confusion matrix for prediction of objects

True positive	False positive
False negative	True negative

Table 2 Analysis of object detection methods through precision, recall, and accuracy

	Background subtraction	Frame differencing	Optical flow	CNN
Precision	0.80	1.00	0.85	0.98
Recall	1.00	0.65	0.89	1.00
Accuracy	0.80	0.65	0.77	0.98

False Negative: An object is not available in the original frame; however, it is able to detect with the method applied on frame.

True Negative: An object not exists in the original frame, and it is not able to detect with the method applied on frame.

From the output of ground truth video of all object detection methods, all the values are calculated as per the number of objects detected correctly, partially, and wrongly and then precision, recall, and accuracy are traced out as per given table. Here, precision and recall are formulated by Eqs. (2) and (3).

$$Precision = \frac{Actual\ Object\ Dotototod}{Actual\ Object\ Detetcted + Object\ not/Partially\ Detetcted} \quad (2)$$

$$Recall = \frac{Actual\ Object\ Detetcted}{Actual\ Object\ Detetcted + False\ Detected\ Object} \quad (3)$$

Precision shows only relevant detection of objects from the image, whereas recall shows all detected objects including relevant or non-relevant objects. As per Table 2, the observation shows that the precision and recall of CNN are greater than the traditional methods, which means the object detection results of CNN are more relevant.

Another observation is that the accuracy of CNN algorithm is 0.98 which is approximately close to 1. After the CNN, background subtraction is also showing better accuracy than frame differencing and optical flow method, but its output is detected in terms of binary image with 0 and 1 values. It has only one channel. Detected objects represent white portion, and the remaining image has a black portion in the background. CNN gives the most appropriate identification of objects using three channels of RGB image. Accuracy of all methods is shown in Fig. 4.

4 Qualitative Analysis

The video used for this experiment is "vtest.avi" having 795 frames [14], which is given as input to all object detection methods. Video is about many people walking on the road. Following Figs. 5a–d show the output of conventional methods and convolution neural network object detection and tracking methods.

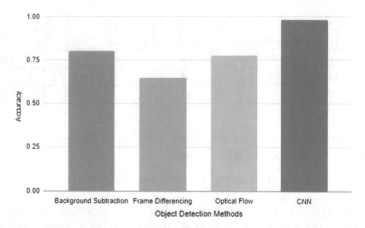

Fig. 4 Accuracy graph of object detection methods

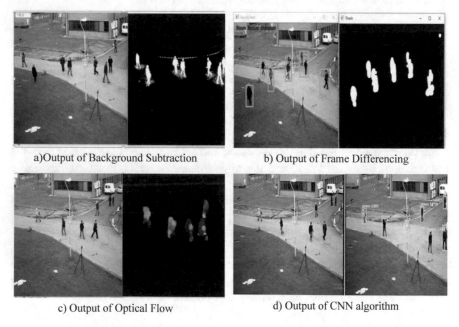

a)Output of Background Subtraction

b) Output of Frame Differencing

c) Output of Optical Flow

d) Output of CNN algorithm

Fig. 5 Methods of object detection. **a** Background subtraction method. **b** Frame differencing method. **c** Optical flow method. **d** CNN-based object detection algorithm

As per the implementation of object detection and tracking methods shown in Figs. 5a, b, objects are detected according to the shape of persons with their foreground mask. Output of background subtraction shows the binary image. Objects are detected with the shadow having a foreground mask. In this method, some objects are detected partially. Occluded objects are detected, but their detection is not very

Table 3 Comparison of object detection and tracking methods

Parameter	Object detection method			
	Background subtraction	Frame differencing	Optical flow	CNN
Object detected	✓	✓	✓	✓
Classification of objects	✗	✗	✗	✓
Selection of regions of interest with bounding box	✗	✓	✗	✓
Result in RGB image with object detection	✗	✗	✗	✓

clear. Frame difference method also shows the binary image as an output. Object has a bounding box in the original frame. Objects are detected without shadow. In this method, occluded object detection does not seem to be appropriate as well as objects are detected more instead of actual total objects present in the frame. The optical flow method has detected and visualized the object through HSV color code shown in Fig. 5c. In this method, some objects are not detected as per the presence of total objects in the input frame.

Convolution neural network object detection method detected all objects correctly. The occluded objects are also detected properly in the output video. Each object has been identified by object localization, and it also classifies the object as per its domain. All classified objects are fitted into the bounding box according to the region of interest of that object with its prediction scores shown in Fig. 5d.

Table 3 shows comparison of object detection and tracking methods with the parameter of object detection, classification of objects with specific domain, bounding box formation according to the region of interest of the object, and the ground truth video having all mentioned parameters with a colored RGB image.

5 Conclusion

In security and surveillance systems, object detection and monitoring are a very significant activity. This paper explores both traditional and deep learning approaches for object detection and tracking. After implementation, the output of those techniques has been discussed.

The quantitative analysis shows precision, recall, and accuracy of all the methods, where the convolution neural network gives more accurate detection of objects compared to background subtraction, frame difference, and optical flow. The conventional approach offers only a foreground mask to distinguish objects, which is not efficient every time for applications where detailed information of objects is required.

Object localization and classification are also important parameters for the process of object identification. This goal is achieved when convolution neural networks are used for object identification; also, bounding boxes appear in ground truth video while performing the method on video. Quantitative and qualitative analyses prove that a convolution neural network is a more promising approach for object detection and tracking. This approach can be further used in video inpainting or video restoration to build efficient and robust mechanisms.

References

1. A. Raghunandan, P. Raghav, H.V. Ravish Aradhya, Object detection algorithms for video surveillance applications, in *2018 International Conference on Communication and Signal Processing (ICCSP)*. IEEE (2018)
2. A. Vahab, et al.,Applications of object detection system. Int. Res. J. Eng. Technol. (IRJET) **6**(4), 4186–4192 (2019)
3. S.A. Chavan, N.M. Choudhari, Various approaches for video inpainting: a survey, in *2019 5th International Conference on Computing, Communication, Control and Automation (ICCUBEA)* (IEEE, 2019)
4. D.H. Parks, S.S. Fels,Evaluation of background subtraction algorithms with post-processing, in *2008 IEEE Fifth International Conference on Advanced Video and Signal Based Surveillance* (IEEE, 2008)
5. A. Shahbaz, J. Hariyono, K.-H. Jo, Evaluation of background subtraction algorithms for video surveillance, in *2015 21st Korea-Japan Joint Workshop on Frontiers of Computer Vision (FCV)* (IEEE, 2015)
6. L.Y. Siong, et al., Motion detection using Lucas Kanade algorithm and application enhancement, in *2009 International Conference on Electrical Engineering and Informatics*, vol. 2 (IEEE, 2009)
7. G. Farnebäck, Two-frame motion estimation based on polynomial expansion, in *Scandinavian Conference on Image Analysis* (Springer, Berlin, Heidelberg, 2003)
8. P.H.L. Shilpa, M.R. Sunitha, A survey on moving object detection and tracking techniques. Int. J. Eng. Comput. Sci. **5**(4) (2017)
9. G. Thapa, K. Sharma, M.K. Ghose,Moving object detection and segmentation using frame differencing and summing technique. Int. J. Comput. Appl. **102**(7), 20–25 (2014)
10. S.H. Shaikh, K. Saeed, N. Chaki, Moving object detection approaches, challenges and object tracking, in *Moving Object Detection Using Background Subtraction* (Springer, Cham, 2014), pp. 5–14
11. N. Singla, Motion detection based on frame difference method. Int. J. Inf. Comput. Technol. **4**(15), 1559–1565 (2014)
12. J. Huang, et al.,Optical flow based real-time moving object detection in unconstrained scenes. arXiv preprint arXiv:1807.04890 (2018)
13. S. Mane, S. Mangale,Moving object detection and tracking using convolutional neural networks, in *2018 Second International Conference on Intelligent Computing and Control Systems (ICICCS)* (IEEE, 2018)
14. vtest.avi Video. Available at https://github.com/opencv/opencv/tree/master/samples/data. Accessed on 11 Sept 2021

A Novel Area Efficient GF(2^m) Multiplier

Asfak Ali, Ram Sarkar, and Debesh Kumar Das

Abstract Galois field multiplication has received a lot of attention from researchers due to its use in encryption, channel coding, and digital signal processing. This paper proposes an area-efficient Galois field multiplier. A sequential approach is adopted here to implement the multiplier in field programmable gate arrays (FPGA). The proposed design gets a 64.58% improvement in an area with respect to 16-bit combinational implementation and 48% with respect to 16-bit sequential implementation. The proposed method also implemented using fixed polynomial, where the design gets a 73.9% improvement in an area with respect to 16-bit combinational imple mentation and 55% with respect to 16-bit sequential implementation. In comparison to early methods, the time delay is also decreased in most cases.

Keywords Galois field · Multiplier · Error control coding · Finite field multiplication

1 Introduction

Efficient polynomial base GF(2^m) multipliers are needed due to the continuous improvement in the field of cryptography [1], channel coding [2], and digital signal processing [3], where m is number of bits. Since the emergence of Internet of things (IoT) in 1999, a huge amount of data is generated every second, for which security must be ensured. For this purpose, the basic block of crypto technology which is nothing but GF multiplier must be area-efficient.

A large number of architectures have been proposed for the Galois Field (GF) multiplier; some of them are combinational, and some are sequential [4–9]. Tang et al. [4] proposed two methods for computation using a pre-computed lookup table.

A. Ali (✉)
ETCE Department, Jadavpur University, Kolkatta, West Bengal, India
e-mail: asfakali.etce@gmail.com

R. Sarkar · D. K. Das
CSE Department, Jadavpur University, Kolkatta, West Bengal, India

© The Author(s), under exclusive license to Springer Nature Singapore Pte Ltd. 2023
Y.-D. Zhang et al. (eds.), *Smart Trends in Computing and Communications*, Lecture Notes in Networks and Systems 396, https://doi.org/10.1007/978-981-16-9967-2_45

Imaña et al. [5] proposed a transpositional method of GF multiplier only for trinomial polynomial. Matsumoto et al. [6] introduced a linear polynomial reduction (LPR) algorithm to multiply two m-bit field elements, where elements were passed through a multiplier array which generates $(2m - 1)$ bit partial prodcuts and divided by primitive polynomial of the field to evaluate GF multiplication. Iliev et al. [7] introduced a parallel programmable Galois field multiplier in which users can use different primitive polynomial as input. Chen et al. [8] proposed a combinational-based $GF(2^m)$ multiplier. They used AND and XOR gates to develop a special cell, and this cell is used in a systolic array to compute the $GF(2^m)$ multiplication, which can be reconfigurable in different sizes. Sangole and Ghodeswar [9] proposed a sequential-based $GF(2^m)$ serial-parallel multiplier. They used aggregation operation using T-FF and D-FF hooked loop structure along with XOR gates to compute the $GF(2^m)$ multiplication.

In this paper, we propose a novel GF multiplier, we consider also its implementation considering fixed polynomial. This design is proved to be better than other methods in terms of area. Besides, in terms of delay also, the proposed design performs well in most of the cases.

2 Fundamental of Galois Field Multiplication

The GF is a set of finite elements denoted as $GF(q)$, where q represents the number of elements in the Galois field. It is also extended using irreducible polynomials of degree strictly less than m over $GF(p)$ as $GF(p^m)$ with respective polynomials shown in Table 1, where p is a prime number, and m is degree of the polynomial. Irreducible polynomials also known as primitive polynomial in a field $GF(p^m)$. For binary fields, we consider a field $GF(2^m)$. Let us consider two elements on $GF(2^m)$ be expressed as:

$$A(x) = a_{m-1}x^{m-1} + a_{m-2}x^{m-2} + \cdots + a_1x + a_0, \quad a_i \in GF(2) \tag{1}$$

$$B(x) = b_{m-1}x^{m-1} + b_{m-2}x^{m-2} + \cdots + b_1x + b_0, \quad b_i \in GF(2) \tag{2}$$

where $GF(2) \in (0, 1)$.

For given primitive polynomial or field polynomial $P(x)$ over $GF(2^m)$, the product of two elements is normal multiplication followed by modulo $P(x)$ and defined as:

$$S(x) = A(x)B(x) \mod P(x) \tag{3}$$

where,

$$P(x) = x^m + p_{m-1}x^{m-1} + \cdots + p_1x + p_0, \quad p_i \in GF(2) \tag{4}$$

Table 1 Field polynomial

m	Polynomials
4	$1 + x + x^4$
8	$1 + x^2 + x^3 + x^4 + x^8$
15	$1 + x + x^{15}$
16	$1 + x + x^3 + x^{12} + x^{16}$

This $S(x)$ becomes a polynomial of degree $m - 1$ as follows

$$S(x) = s_{m-1}x^{m-1} + s_{m-2}x^{m-2} + \cdots + s_1x + s_0, \quad s_i \in \mathrm{GF}(2) \tag{5}$$

3 Proposed Architecture

In this section, algorithm and architecture of the proposed multiplier are discussed.

3.1 Galois Field Multiplier

Algorithm : 1

Input: $A(i)$, $B(j)$, $P(j + 1)$ **Output:** $S(i)(i, j = 0, \ldots, m - 1)$

Step 1: GF Multiplication
$for\ i = m - 1\ down\ to\ 0\ loop$
$for\ j = m - 1\ down\ to\ 0\ loop$
(a) $M_j = B(j)\ AND\ A(i)$
(b) $D(j) = (M_j\ AND\ P(j))\ XOR\ Q(j - 1)\ XOR\ Q(m - 1)$
//D and Q are i/p and o/p of a Delay flip-flop respectively
$endloop\ endloop$

Algorithm The multiplier takes m number of clock cycles to compute GF(2^m) multiplication using Algorithm 1. In step 1.a, it calculates bit-wise multiplication as

$$M_j = A(x_i) \cdot B(x_j) \tag{6}$$

Polynomial reduction is done in step 1.b. Firstly, AND operation between primitive polynomial and multiplication output of every iteration is calculated, then XOR between multiplication result, previous multiplication output and MSB of previous multiplication output is calculated for polynomial reduction as,

$$D(x_i) = (M_j \text{ AND } P(j)) \oplus Q(x_{i-1}) \oplus Q(X_{m-1}) \tag{7}$$

$A(x_i)$ is taken bit by bit from MSB to LSB. Multiplication result is stored in a D flip-flop. Result shifts one bit from LSB to MSB in every iteration. If multiplication output has "1" in MSB, then only it performs polynomial reduction operation in the next iteration. It happens due to the following reason. We know every primitive polynomial has MSB as "1". Hence, when multiplication output has "1" in MSB, only then the modulus operation is performed, and the result is shifted one bit. Moreover, while doing this, we may skip XOR operation of this "1" with $P(x_m)$. After m iteration, results are loaded in the output register $S(j)$ from output of D flip-flop $Q(j)$ in step 1 (b). A top-level view of the proposed model is shown in Fig. 1. Sequential Galois field multiplier structure with variable polynomial is shown in Fig. 2.

Fig. 1 Top-level view of Galois Field multiplier structure with polynomial input

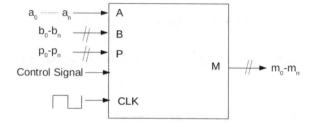

Fig. 2 Sequential Galois Field multiplier structure with variable polynomial

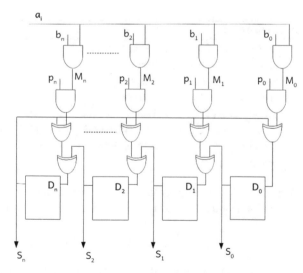

Fig. 3 Sequential Galois field multiplier structure for $m = 4$ with fixed polynomial $x^4 + x + 1$

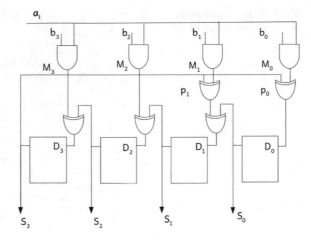

3.2 Fixed Primitive Polynomial Multiplier

Next, we consider the primitive polynomial as fixed. The serial-bit multiplier is developed using the steps shown in Algorithm 2.

Algorithm : 2

Input: $A(i)$, $B(j)$, $P(j + 1)[Fixed]$ **Output:** $S(i)(i, j = 0, ..., m - 1)$

Step 1: GF Multiplication $for\ i = m - 1\ down\ to\ 0\ loop$

$for\ j = m - 1\ down\ to\ 0\ loop$

$(a)\ M_j = B(j)\ AND\ A(i)$

$(b.i)\ if\ p(j) = 1\ then$

$D(j) = M_j\ XOR\ Q(j - 1)\ XOR\ Q(m - 1)$

//D and Q are i/p and o/p of a Delay flip-flop respectively

$(b.ii)else$

$D(j) = M_j\ XOR\ Q(j - 1)$

$endif$

$endloop$

Algorithm The multiplier takes m number of clock cycles to compute GF(2^m) multiplication using Algorithm 2. This algorithm is similar to Algorithm 1, except the step 1.b is designed in two parts. In step 1.b, it calculates bit-wise addition as well as polynomial reduction with respect to $P(x)$. If $P(j)$ is "1", then only it calculates modulus as well as addition; else, it only calculates addition operation is i.e when $P(X_i) = 1$ (step 1.b.i)

$$D(x_i) = M_i \oplus Q(x_{i-1}) \oplus Q(X_{m-1}) \tag{8}$$

Fig. 4 Area comparison plot
of Galois field multiplier

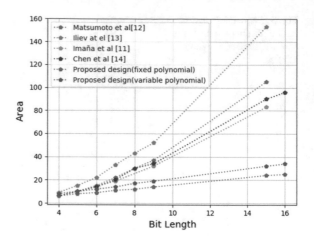

And, when $P(X_i) = 0$ (step 1.b.ii)

$$D(x_i) = M_i \oplus Q(x_{i-1}) \tag{9}$$

After m iteration, results are loaded in the output register $S(j)$ from output of D
flip-flop $Q(j)$. For $m = 4$ considering fixed polynomial $x^4 + x + 1$, the multiplier is
shown in Fig. 3.

4   Experimental Results

We compare our work with both combinational [4–8] and sequential [9] GF multipliers. For delay evaluation, the authors of [5–7] and [8] used Xilinx-virtex-V100FG256.
In [4], authors used Altera-Stratix-II-EP2S130, in [8] in addition to Altera-Stratix-II-
EP2S130, another device Xilinx-virtex4-VXC4VSX55 was used. However, the work
[10] showed that Xilinx-virtex4-VXC4 VSX55 has close performance to Altera-
Stratix-II-EP2S130. To compare our results, we use the Xilinx devices used in many
previous papers. Time delay of the combinational architectures [4–8] and the proposed design is recorded in Table 2, where the first column represents the different
designs, second represents the device name. The third, fourth, and fifth columns,
respectively, represent the time delay of the GF(2^4), GF(2^8) and GF(2^{16}) multipliers
for different designs. From this table, we can clearly see the maximum path delay in
the proposed design gets improved than those of past methods.

For area calculation, the authors of [5–8] used Xilinx-virtex-V100F G256 which
are indicated in Table 3, where the first column shows GF multiplier with the different
bits; second, third, fourth, fifth, sixth, and seventh columns represent the area for
the designs mentioned in [5–8] proposed work without and with fixed polynomial,
respectively.

Table 2 Comparison of time delay (ns) with combinational architecture

Work reference	Device	GF (2^4)	GF (2^8)	GF (2^{16})
Matsumoto et al. [6]	Xilinx-Virtex V100FG256	12.314	17.516	–
Iliev et al. [7]	Xilinx-Virtex V100FG256	12.794	24.311	–
Tang et al. [4] GLUT	Altera-Stratix II EP2S130	11.106	–	–
Tang et al. [4] GPB	Altera-Stratix II EP2S130	19.133	–	–
Chen et al. [8]	Xilinx-Virtex V100FG256	12.215	18.146	–
	Xilinx-vertix4 VXC4VSX55	6.300	8.168	–
Proposed design with variable polynomial	Xilinx-Virtex V100FG256	7.418	7.760	8.462
	Xilinx-vertix4 VXC4VSX55	3.810	3.874	3.955
Proposed design with fixed polynomial	Xilinx-vertix4 V100FG256	7.184	7.418	7.418
	Xilinx-vertix4 VXC4VSX55	3.806	3.810	3.810

Table 3 Comparison of area (slice) with combinational architecture in Xilinx-virtex-V100FG256

Method	Matsumoto et al. [6]	Iliev et al. [7]	Imaña et al. [5]	Chen et al. [8]	Proposed design with variable polynomial	Proposed design With fixed polynomial
GF(2^4)	6	9	–	6	8	6
GF(2^5)	10	15	10	10	10	8
GF(2^6)	15	22	–	14	12	9
GF(2^7)	22	33	19	20	14	11
GF(2^8)	30	43	–	30	17	12
GF(2^9)	37	52	32	34	19	14
GF(2^{15})	105	153	83	90	32	24
GF(2^{16})	–	–	–	96	34	25

Table 4 Comparison of sequential architecture in Spartan3 xc3c50-5pq208

Family: Spartan3

Target device: xc3c50-5pq208

S. No.	Parameters	Conventional design	Sangole et al. [9]	Proposed design with variable polynomial	Proposed design with fixed polynomial
1	No. of bits	8	16	16	16
2	No. of clocks	No. of bits	No. of bits	No. of bits	No. of bits
3	Delay (ns)	7.31	6.280	6.878	6.318
4	No. of slice	26 out of 786	20 out of 786	18 out of 786	9 out of 786
5	No. of slice flip-flop	34 out of 1536	31 out of 1536	16 out of 1536	16 out of 1536
6	No. of 4 input LUTs	59 out of 1536	53 out of 1536	31 out of 1536	16 out of 1536
7	No. of bonded IOBs	36 out of 124	36 out of 124	50 out of 124	34 out of 124

It can be seen that in the proposed design, the area decreases for number of bits >5 in comparison to other works. For $GF(2^{16})$, the number of slices of the proposed design is decreased by 64.5% with respect to [8] which increases to 73.9% for fixed primitive polynomial. An illustration of area comparison of the different GF multiplier architectures is shown in Fig. 4, where the x-axis represents the number of bits, and the y-axis represents the area in the number of slices. From this, it can be clearly seen that the result of the proposed design outperforms the past results.

Comparison with sequential architecture of [9] is shown in Table 4 using Spartan3-xc3c50-5pq208. From this table, we observe that our design is always better than conventional combinational design. In comparison to [9], although time delay increases by 10% number of slices is decreased by 10%, number of flip-flop slices by 48%, number of 4 input LUTs by 42%. For fixed polynomial, with the increase of delay by only 0.6%, number of slices, number of flip-flop slices, number of 4 input LUTs are decreased by 55%, 48%, 70%, respectively. Number of bounded IOBs is also decreased.

5 Conclusion

In this paper, we have presented a new multiplier hardware architecture over the $GF(2^m)$ field based on their primitive polynomials. Given a primitive polynomials, our design become fixed depends on that polynomial. We have used D flip-flop, XOR, and AND gates to implement $GF(2^m)$ multiplier. We showed that area and path delays are reduced in the proposed design. As the slice and number of LUTs are reduced, the power consumption would also reduce.

Acknowledgements A part of this work is funded by DST(ICPS) CPS-Individual/2018/403(G).

References

1. N. Koblitz, Hyperelliptic cryptosystems. J. Cryptol. **1**(3), 129, 150 (1989)
2. R.E. Blahut, *Theory and Practice of Error Control Codes* (Addison Wesley, 1983)
3. R. Lidl, H. Niederreiter, *Introduction to Finite Fields and Their Applications* (Cambridge University Press, Cambridge, 1994)
4. H. Yi, S. Tang, L. Xu, *A Versatile Multi-input Multiplier over Finite Fields* (International Association for Cryptologic Research (IACR), Cryptology ePrint Archive, 2012), p. 545
5. J.L. Imaña, Reconfigurable Implementation of Bit-Parallel Multipliers over $GF(2^m)$ for Two Classes of finite Fields, in *International Conference on Field-Programmable Technology (ICFPT)*, pp. 287– 290 (2004)
6. M. Matsumoto, K. Murase, Multiplier in a galois field. U.S. Patent, No. US4918638 A (1990)
7. N. Iliev, J.E. Stine, N. Jachimiec, Parallel programmable finite field $GF(2^m)$ multipliers, in *Proceedings of IEEE Computer Society Annual Symposium on VLSI Emerging Trends (ISVLSI 04)*, pp. 299–302 (2004)
8. R.J. Chen, J.W. Fan, C.H. Liao, Reconfigurable Galois field multiplier, in *2014 International Symposium on Biometrics and Security Technologies (ISBAST)*, pp. 112–115 (2014)
9. A.M. Sangole, U. Ghodeswar, Design of bit serial parallel multiplier using finite field over $GF(2^p)$, in *2015 International Conference on Advances in Computer Engineering and Applications (ICACEA)*, IMS Engineering College, Ghaziabad, India, pp. 961–965 (2015)
10. Altera, Stratix II vs. Virtex-4 Performance Comparison. White Paper, ver. 2.0 , Altera, Inc (2006)

Virtual Testbed of Vehicular Network for Collision Simulation and Detection on SUMO and OMNeT++ Simulators

Tatini Shanmuki Krishna Ratnam and C. Lakshmikanthan

Abstract Vehicular safety technologies play a vital role in preventing or minimizing the impact of vehicle collisions to reduce life-threatening injuries and keep down vehicle collision-related casualties. One such application is connected vehicles, powered by vehicle-to-infrastructure (V2I) technology to enhance safety on road. It enables all the vehicles on a road within its range to communicate their speed, position, and heading direction to roadside unit (RSU) through cooperative awareness messages (CAM). This process needs three major operations. The first one is receiving the data from the vehicles, the second one detects the collision, and the third one communicates it with the vehicle in case of an impending collision. In this study, we developed a sophisticated algorithm to detect collisions. On detection of the impending collision, RSU sends a warning message to the concerned vehicle. This alerts the driver to take control measures like brake and speed limiting. Here, we implemented the intersection and rear-end collision scenarios using simulation of urban mobility (SUMO) traffic simulator and developed vehicular network (VANET) on network simulator OMNET++ . Veins framework combines both traffic and network simulator. Now using this computerized testbed, we can simulate the collision scenarios on the connected network and evaluate the timeline and data delivery rate with which the latter received the signal in order to take control actions like brake or halt the vehicle.

Keywords Simulation · Testbed · Collision · Detection · SUMO · OMNeT++ · VEINS · Vehicular network · V2I communication · Roadside unit

1 Introduction

According to the World Health Organization (WHO), nearly, 1.35 million people are dying every year due to road traffic crashes. These crashes cost 3% of the gross domestic product for most countries. WHO's global status reports on road safety

T. S. K. Ratnam (✉) · C. Lakshmikanthan
Department of Electronics and Communication Engineering and Department of Mechanical Engineering, Amrita Vishwa Vidyapeetham, Coimbatore, India
e-mail: cb.en.p2ael19023@cb.students.amrita.edu

© The Author(s), under exclusive license to Springer Nature Singapore Pte Ltd. 2023   485
Y.-D. Zhang et al. (eds.), *Smart Trends in Computing and Communications*, Lecture Notes in Networks and Systems 396, https://doi.org/10.1007/978-981-16-9967-2_46

emphasize the need for strong policies, smart road design, enforcement, and technological developments as preventive measures to help in reducing road incidents [12]. Collision detection is one of the most significant as well as the challenging safety application. The common causes are over speeding and exceeding safety distance. In the present time, VANET applications enable the rapid exchange of awareness and safety messages among the connected vehicles, help to reduce collisions [8].

VANET can connect vehicle-to-infrastructure (V2I), vehicle-to-vehicle (V2V), and vehicle-to-everything (V2X) [2]. So to address its complex nature, a simulation testbed is needed to provide precision and repeatability of the test cases mimicking the real-time events and study their capacity and capabilities. Simulation testbed not only reduces development and testing periods but also very economical and boosts rapid development cycle. These computerized simulations for communication networks are offered by softwares such as network simulator-1, ns-2, ns3, and OMNeT++; driving testbeds are CORSIM, PARAMICS, VISSIM, and SUMO. Here, collision scenarios are simulated on the SUMO software and collision detection algorithm designed under the supervision of vehicular network on OMNeT++. We choose SUMO and OMNeT++ simulators as they are open-source applications with robust performance [9].

2 Literature Survey

Vehicular network research started around 2000 in the scope of increasing traffic safety. Connected vehicular networks are an extensive part of building an intelligent transportation system. Author Validi's paper focused on study of VANETS in combination with traditional vision-based sensors for quick and reliable safety features [7]. In urban areas, with increased traffic flow, there is a need for smart methods for handling traffic to avoid accidents. So, author 'Senthil Kumar Mathi' proposed a model to monitor and regulate the traffic flow at intersections through RSU to alert the driver with upcoming dangers like collisions [1].

Author 'Rajeswar Reddy' developed traffic simulation in SUMO and communication network in OMNeT++, so vehicles can communicate with a RSU in a V2I fashion. He concluded that, with increase in density of traffic the communication network will be busy, which may lead to loss of data and delay in communication [3]. Author 'Sivraj P' studied Wi-Fi, DSRC, and modified DSRC communication protocols for VANETs on SUMO and NETSIM. He analyzed that among the other variants modified DSRC with IEEE 802.11 g is having good performance [2, 10].

3 Methodology

The papers mentioned in the above literature survey section focused mainly on the study of comparison on various technological solutions for vehicular networks for enhancement of safety and comfort of the vehicle and its occupants. The work done to understand the specific road events such as different collision scenarios and the system behavior for these collisions at various speeds, traffic conditions, and road conditions are minimal. In this work, we used V2I communication through DSRC protocol to establish a vehicular network. RSU connected automatically to all the vehicles entering the vicinity of this communication range. This RSU is part of infrastructure that receives the CAM from vehicles showing their speed, position, and heading direction. We implemented a sophisticated collision detection algorithm in RSU to detect the collision chances using CAMs. And recorded the time taken for detection of collision to generate an alert or control action halt. We identified the collision scenario where the collision detection is fastest. This whole scenario is to be simulated using the below software tools [1, 6, 9].

1. Python 3.8,
2. Simulation of urban mobility simulator (SUMO),
3. Veins (integrated solution),
4. OMNeT++ (network simulator).

SUMO for Traffic Simulation

In SUMO, we imported real-world road maps and simulated traffic conditions to replicate the roads and road infrastructure like traffic lights, one-way, and two-way lanes that can accommodate denser and less dense traffic conditions, also integrated with other network simulator OMNeT++ and an integration simulator veins to provide a vehicular network for simulating connected vehicles on realistic traffic demands and incidents. To simulate traffic mobility in SUMO, we needed two inputs, they are street network, i.e., map and traffic demand, i.e., vehicle routing as shown in Fig. 1.

In SUMO, there are three ways to generate these road maps, they are possible by

(i)   Using 'netgenerate' command, we compiled the subscripts that contain information of lanes, junctions, etc., to generate road network file.
(ii) 'NETEDIT' is a visual network editor; offers powerful select, edit, and highlight capabilities for user defined road and traffic designing.
(iii) Using 'netconvert' command, we converted real road maps imported from external source like OpenStreetMap database into network files (Fig. 2).

We generated the road network and added traffic flow to create an active map with moving vehicles by compiling the network file(net.xml), demand file(.route.xml or trips.xml), and poly files(.poly.xml) to generate a SUMO executable configuration file with extension.sumo.cfg. We can visualize it by running it on the SUMO-GUI interface. Figure 3 depicts the real-world roads with traffic lights [11].

Collision Simulation

Through the SUMO application, we designed road maps and generated the traffic entities. Now, we need to simulate the collision events to test the communication protocol on network simulator OMNeT++ software. SUMO software is a controlled

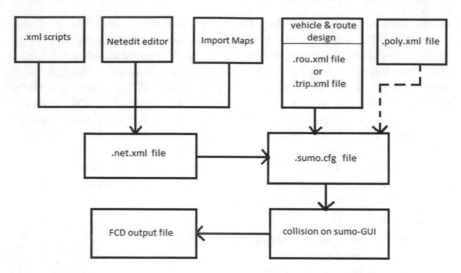

Fig. 1 Flowchart for simulation of road traffic

Fig. 2 Creation of network file—defines road map

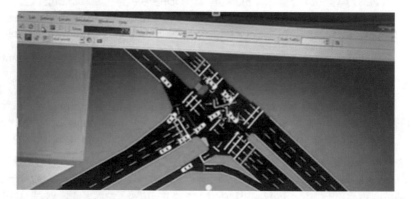

Fig. 3 Real-world traffic network

environment that automatically detects any malicious events and controls the traffic to make it free of collisions. Therefore, the following operations we achieve vehicular collisions:

(i)　　Halt vehicle on edge (lane) and route another vehicle on the same route, cause a rear-end collision

(ii)　　Add variable speed signs to the edges (lanes), causing ambiguity in vehicle routing and leads to accident scenarios like intersection collisions, etc.

(iii)　Add variable speeds to vehicle definition and make overtaking scenarios that lead to a collision.

(iv)　Set the minimum gap parameter of SUMO, so whenever the gap between the vehicles reduced below the minimum gap, collision is detected.

After identifying the collision, SUMO provided a scope to take actions like warn, stop, and remove a vehicle to identify the collision. In Fig. 4, vehicle collision is indicated by a warning message in red color. This warning contains information about vehicles involved in a collision, minimum gap between them, lane identity, and type of collisions. We are able to succeed in simulating intersection and rear-end type collisions. In the warning message, rear-end collision type is specified as a move for stage parameter and intersection collision is identified with the same name [8].

To implement a collision-free algorithm in network simulator, we need vehicles information. As SUMO can also connect to the other software through traffic control interface (TraCI), it helped in controlling the simulation of road traffic and getting

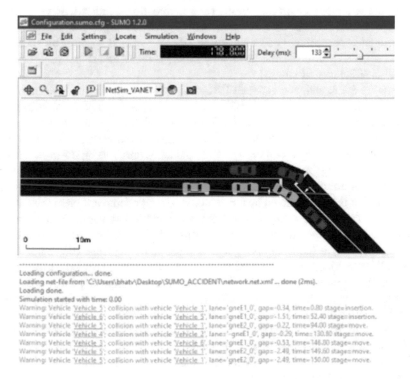

Fig. 4 Collision events in SUMO-GUI interface

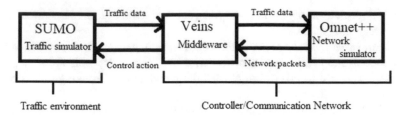

Fig. 5 SUMO and OMNeT++ interface with veins

FCD data during runtime to share it with the network simulator through a middleware framework called veins [6].

OMNeT++ for Network Simulation

OMNeT++ is an event-based network simulator, and it can prototype vehicular networks to communicate with road traffic simulated on SUMO using the veins a middleware framework. Veins will enable SUMO's traffic network models to work under the influence of OMNeT++ communication network as shown in Fig. 5. OMNeT++ offers a platform for implementing and re-using existing communication protocols, and it also supports automotive communication standards. So we can build a compatible and reliable vehicular network. [8] We developed a wireless network with dedicated short-range communication (DSRC) in V2I fashion by specifying medium access control (MAC) layer with IEEE 1609 standard and physical layer with IEEE 802.11p standard [3].

Messages are exchanged between RSU and vehicles within the maximum interference distance range of 2600 m. RSU will get the speed and position information as cooperative awareness message (CAM) from vehicles. The algorithm shown in Fig. 6 checks the chances for collision through current vehicle's position and speed as inputs, represented by vectors $\times 0$ and $v1$, and takes previous inputs in B. Initialize the set C for holding the vehicles that are to be collided with the current vehicles and calculated the current vehicle's position for its changeover time. Every vehicle generated CAM that is to be received by the detector. The algorithm assigned this CAM to CAM $b \in B$ (mapping messages with current vehicles). The vehicle's position over time and calculated the difference between this vehicle and the current vehicle $d(t)$. We computed $d(t)$ as scalar by taking it as squared value of distance $d(t)$, i.e., $D(t) = |d(t)|^2$. We must look for $D(T)$ minimum value over time t; now take time t where $D(t)$ is at its minimum value. If $t < 0$, vehicles are moving away from each other, it means there is no chance of collision. Else the vehicles are moving toward each other. Then, take distance d between vehicles. Check if the d value is lower than the $d$min threshold. If so, then add b to C, i.e., mapping alert to the collision vehicles set, b identifies the vehicles to send alert message, that will be involved in the collision and sends an alert before the collision happens. In our scenario, we are sending a message to halt vehicles for 50 s [5, 6, 9].

Fig. 6 Flowchart of collision detection

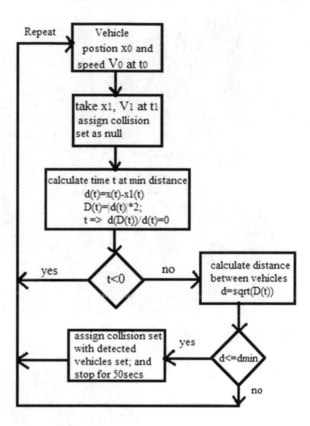

4 Results

Veins support parallel execution of traffic mobility on SUMO and vehicular network on OMNeT++. Red signals on vehicles indicate the threat of collision as depicted in Fig. 7 and 8.

On detection of a collision, RSU sends a message to vehicles and halts for a time duration of 50 s, at their current position to avoid accidents. Figure 9 shows the position of vehicles for time graph. Constant straight lines depict that the vehicles are in a halt state.

For any communication network, its reliability is highly dependent on data (packets) delivery time and data transfer, without any losses and delays. Here, we measured network reliability through packet delivery ratio (PDR), which tells how much percent of data packets received out of all the data packets transferred [4].

$$PDR = \frac{\text{data packets received} * 100}{\text{data packets transmitted}}$$

Fig. 7 Vehicular network on OMNeT++

Fig. 8 Mobility model on
SUMO-GUI at run time

We got a maximum of 86% PDR. We calculated it by accessing the network performance data generated after complete execution. In Fig. 10, bar chart of the total received packets by individual nodes (vehicles) is plotted. In Fig. 11, bar chart of total lost packets by individual nodes (vehicles) is plotted. We took average values of all those parameters to compute PDR.

The average time delay for our vehicular network is calculated by taking the variation in total busy time and channel busy time between RSU and individual nodes (vehicles). The average result is 200 ms. Figure 12 illustrates the bar chart of a busy time for the vehicles.

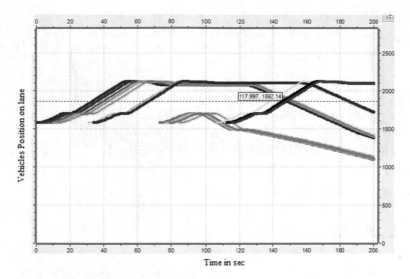

Fig. 9 Vehicles position versus time plot

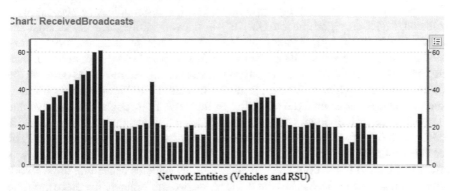

Fig. 10 Bar chart of total received packets

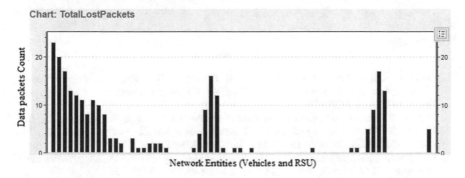

Fig. 11 Bar chart of total packets lost by vehicles

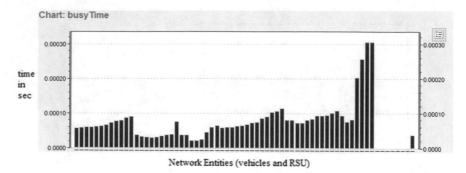

Fig. 12 Bar chart of busy time for vehicles

5 Conclusion

This paper aims to prevent or minimize the impact of vehicle collisions using vehicle-to-infrastructure fashioned VANET technology. We simulated traffic collision scenarios on traffic mobility application SUMO, communication network with collision detection algorithm on OMNeT++ network simulator, and created the VANET functionality by integrating the OMNeT++ communication network to simulated traffic environment on SUMO via veins. We achieved supervisory control over the vehicles within the network range through an exchange of messages between the roadside unit and vehicles. We used standard wireless DSRC protocol to communicate the data. We computed the reliability of the network through PDR and average time delay for an individual vehicle to the roadside unit channel. We attained 86% PDR and 200 ms for average individual time delay for vehicles. We improved PDR for this proposed VANET more than the existing methods [4]. This scenario was effective for short to medium range applications, as vehicles exceeding the network range, communication is lost. To extend the range, we can strategically place multiple RSUs on lane starting and ending based on high-probability accident zones.

References

1. Dr. M. Senthil Kumar, E. Joseph, S, D., V, M. Karthik, S, H., "Design and Implementation of message communication to control traffic flow in vehicular networks", Int. J. Eng. Adv. Technol. (IJEAT), **9** (2019)
2. M. S., S. P., "Performance comparison of communication technologies for V2X applications," in *2020 5th International Conference on Communication and Electronics Systems (ICCES),* pp. 356–362 (2020). https://doi.org/10.1109/ICCES48766.2020.9137879
3. G. Rajeswar, R. Ramanathan, An empirical study on MAC layer in IEEE 802.11p/WAVE based vehicular Ad hoc networks. Procedia Comput. Sci. **143**, 720–727 (2018)
4. A. Mukhopadhyay, V. A. Bharadwaj, "V2X based road safety improvement in blind Intersections,"*2020 Second International Conference on Inventive Research in Computing Applications (ICIRCA),* (2020), pp. 964–968. https://doi.org/10.1109/ICIRCA48905.2020.9183253

5. M. Schratter, M. Hartmann, D. Watzenig, Pedestrian collision avoidance system for autonomous vehicles. SAE Int. J. of CAV **2**(4), 279–293 (2019). https://doi.org/10.4271/12-02-04-002
6. G. Avino et al., "Poster: a simulation-based testbed for vehicular collision detection," 2017 IEEE Vehicular Networking Conference (VNC), (Torino, 2017), pp. 39–40. https://doi.org/10.1109/VNC.2017.8275655
7. A. Validi, T. Ludwig, C. Olaverri-Monreal, "Analyzing the effects of V2V and ADAS-ACC penetration rates on the level of road safety in intersections: evaluating simulation platforms SUMO and scene suite," *2017 IEEE International Conference on Vehicular Electronics and Safety (ICVES)*, (Vienna, 2017), pp. 38–43. https://doi.org/10.1109/ICVES.2017.7991898
8. F. Lyu et al., "Towards rear-end collision avoidance: adaptive beaconing for connected vehicles," in IEEE Conference, Transactions on Intelligent Transportation Systems. https://doi.org/10.1109/TITS.2020.2966586
9. M. Malinverno, J. Mangues-Bafalluy, C.E. Casetti, C.F. Chiasserini, M. Requena-Esteso, J. Baranda, An edge-based framework for enhanced road safety of connected cars. IEEE Access **8**, 58018–58031 (2020). https://doi.org/10.1109/ACCESS.2020.2980902
10. "Documentation-Veins." [Online]. Available: https://veins.car2x.org/documentation/. Accessed 25 May 2020
11. "SUMO User Documentation -Sumo." [Online]. Available: https://sumo.dlr.de/wiki/SUMO_User_Documentation#Introduction. Accessed 06 Jun 2020
12. "WHO Road Safety Status Report" https://www.who.int/violence_injury_prevention/road_safety_status/report/en/. Accessed 2 Aug 2020

Artificial Intelligence-Based IoT Clustered Smart Vending Machine

Ravivanshikumar Sangpal, Suhas Khot, Pratiksha Vallapure,
Rajkumar Mali, Rasika Kumbhar, and Aparna Hambarde

Abstract Vending machines are a vital part of normal people's everyday life in countries like Japan or the USA. India is lacking in the use of effective methodologies in vending machine space. The smart vending machine is designed to work ambiently with IoT hardware and the proprietary design architecture of the physical machine. This vending machine is designed to reduce the mechanical complications present in the current machinery and use technically advanced systems based on IoT; which will help in remotely interfacing with the machine as well as intelligently compiling the data which our artificial intelligence system interprets, giving us predictions for customers as well as the vendor. The machine will be connected to the cloud, responsible for comprehensive data collection and processing. Our system will be having a cluster of vending machines that will be interconnected to each other. The singular vendor can view the data generated by all these machines on a single console and will be able to control and monitor the aspects such as product management, machine state check, and many more from the console itself. The machine will not be having any physical interface for users, and we have designed an utterly end-to-end architecture for this system which will revolutionize the vending machine space in the industry.

Keywords Vending machine · IoT · Artificial intelligence · Cloud computing · IoT cluster · Embedded controller systems

1 Introduction

The vending machines we are considering in this monograph are the product vending machines, i.e., snacks or beverage vending machines which are used in public exposure. The spectrum of the market for these product vending machines is quite wide, and we will be focusing on the corporate usage for explaining in perspective. These vending machines are mostly imported from other countries, we consider recent

R. Sangpal (✉) · S. Khot · P. Vallapure · R. Mali · R. Kumbhar · A. Hambarde
Department of Computer Engineering, KJ College of Engineering and Management Research,
Savitribai Phule Pune University, Pune, Maharashtra 411048, India
e-mail: programmer0608@gmail.com

© The Author(s), under exclusive license to Springer Nature Singapore Pte Ltd. 2023   497
Y.-D. Zhang et al. (eds.), *Smart Trends in Computing and Communications*, Lecture Notes
in Networks and Systems 396, https://doi.org/10.1007/978-981-16-9967-2_47

market drift, and we can observe some manufacturers which are locally based in India. The machines manufactured by these manufacturers are quite effective in terms of general use cases, but they have not addressed all the issues in the existing systems. To understand our solution, let us just understand the present condition of snacks/beverage vending machines.

1.1 Introduction to Standard Machinery

Vending machines in general are quite bulky. The purpose of those machines was to reduce the need for a physical vendor at the place. If you look at some of these machines, they are very identical to the beverage fridge, and we see at local stores. This machine has three major parts or sections namely:

Controller Board: The controller board is the brain of the system which integrates every mechanical component and system to function properly. This board is having controller logic coded in and is mainly used by the number-pad present at the front of the vending machine.

Product Storage: This phase of the machine is where every product is stored for users to buy. A very generic mechanism known as spring mechanism is used as the main dispatching mechanism which pushes the product indicated by the controller. This spring mechanism has its own set of disadvantages.

Dispatchment Block: This part of the machine is where the products get relocated so that users can pick them up. Usually, the products which are pushed by the spring mechanism fall in this area. This block is open access, and users can easily pick their items as they fall.

1.2 Problems Derived from Existing Systems

The product vending machines available in the current market are designed and built considering the old and traditional design followed globally. These machines now need to be cashless considering the Digital India transition and also the COVID-19 restrictions [1]. Cashless vending machines are available in the market, but they are having some issues which need to be re-addressed: (a) Those machines do not provide adequate feasibility for the user to use them fluently. (b) It needs physical intervention for the transaction to be completed and the purchase to be made by the user. (c) The complete process is not automated, which makes the user think that directly providing cash is better than these cashless vending machines. These were only some of the issues faced by the pre-existing machines. This pre-existing system creates a lot of issues for the people using them on a regular basis.

Usually, these machines are to be seen in a commercial environment surrounded by people who work there; the small issue like you need to physically touch the machine to complete the purchase of the product is now going to have a lot of trouble considering this COVID-19 situation [2, 3]. Also, these machines work on a singular unit basis, i.e., it is a single machine having the program installed on it and provides all the data through that machine itself. This generates issues in a larger corporate environment where a lot of machines are needed in that campus itself. These machines need to be based on an interconnected server-client system which will ensure centralized data is used throughout the range of machines used in that campus.

This will ensure that the company does not need to take note of every machine placed, separately. Instead, they can just update the central database and the changes will be automatically reflected to all the machines under that network. Also getting to the point that the user needs to physically intervene in the system for the process to complete is going to be changed. Users need to touch the small display of the machine to select what product they have to buy and also what quantity is needed by them to be bought from the vending machine. This will no longer be true because our system makes this process completely automated and you can stay away from the machine by not going near or touching it to get your purchase done. We will create a secondary interface available to you so that interact with that 2nd interface to get your things done rather than touching the machine. This interfacing methodology is elaborated in subsection 2.1.

1.3 Approach to Smart Vending Machine

We have completely redesigned the overall infrastructure of these vending machines in perspective of how they are used by the user, vendor, and the client as well. We have considered all the aspects of the points stated above and designed this system which will revolutionize the vending machine market.

The smart vending machine is built with the integration of IoT. The vending machine will have controller boards that will be connected to the cloud and completely synchronized with the server. The IoT implementation is not what highlights this project but the use of IoT clustering for managing multiple machines under the same network is the breakthrough in the design. The use of clustering makes sure that all the machines in the cluster or network stay connected with each other and pass the data to the cloud. This will not affect drastically for the end-user but the vendor will be having a tremendous advantage in managing all those machines from the cloud console. The smart vending machine (SVM) also has mechanical design changes which will eliminate the bulky part of the current vending machines and make our resulting product to be modern and compact. The central architecture is having three components as depicted in Fig. 1.

The first module is the physical machine itself which is having a controller board connected to IoT sensors. The machine will have all the data generated and pass it

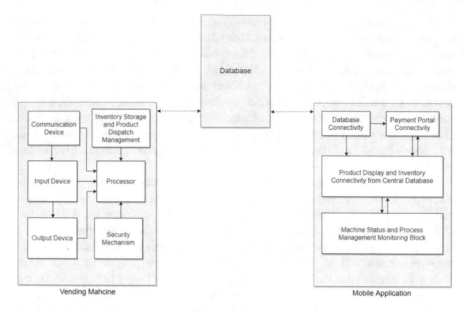

Fig. 1 Central architecture of smart vending machine

to the database on the cloud. The cloud acts as a mucilage of the system. The final component of the system is the mobile application. This application will be used to interact with the vending machine without touching it. The secondary interface is the mobile application that will fetch data from the cloud, and the user will update the requirement on which the machine will trigger a specific item to be dispatched. The dispatched item will then be collected by the user to complete the purchase.

SVM will use online transactions for all the purchases made by the users. The user will select the desired item on the mobile application, and the transaction needs to be completed from the mobile using any convenient method such as UPI, net banking, debit, or credit cards. The completion of the transaction will trigger the mechanical machine, i.e., once the transaction is successful, the authorization token is collected by the server, and the token authenticity is maintained in the database. Once this token is received, the trigger will be launched which will result in triggering the product dispatching mechanism.

2 The Architecture of Smart Vending Machine

As explained in the above subsection, the central architecture of the vending machine is a combination of IoT-based hardware, interactive mobile application, and cloud systems for data collection and interpretation. Let us understand the flow of the system:

The user approaches the vending machine to purchase a specific item from the machine. The machine does not have any kind of manual input, and the user does not need to touch anything. The SVM will have a QR code on the front glass of the machine. The QR code is responsible for machine identification and product information fetching. Once the QR code gets scanned, the user is redirected to our application in which he/she can view all the products present in the machine.

All the data will be visible as if you are on an e-commerce Website. You can easily select the products you intend to purchase and proceed with payment. Once the payment is successful and the token is received on our cloud, the product dispatchment triggers, and your product will be accessible to you in the bottom section of the machine. This complete process runs for all the machines present in the cluster, and each bit of data is stored on the cloud for further computation.

2.1 Software Architecture and Integration

The software end of the system is where the magic happens. The computing of all the data is done in a cloud server. The base of this system is the cloud which will be provided by services such as AWS, Azure, or Google Cloud. The base of all the software is the database which contains the data of the complete system. The schematics of the database are very crucial for other components of this system.

The second layer on top of the barebones database schema is the software stack known as the MERN stack [4]. The application will be designed using the frameworks present in the MERN stack, i.e., MongoDB for database, express, and Nodejs for logical programming in the system, and React.js + REACT Native for Web application and native mobile application designed for vendor and end-user, respectively.

Database Schematics: The database is the most important component of this software system as all the data will be originating from here and the machine-generated sensor data will be stored here. The base schema of the database includes primary data such as (a) number of machines present in the cluster, (b) products listed with details provided by the vendor, (c) the number of products present in any specific machine (inventory data), and (d) the shell structure of data going to be generated in future such assensor data, transactional data, user logs, and product update logs.

User Application: This application will be designed to display the data fetched from the server. The application will have a QR code reader to identify the target machine in front of the user. After that, the interface will be quite simple and intuitive so that users will not face any issues while using the application [5]. We have tried to eliminate the learning curve for this application as it is based on UI trends followed by top applications such as Amazon and Flipkart.

This UI will further initiate the process of the transaction. The transactions will be handled by an API in JavaScript, and the data will securely pass through the cloud server. Once the Auth token is received in the database, the trigger will be launched which will communicate with the controller board on the machine and pass the signal to dispatch the specific product asked by the user.

Vendor Console: This is the most interesting end of the system. This console will have the same UI as the mobile application and will contain administrative privileges so that intricate data can be accessed on this console. The vendor can view all the transactions done on any specific machine and the inventory condition of the products in all present machines in the cluster. This console will view statistical data reports of any period selected depicting the growth or downfall in sales on any single product or a complete range of products.

The statistical data will also display the health of the machines which will be calculated based on the sensor data provided by the machine itself. The data will also depict the need for a specific vending machine in a certain location.

2.2 Distinctive Approach

The software architecture of the system is very peculiarly designed with consideration of nearly a year of granular research on multiple parameters of the vending machine ecosystem. The MERN stack approach referred to above is selected based on this contemporary research. The other options or approaches of the software stacking process are MEAN, LAMP, Meteor stack, and many more. The reason for choosing the MERN over the others is that the MERN stack is the most up-to-date and agile framework structure which enables users to experience a lag-free and intuitive application process. Also, the chosen software framework will ensure the feasibility, scalability, atomicity for transactions, and highly encrypted data flow structure so that the system stays robust.

Looking forward to the hardware perspective of the SVM, this design and methodology are pioneered by us. The existing systems do integrate technologies such as IoT and smart connectivity but the balance which we have created in the feature set for customers, and vendor is a phenomenal achievement. We have focused on the elimination of bulky, traditional designs and have adopted modular and modern material choices. This system will surely change the thought of vending machines in the eyes of the world.

2.3 Hardware Architecture and Integrations

The majority of the data is gained from this hardware system. Let us dive deeper into the intricacies of it. The hardware of the system is the actual vending machine. The vending machine will be fully functional but as stated before, most of the complicated mechanisms are eliminated from the traditional design because the use of those mechanisms is no longer needed and it would increase the bulk of our system unnecessarily. The SVM will have its brain as controller boards such as Arduino, ESP32, Node MCU, etc. [6], for purposes such as main computation and control, connectivity, and wireless communication, respectively.

The other components of the system are the motors which control the tray system holding the products. The tray system is also decreased in size and materials so that the weight of the overall chassis is drastically reduced. Moving on to the circuitry of the main system, the below figure depicts the architecture of the hardware integrated into the smart vending machine (Fig. 2).

The components in this system are interconnected to each other for the completion of the designed modules. Each module is interconnected to each other in such a way that data that is generated in a single module can be passed to other components with minimum latency so that machine is fast and fluent when a transaction is triggered by the user.

Every product dispatched and transaction made will generate the data which will, in turn, help the vendor to manage the elements of this system in order of his profit. The machine's overall construction will not be done using the general aluminum chassis but with a new approach.

The complete chassis and external paneling of the vending machine will be constructed using composite materials made out of recycled plastic. This has cut down our cost by miles, and the structural integrity of the system is also not hampered

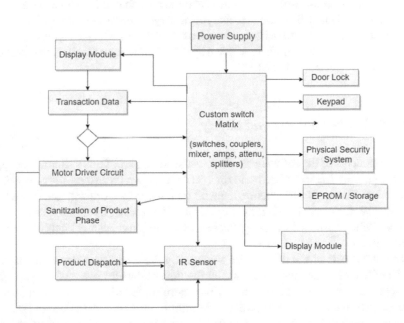

Fig. 2 Hardware integration architecture

as the internal bulk is eliminated by us previously. This complete approach leads us to a cheap to manufacture and easy-to-control and access system of smart vending machine.

3 Safety and Security

As these machines are intended to be placed in public environments, the safety and security of the products in the vending machine, as well as the vending machine itself, are very important.

3.1 Hardware Security

All the vending machines will be equipped with sensors that transmit the data to the cloud. This enables a new gateway to SVM security. We have efficiently placed several sensors in the vending machine which will ensure the safety of the vending machine and the products in them. If any sensor detects an intrusion or abnormal activity in the general workflow of the machine, an immediate alert will be sent to the vendor [7]. This will ensure that everything stays under the control of the vendor. The vendor can also set up periodic checks of the systems which ensures everything stays safe and maintains a sense of security. The below figure depicts the structure of the array of sensors in the SVM.

3.2 Software Security and Safety

The components of the software are the most secured part of this system. The application is built on REACT Native and runs on Android which ensures the safety of the application. The application before launching on the Play Store will undergo a brief security check by the Google Play Store team. The passing of our application from the Play Policy ensures that our application does not intercept any kinds of malware or viruses.

The cloud part of the data is also secured by the cloud provider which will be AWS or GCP or Azure but if we consider the flow of data through these layers, then it will be encrypted in a 256-bit encryption algorithm for which the key will be local for the individual user and the server configuration. We assure the safety and security of transactional data as there are multiple layers of security through which the data will be passed into the system.

Talking about the physical safety of end-users, we have taken into consideration the norms which are implicated by the Government of India following COVID-19 and sanitization regulations. The system will be having a UV lamp-based sanitizer in-built so that every product will stay sanitized throughout the process.

4   Artificial Intelligence Integrated with IoT

On top of all these integrations among IoT hardware and software stack, our system adds a layer of artificial intelligence which proves to be the cherry-on-top for all vendors and users.

The AI system will be running in the cloud and will predict certain parameters which are useful for the vendor [8]. As all the data collected by the sensors is stored, we have our private data set on which our AI model will predict parameters such as (a) the predictive growth and fall in terms of sales of products, (b) the list of products which are top-selling and should be stocked more, (c) the list of products which are not sold often and hence need, less amount of stock, (d) the prediction which states that which machine should be turned down for small regular intervals based on the timestamps of the transactions so that components remain healthy, and (e) list of products which are stocked in the machine for some amount of days but are about to expire which includes products based on dairy and baked goods. This data will help the vendor to efficiently manage the overall machines and products and gain profit.

Also, an AI model is designed for users as well which predicts the products you intend to buy based on the past transactions you have done. This is similar to suggested products on Amazon. This will help users to reduce the steps while buying anything as they see their regular products right on top of the others.

5   Fundamental and Supplemental Enhancements

Summarizing the overall improvements which are made in smart vending as compared to other proprietary or generic product vending machines are:

5.1   Customer-Focused Enhancement

In older vending machines, users needed to carry money in form of cash or coins, or tokenized pieces. Whereas in SVM, they do not need to carry any kind of money with them. In older machines, the users would use manual input for ordering the product and in SVM, and they do not need to touch the machine to get the product. This vending machine can be considered as a step forward in the direction of the digital era with ubiquitous computing.

5.2   Vendor-Focused Enhancement

Vendors are having the most precious advantage in this system of a console control. The one-stop portal where a vendor can access any information in terms of statistical data or conditional report of any specific machine or inventory status of all products, etc. Also, vendors can access this data remotely, if the vendor decides to leave for a weekend to a holiday, he can keep an eye on the console which will provide all the needed information. Also, the insights provided by the AI model stand helpful for getting the most out of the vending machine infrastructure in terms of profits, savings, inventory, etc. The vendors are in fine fettle if they use our smart vending machine system.

6   Circumscription of Smart Vending Machine

As the complete architecture of this machine is based on the Internet, the absence of the Internet will leave the process unrendered. The location of the establishment of SVM is suggested to include an in-house WI-FI or local server connection for annexing the machines available in the network. The workaround for this issue is that the vending machines will need a similar software with local system coded so that the vending machines will access the data when necessary. But the problem of transactional data parsing will still exist in such cases. This leads to attenuation of the SVM.

Even if the structural integrity of this work is acclaimed to be better than the general metal chassis designs, deeper investigations regarding long exposure to external environments and result analysis are needed to be done so that the structural integrity is maintained. Also, if a person is not carrying a smartphone or an Internet-connected device or the device they have does not have adequate camera resources; the user will be unable to use the machine at all. The vending machine is also having multiple parameters which are required to be tested for long periods like the collection of data affecting the server scaling, cloud stability, and sustainability in the longer run, coherence among the IoT hardware in the system is also needed to be tested for a longer duration for assuring the quality and feasibility of this work.

7   Conclusion and Future Work

We can conclude on an amicable note that the smart vending machines can be used in a wide variety of commercial and non-commercial infrastructures benefitting the larger group of people including both sides of the coin, users, and vendors. We believe that this architecture will be a revolutionary start to a new normal in our Indian markets. We also have some design improvements to be considered before wrapping up this monograph, but we need to invest a tremendous amount of testing and quality analysis before deploying this vending machine in the open market.

This research and development are performed under Prof. Aparna Hambarde. We would like to thank her for assistance with documentation, intellectual aspects, and for comments that greatly improved the manuscript. This research is sponsored by our college, i.e., K J College of Engineering and Management Research, Pune; for which we would like to thank our principal sir Dr. Suhas S. Khot for helping us in multiple aspects of this research. We would also like to show our gratitude to them for sharing their pearls of wisdom with us during the course of this research, and we thank our reviewers for their so-called insights. We are also immensely grateful to our parents for their comments on an earlier version of the manuscript, although any errors are our own and should not tarnish the reputations of these esteemed persons.

References

1. D.T. Wiyanti, M.N. Alim, Automated vending machine with IoT infrastructure for smart factory application. J. Phys.: Conf. Series. https://doi.org/10.1088/1742-6596/1567/3/032038
2. N.Y.N. Izuddin, H.A. Majid, N.M.A. Al-fadhali, IoT monitoring system for vending machine. Progress Eng. Appl. Technol. **2**(1), 514–521 (2021). https://doi.org/10.30880/peat.2021.02.01.051
3. T. Dahanayaka, D. Chinthaka, A. R. Lokuge, I. Atthanayake, Cash sanitization machine. Euro. J. Adv. Eng. Technol., **7**(11), 24–25 (2020), ISSN: 2394–658X
4. M. Mohammadi, *How to Integrate React Native with NoSQL Database (Hosted on Azure)* (LAHTI UNIVERSITY OF APPLIED SCIENCES, Napapiiri Jukola, 2020)
5. K. Roy, A. Choudhary, J. Jayapradha: Product recommendations using data mining and machine learning algorithms. ARPN J. Eng. Appl. Sci., **12**(19) (2017), ISSN 1819–6608
6. O. H. Yahya, H. S. ALRikabi, I. A. Aljazaery, Reducing the data rate in internet of things applications by using wireless sensor network. iJOE **16**(3) (2020)
7. A. Hendra1, E. Palantei, Syafaruddin, M. S. Hadis, N. Zulkarnaim, M.F. Mansyur: wireless sensor network implementation for IoT-based environmental security monitoring. The 3rd EPI International Conference on Science and Engineering 2019 (EICSE2019). https://doi.org/10.1088/1757-899X/875/1/012093
8. P. Behera, S. Sanjay, B. Prathima, P. Singh, D. M. Sunil, Machine learning based drought prediction system using cloud and IOT. Int. J. Modern Agricult., ISSN: 2305–7246 **10**(2) (2021)

Artificial Intelligence-based Vehicle In-Cabin Occupant Detection and Classification System

M. Tamizharasan and C. Lakshmikanthan

Abstract Vehicle safety is the primary and necessary aspect of the automobile industry. An airbag is one of the passive safety systems available in an automobile. However, the airbag deployment needs to be controlled to avoid accidents due to it. In low-velocity crashes, the injury caused by airbag deployment is higher than the impact inside the vehicle. Children are more vulnerable to airbags when they sit near airbag housing without proper seatbelts or child seat arrangements. Deployment of the airbag when no occupant is sitting on a seat is unnecessary. So, it is important to detect the occupants' presence in a seat and their classes such as a child or an adult. The primary aim of the paper is to detect the occupants' presence and classify them into different classes. The occupant classes used are child and adult. In this project, we developed a technique to identify the occupants' presence and verify the data from one sensor using another. We collected the image data of the occupants using a camera in a sedan and hatchback vehicles. We analyzed the images using a deep learning algorithm. The output classified the occupants as child and adult. A load cell sensor mounted on the seat was used to measure the weight of the passenger. This data was used to confirm the occupant classification. We evaluated the model detection and classification performances with the parameters such as precision 0.95, recall 0.97, and F1-score 0.96 for image dataset, and we got 0.73 as a classification accuracy for load cell dataset. Finally, we compared both the model performances.

Keywords Detection · Classification · Deep learning · Machine learning · Bounding boxes

M. Tamizharasan (✉)
Electronics and Communication Engineering, Amrita Vishwa Vidyapeetham, Coimbatore, India
e-mail: tamilarasanukl@gmail.com

C. Lakshmikanthan
Department of Mechanical Engineering, Amrita Vishwa Vidyapeetham, Coimbatore, India
e-mail: c_lakshmik@cb.amrita.edu

1 Introduction

Improvement in the safety of vehicle occupants is an important domain in the automotive industry. There are many safety systems available in a vehicle such as automated emergency braking (AEB), traction control system (TCS), electronic stability control (ESC), roll stability control, tire pressure monitoring system (TPMS), and adaptive cruise control (ACC). Each system has its function and purpose concerning safety. Sensors are used to measure or detect the presence of traffic lights or speed bumps, vehicles, pedestrians, and even road edge detection. Image processing algorithms are used for these applications. For example, image processing algorithm for unmarked road edge detection [1] was performed for vehicle safety. For vehicle comfort, automatic seat adjustment using face recognition [2] was performed using image processing technique. For both safety and comfort, real-time detection of speed bump and distance estimation [3] was performed using image processing technique. Typically, we classify the safety systems into two types, such as active safety system and passive safety system. An active safety system prevents the vehicle from a crash. Passive safety systems protect the occupants of a vehicle if a crash occurs. Among all, airbags are primary and passive safety systems. They typically make airbags of nylon fabric, and it is a deflated balloon that is folded inside the steering wheel in such a way that whenever there is a collision, it inflates extremely fast in about 200 miles per hour [4]. We installed a deceleration sensor in different areas, such as the front guard or motor firewall. Setting off the sensor enacts a pyrotechnic device that contains the chemical called sodium azide that touches off and delivers an enormous amount of nitrogen gas after that, it blows-up the airbag. Airbags can reduce the number of deaths or injuries that happened because of crashes. However, it can produce a range of wounds. The usually noticed wounds are slight, scraped spots especially above the chest such as neck, face, and upper arms. A portion of the primary reports was eye wounds. The distribution of airbags might separate. These wounds might be more regrettable to the patients who are wearing eyeglasses [5]. Airbags can reduce serious injuries and death rates for adult occupants who are sitting in the front seat passenger side. Occupants who are sitting too close to the airbag are more vulnerable. It recommended children who are under 13 years old to not sit in the co-driver seat [6]. Airbag deployment will strike the child's head. Besides that, it can crush the head of an infant who is sitting in the passenger seat. Suddenly stopping the vehicle causes the child to slide forward, and becoming closer to the airbag housing because of that, the deployment of an airbag affects the child's face and causes critical spine damages [7].

The primary aim of this project is to detect each occupant sitting in a vehicle and classify them into two different categories, such as adult and child. We have used the image dataset and load cell dataset. We have used YOLO version 3 object detection algorithm for detection and classification for image dataset, whereas we used support vector machine (SVM) algorithm for load cell dataset values. Finally, we studied and analyzed both model's performances.

2 Literature Survey

Author K Jost has implemented an occupant detection for airbags and the passive occupant detection system [8]. This depends on a pressure sensor fixed to the seat, bladder and connected to electronic control unit (ECU). Limitation: This system recognizes heavy objects placed on a seat and detects them as a child and deploys an airbag during a crash.

In-vehicle occupancy detection with CNN on thermal images, authors Nowruzi and Ghods have introduced a new thermal image dataset to detect and count the occupants inside the vehicle using convolutional neural networks [9]. Limitation: This method can only count the number of occupants sitting in the vehicle and not able to classify whether the occupant is an adult or a child. Since this method used thermal sensors and a thermal dataset, the cost of the project was expensive when compared to the pressure sensor.

Low-cost power in-vehicle occupant detection with mm-wave FMCW radar, authors are Alizadesh and Abedi [10]. Limitations: There was no classification part done to identify whether the occupant is a child or an adult. The cost of the radar was also expensive.

In-vehicle occupancy detection and classification using machine learning, authors Malneedi Vamsi and K. P. Soman have introduced a method to detect and classify the occupant into an adult or a child using a machine learning algorithm [11].

Limitations: This is limited only to hatchback cars and not for sedan cars, and this is not suitable when occupants are exposed to bright light conditions. They used single sensor, size of the dataset was small, and also false positive rate is also high for their machine learning algorithm.

In this project, we have used two different vehicles (sedan, hatchback) to make sure that it is not limited only to a specific type of car. We also have performed a machine learning algorithm for load cell dataset and calculated accuracy, precision, recall, and F1-score. Finally, we compared the performance of both the models.

3 Methodology

The major part of the project used an artificial intelligence technique to perform the classification of an occupant in a vehicle. Artificial intelligence is the overall technique of anything that requires some kind of decision making to take place. A part of a subset is called machine learning, and inside the machine learning, there is a subset called deep learning.

Figure 1 shows the overall idea of the project. It comprises two parts. They are based on camera and weight sensor. We have used two different artificial intelligent

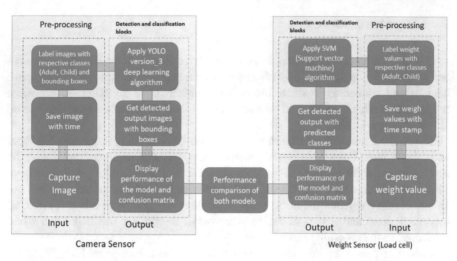

Fig. 1 High-level block diagram of proposed method

models to perform classification tasks. Finally, we compared the performance of both the models.

Occupant Detection and Classification Using Image Dataset

Typically, we can find deep learning applications in image processing, videos, texts, and audio. We use the deep learning technology in this project since the project dealt with an image dataset as an input. This process involved dataset collection, image pre-processing, feeding the dataset to deep learning neural network algorithm, and finally, detection and classification task.

Occupant detection and classification system using images are shown in Fig. 1. It started with input block followed by pre-processing, detection and classification blocks, and finally, output blocks. We have captured and fed the occupant images as an input to the model. We have labeled each occupant's face in an image for class names (adult, child) before feeding into a deep learning algorithm. We have trained the images using the YOLO version 3 algorithm. Finally, we got the output bounding boxes with respective class names and a confident score on top of each image.

Image dataset collection

We have used hatchback and SUV vehicles to collect the dataset. We collected images with many combinations, such as with and without occupants. There were 16 scenarios in which occupants can sit and travel on a 5-seat vehicle. The occupant's seating arrangements are driver (D), co-passenger (CoP), passenger behind the driver (P1), passenger sitting in the middle of the rear seat (P2), and passenger sitting behind the co-passenger (P3).

We captured the images as shown in Fig. 2 from the hatchback and sedan vehicles. We have captured each image and saved it with different names at different times. We have collected datasets in different lighting conditions with variations in

Fig. 2 Sample images from the dataset

gender, variations in seat positions, and different age groups of people. The number of windows varies, which causes different lighting conditions. Further, the camera position and orientation vary, which results in a different perspective.

Data pre-processing

Once we are done with the dataset collection, the next step is to put bounding boxes on top of these images. Bounding boxes are rectangular boxes drawn on every face image. On top of each bounding box, there will be a class name associated with it. There are multiple tools available to perform bounding boxes and label the images. We have used labeling tool to do bounding boxes and labeling for each class. We have collected 1280 images and annotated using labeling tool as shown in Fig. 3.

The Algorithm Used for Image Dataset

The primary aim is to detect the location of an occupant and predict its respective class with bounding boxes for each class. To perform detection and classification, we have used the algorithm called YOLO version 3 [12] which is a part of the deep learning. YOLO applies single neural networks to an entire image, and it divides an image into grid cells. Each grid cell gives the probability values. Finally, it predicts the bounding boxes for each region. YOLO uses convolutional neural networks (CNNs).

The structure of the YOLO algorithm contains several convolutional layers and pooling layers and two fully connected layers at the end. YOLO can distinguish unique items within the picture with the corresponding area. YOLO is faster than other deep learning algorithms such as R-CNN and faster and fast R-CNN, and it applies neural networks to the entire images. It divides the pictures into boxes or

Fig. 3 Sample of an annotated image using labeling tool

regions and gives probability values for each area and picks the best one as per the probabilities. The model Darknet53.conv.74 [13] is used as a transfer learning model to train on top of our classes.

Transfer learning model: Transfer learning is a method where we reuse the learned knowledge from one task to another related task. Transfer learning is an approach to use an available model which is pre-trained for some sort of classes. Darknet 53 acts as a feature extractor in the YOLO version 3 algorithm. Darknet 53 is a convolutional neural network that has got 53 layers stacked one after the other. YOLO version 3 neural network has three types of YOLO layers at the last. We configured the last three YOLO layers for own dataset. Finally, we trained the model with the custom dataset.

Occupant Classification Performed Using Load Cell Dataset

The load cell is a cantilever beam-type load cell that is used to measure the force, tension, or pressure applied to it. Load cell contains strain gauge in a Wheatstone bridge configuration. One side of the beam is fixed, and the pressure is applied on the other side of the beam. Whenever there is a pressure applied strain gauge, it deforms and produces an electrical signal. The electrical signal produced by the load cell is directly proportional to the amount of pressure applied to it.

We have used a 200 kilogram load cell sensor for this project. The load cell setup block diagram is shown in Fig. 4.

The load cell is connected to the HX711 load cell amplifier module. HX711 is an analog to digital converter which converts the load cell value in digital, and the amplified output value is given to Arduino. The Arduino board calculates and converts the weight values in terms of grams and displays the output on the computer.

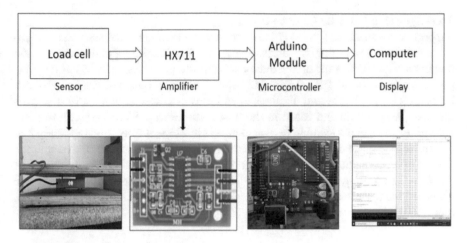

Fig. 4 Load cell setup block diagram

We positioned the load cell in between two wooden plates in such a way that any person can sit on top of it.

Collection of dataset

The load cell is connected to HX711, HX711 is connected to Arduino, and the final output is displayed on a computer. The load cell output weight values are logged in a CSV file with a timestamp. We have concentrated and placed the load cell sensor only on the front passenger seat and collected the data as shown in Fig. 5.

According to national transportation safety board, children who are below 29 kms are not supposed to sit in the front passenger seat so when the occupant weights below 29 kms were considered as a child class and weights which are above 29 kms were considered as an adult class. Overall, we collected 500 load cell values using sedan and hatchback vehicles.

Fig. 5 Load cell complete setup

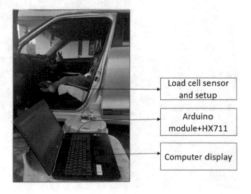

The Algorithm Used for Load Cell Data

Support vector machine (SVM) [14] is a supervised learning and classification algorithm for machine learning. Supervised learning has a labeled input data for specific output classes. SVM is the model that best splits the two classes. SVM has a hyperplane, which is the widest margin that separates the two classes. So, the distance between the two classes and the lines is as far as possible. In this dataset, it splits the two classes (child and adult) in the best possible way. SVM varies from other algorithms in machine learning in a manner that it selects the decision boundary that maximizes the distance from the nearest data points of all classes. We have split the dataset into two parts. One is testing, and the other one is training. We used 30% of the dataset for testing and the remaining 70% of the dataset for training. Finally, we evaluated the performance of the model and calculated confusion matrix, precision, recall, and F1-score values.

4 Results

Image dataset results

We trained and tested the images in the dataset using the YOLO version 3 algorithm and received detections with a confident score per bounding box as an output, as shown in Fig. 6.

Fig. 6 Sample detection and classification output images

Bounding boxes tell where the occupant or a person is present in an entire image. The confident score shows how precious the model believes that the predicted bounding box contains the actual image.

Confident score can be calculated using formula:

$$C = \text{Pr(object)} * \text{IoU},$$

IoU is a shortened form of Intersection over Union. IoU is a ratio between the area of overlap and the area of union. Whenever there are no objects in a cell, the value of a confident score will be zero. We can evaluate the performance of the model using many parameters which are IoU, precision, F1-score, and recall. The average IoU value is 82.67%. mAP is a shortened form of mean average precision. mAP value we obtained is 99.70%.

Precision is the total ratio between current predictions and total predictions, and it tells how good the model at whatever is it predicted. We obtained 0.95 as a precision value. The recall is the ratio between correct prediction and the total number of correct items in the seat. We obtained 0.97 as a recall value and 0.96 as a F1-score.

The graph shown in Fig. 7 represents the losses for each iteration. The blue curve shows the loss values, and it started from a random point and continued to move in

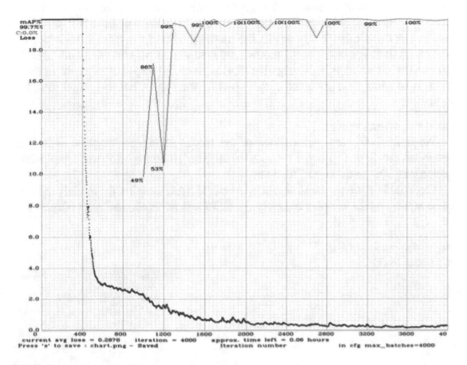

Fig. 7 Loss value and mAP value graph

```
calculation mAP (mean average precision)...
Detection layer: 16 - type = 28
Detection layer: 23 - type = 28
164
detections_count = 689, unique_truth_count = 561
class_id = 0, name = Adult, ap = 99.95%          (TP = 191, FP = 26)
class_id = 1, name = Child, ap = 99.45%          (TP = 353, FP = 0)

for conf_thresh = 0.25, precision = 0.95, recall = 0.97, F1-score = 0.96
for conf_thresh = 0.25, TP = 544, FP = 26, FN = 17, average IoU = 82.67 %

IoU threshold = 50 %, used Area-Under-Curve for each unique Recall
mean average precision (mAP@0.50) = 0.997014, or 99.70 %
Total Detection Time: 6 Seconds
```

Fig. 8 Performance of image dataset model

the direction where the losses are minimal. Finally, at the end of 4000 iterations, the average loss value we obtained was 0.2876.

Class ID=0 and Class ID=1 represent adult and child class, respectively, as shown in Fig. 8.

True positive (TP) is predicted positive, and it is true. For example, the model predicted the class as an adult, and its ground truth is an adult or the model predicted the output class as a child, and its ground truth is a child. False positive (FP) predicted positive, but it is false. For example, the model predicted the class as an adult, but its ground truth is a child, or the model predicted the output class as a child, and its ground truth is an adult. False negative (FN) is predicted output class as not an adult, but it is an adult, or it predicted output class as not a child, but its ground truth is a child.

Load cell dataset results

We have shown the performance of the model in Fig. 9. We have calculated precision, recall, and F1-score for both the classes. There were 150 test sample data, which was 30% in the overall sample.

From Table 1, we can observe that in overall 150 test samples, the model predicted 78 samples and 32 samples correctly for child and adult classes, respectively. The

Fig. 9 Performance of load
cell dataset model

```
[[78  0]
 [40 32]]

                   precision   recall   f1-score   support

              0       0.66      1.00      0.80        78
              1       1.00      0.44      0.62        72

       accuracy                           0.73       150
      macro avg       0.83      0.72      0.71       150
   weighted avg       0.82      0.73      0.71       150
```

Table 1 Confusion matrix of the model

	Predicted child (0)	Predicted adult (1)
Actual class child (0)	78	0
Actual class adult (1)	40	32

Table 2 Performance comparison

S.No	Predicted output with probability values		Actual output
	Image dataset model	Load cell dataset model	
1	Adult (1.0)	Adult 0.98)	Adult
2	Adult (1.0)	Adult (0.98)	Adult
3	Adult (1.0)	Adult (0.98)	Adult
4	Child (0.95)	Child (0.50)	Child
5	Adult (1.0)	Child (0.50)	Adult
6	Child(1.0)	Child (0.78)	Child
7	Adult (0.99)	Child (0.50)	Adult
8	Adult (1.0)	Adult (0.96)	Adult
9	Adult (1.0)	Adult (0.64)	Adult
10	Child(0.84)	Child (0.58)	Child

remaining 40 samples belong to the adult class, but the model misclassified and predicted as child classes. The overall accuracy value obtained is 0.73.

Occupant classification performance comparison

Table 2 shows the difference in the performance of both models. We can verify the output detections using image dataset of front passenger seat occupants with load cell dataset outputs. The image dataset model trained with the YOLO version 3 algorithm performs better than the model trained with the load cell dataset. In ten samples, the model trained with load cell dataset misclassified two samples. Based on the analysis, we can conclude that the image dataset is more reliable and gives better performance.

5   Conclusion

We achieved occupant detection and classification task using an artificial intelligence algorithms. We captured in-vehicle occupant images using a camera connected with the Raspberry Pi module. We performed bounding boxes for each image and classification for each bounding box using labeling application. We used sedan and hatchback vehicles to capture the occupant images and classified them into two classes (child, adult) using you only look once (YOLO) deep learning algorithm.

We displayed detection results as an output with bounding boxes for each class. Finally, we evaluated the model performance using IoU, mAP, precision, recall, and F1-score values. We performed a machine learning algorithm for load cell dataset and calculated accuracy, precision, recall, and F1-score. Finally, we compared the performance of both the models.

6 Future Scope

This project can be extended by adding more classes to the system. The classes can be an empty seat, infant, infant in infant seat, child, child in child seat, and everyday object available in the vehicle. The dataset can be trained with other algorithms, and performance can be compared. The future score of the project is to perform a sensor data fusion technique. This can be achieved by fusing both image dataset and load cell dataset values.

References

1. J. Annamalai, C. Lakshmikanthan, *An Optimized Computer Vision and Image Processing Algorithm for Unmarked Road Edge Detection: Methods and Protocols* (2019). https://doi.org/10.1007/978-981-13-3600-3_40
2. M. Vamsi, K.P. Soman, K. Guruvayurappan, Automatic seat adjustment using face recognition. Int. Conf. Inventive Comput. Technol. (ICICT) **2020**, 449–453 (2020). https://doi.org/10.1109/ICICT48043.2020.9112538
3. V.S.K.P. Varma, A. Sasidharan, K. Ramachandran, B. Nair, Real time detection of speed hump/bump and distance estimation with deep learning using GPU and ZED stereo camera. Procedia Comput. Sci. **143**, 988–997 (2018). https://doi.org/10.1016/j.procs.2018.10.335
4. C. Woodford, Airbags (2008/2020). Retrieved from https://www.explainthatstuff.com/airbags.html
5. K. Cunningham, T.D. Brown, E. Gradwell et al., Airbag associated fatal head injury: case report and review of the literature on airbag injuries. Emergency Med. J. **17**, 139–142 (2000)
6. J.A. Saveika1, C. Thorogood, Airbag-mediated pediatric atlanto-occipital dislocation. Am. J. Phys. Med. Rehabil. **85**(12), 1007–10 (2006). PMID: 17117005. [PubMed] [Read by QxMD]
7. A1. Quiñones-Hinojosa, P. Jun, G.T. Manley, M.M. Knudson, N. Gupta, Airbag deployment and improperly restrained children: a lethal combination. J. Trauma. **59**(3), 729–33 (2005). PMID: 16361920. [PubMed] [Read by QxMD]
8. K. Jost, "Delphi occupant detection for advanced airbags. The passive occupant detection system (PODS) from delphi automotive systems." Autom. Eng. Int. **108**(2000)
9. F. Nowruzi, W. El Ahmar, R. Laganiere, A. Ghods, "In-vehicle occupancy detection with convolutional networks on thermal images," 2019 IEEE/CVF Conference on Computer Vision and Pattern Recognition Workshops (CVPRW) (Long Beach, CA, USA, 2019), pp. 941–948
10. M. Alizadeh, H. Abedi, G. Shaker, Low-cost low- power in-vehicle occupant detection with mm-wave FMCW radar (2019)
11. M. Vamsi, K. P. Soman, "In-vehicle occupancy detection and classification using machine learning," 2020 11th International Conference on Computing, Communication and Networking Technologies (ICCCNT) (2020), pp. 1–6. https://doi.org/10.1109/ICCCNT49239.2020.9225661

12. J. Redmon, A. Farhadi, An incremental improvement. arXiv (2018)
13. J. Redmon, Darknet: open-source neural networks in c. http://pjreddie.com/darknet/ (2013–2016)
14. C. Cortes, V. Vapnik, Support-vector networks. Mach. Learn. **20**(3), 273–297 (1995)

A Hybrid Approach Towards Machine Translation System for English–Hindi and Vice Versa

Shraddha Amit Kalele, Shashi Pal Singh, Prashant Chaudhary, Lenali Singh, Ajai Kumar, and Pulkit Joshi

Abstract With the rapid progress in the technology and data in the public domain, the machine translation and data science have made remarkable progress. In this paper, we discuss our specific use case of developing machine translation system for English to Hindi and Hindi to English language translation. For this system, we have used the daily proceedings of the Lok Sabha as data and developed NMT-based machine translation system on the top of already available rule-based machine translation system. Developed system has been evaluated using bilingual evaluation understudy (BLEU) as well as the human evaluation metrics using comprehensibility and fluency. In machine translation (MT), there is the trend of measuring post-editing time, and thus, we have also evaluated our system by measuring post-editing time using open-source tool.

Keywords Machine translation · Neural network · Hybrid · Parliamentary domain · Evaluation · BLEU scores · Human evaluation

S. A. Kalele (✉) · S. P. Singh · P. Chaudhary · L. Singh · A. Kumar · P. Joshi
AAI, Centre for Development of Advanced Computing, Pune, India
e-mail: shraddhak@cdac.in

S. P. Singh
e-mail: shashis@cdac.in

P. Chaudhary
e-mail: cprashant@cdac.in

L. Singh
e-mail: lenali@cdac.in

A. Kumar
e-mail: ajai@cdac.in

P. Joshi
e-mail: pulkitj@cdac.in

523

1 Introduction and Background

With the advancement in machine learning technologies, the language technology has reached at an advanced level of maturity, wherein it is impacting everyone's life in the world irrespective of the language. Language technology has made the work of translation easy and less time consuming. Many machine translation tools and systems are available for use, but to best our knowledge, none of them assists for translation from English to Hindi and vice versa for parliamentary domain. In this paper, we share our experience in developing hybrid machine translation system for translating from English–Hindi and vice versa for daily proceedings in parliamentary domain.

Daily processing of the Lok Sabha produces many documents in various domains, namely papers to be laid on the table (PLOT), list of business (LOB), Bulletin Part I and Part II, synopsis, etc. Papers to be laid on table includes reports and notifications from the institutes, companies, etc. which comes under government or financed by government. Parliamentary Bulletin Part I contains a brief record of the items taken up daily like question, papers to be laid, bills introduced, etc. Parliamentary Bulletin Part II includes items of general interest to members like changes in responsibility of members, progress of Bills and matter related to security, etc. Synopsis contains a brief record of speeches of members made during the session, etc. Bulletin Part I and Synopsis are published daily, while Parliamentary Bulletin Part II is published during the session and also in intersession as well. All English data produced during the working has to be translated to Hindi, the same day for further processing. To assists the translation of daily proceedings, a hybrid system has been developed. The system uses both rule-based and NMT approach for translation.

1.1 Background

Until 2013, rule-based, statistical and hybrid approaches were used to develop machine translation systems from English to Indian languages. The system that works on pseudo-target approach is Anglabharti [1] (IIT, Kanpur, India); other was example-based approach AnglaHindi (IIT, Kanpur, India) [2] that translate frequently occurring noun and verb phrases. Anusaaraka [5, 6], a Paninian grammar-based system, was developed by Indian Institute of Technology, Kanpur. Another system MANTRA-Rajya Sabha [3] developed by CDAC Pune, India, is a machine-aided translation tool primarily designed to translate Rajya Sabha proceedings from English to Hindi language. Dave et al. [4] proposed a machine translation approach based on universal networking language (UNL). Apart from this, many SMT system has been developed for Indian language translation.

Neural machine translation is an emerging approach in machine translation, which is recently proposed [9, 10] and [11]. Unlike the traditional phrase-based translation system [12] that is made of many small sub-parts that are tuned separately, neural machine translation attempts to build and train a single, large neural network that reads a sentence and outputs a correct translation [13].

Recently, the application of deep neural networks is making tremendous improvement in machine translation. Google came up with a very good translation system called Google neural machine translation (GNMT) system [7, 8], which used millions of samples for translation. It consists of LSTM networks. They have also implemented attention mechanism for dealing with contextual dependencies of the text and also handled the occurrence of unknown words by dividing the words into finite set of common words.

The primary problem for the machine translation research in Indian languages is unavailability of good-quality parallel sentences in abundance. For this reason, our first objective was to create a parallel corpus, which covers sentences from an untouched domain and that too of good quality and in large number. The motivation behind this was to support machine translation research by creating aligned parallel corpus. Another problem with available open-source corpus was impurities such as spelling mistakes, presence of unknown characters, data in different fonts and varied formats that makes it unusable for training or other research work. Thus, to make use of such data, it has to be pre-processed properly. The pre-processing includes various stages along with the cleaning, vetting, aligning, etc. The process of data creation is discussed in next section.

2 Data Creation

A corpus is a collection of authentic text organized into datasets. 'Authentic' in this case means text written is created by a native of the language. These datasets are important for development of any NLP application. Generally, for neural machine translation, a larger size of corpus is better. But, one cannot prioritize quantity over quality. A high-quality training data is required for good training results. This can be achieved by ensuring accuracy, completeness and timeliness. The process of data creation comprises of following steps.

2.1 Identification of Data Source

A good-quality parallel and aligned data plays a crucial role in the development of any machine learning application. For development of neural machine translation system, a huge and accurately aligned parallel corpus in Unicode is required. Many open-source datasets are available on web, but these are often problematic in terms of quality. Thus, we decided to create our own dataset for MT application. So, the obvious challenge was not only to identify the website for data creation but also handling the inconvenience and messiness of the scraping and assembling process.

For development of this system, we have chosen the Lok Sabha website (Parliament of India, Lok Sabha) for data collection. This website contains data in the form for bilingual (i.e. English and Hindi) PDF files. These files are the records of the daily proceedings of Lok Sabha, discussed in introduction and background.

2.2 Preparation of the Parallel Corpus

A parallel corpus is a corpus that contains a collection of original texts in one language and their translations into another languages. The corpus can be mono- or bi- or multidirectional in nature. English and translated Hindi files at the chosen website are kept year-wise in each subdomain separately. The data files are not aligned or parallel. The data from identified domains was downloaded.

Data transformation

After the mapping was established, the PDF data was converted to doc or txt format using optical character recognition (OCR). It was then found that the majority of the Hindi data was not in Unicode, and after doing OCR, many characters were lost or converted into a junk. In many cases, where the data was in tabular format, formatting got lost and it resulted in corrupted data. Thus, converted data required manual and semi-automatic validation after which this data was then further aligned for generation of parallel corpus.

Alignment of data

The alignment of corpus refers to storing the text of the source language and the corresponding translation and aligning the two texts at paragraph, sentence or phrase or word level. Here, we have aligned the data at sentence level. Each parliamentary subdomain has its own style of writing. In PLOT and LOB, single paragraph represents a single sentence (length up to 30 words). This may also contain embedded tables with numeric and text data as well. In synopsis domain, a single paragraph contains multiple sentences with generally no tables or other numeric data. The text up to length four was removed and was later processed and used for lexicon building. The table data was removed manually by analysing the source PDF files. The paragraphs in synopsis domain were broken into valid sentence and later were aligned with Hindi corresponding data manually. This resulted in approx. 5 lakhs of aligned parallel sentences and 20 thousand lexicon and multiword domain terms.

Data compilation

The compilation of the converted and corrected data was done with following points in the mind.

– Language Authenticity: During the process, only valid language data was taken, and other data was deleted during the validation and cleaning process.

Table 1 Showing the examples of corrupted data

Actual Data	OCR Data
सी.आर. चौधरी	सी-आर चधरी, सी-Qआर चिघरी, च धरी चरण सिंह
सं. ए-12011/1/2017 –स्था. (खंड-1), एफ सं. 10-8/2016–बीबिएंडपीए	संट ए-1201101020 17-स्था" (खंड-1), एफ सं 10-82016–बीबिएंडपीए
राष्ट्रीय कृषि विपणन संस्थान	राप्रिय कषि विपणन संस्थान
का.आ. 242 (अ); का.आ. 490(अ)	का-Qआ2 242(अ); का:आं 490(अ)

- Data sampling: The validated, aligned and parallel data was bucketed domain-wise; the frequency of each sentence was counted, and the underlying structure types of sentences were identified using the in-house developed tool for sentence structure identification.

Challenges in Data preparation

The whole process of parallel corpus creation posed many challenges. Few of which are discussed below.

- *Similarity in the names:* The names of downloaded files were similar, and there is no identifier to establish a mapping between the English and its corresponding Hindi data. This mapping has to be established manually to create a link, by analysing content of each data file.
- *Missing files*: In certain cases, only the monolingual (English or Hindi) files are available. These single files need to be identified and marked separately. This was done by analysing the content of each file and mapping of downloaded files.
- *Corrupted data*: During the process of conversion, the Hindi characters were lost at many places or wrongly converted to some other character. The examples of which are given in Table 1.

Sentence breaks at wrong place: The English sentences were broken at wrong places leading to the mismatch of English and its corresponding Hindi data.
Stylistic variation in translation: In synopsis domain, the translators take liberty in following cases: (i) translate two English sentences together in to a single Hindi sentence; (ii) one English sentence into two corresponding Hindi sentence; (iii) translates only meaning of the English sentences without capturing the sentence structure.

All this makes the alignment of sentences much difficult and time consuming. The examples of above discussed stylistic variations are given in Table 2.

- *Repetition:* In some domains like list of business, etc., there were lot of repetition in sentences, and the sentence length and complexity of sentence were more than in other subdomain. This led to more efforts and less data in terms of unique corpus.

Table 2 Examples of stylistic variations

English	Hindi
Industrial development will push the State on the road to development automatically.	राज्य में उद्योग लगाइए विकास स्वतः होगा।
There are 35-36 sugar mills in Bihar which are lying closed today.	बिहार में 35-36 चीनी मिलें बंद हैं।
If a juvenile reports any crime to the police, it should be taken seriously and action should be taken if a police man ignores his complaint.	यदि किशोरावस्था का बालक किसी भी पुलिस स्टेशन में जाकर अपराध की जानकारी देता है तो उसे तुरन्त ही गंभीरता से लिए जाने की जरूरत है। यदि कोई पुलिसकर्मी उनकी शिकायतों की अनदेखी करता है तो उसके खिलाफ कार्रवाई किए जाने की जरूरत है।

Addressing the Challenges

All the challenges mentioned above were handled efficiently and effectively to produce a good-quality, unique and aligned parallel data.

– File mapping issues were resolved by manually checking the content of each file and bucketing it in correct domain.
– The conversion errors were correct by semi-automatic method, wherein first few files were corrected manually which led to with the identification of the pattern or specific issues, and later, the data was corrected with semi-automatic methods and through programme.
– The breaking of English sentences during the alignment process was done both by manual efforts and sometimes with the help of available text editors.
– The process of validation and final alignment was done manually.

The data preparation took more time than expected due to the challenges mentioned above. All the efforts yield 5 lakhs of cleaned and aligned parallel sentences which was used to develop the machine translation system and can be used in many other NLP application further.

3   Development of the System

We have developed the hybrid model for machine translation from English to Hindi and vice versa. The hybrid includes the rule-based, i.e. tree adjoining grammar (TAG) and neural-based machine translation system. The TAG consists of multiple and many elementary trees, and the substitution and adjunction operations are performed on the tree to parse the source sentences to target. TAG largely handles the domain of the sentences to give better accuracy as its formalisms and dependencies defined by the elements of grammatical rules. For example, for parsing the English sentences, 'the opinion poll does not have a single question on whether the respondent received any food support from the Government or other sources (Fig. 1)'.

Neural machine translation (NMT) is method of translation source to target language based upon the artificial neural network to envisage the likelihood or the

Fig. 1 Parsed tree Source:
http://nlp.stanford.edu:8080/
parser/index.jsp

```
Parse [14]
(ROOT
  (S
    (NP (DT The) (NN opinion) (NN poll))
    (VP (VBZ does) (RB not)
      (VP (VB have)
        (NP
          (NP (DT a) (JJ single) (NN question))
          (PP (IN on)
            (SBAR (IN whether)
              (S
                (NP (DT the) (NN respondent))
                (VP (VBD received)
                  (NP (DT any) (NN food) (NN support))
                  (PP (IN from)
                    (NP
                      (NP (DT the) (NN Government))
                      (CC or)
                      (NP (JJ other) (NNS sources)))))))))))
    (. .)))
```

probability of a given sequence of words/token, classically modelling whole input sentences in a sole integrated model. It is based upon many factors such as.

3.1 Model Building

Language model is built using the transformer deep learning model. Transformer uses the attention mechanism and weighs the input data differently according to its significance. Attention mechanism helps top detect the relationship between the words of sentence and weigh their significance separately while building model rather than taking phrase-based approach unlike its predecessor statistical-based machine translation (SMT).

3.2 Pre-processing Data for Model Building

To prepare the data for model building byte pair encoding were applied and then vocab files were generated.

3.3 Encoder–Decoder Model and Beam Search for Inference

Transformer-based encoder and decoder [16] are used to build model. Alternatively, RNN and LSTM also can be used, but transformer provides the characteristics like positional embedding, non-sequential processing and most importantly self-attention which makes the model more dynamic as compared to RNN and LSTM. Beam search [17] takes each position separately and looks for optimum result at that position and then moves to next position repeating the same for each position until it reaches the end of the sentence. At the end of the beam search completion, it returns the output with the best result possible based on the trained model.

Adam optimizer is used, and various combinations of parameters were manipulated to get the best possible result.

4 Evaluation and Results

The developed machine translation system was evaluated using both automatic and subjective evaluation methods. For evaluation, the test data a sample data of varying length covering various sentence structures from different parliamentary domains was extracted. The BLEU score for English–Hindi translation system is 21.18 and for Hindi–English translation is 41.47. Blind and open methods were used for subjective evaluation. In case of blind evaluation, the evaluators were only provided with the machine translated output, while in the open-source sentences, it was also provided along with MT output. In both open and blind evaluations, each sentence was given score for the sentence comprehensibility and fluency on minus 1 to 4 scale. Later, the overall comprehensibility and fluency were calculated using formula. Blind and open evaluation scores are given for the systems below (Table 3).

Apart from this, the post-editing time was also calculated using the open-source tool [15]. The post-editing time required for sample data was 3.34 hours for Hindi output, but for English output, it was 4.68 hours.

Table 3 Scores of subjective evaluation

Subjective evaluation				
System	Comprehensibility		Fluency	
	Blind	Open	Blind	Open
English–Hindi	60.13	52.75	46.08	42.27
Hindi–English	96.99	94.13	92.25	79.39

5 Conclusion and Future Work

In this paper, we have discussed our approach for development of machine translation system for English to Hindi and vice versa for specific domain (parliamentary). The results that we got were encouraging. Though advanced AI models are available and proven effective for other languages (English), data for Indian languages still remains a challenge. Much of Indian language data is available over the web, but none of it is readily usable due to various issues such as data formats, data encoding and unavailability of parallel corpora. Thus, a lot of manual effort is required for creation of the aligned, parallel corpus. The focus of future work is on developing tools for the pre-processing and preparation of data.

Acknowledgements We are thankful to Ministry of Electronics and Information Technology, Government of India, for supporting our work through the pilot Bahu-Bhashak project under Natural Language Translation Mission. We are also thankful to GIST Group of CDAC Pune, to carry out evaluation under CoE Bahu-Bhashak project under Natural Language Translation Mission.

References

1. R. M. K. Sinha, K. Sivaraman, A. Agrawal, R. Jain, R. Srivastava, A. Jain, ANGLABHARTI: a multilingual machine aided translation project on translation from English to Indian languages, in: Systems, Man and Cybernetics. Intelligent Systems for the 21st Century, IEEE International Conference on, 2 (IEEE, Vancouver, Canada, 1995), pp. 1609–1614
2. R. M. K. Sinha, A. Jain, AnglaHindi: An English to Hindi Machine-aided Translation System (MT Summit IX, New Orleans, USA, 2003), pp. 494–497
3. CDAC-MANTRA, Available from: https://www.cdacindia.com/html/aai/mantra.asp. Last updated 07 Aug 2017
4. S. Dave, J. Parikh, P. Bhattacharyya, Interlingua-based English–Hindi machine translation and language divergence, Mach. Transl. **16**, 251–304 (2001). https://doi.org/10.1023/A:1021902704523
5. A. Bharati, V. Chaitanya, A. P. Kulkarni, R. Sangal, Anusaaraka: machine translation in Stages, arXiv preprint cs/0306130 (2003)
6. S. Chaudhury, A. Rao, D. M Sharma: Anusaaraka: an expert system-based machine translation system, in: Natural Language Processing and Knowledge Engineering (NLP-KE), 2010 International Conference on, (IEEE, Beijing, 2010), pp. 1–6
7. M. Johnson, M. Schuster, Q. V. Le, M. Krikun, Y. Wu, Z. Chen, N. Thorat, F. Viégas, M. Wattenberg, G. Corrado, M. Hughes, J. Dean, Google's multilingual neural machine translation system: enabling zero-shot translation, arXiv preprint arXiv:1611.04558 (2016)
8. Y. Wu, M. Schuster, Z. Chen, Q. V. Le, M. Norouzi, W. Macherey, M. Krikun, Y. Cao, Q. Gao, K. Macherey, J. Klingner, A. Shah, M. Johnson, X. Liu, L. Kaiser, S. Gouws, Y. Kato, T. Kudo, H. Kazawa, K. Stevens, G. Kurian, N. Patil, W. Wang, C. Young, J. Smith, J. Riesa, A. Rudnick, O. Vinyals, G. S. Corrado, M. Hughes, J. Dean, Google's neural machine translation system: bridging the gap between human and machine translation, arXiv preprint arXiv:1609.08144 (2016)
9. N. Kalchbrenner, P. Blunsom: Recurrent continuous translation models. In: Proceedings of the 2013 Conference on Empirical Methods in Natural Language Processing, Association for Computational Linguistics, Seattle (2013), pp. 1700–1709

10. I. Sutskever, O. Vinyals, Q. V. Le, Sequence to sequence learning with neural networks. CoRR abs/1409.3215. http://arxiv.org/abs/1409.3215
11. K. Cho, B. Van Merrienboer, D. Bah-danau, Y. Bengio, On the properties of neural machine translation: encoder-decoder approaches (2014b). arXiv preprint arXiv:1409.1259
12. P. Koehn, F. J. Och, D. Marcu, Statistical phrase-based translation. In: Proceedings of the 2003 Conference of the North American Chapter of the Association for Computational Linguistics on Human Language Technology, vol 1 (Association for Computational Linguistics, Stroudsburg, PA, USA, NAACL '03, 2003), pp. 48–54. https://doi.org/10.3115/1073445.1073462
13. B. Dzmitry, K. Cho, Y. Bengio, Neural machine translation by jointly learning to align and translate (2014a). CoRR abs/1409.0473. http://arxiv.org/abs/1409.0473
14. Stanford POS tagger, http://nlp.stanford.edu:8080/parser/index.jsp
15. http://www.clg.wlv.ac.uk/projects/PET/#:~:text=PET%20is%20a%20stand%2Dalone,time%20amongst%20other%20effort%20indicators
16. A. Vaswani, N. Shazeer, N. Parmar, J. Uszkoreit, L. Jones, A. N. Gomez, L. Kaiser, I. Polosukhin, Attention is All You Need. https://arxiv.org/abs/1706.03762. Last modified 06 Dec 2017
17. Beam Search, https://en.wikipedia.org/wiki/Beam_search. Last edited 28 Oct 2021

Gender Issues and New Governance: Managing Menstrual Health and Hygiene with Emerging Multimedia Options for a Progressive Economy

Suparna Dutta and Ankita Das

Abstract A nation-building process needs to be comprehensive and progressive with strong backward and forward linkages. Hence, insight into the evolution and foresight driven endeavors to change, grow and develop further become the cornerstone for sustainable development. Women-centric issues, especially those related to health, hygiene and sanitation will play a cataclysmic role in this broad context. Here the design and implementation of an effective outreach communication for social mobilization based on substantial 'reversal of learning' is expected to lead toward more responsive, potent and sustainable governance. We are witnessing a progressive global interest in the domain and much of so in India. What is still missing is an overarching, orchestrated mechanism, framework and governance which will curate novel initiatives and ideas, filter redundancy and repetition and harness emerging multimedia options by tapping the state-of-the- art cognitive and immersive technologies to loop the population into a dialogic communication for inclusive growth. This innovative, customized, mediate communication paradigm will disseminate knowledge to spread awareness for better menstrual management practices across the nation to promote a healthy and efficient workforce to support a healthy economy.

Keywords Menstruation · Health · Hygiene · Usage and disposal of FHPs · Insinuators · Awareness · Attitudes · Taboo · Practices · Prototypes and stereotypes · Knowledge · FHPs · Outreach · Communication design · Reversal of learning · Multimedia · Cognitive and Immersive tools

1 Introduction

Menstruation is a characteristic unavoidable truth and a month-to-month event for the 1.8 billion young girls, women, transgender men and *non-binary* people of conceptive age. However, a huge number of the menstruating individuals across the world are

S. Dutta (✉) · A. Das
Birla Institute of Technology, Mesra, Ranchi (Off campus- A-7, Sector 1, Noida, India
e-mail: s.dutta@bitmesra.ac.in

© The Author(s), under exclusive license to Springer Nature Singapore Pte Ltd. 2023   533
Y.-D. Zhang et al. (eds.), *Smart Trends in Computing and Communications*, Lecture Notes in Networks and Systems 396, https://doi.org/10.1007/978-981-16-9967-2_50

denied the privilege to deal with their menstrual cycle in a simple, healthy and a digni-fied manner. Sexual orientation imbalance, prejudicial social standards, social taboos, destitution and absence of fundamental intervention from the governing authorities frequently neglect the needs felt by the bleeding individuals. Girls from any low and middle-income countries (LMIC) including India enter puberty with huge knowl-edge gaps, prejudices and misconceptions about menstruation and hence remain unprepared to not only cope with it but also remain unsure of when and where to seek help from [1]. Kaur also sites gender inequality as the main propellent for this social malaise. Unequal rights usurped traditionally by men has resulted in downing the female cry for help and protest despite being the home maker, the silent work-force, and the most alert and conscious beneficiary of development initiatives. Nidhi Mallik (2013), in an exploratory study, identified that 'the awareness and knowledge about menstruation, most of the girls were dependent upon their mothers and other family members. The fact that even educated girls were not discussing these problems with their parents for reproductive health problems reflects the poor communication between them.'

'It is now common knowledge that men can easily (and have already begun to) support and influence women and girls in managing menstruation at home, schools, work, and at the community level by assuming many socio-economic roles as husbands, fathers, brothers, students, teachers, colleagues, leaders and policy-makers.', [2]. Whereas, the stark ground reality shows (Namrita Rai 2019) that 'Around 63.7% women still experienced embarrassment due to the lack of facil-ities to dispose sanitary pads, while 72.6% found it problematic to find bins for disposal, and 80.1% claimed that their bins carrying waste were not emptied on a regular basis. And a significant majority of 84.6% women struggled to conceal the sanitary napkin while throwing it in a bin so that the male members do not get to see what they were throwing.' Whereas, Syed Uzair Mahmood (2019) shared through a survey that 'These sanitary napkins are not disposed in dustbins but are littered on the roads, free plots or burnt openly creating risk of contamination and disease transmission. In rural-urban divide, 46.89% of the urban and in 60.96% of the rural girls, burnt their soiled FHPs but 45.23% of the urban girls and 12.33% of the rural girls disposed soiled FHPs by throwing it with the routine waste. However, it was rather interesting to note that semi- skilled and semi-literate men were more comfort-able in discussion issues related to contraception and pregnancy to a great degree, and slightly less on childbirth but certainly not shut toward menstruation (Singh and Singh 2014). Gragi Abhay Mudey [3] also stated in her research that 'It was found that 37.33% girls disclose only to their mothers regarding menstruation, whereas 28.67% were comfortable about disclosing to all. The various reasons given for not wanting to disclose were; irritation caused due to unnecessary remarks especially for not attending school and functions (46.26%), disgusting feeling (15.89%) and 9.35% to avoid irritation when others talk about it.'

In the Menstrual Health Management (MHM), disposal of the soiled FHPs play as critical as critical role as it is to manage the bleeding. In the absence of a state managed supportive system and ambience, women have had little choice but invent their very own ingenious methods and ways of creating crude and accessing proper

FHPs, using, changing and disposing them during their active menstruation in accordance to their individual awareness, education, economic status and social norms and ambiance. Transgender men and non-double people who discharge too regularly face more damaging segregation due to their sexual orientation and menstrual discharge in them is looked up as demonic and unnatural. This keeps them away, like menstruating females from getting access to the materials, offices, schools and services that they need urgently. Often, such exclusion and disadvantage cause serious and extensive adverse psychosomatic effects on the lives of the individuals (Aleena Baig 2019). Similar observations have come from similar researchers and scholars like [4], Naveeta Kesharwani (2010), Sushmita Malaviya (2019) and (Sumpter and Torondel 2013). Surveys and studies also reveal that most frequently the plastic used in sanitary napkins are not bio-degradable and therefore the improper, hapless littering of the soiled FHPs lead to health and ecological hazards. Stray animals scavenging this filth carry and spread the germs far and wide which few notice and often it is the unsuspecting rag pickers and street urchins and who become the first line of human contact with this virulent vector force to spread contamination and dreadful diseases. Fortunately, this dreadful, unmentored, untrained and disorderly method of metropolitan waste administration in the cities, mushrooming residential areas, markets, towns and even the villages now are progressively getting noticed by our civil society agents and getting articulated at various platforms [5].

1.1 Mediated Communication

Immersive Technology has for some time been the hottest cafe for socialization across the world. The pandemic and the consequent lockdown have only emphasized this global phenomenon provoking online chattering to the next level. The language was the god gone astray in flesh (Frantz Fanon) in the age of paper and ink and it may now be said that social media, the "hashtags" and the ever-emerging multimedia communication options have become the near equivalent of the same in the age of these virtual narratives of our times. These immersive vicarious experiences which are driven by our needs, taste and options to see and share content is propelled greatly by the evolving Cognitive and Immersive technologies. The addictive potency of such technology is arming this fast-spreading medium with the enormous power of advocacy and influence. Consequently, such mediated communication is creating the opportunity to create sharper subgroups based on interest and deliberate content. Hence, much of the critical ideological and socio-emotional matter like gendered content is shared more freely among the members making it the indispensable point of conjunction, a platform and the coming together for consumers who are keen to chat, share, mingle and do business together and believe spontaneously and consciously. This is where, the medium is gaining the power to influence and has become a tool for both advantage and ace in contentious and sensitive social communication drives and initiatives. In fact, smart governance should ideally take advantage of these online community clubs to disseminate critical social messages for greater attention and retention by the audience. But for sharper impact, content creation for such outreach communication should ideally be based on participatory communication. Dialog and discussion are the two holy wheels of such targeted participation and such participation is primarily possible with help of innovative mediated communication.

1.2 Objective

This study is based on a field survey conducted in the year 2019–2020. The scope of the survey was limited to the metropolitan city of Bengaluru and its suburbs in the south of India. The participatory action research used qualitative research methodology and tools to explore and comprehend the worries and propensities of menstruating individuals and people connected with them in this, demystify myths on menstruation, augment self-adequacy and esteem in women, assist vulnerable menstruating individuals with capacity building aptitude and skills to overcome obstacles to their wellness, freedom, privacy, sexual orientation-based violence during this vulnerable period, menstruation induced discomfort, promote proper disposal of soiled menstrual waste and design a communication framework that seamlessly integrate into the e-governance ecosystem of urban, semi-urban and rural areas in India keeping the plural hyper-local context of our nation.

1.3 Background

Menstruation from menarche to menopause associates with new changes and vulnerabilities that needs to be collectively dealt with at all levels. Menstruating individuals require support and some more so through their menstruating life till menopause irrespective of class and education. But especially vulnerable in this are the base-of the pyramid population, the low middle class and the destitute. It is well understood today that menstrual well-being and hygiene interventions can augment gender transformative skills, performance abilities and professional competence. It also well understood that there is a need to design a dynamic ecosystem of trusteeship for this where all the stakeholders would align to create an overarching inclusive framework based on a participatory model that initiates sustainable dialog to not only ensure 'reversal of learning' but also the integration of this dynamic, dialog-based ecosystem of progressively informed trustees into the emerging e-governance ambit in India [6]. Though issues related to menstrual health and hygiene has to become a prominent and integral part of all governance to ensure the development to be glitch free and the inclusion has to be universal [7] yet, unfortunately, even the survey conducted reiterates that menstrual health, hygiene and awareness still remains a causality in our country especially in the marginalized, deprived and the lower middle class (LMC) segments of the population. The situation is not very encouraging and healthy in the middle class population as well specially in the tier 2 and tier 3 cities. However, the upper middle class and affluent families of the metropolitan cities are much more fortunate in the sense that they have overcome much of the taboos and ritualistic Indian traditional approach toward menstruation. So prompting and sensitizing a responsive digital governance for a sound and comprehensive inclusive development becomes the responsibility of all.

1.4 Current Scenario

It is the Community-based menstrual health tips which help to battle and give succor to the menstruating individuals. So, a network of competent stakeholders from school teachers, village midwives, primary healthcare centers both in urban and rural areas, government dispensaries, dedicated district (preferably block level) administrative machinery and so on need to come out as the talkative interlocuters. But here, it is the communication paradigm used that is going to make all the difference [8]. The communication principle of 'seeing is better than hearing, but it is best to experience...' is what needs to be pegged on in this context of targeted outreach. The governance and its partners need to move out of printed flyers, and garrulity and resort to the magic of cost-effective multimedia communication. Borne by ICT, the emerging immersive and cognitive technologies like 3-D Animation, Augmented, Virtual and Mixed Reality and the virtual games are the new magic wands of effective communication. They customize and they personalize the content and they also can cater to the individual and the hyper-local through the social media. And the potency of the immersive technology can attract and impinge on the minds of the end user through the incessant use of content packed in the infotainment mode. The new governance across the country must harness this and developing content will not be difficult and there is a steady progress of academic institutions and Institutions of Higher Educations (IHE) who are showing interest in offering degree diploma program in Animation and Multimedia Studies. Working in tandem with such IHEs, the entertainment or the Animation Industry, the government can easily design such modules for their outreach communication.

1.5 Reversal of Learning

Governance and policy making share a hand and glove relationship. Unfortunately, policy decisions are made from the federal offices which are frequently cut-off from the ground reality. Of course, this does not indicate lack of competence and skill in the official functionaries who are mostly selected through the rigors of a highly competitive and complex selection procedure. But what mostly go lacking is the familiarity with the hyper-local context, traditions and nuances. Fortunately, the ongoing pandemic has brought this hyper-local to the forefront and hence addressing it will no longer remain a difficult domain for anyone. This familiarity is expected to bring the policy makers will come in close contact with the local ethnicity, culture and typical traditions, knowledge, worldview and customs developed thereof and push them toward a deeper comprehension and develop a more holistic perception of the factors that promote most of such social malaise. It is indeed disappointing that even after 7 decades of independence and self-governance, we in India still need to frame popular gender and family-planning slogans where as in contrast, the governance has been robust and progressive enough to make India become a signatory nation to the United Nations Sustainable Development Goals that idealize a gender bias free and neutral global society at all levels and locations. So, doubting the intentions of governments is not always fair, nor is it always correct to distrust the wisdom of the government. What normally remains missing in such socio-economic evaluation context is the awareness and acceptance of the hyper-local reality and its dynamics by the educated policymakers. It is a kind of 'traditional disconnect' and a 'dissociation' that stems from ignorance often stemming from the conceit born from the modern university driven education of the policymakers. Once this 'connect' between the hyper-local context and the governance happens, the latter is expected understand, learn and accept both the subtle local social matrix that change every 100 km with the dialect in India, the local wisdom as well as the local stereotypes with all its idiosyncrasies. This learning and acceptance is of critical value for the lawmakers because it is here where 'local resistance' and thereafter the popular disconnect with laws, policies and the government itself begins. The sloth and power of local tradition in which most of such eccentricities are meshed in and borne by should not be ignored while framing our rules, policies and political agendas. Behavior change communication relies on such 'typical knowledge' of tradition, culture and worldview. So, when policies are framed, it is done with reference to this knowledge of the local order. But most importantly, it is when communication modules are designed that this knowledge of the local came handy in customization of the content which is critical for behavior change communication and thereby inclusive growth.

2 Data Analysis Derived from Survey Par Conducted in Bangaluru in 2019–20

2.1 Menstrual Health

There were significant number of respondents who knew of and suffered some of the vaginal and health diseases related and caused by improper menstrual hygiene, but they still did not have specific knowledge of those dreadful diseases like the Urinary Tract Infections (UTI) and Reproductive Tract Infection (RPI). However, almost none connected such recurring infections with FHPs and their hygienic use and disposal. The survey also found individuals who themselves or their friends who suffered from 'yeast infections' and all they understood of it was that an odor filled discharge was to be expected. It was less than 10% respondents who realized the connection between appropriate menstrual practices sound vaginal and menstrual health, with recurring problems like irritation, fever or some other serious health issues that even lead to chronic conditions like cancer. The survey did not come across individuals who mentioned cervical cancer, hepatitis B and the like on their own while talking about menstruation and MHM.

2.2 Knowledge and Awareness

An overwhelming majority of the female respondents shared that they first got to know of menstruation from their mothers and from information and experiences shared by their friends and other classmates. Less than 20% of the female respondents admitted to learnings from menstrual health camps in school or biology subject they studied as part of their curriculum. These were also young girls who spoke about their proactive schools and teachers who talked to them and shared relevant information. Some also talked of having met some educated, emancipated relatives who introduced them to novel FHPs and talked to them of hygiene and proper disposal of FHPs. However, none of the respondents knew about insinuators and were quite startled and impressed by the smokeless mechanism that was capable of destroying regular domestic waste and the FHPs easing their challenge of waste disposal. Though less than 15% of the female respondents correlated a clean, green pollution environment with MHM, an overwhelming 63% of them agreed that schools, residential areas, transport stations and hubs, markets, etc. need insinuators.

2.3 Social Taboo

The survey revealed a varied and probably a hint of a progressive change. Mostly, menstruation was an avoidable topic for discussion as it was looked upon as natural and regularly occurring and hence to be taken in stride. Heavy bleeding and related discomfort are still a matter of individual concern. Less than 20% of the sisters and mothers were overly concerned about their young daughters bleeding heavily. Majority did not have or looked for any specific information from anyone or source on heavy bleeding and its cure nor felt the need to do so. Over 76% of the female respondents did not feel local administration had anything to do with heavy bleeding and FHPs, but they were all surprised with the workings of an insinuators. Regarding talking about menstruation. 40 percent of the female respondents spoke to each other, their mothers and neighbors and purchased and shared FHPs. More than 20 % of the male respondents admitted to purchasing the FHPs for their wives and even their sisters and other female relatives at home. These were also male respondents who were aware of the pain, fatigue and discomfort caused by menstruation and appreciated insinuators. 13% of the male respondents were open to health camps and support from local administrations. But more than 87% of all the women still would avoid places of worship and treat menstruation as ritually polluting. There were many local customs of prohibition of both consuming and serving of food during this period and men felt no compunction to cooking to take such taboos forward. Not attending school was another overwhelming trend among the lower income families. But females, generally went back to work the moment the heavy bleeding stopped irrespective of exhaustion, weakness or irritation which never was to be considered.

2.4 Disposal

The survey revealed a universal exposure and knowledge of improper and ugly disposal of soiled FHPs. Littering was again a universal annoyance which was prosaically accepted by all. The resulting stink and animal menace though a source of embarrassment was still not overtly connected with health hazards. It is still very common to through and litter unwrapped soiled FHPs from widows, moving trains, and in public places with little guilt and responsibility. The demand for administrative support for such waste management is still less than 18% and people are still unable to differentiate between dry and wet waste. Significantly, there is still no popular protest and opinion against callous disposal of city garbage where used FHPs are crudely gathered, visible transported with other waste material. Such menstrual waste unsurprisingly is commonly found in public unisex toilets, college washrooms and dumped trough ducts into activity and public space in schools, colleges and work areas. Yet, no warning or public instructions were seen for the proper disposal of FHPs from any authority.

3 Conclusion

India as an economic super power suffers a few paradoxes which needs immediate attention. What is of advantage here is the progressive unison in accepting gender challenges at all levels and the realization that despite having reached the orbit of Mars in a maiden venture, we have still left our womenfolk in the more distant Venus. This is especially true of our Menstrual Health Management (MHM) practices. Beyond debate, it is individuals from the middle and lower social rung that suffer most and are not only affected with low self-esteem and suffer socially debilitating consequences for this required normal biological condition but are thence affecting the nation's progress to level next by spreading diseases and remaining unproductive at work and school. A simple solution to this is the governance taking recourse to multimedia communication where an overarching communication framework is created that would induce spontaneous participation of the people into open and free conversation with experts from the field. Doctors, counselors, administrators should create their specific knowledge driven content in infotainment packages and make them available to the people to answer their questions and alley their fears and doubts in effective dialogic communication. But what needs to be seen is that both privacy, security and authenticity should be kept at the premium. Keeping in mind the open architecture of the immersive multimedia technology, security of the content uploading and the end users should be constantly kept in mind. At the same time, the potency of the emerging animation and multimedia communication should be harnessed for greater outreach, sharper reception, deeper learning and content creation and content delivery. The appeal, the playful, recreative ambience and info-tainment potential that such cognitive and immersive technologies are capable of delivering should be used to create participation of the end users. What is to kept in mind is the dialogic format of communication where questions, doubts and fears could be expressed and allayed and addressed by the experts as an integral part of effective governance that is itself aware of not only the problems but the cause and context of the problem and its solution as well without bias and arrogance and insularity of education and power.

(The Authors did not include the Menstruating Third Gender within the Scope of this Study).

References

1 R. Kaur, K. Kaur, R. Kaur, Menstrual Hygiene, management, and waste disposal: practices and challenges faced by girls/women of developing countries. J. Environ. Public Health **2018**, 1–9 (2018). https://doi.org/10.1155/2018/1730964

2 A. Dasgupta, M. Sarkar, Menstrual hygiene: how hygienic is the adolescent girl? Indian J. Commu. Med.: Official Publ. Indian Assoc. Preventive Soc. Med. **33**(2), 77–80 (2008). https://doi.org/10.4103/0970-0218.40872

3 M.A. Bhausaheb, N. Kesharwani, G.A. Mudey, R.C. Goyal, A Cross- sectional Study on Aware-ness Regarding Safe and Hygienic Practices amongst School Going Adolescent Girls in Rural Area of Wardha District, India (2010)

4 S. Puri, S. Kapoor, Taboos and Myths associated with women health among rural and urban adolescent girls in Punjab. Indian J. Community Med., **31**, 168–70 (2006). [Google Scholar]

5 S. Ahn, K. Cho, Personal hygiene practices related to genito-urinary tract and menstrual hygiene management in female adolescents. Korean J. Women Health Nurs. **20**(3), 215 (2014). https://doi.org/10.4069/kjwhn.2014.20.3.215

6 N. Mehta, S. Dutta, Multimedia for effective communication. 2016 International Conference on Signal Processing, Communication, Power and Embedded System (SCOPES) (2016). https://doi.org/10.1109/scopes.2016.7955510

7 A. Kumar, K. Srivastava, Cultural and social practices regarding menstruation among adolescent girls. Soc Work Public Health., **26**, 594–604 (2011). [PubMed] [Google Scholar]

8 S. Dutta, N. Mehta, R. Pratik, Humane Digital Route to Customize Communication for Sustainable Development. INTED2013 Proceedings, (2013), pp. 3090–3096

An IoT-Based Smart Parking System for Smart Cities

Venkatesh Mane, Ashwin R. Kubasadgoudar, Raghavendra Shet, and Nalini C. Iyer

Abstract There has been an exponential growth in the vehicle population in recent times, adding to the metropolitan traffic. This increase in the number of vehicles has led to the problem of inadequate parking spaces resulting in traffic congestion. To address this challenge, a smart parking system is proposed in this paper, which makes use of TIME RESOURCE SHARING to effectively utilize the parking spaces based on peak demand time and enables prior identification and reservation of parking space with the help of unique identification. The system periodically updates the parking status. The interaction of the driver and the owner of the parking space takes place through the application connected through the cloud. The related work in this area has also been referred. The proposed system has the potential to transform the current parking method and alleviate the traffic congestion caused by insufficient parking space.

Keywords Time-shared parking · Unique code · Traffic congestion · Parking space · Arduino

1 Introduction

Smart city development is one of the major areas that need to be addressed. Smart city development refers to the efficient utilization of the available resources with the active interaction with the community leading to the improvement in the quality of life and in the overall development of the nation. In present times, with the ever

V. Mane (✉) · A. R. Kubasadgoudar · R. Shet · N. C. Iyer
KLE Technological University Hubballi, Hubballi, India
e-mail: mane@kletech.ac.in

A. R. Kubasadgoudar
e-mail: ashwinrk@kletech.ac.in

R. Shet
e-mail: raghu@kletech.ac.in

N. C. Iyer
e-mail: nalini.c@kletech.ac.in

© The Author(s), under exclusive license to Springer Nature Singapore Pte Ltd. 2023   543
Y.-D. Zhang et al. (eds.), *Smart Trends in Computing and Communications*, Lecture Notes in Networks and Systems 396, https://doi.org/10.1007/978-981-16-9967-2_51

increase in vehicle population for the ease of transportation, there arises a serious problem of inadequate parking spaces. The available parking spaces prove to be insufficient to meet the growing demands. Searching for a vacant parking space becomes a tedious, time-consuming task for the drivers leading to traffic congestion and causing hindrance to the flow of traffic [1]. It also contributes to increasing the extent of pollution and increasing the risk of accidents. Hence, this issue of inadequate parking space should be addressed to achieve smart city development.

'To address the issue, many new parking lots are being built. But this means new constructions and utilization of extra space, which is again a big hassle [2]. To reduce the traffic congestion caused by inadequate parking spaces, many smart parking systems have been proposed. The conventional method adopted is that the driver simply searches for a parking space. They continue to do so until he finds a vacant space [3]. Some systems check for the availability of parking space and broadcast it to the drivers [4]. In this, there is no reservation facility, and sometimes instead of reducing traffic congestion, they worsen it. Therefore there exists a greater demand to address this issue with an effective strategy.

Our contribution to meet the demands of parking space efficiently is as follows. A smart parking system with time-shared facility is proposed, which adds a new dimension to the existing parking space by introducing the concept of time sharing, which makes use of city resources efficiently, which excludes new constructions and major investments for parking spaces. Shared parking refers to the parking facility that is used jointly by several sectors, maybe offices or residential flats, and designed to take advantage of differences in peak parking demand times, which can vary by day, week, and season [5] For instance, apartment parking space remains vacant during the weekdays when the owner goes for work and the parking space allocated to offices remain vacant during weekends [8, 9].

The proposed system has a unique device at the parking space for the vehicle identification and monitoring of parking status. The system also caters to the collection of parking fees based on the time duration and request from the customer [6, 7]. The interaction between the customer and the owner of the parking space takes place with the help of an application wherein all the information is stored in the cloud.

The organization of the rest of the paper is as follows. Section 2 describes the details of the proposed methodology. Section 3 describes the scope of implementation and the conclusion of the article in Sect. 5.

2   Proposed Design

The proposed system aims at providing a TIME-SHARED RESOURCE facility that makes use of existing parking spaces without having to build new constructions or new parking facilities to address the problem of inadequate parking spaces. Timed Shared Resource Parking basically aims at providing parking spaces during peak hours of the day. Apartment parking space remains vacant during the weekdays when the owner goes for work, and the parking space allocated to offices remains

Fig. 1 Overview of the system

vacant during weekends. Thus there exists a need to identify and utilize the spaces available in an efficient manner. Therefore an integrated system providing a solution on a time-shared basis is illustrated in Fig.1, along with the operational details.

The system mainly focuses on the effective utilization of parking space through the interaction between the user and the owner, connected through an application via a unique device.

Every parking space is connected to the Internet through the proposed application. This enables the communication between the user and the owner of the parking space. This information of vacancy or occupancy is sent to the user through the application. The user then reserves a parking space from the available spaces. The owner determines the availability and the time duration for which the parking space is available for the user to be utilized. Then the same data/information is updated in the database.

Each parking lot is deployed with a unique ID for the validation/confirmation of the user's identity and booking. The system acts as a centralized body that dynamically allocates parking spaces on a time-shared basis. When the user reaches the parking spot, the user is identified by the unique ID given at the time of signing up. Since the information of each parking lot is stored in the database of the system, the system is able to synchronize all requests from the user. Therefore, whenever a user books a parking space, that space is reserved and is unavailable for other users.

The proposed system is a combination of hardware and software modules. Details of each are detailed in the further sections.

3 System Architecture-Hardware

The functional details of the system are shown in Fig. 2.

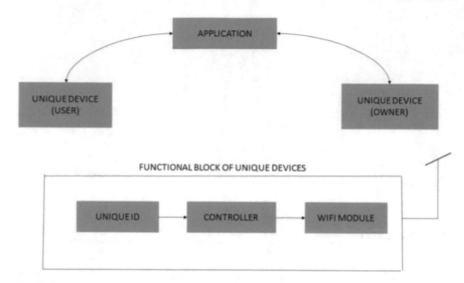

Fig. 2 Functional block

The system basically comprises a unique identification device and an application. The interaction between these unique devices takes place through the application.

Each parking space is deployed with a hardware module, and every vehicle will have a unique identity card, which is basically an RFID. Each such hardware module comprises electronics and embedded circuits that provide unique identification determination based on the parking slot booking. The module consists of a controller which does all the computations and processing of the data. The controller is able to access the Internet through the help of a Wi-Fi module. This module enables the hosting of the application. It also consists of a real-time clock (RTC) that helps to keep track of time duration.

4   System Architecture-Software

The flow chart for the application running between the two dedicated devices is shown in Fig.3.

The software module consists of an application designed on an android platform. The user has first to create an account to be able to use the services. Once the account is created, the user can log in with his username and password. Users can then select the required destination and check the availability of parking spaces in the vicinity. If free spaces are available, then the user can reserve the parking space. One user is allowed to reserve only one space. For booking, the user has to enter his details, including his unique identification number. Once the Parking space is reserved, the database is automatically updated from vacant to occupied hence no other user is

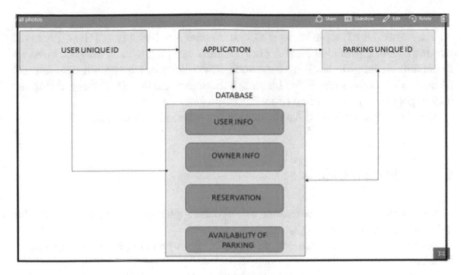

Fig. 3 Software architecture

allowed to park in that space. Prior notifications are sent to the user's phone to indicate that reservation time is about to expire.

The owner has to sign-up for the first time and then provide the required information. The owner is also given a unique identification number. When the owner logs in to his account, he has an option to rent his parking space for a particular time interval. This information is used by the user to enable time-shared parking.

5 Implementation

The proposed system mainly focuses on assigning parking spaces on a time-shared basis. Availability of parking space before reaching the destination is made available to the user with a booking facility. As a result, this reduces the "multiple-car-chase-single-slot" phenomenon; hence the system reduces the traffic congestion, thereby leading to the decrease in the extent of air pollution.

Other salient features of the module include a number plate tracking system, Bluetooth connectivity, tracking facility, unique identification, which helps in achieving the aim proposed. The software module consists of the various locations available for parking in the vicinity, which can be monitored using GPS modules. These free spaces can be identified, booked, and utilized in a feasible way. The integration of both of these modules paves the way for the achievement of a smart parking system. Even vehicle identification, in case of theft, is also addressed through a unique ID given to each vehicle. In case of a change in a number plate, the unique ID remains the same and hence addresses the issue of identification of the stolen vehicle.

As the system involves user and owner communication through the application, we may embed the Google navigation system onto it to better direct the users to the parking space. The unique identification number assigned to each user can be made to contain all the details of the user, such as his license, registration number, and address, so that the user can be identified through a unique ID. Hence, paving the way to paperless identification in a smarter way.

Details of the application developed are described in further sections.

6 Application Details

The user and the owner are connected through the application. This application helps to know the status of parking spaces.

Figure 4 shows the home screen, which has a sign-up and login option. The sign-up page asks for the details.

After signing up, a unique ID is assigned. After logging in, an options page is displayed, as shown in Fig. 5.

The owner has an option to rent the space, and the customer has an opportunity to book a parking space, as shown in Fig. 6.

Fig. 4 Application home and sign-up page

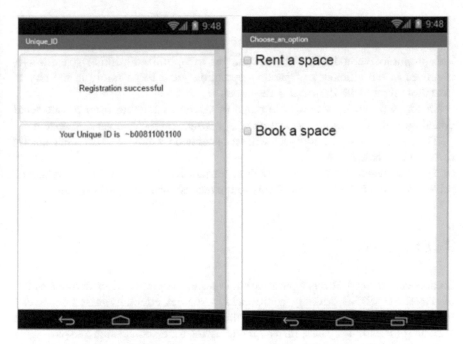

Fig. 5 Unique ID and user options

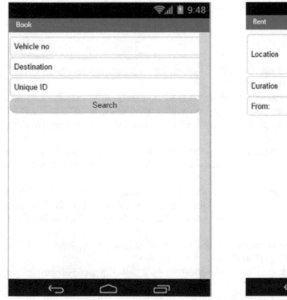

Fig. 6 Booking and renting page

7 Discussion on Results

The proposed method offers the advantages of optimized parking space usage, improves in the efficiency of parking operations, and helps traffic in the city to flow more freely with the existing resources.

Our system works with a wide range of operators ranging from private retail operators to the government or private sector-owned large-scale parking systems.

The owner of the parking space will be benefitted by getting maximum capacity from the available space.

Space management was easier with an instant real-time reporting system. The drivers were guided quickly and easily to the nearest available parking space.

8 Conclusion

To address the need of inadequate parking space, the proposed system makes use of TIME SHARING parking facility, which enables effective utilization of city resources along with active interaction of the community. This reduces the time spent searching for parking spaces, thereby reducing the traffic congestion and paving the way to the development of the smart city.

References

1. R. Shet, N.C. Iyer, Y. Jeppu, Fault tolerant control system for autonomous vehicle: a survey. J. Adv. Res. Dyn. Control Syst. **12**(8), 813–830 (2020)
2. P.S. Pillai, R. Shet, N.C. Iyer, S. Punagin, Modeling and simulation of an automotive RADAR. In *Information and Communication Technology for Competitive Strategies (ICTCS 2020)* ed. by. A. Joshi, M. Mahmud, R.G. Ragel, N.V. Thakur. Lecture Notes in Networks and Systems, vol. 191. (Springer, Singapore, 2022)
3. N.C. Iyer et al., Virtual Simulation and Testing Platform for Self-Driving Cars. In: *ICT Analysis and Applications. Lecture Notes in Networks and Systems*, ed. by S. Fong, N. Dey, A. Joshi, vol 154 (Springer, Singapore, 2021)
4. R.M. Shet, N.C. Iyer, P.C. Nissimagoudar, A. Kulkarni, J. Abhiram, S.K. Amarnath, Motion Control and Sensor Fault Diagnostic Systems for Autonomous Electric Vehicle. In: *ICT Analysis and Applications. Lecture Notes in Networks and Systems*, ed by S. Fong, N. Dey, A.Joshi, vol. 154 (Springer, Singapore, 2021)
5. P. Shastri, A. Chauhan http://timesofindia.indiatimes.com/city/ahmedabad/Ahmedabad-lets-get-smart-link-fee-to-parking-demand/articleshow/50885898.cms
6. A. Roy. http://timesofindia.indiatimes.com/city/nagpur/New-bldgs-will-require-double-the-parking-space/articleshow/51001996.cms
7. P. White, *No Vacancy: Park Slopes Parking Problem and How to Fix It*, http://www.transalt.org/newsroom/releases/126
8. http://fresno.ts.odu.edu/newitsd/ITS_Serv_Tech/park_sys_tech/parking_systems_tech_report_print.htm
9. Parking Management for Smart Growth by Richard W. Willson

Agriculture Stakeholder's Information Need—The Survey in Gujarat, India

Axita Shah and Jyoti Pareek

Abstract Agriculture plays an important role in Indian economy. In recent era applications and technology usage is drastically increased. To fulfill the demand in Agriculture, it is important to know the information need of the farmer. Farmer's Information Need can be satisfied by the varied human stakeholders in Agriculture or by the technology use. We have prepared Agriculture Questionnaire to know the farmer's information need. Our Agriculture Survey is filed as copyright, detailed in [1], We have conducted survey to the varied demography of the farmer on different time and at different location. To improve the accuracy of the survey, we have interviewed the stakeholders on one-to-one basis directly. Our survey to the agriculture stakeholders of Gujarat also enables and facilitate us examining the channels of information communication, gathering information need and knowing varied need in natural Guajarati language. Facts and figures of the survey leads to the necessity of intelligent retrieval system.

Keywords Agriculture survey · Farmer information need in agriculture · Indian agriculture questionnaire

1 Introduction

According to World Bank Data,[1] 70% of the world's poor who live in rural areas and are poorly educated, Agriculture is the main source of income and employment. 64.93% of land area is used for Agriculture where Crop Production Index is also increasing exponentially, i.e., 143.6 in 2020 (Base 2016–2017) which proves that Agriculture plays an important role in Indian Economy. According to Agricultural

---

[1] http://data.worldbank.org/—The World Bank Group is one of the world's largest sources of funding and knowledge for developing countries.

A. Shah (✉) · J. Pareek
Department of Computer Science, Gujarat University, Ahmedabad, India
e-mail: axitashah@gmail.com

J. Pareek
e-mail: jspareek@gujaratuniversity.ac.in

© The Author(s), under exclusive license to Springer Nature Singapore Pte Ltd. 2023
Y.-D. Zhang et al. (eds.), *Smart Trends in Computing and Communications*, Lecture Notes in Networks and Systems 396, https://doi.org/10.1007/978-981-16-9967-2_52

Statistics at a Glance[2] 2020 [2]. In 2011 68.9% of India's population lives in rural India for their livelihood. 263.1 million Population from total 1210.9 million population works in Agriculture from that 45.1% are cultivators and other 54.9% are Agricultural laborers. There are 8 union territories and 28 states in India. Average 57.8% of Agricultural households to rural households where in Gujarat, it is 66.9%.[3]

Farmer's Information Need can be satisfied by the varied human stakeholders in Agriculture or by the technology use. Agriculture expert stakeholder who has knowledge of farming help farmers to resolve their problem. But Several Farmers do not have an access of other human expert. Hence, Technology usage is blessings for them. At one side poor people are poorly educated while the other side Literacy rate in Adult and youth is also increasing, i.e., 74%. Literacy Rate of Gujarat is also 79.31% that shows increase in the usage of existing online and offline knowledge. In Gujarat farmers are getting the benefit of Directorate of Agriculture to increase and protect the crop production.

There are problems in getting the data/information like Static Information, More than one information source, unavailability of an expert, etc.... which cannot be resolved by using Traditional App or Expert's direct view [3]. Hence, this information leads to the necessity of an agriculture expert system which retrieves from diverse sources. We have explored existing agriculture data sources which is useful to resolve farmer's query. Sources of Information from trusted web sites and publication by agriculture expert include but are not limited to Agriculture crop cultivation textual knowledge, statistical data etc., by Ministry of Agriculture, Department of Agriculture, Government of India, Indian Agriculture Universities, Agriculture Companies and so forth publishes numerous online and offline informative documents for the welfare of the farming individuals. This is the place where data of agriculture exists on Internet in a large volume. The client needs different types of information for crop cultivation. We came to know about it from our agricultural survey [1]. And it is very important to have data to provide the information that the stake holders need. We have researched, reviewed and stated the sources of such data and the information in the *Agriculture Content-Set* section available from the government and other institutes that is required for direct access or as part of technical retrieval system research. This data can be useful to guide the farmer about crop cultivation system. These data are available on Internet for the usage of the farmers that is definite knowledge seeker.

Researchers in this field have developed technical application to fulfill the demand of information and knowledge seeker. Information and Communication Technology systems and applied artificial intelligence techniques are available at farmer's doorstep to sort out farmer's information need. We have reviewed and summarized literatures of applied AI in Agriculture and stated in *Research status of*

---

[2] Remark: Agricultural Statistics at a Glance is a flagship publication of the Directorate of Economics and Statistics, Department of Agriculture and Cooperation, Ministry of Agriculture, Government of India containing data on a wide range of economic indicator, particularly in agriculture sector. https://eands.dacnet.nic.in/PDF/Agricultural%20Statistics%20at%20a%20Glance%20-%202020%20(English%20version).pdf.

[3] https://en.wikipedia.org/wiki/2011_Census_of_India.

applied AI in Agriculture section. It will be better if we provide this knowledge to its users on one platform.

Information need of Agriculture stakeholders mostly achieve via friend advise, expert adviser, Mobile SMS, Helpline no., Cable tv, Newspaper, Expert Adviser, Training and Demonstration, Chat, WhatsApp, video or Audio conference. To understand farmer's information need, we have surveyed few of the agriculture stakeholders of Gujarat and collected information related to their crop cultivation. Our Survey by Questionnaire is explained, summarized and analyzed in *Analysis and Interpretation of Agriculture Crop Cultivation System Survey* section. From the other established sources and from the survey, we have collected 2848 user's query in natural language. We have also prospected stakeholders' information need in a form of natural language query as a part of survey.

2 Agriculture Content-Set

Informative web access provided by various autonomous entities in Agriculture includes but are not limited to, Department of Agriculture, Cooperation and Farmers Welfare,[4] Ministry of Rural Development,[5] Area and Production Statistic by Ministry of Agriculture and farmers welfare,[6] Government Agriculture Data,[7] The Indian Council of Agricultural Research (ICAR)[8]—autonomous organization under the Department of Agricultural Research and Education (DARE), Ministry of Agriculture, Government of India, Directorate of Economics and Statistics, Department of Agriculture and Cooperation, Ministry of Agriculture, Government of India At a Glance Report,[9] Indian Agricultural Statistics Research Institute (IASRI),[10] Department of Agriculture and Cooperation, Ministry of Agriculture, Government of India,[11] National Institute of Agricultural Extension Management,[12] Food and Agriculture Organization of the United Nations,[13] The International Crops Research Institute for the Semi-Arid-Tropics,[14] Indian Agriculture College/Universities News,[15]

---

[4] http://agricoop.nic.in/.

[5] https://rural.nic.in/.

[6] https://aps.dac.gov.in/Home.aspx?ReturnUrl=%2f.

[7] https://data.gov.in/search/site?query=agriculture.

[8] http://www.icar.org.in/.

[9] http://eands.dacnet.nic.in/Default.htm.

[10] https://iasri.icar.gov.in/.

[11] http://agricoop.nic.in/.

[12] https://www.manage.gov.in/default.asp.

[13] http://www.fao.org/index_en.htm.

[14] http://eprints.icrisat.ac.in/information.html.

[15] http://agricollegenews.com/.

Agriculture Technical News,[16] Indian Agriculture Retailer News,[17] Agricultural and Processed Food Export Development Authority,[18] Agriculture Insurance,[19] Farmer's Portal,[20] AGMARKET,[21] mKisan—SMS Portal for Farmers by Department of Agriculture and Cooperation,[22] Indian Agriculture Jobs,[23] Consortium for e-Resources in Agriculture,[24] Indian Agricultural Research Institute (IARI)—Journal,[25] e-learning portal on Agricultural Education[26], Information in Agriculture Science,[27] Bhartiya Kisan Sangh gujarat,[28] Data bank on Agriculture and allied sector,[29] four Agriculture Universities of Gujarat,[30,31,32,33] Online Agriculture Community[34], Agriculture Research Article Search Websites[35], Agriculture Knowledge Discovery Websites[36,37], Agricultural Census[38], Agriculture Time series of India's economic indicator by EPW Research Foundation[39] and other Indian Agricultural Web Sites indiaagristat[40], khetiwadi[41], kisan[42], ikhedut.aau[43], nabard[44], agriwatch[45], etc....

[16] http://agtechnews.com/.

[17] http://agriretailers.com/index.htm.

[18] http://apeda.in/apedawebsite/.

[19] http://agriinsurance.com/.

[20] http://farmer.gov.in/.

[21] http://agmarknet.nic.in/agmarknetweb/default.aspx/.

[22] https://mkisan.gov.in/.

[23] http://www.indiaagronet.com/Agriculture-Jobs/.

[24] http://www.cera.iari.res.in/index.php/en/.

[25] http://epubs.icar.org.in/ejournal/.

[26] https://ecourses.icar.gov.in/.

[27] https://www.agriscienceindia.in/.

[28] http://bharatiyakisansangh.org/.

[29] https://data.worldbank.org/indicator/NV.AGR.TOTL.ZS.

[30] https://nau.in/index.

[31] http://jau.in/.

[32] http://www.aau.in/.

[33] http://www.sdau.edu.in/.

[34] http://www.agricultureinformation.com/.

[35] http://oajse.com/subjects/agriculture.html.

[36] http://www.agnic.org/.

[37] http://www.sciseek.com/Agriculture/.

[38] http://agcensus.nic.in/.

[39] http://www.epwrfits.in/Agriculture_All_India_State.aspx.

[40] http://www.indiaagristat.com/.

[41] www.khetiwadi.com.

[42] www.kisan.net.

[43] http://faq.ikhedut.aau.in/.

[44] www.nabard.org.

[45] www.agriwatch.com.

Agriculture and Horticulture Data Sources are also available for Retrieval as per the user's demand. There are around 1200 freely available textual and multimedia resources on government and private institutes websites categorized and listed as dag,[46] atma gujarat,[47] sdau,[48] aau, [49] icar,[50] iari,[51] nau,[52] jau,[53] gov farmer,[54] ikhedutaau,[55] ikhedut gujarat,[56] agricoop,[57] soilhealth.dac.gov,[58] dacnet,[59] prsindia,[60] doh.gujarat,[61] gujecostat,[62] gsfcagrotech,[63] nhb,[64] agriculturetoday,[65] youtubechannels on Agriculture[66] and topical Search engine as google.[67]

Agriculture information is published weekly or monthly in the newspapers/magazines such as National Agriculture Today, Kisan samaj in Hindi from Indor, krushik jagat[68] in Hindi from Bhopal, Krushi Jagran,[69] krushi prabhat,[70] Agro Sandesh,[71] Krushi Vignan,[72] Real Target, Krushi Govidhya by Aanand University, Krushi Goldline in Hindi from Jaypur, Kisan Sandesh, Fasal Kranti,[73] Digital agrimedia film, agro house publication.[74]

---

[46] https://dag.gujarat.gov.in/images.

[47] https://atma.gujarat.gov.in/.

[48] http://www.sdau.edu.in/.

[49] http://www.aau.in/sites/default/files/.

[50] https://icar.org.in/sites/default/files.

[51] https://www.iari.res.in/files.

[52] http://old.nau.in/files.

[53] http://www.jau.in/attachments.

[54] https://farmer.gov.in/.

[55] http://agri.ikhedut.aau.in.

[56] https://ikhedut.gujarat.gov.in/.

[57] http://agricoop.gov.in/.

[58] https://soilhealth.dac.gov.in/Content/FAQ.

[59] https://eands.dacnet.nic.in/.

[60] http://prsindia.org/sites/default/files/.

[61] https://doh.gujarat.gov.in/.

[62] https://gujecostat.gujarat.gov.in.

[63] http://www.gsfcagrotech.com/.

[64] http://nhb.gov.in/.

[65] http://www.agriculturetoday.in/.

[66] https://www.youtube.com.

[67] https://www.google.com/.

[68] https://www.krishakjagat.org/Web/single_epaper/486.

[69] https://subscription.krishijagran.com/.

[70] http://www.krushiprabhat.com/epapers-listing.

[71] http://sandeshepaper.in/edition/57311/agro-sandesh.

[72] https://krushivigyan.blogspot.com/2018/12/scintilla.html.

[73] http://fasalkranti.com/news.

[74] http://agrohouse.in/agrohouse/shop_book.php#.

Agriculture industry publishes the facts and figures with statistics by magazines, articles, periodicals, textbooks as in textual or multimedia format as data and information. Researchers from universities, industries and government institutes working on Agriculture generate or collect knowledge in database as an example, Soil Health Card,[75] faolex[76] and data warehouse. Healthy e-mail/message communication of community expert is also available as a community data. Generated predicted knowledge is also helpful to the Farmer. Indian farmer can use this available form of information via technical application to increase productivity of farming and generate revenue.

3   Research Status of Artificial Intelligence in Agriculture

Vast research in Academia has been carried out for the applications of artificial intelligence in Agriculture. Applied Agriculture scopes a variety of topics specified in [4–9] includes but are not limited to, Intelligent retrieval, Information system, expert system, IOT-based system, machine learning, geographic context-based expert system, Question-answering system, Agri Robot, Sense-based expert system, Rule-based expert system, Fuzzy logic-based expert system, etc. Private Industry and the government also inflate its agriculture framework by providing information via web portal and mobile applications. Existing providers' study has been researched, reviewed and submitted it as an article in our submitted paper [10] that talks about research status of applied artificial intelligence approaches in Agriculture and the research gap.

Various researchers have worked in the retrieval process from Agriculture content. Techniques such as Information Retrieval (IR) [11], Question-answering (QA) system [12–14], Search engine [15–17], and Natural language and keyword-based interfaces are developed to get solution of user's information request. They have worked in the optimization of crop cultivation system, from the seed plantation till the harvesting process. These techniques are implemented using data, rules, thesaurus and ontology-based knowledge. Retrieval is improved by understanding user's request and retrieval sources such as documents in IR, and QA system and web pages for search engine.

Based on our study, we hypothesized the requirement of an intelligent system that retrieves from diverse government data sources and that requires to fulfill the information need of the farmer. To know the information need, we have prepared the survey, detailed in [1], conducted the survey and its interpreted analysis and results are as follows.

---

[75] https://www.soilhealth.dac.gov.in/.

[76] faolex.fao.org/faolex/index.htm.

4   Analysis and Interpretation of Agriculture Crop Cultivation System Survey

To fulfill the information need of the Agriculture Stakeholder, specifically Farmers in Agriculture Crop Production System, we had conducted a Survey via well-structured Agriculture Questionnaire as copyright Filed for "Survey-Questionnaire of Indian Agriculture Stake Holder's Information Need for Agriculture Crop Cultivation Information System of India". Detail Survey Questionnaire is given in the filed copyright—[1]. Objective of the survey is to know the varied user's natural language request as a part of Man–Machine Interaction, to gather the information needs of farmer community in Gujarat, India and to examine the channels of information communication and sources of information dependence used by the farmers.

It is an agriculture survey for Agriculture Crop Production Information System of Gujarat, India in Questionnaire form. We have prepared Questionnaire in both English and Gujarati languages. This Survey is distributed among 5 categories, i.e., personal information of stakeholder, Questions regarding Business/Job in Agriculture Crop Production System, Queries for Agriculture Crop Production and Role of IT in Agriculture. Agriculture questionnaires collect information on a core set of indicators such as Basic crop production of main crops, with emphasis on improved measures of Quantification of production and Plot size, Land holdings, Farming practices, i.e., Mechanization, Soil and environmental management, Water management, Input use and technology adoption, Use of technology and farming implements, Seed varieties, Fertilizer, pesticides/herbicides applications, etc....We have referred several Indian government agriculture websites and Indian council of agriculture research (ICAR) benchmark book published by agro house [18] and Agriculture Categories by EPW Research Foundation[77] to prepare this Questionnaire.

Kindly refer Survey-[1] to get in detail considerate category wise questions. We have conducted this survey by visiting AgriAsia-2016, 2017 and 2019, (Companies Visited in AgriAsia: Agrostar, reliance foundation, vimax rop science ltd, gnfc, Agriculture Farmers welfare and cooperation department—Government of Gujarat), Gujarat Vidhyapith KVK Seminar and Department of Agriculture, Horticulture and Statistics, Government of Gujarat. There are around 308 individuals' involvement with 272 survey responses from AgriAsia, and 36 opinions by conducting an interview of Experts, Advisors, Government Domain Employees for knowing their viewpoint in Agriculture research.

---

[77] http://www.epwrfits.in/Agriculture_All_India_State.aspx.

5   Results and Discussion: Analysis and Interpretation of Agriculture Crop Cultivation System Survey

Our survey analysis, after data collection and preprocessing is as follows: Diverse Age group of participants in survey below 20 Yrs is 4.5%, 20–40 is 42.5%, 40–60 is 48.5% and 4.5% are above 60. From them 88.8% are Male and 11.2% are female. Around 94% from them own the farming unit.

Here, in Fig. 1 depicts personal facts and figures of survey participant. A of Fig. 1 expresses the speech ability of local, national and international languages. More participants know Gujarati compared to Hindi, English and other languages. More than 60% participants are having basic knowledge of technology usage such as use of computer, mobile phone and application, summarized in B of Fig. 1. Varied Education Qualification of respondents in C and respondent's agriculture role in D are specified in Fig. 1

A to E of Fig. 2 depicts figurative summary of respondent's agriculture crop cultivation system practice. Most of the respondent's crop cultivation frequency in a year is twice shown in A of Fig. 2. Total counts of types of soil at farm in B of Fig. 2, total categories of crops which you are producing in farm in C of Fig. 2, total count of preferred medium of information sources in C of Fig. 2, Sources of farmer's help need summary count in D of Fig. 2 and Requirement of information need's category summary in E of Fig. 2 are demonstrated in relevant chart. These

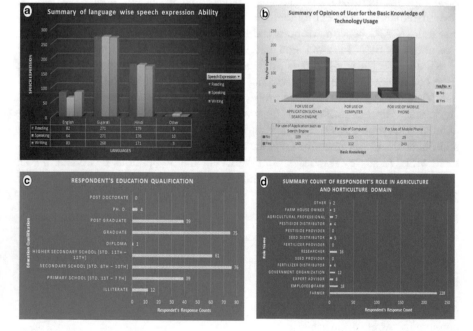

Fig. 1   Personal informative analysis of survey participants

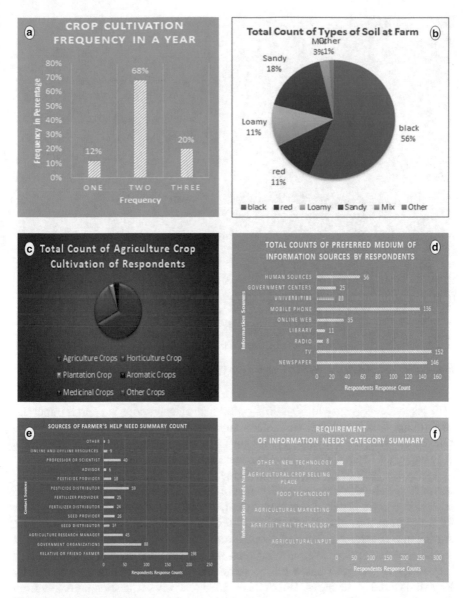

Fig. 2 a–f summary of agriculture crop production system usage by survey participants

analyzed, interpreted and demonstrated summary of Figs. 1 and 2 indicates diversified background of survey respondents.

Figure 3 identifies varied application usage by the respondents in the Bar chart. Out of these, Agrostar, Agrimedia, Mandi and mkrishi are maximum used among the community. Application usage may get changes depending on the current availability of application, latest functional and error-prone applications in the market. Around

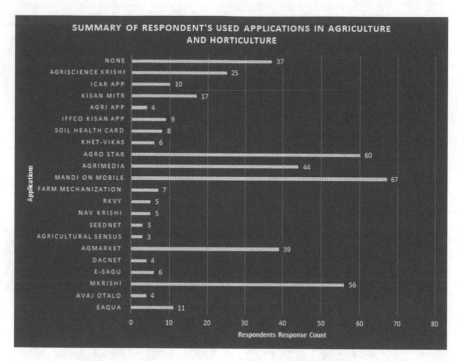

Fig. 3 Bar chart of usage of technical application by participants in agriculture

70% respondents agree with 50–100% varied span of technology support is important in farming. 83% agree that technology supports have definite role in Agriculture where 10% disagree and other 7% do not know about it. And around 57% farmers who are using Kisan Call Center Service are satisfied with the expert's advice.

Query Set in Natural Gujarati Language: We have collected around 1339 agriculture's stakeholder's request query. Out of 1339, 566 is from AgriAsia 2016, 506 is from AgriAsia 2019, 199 is from scientists at Kisan Vigyan Kendra, and other 68 from companies like GNFC, Kisan Call Center, Reliance Foundation, Agrostar, Agriculture farmers welfare and cooperation department-government of gujarat at AgriAsia via interview. We have gathered around 2848 agriculture's stakeholder's request query analyzed based on the frequency of keywords and represented by Keyword Cloud in the Fig. 4. Out of 2848, 1339 as a part of questionnaire, 299 from Agriculture FAQ Government Book, 625 is from Agrimedia mobile app and 585 is from IKhedut FAQ. Our analysis identifies mostly asked types of the agriculture queries listed as in Table 1.

Our result of conducted survey toward Agriculture Stake holders cognize farmer's context and situation in Gujarat and their need as a part of natural language queries. It also identifies the necessities of tools and technologies as well as agriculture scientific tips.

Fig. 4 Keyword cloud of agriculture stake holder's requested query

Table 1 Analyzed categories of agriculture queries

Mostly Asked Agriculture Queries
ફૂલો અને શાકભાજી ઉગાડવાની રીતો
ખાતર ને લગતા પ્રશ્નો, ખેતીમાં રોગને લગતી માહિતી
પાક માટે દવા અને પાણી નાં છટકાવને લગતા પ્રશ્નો
જીવાત અને ઢોરથી પાકોનુ રક્ષણ કરવાના ઉપાયો
પાકની લણણી અને કાપણી કરવાની વિવિધ રીતો
સરકારની વિવિધ યોજનાઓ જેમકે વીમા યોજનાઓ
જંતુનાશક અને ફૂગનાશકના છંટકાવ ની માહિતી
સુધારેલી પાકની જાતના ફાયદા
પાકની નીંદણ માટેનો વૈજ્ઞાનિક અભિગમ
વિવિધ પ્રકારની જમીનો ની માહિતી
ઔષધીય ખાતર તૈયાર કરવાની પધ્ધતિ
ઋતુ ને લાયક વરસાદ અને તાપમાન ને અનુરૂપ પાક ની માહિતી
વિવિધ પાકના સ્તરે ટેકનોલોજી નો ઉપયોગ
સિંચાઈ ની સગવડ ને લગતા પ્રશ્નો

6 Conclusion

Analysis of undertaken survey shows immense potential to standardize, regulate and automate the agricultural processes and solve the problems in Indian Context in a form of search and retrieval-based artificial intelligent system. Our current analysis escort toward viable, efficient and cost-effective solution of obtaining wisdom through the working with independent individual. Our Survey to the Agriculture Stake holders cognize us about the information need of varied farming practices to diverse demographics. Even our future research will analyze farmer's Information need that is in Natural Language Query. We will extensively explore, classify, process and analyze natural language query.

Diverse available data and information resources in natural language should be utilized on one platform via an agriculture expert information system. Even Agriculture Stakeholder is not required to know the technicality context of the resource. In future, we will propose Retrieval system that fulfill the above specified mandatary request. This Retrieval system will provide easy to access, user friendly interface, one-point access mechanism for diverse information needs, even accessible by non-technical persons, and provide relevant solution to the user's request. The complexity of this implementation would be diverse storage and access mechanism of diverse computer resources, natural language ambiguity, difficulty in defining knowledge and inference for the same.

We hope that this work will help researchers in the field to acquire the necessary information and technologies to further conduct more advanced research.

References

1. A. Shah, J. Pareek, Survey-Questionnaire of Indian Agriculture Stake Holder's Information Need for Agriculture Crop Cultivation Information System of India. Literary/Dramatic works Copyright Certified—L-113297/2022. (2021)
2. G. India, *Of: Agricultural Statistics at a Glance—2018* (Controller of Publication, New Delhi, Government of India, 2018)
3. A. Rafea, *Expert System Applications: Agriculture. Central Laboratory for Agricultural Expert Systems*, PO Box. 100, (1998)
4. B.T. Bestelmeyer, G. Marcillo, S.E. McCord, S. Mirsky, G. Moglen, L.G. Neven, D. Peters, C. Sohoulande, T. Wakie, Scaling up agricultural research with artificial intelligence. IT Professional. **22**, 33–38 (2020)
5. Y. Awasthi, Press "A" for artificial intelligence in agriculture: a review. JOIV : Int. J. Inf. Visual. **4** (2020). https://doi.org/10.30630/joiv.4.3.387
6. A. Gambhire, B.N. Shaikh Mohammad, Use of artificial intelligence in agriculture. SSRN Electron. J. (2020). https://doi.org/10.2139/ssrn.3571733
7. S. Chaterji, N. DeLay, J. Evans, N. Mosier, B. Engel, D. Buckmaster, R. Chandra, Artificial intelligence for digital agriculture at scale: techniques, policies, and challenges. arXiv preprint arXiv:2001.09786 (2020)
8. M. Pathan, N. Patel, H. Yagnik, M. Shah, Artificial cognition for applications in smart agriculture: a comprehensive review. Artific. Intell. Agricult. **4**, 81–95 (2020)
9. K. Jha, A. Doshi, P. Patel, M. Shah, A comprehensive review on automation in agriculture using artificial intelligence. Artific. Intell. Agricult. **2**, 1–12 (2019). https://doi.org/10.1016/j.aiia.2019.05.004
10. A. Shah, J. Pareek, Contemporary research status and recommendation of applied artificial intelligence in agriculture. Towards Excellence. 14, (2022)
11. A.S.Sharma, A., A framework for semantics and agent based personalized information retrieval in agriculture—IEEE Conference Publication. In: 2nd International Conference on Computing for Sustainable Global Development (INDIACom) (2015)

12. T. Kawamura, A. Ohsuga, Question-answering for agricultural open data. In: *Transactions on Large-Scale Data-and Knowledge-Centered Systems XVI*. pp. 15–28. (Springer, 2014). https://doi.org/10.1007/978-3-662-45947-8_2

13. S. Gaikwad, R. Asodekar, S.Gadia, V.Z. Attar, AGRI-QAS question-answering system for agriculture domain. In: *2015 International Conference on Advances in Computing, Communications and Informatics, ICACCI 2015*. pp. 1474–1478. Institute of Electrical and Electronics Engineers Inc. (2015). https://doi.org/10.1109/ICACCI.2015.7275820

14. M. Devi, M. Dua, ADANS: an agriculture domain question answering system using ontologies. In: *Proceeding—IEEE International Conference on Computing, Communication and Automation, ICCCA 2017*. pp. 122–127. Institute of Electrical and Electronics Engineers Inc. (2017). https://doi.org/10.1109/CCAA.2017.8229784

15. N. Xie, W. Wang, *Research on Agriculture Domain Meta-search Engine System*. Presented at the October 18 (2009). https://doi.org/10.1007/978-1-4419-0211-5_68

16. W. Li, Y. Zhao, B. Liu, Q. Li, *The Research of Vertical Search Engine for Agriculture*. Presented at the October 18 (2009). https://doi.org/10.1007/978-1-4419-0211-5_2

17. C.Y. Sufen, T.P., L.S., S., Study of agricultural search engine based on FAO agrovoc ontology and google API. In: *World Automation Congress*. IEEE (2010)

18. Indian Council of Agricultural Research: Handbook of Agriculture. Directorate of Knowledge Management in Agriculture, New Delhi (2012)

System Design and Implementation of Assistive Device for Hearing Impaired People

P. Nikita, B. H. Shraddha, Venkatesh Mane, N. Shashidhar, P. Vishal, and Nalini C. Iyer

Abstract This paper discusses the design, implementation and development of assistive device for hearing impaired. Sign language is one of the oldest and most natural form of language for communication, but since most people does not know sign language. Finding interpreters for different group of people is difficult. To treat the deaf people as one among the society, a user-friendly assistive device is necessary. This paper provides the details about the implementation of standalone interpreter using transfer learning for finger spelling-based American Sign Language using Raspberry Pi. A Graphical User Interface (GUI) is created and tested for establishing a two-way communication to convert text into sign language. A language barrier is created using an assign language structure that is different from normal text. Hence, the users will now depend on the vision-based communication that will be able to bring normal people, deaf and mute people on the same grounds of interaction. There is a need to build a possible sign language translator, which can take communication in sign language and translate them into written and oral language. Such a translator that uses neural networks for finger spelling-based American Sign Language would greatly lower the barrier for many deaf and mute individuals to be able to better communicate with others in day-to-day interactions. In proposed method, hand gesture is first passed through a filter, and after the filter has applied,

P. Nikita (✉) · B. H. Shraddha · V. Mane · N. Shashidhar · P. Vishal · N. C. Iyer
KLE Technological University, Hubballi, India
e-mail: nikhita_patil@kletech.ac.in

B. H. Shraddha
e-mail: shraddha_h@kletech.ac.in

V. Mane
e-mail: mane@kletech.ac.in

N. Shashidhar
e-mail: shashisn@kletech.ac

P. Vishal
e-mail: vishalbps@kletech.ac.in

N. C. Iyer
e-mail: nalinic@kletech.ac.in

© The Author(s), under exclusive license to Springer Nature Singapore Pte Ltd. 2023
Y.-D. Zhang et al. (eds.), *Smart Trends in Computing and Communications*, Lecture Notes in Networks and Systems 396, https://doi.org/10.1007/978-981-16-9967-2_53

the gesture is passed through a classifier which predicts the class of the hand gestures. This method provides 95.7% accuracy for the 26 alphabets and 0, 1, 2 numerical.

Keywords American Sign Language · GUI · Translator · Interpreters

1 Introduction

One of the five senses necessary for human survival is hearing. Language acquisition, learning to speak, and general knowledge all require the use of the hearing sense. Hearing, above all, allows people to communicate and socialize with others. Hearing loss affects deaf persons in numerous ways; they are unable to communicate or mingle properly with the general public. Furthermore, deaf people's safety is jeopardized in many situations because they are unable to make the correct decision based on the brain's analysis of noises, particularly worrisome ones. Hearing loss is treated differently depending on the severity of the condition. The only disability dumb and deaf people have is related to communication and they cannot use spoken languages hence the only way for them to communicate is through sign language and American sign language is a predominant language. Hands are the essential part of body which will be used by deaf and dumb people to communicate their thoughts with other people by using different gestures. Gestures are represented in non-verbal form where in there would be exchange of messages and these gestures will be understood with vision. This non-verbal communication of deaf and dumb people is called sign language. Sign language is often considered as a visual language. The three major components involved in this language are as follows [4].

Finger spelling: The finger spelling method also known as the manual alphabet is a manual form of communication.

World level sign vocabulary: This method involves collection of large amount of data set. The experimentation will be carried out using multiple deep learning methods to identify the signs depending on the words and performs its evaluation. This method is applicable to large-scale scenarios.

Non-manual features: It is the combination of eye gaze, position and movement of head and upper body. This basically creates a model that predicts the finger spelling-based hand gestures to produce a full word. Figure 1 shows the different gestures that are trained for the proposed methodology and also reflects the American Sign Language symbols.

American Sign Language is the main sign language. Since the only obstacle for deaf mute people is communication, they cannot use spoken language, so the only way to communicate is through verbal gestures [5–8].

Fig. 1 American sign
language symbols

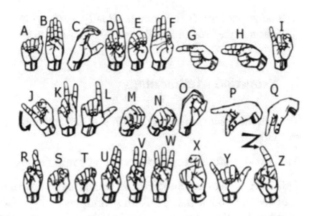

2 Literature Survey

Through the recent years there has been tremendous research carried on the hand gesture recognition which do not ensure reliable results.

- Mahamud and Zishan A hearing aid for the deaf and hard of hearing. "In this study, an assistive gadget for hearing impaired people is proposed, allowing them to communicate freely with anyone who comes close to them". The disabled person must wear a wristwatch-style device and carry an Android phone. This device receives voice and continuously converts it into legible text, which is displayed on the LCD screen. When someone wants to hear something from them, all they have to do is open the application on their phone and go through his watch [1].
- Dhanjal and Singh "Tools and Techniques of Assistive Technology for Hearing Impaired People". Education is the most crucial aspect of any human life since it allows us to advance in life by allowing us to transmit information through speech, writing, or other means. Hearing loss or deafness makes it harder to communicate and comprehend others. Hearing impaired people can communicate with normal people with less stress with assistive technology. Assistive Listening Devices (ALDs), Augmentative and Alternative Communication (AACs), and Alert systems are three types of assistive devices [2]. Various tools and strategies based on parameters like as alarm systems, speech augmentation, SL recognition and learning apps, telecommunication systems, and voice to text are described in this study article. The majority of hearing aids on the market are prototypes and not yet commercially available.
- Ren et al. [3]. "Wireless Assistive Sensor Networks for the Deaf". The purpose of this study is to develop a wireless sensor network-based assistive system (WASN) to assist the deaf or hearing impaired in being more aware of their surroundings. There's also a quick rundown of the latest assistive technology. For two common application scenarios, smart home and smart school playground, the system architecture, components, and specifications are described. To implement the system,

a node platform is being built. Finally, simulation is used to assess the system's performance, which is followed by a feasibility and availability analysis [3].

3 Proposed Methodology

This section discusses about the different design alternatives and the proposed design methodology.

(i)   Hand gesture recognition based on computer vision (Fig. 2):

This method uses computer vision technology to recognize the user's gestures that uses one or two hands to perform specific gestures in front of the camera. It will be connected to the frame of the system, including gestures that use one or two hands in front of the camera. These kinds of systems include various methods of extracting features and classifying gestures as per the application requirement [9].

(ii)   Real-time recognition of gestures through finger segmentation:

The block diagram of real-time recognition of gestures through finger segmentation is shown in the Fig. 3. The process is as follows. Initially perform hand recognition by subtracting the background part of it. The result of hand recognition is converted into a binary image. The principle of segmentation is applied to the fingers and palms to facilitate the finger recognition. Thus, gestures are recognized by a simple rule classifier in these kind of systems.

(iii)   AI hand tracking and gesture recognition:

Gesture recognition provides computer with real-time data for executing user commands. The motion sensor in this device can track and interpret gestures and use

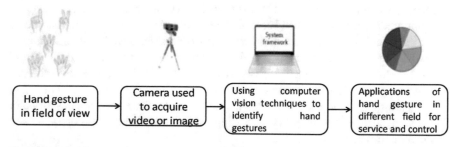

Fig. 2 Hand gesture recognition based on computer vision

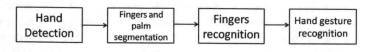

Fig. 3 Block diagram of real-time recognition of gestures through finger segmentation

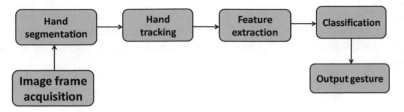

Fig. 4 Block diagram of hand tracking and gesture recognition using AI

it as the main source of input data. Most of the motion control solutions involve the combination of 3-D cameras, depth sensors, infrared cameras and machine learning systems. The machine learning algorithm learns from the marked depth image of the hand so that it can recognize the position of hand and fingers (Fig. 4).

Gesture recognition consists of three basic levels. They are as follows.

Detection: This device uses a camera to detect hand or body movements. With the help of machine learning algorithms, the image is segmented to determine its edge and position of the hand. **Trackiug**. This device tracks the movement frame by frame, record each and every movement and provides the accurate input for data analysis [6]. **Recognition**: This system tries to find the patterns based on the collected data. When the system finds the match and interprets the gesture, it will perform the action associated with the gesture. The feature extraction and classification in the following scheme realizes the recognition function. This completes the discussion of three different design alternatives of assistive systems. The discussion of the proposed design is as follows.

The functional block diagram is shown in the Fig. 5. The functions are expressed as a set of elementary blocks connecting the inputs and outputs. The role of web camera is used to recognize the gestures. They are further applied to the preprocessing block for converting the images to the standard sizes. The system is developed in such a way that converts the gestures into its respective text. Further the images are stored

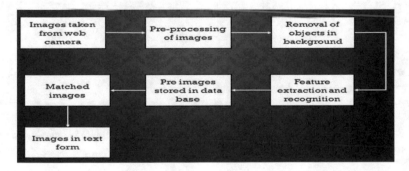

Fig. 5 Functional block diagram of system

in the database. The system uses the appropriate database to convert the image to text. Thus, it can be concluded that this system provides the output data in the form of text which in turn helps to reduce the communication gap between deaf and dumb people.

4 Implementation

With the help of the proposed design, the system architecture is developed and the specification of each component is discussed in this section. It also includes the detailed description of the algorithm and flowchart of the proposed design. **Data Collection**: It is the method of collecting data and information of the variables which will be targeted in a conventional system. It provides the authorization to respond to the questions which will be relevant and further estimates the outcomes. It is often considered as a component of research in all the kind of study fields (including technical and non-technical). The methods may diverge by discipline, the prominence on ensuring precise and sincere collection remains the same (Fig. 6).

The complete data set consists of 14,500 images which are of 60 × 60 pixels. The total number of classes that are available in this data set are 29 [each comprises with 500 images], 26 for the letters A–Z and 0,1,2 numbers. This data is the sole of the user gesturing in ASL, with the images taken from the laptop's webcam. These photos were then cropped, rescaled, and labeled for use. Note the difficulty of distinguishing fingers in the letter E. Five different test sets of images were taken with a webcam under different lighting conditions, backgrounds, and use of dominant/non-dominant hand. These images were then cropped and pre-processed. **Data Augmentation**: Data expansion during data analysis is a technique of expanding the amount of data by adding a slightly modified copy of existing data or newly created synthetic data from existing data. It acts as a regularizer and helps to reduce over fitting when training

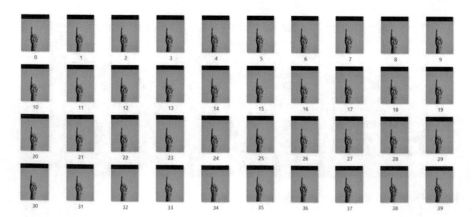

Fig. 6 Data collection

machine learning models. **Transfer learning**: This system uses transfer learning methods to build, train, and test models. Transfer learning is a machine learning technique in which a model trained on one task is redirected to a second related task. Focus on storing the knowledge gained from solving one problem and applying it to another related problem. In practice, few people train a convolution network from scratch (randomly initialized) because they rarely get enough data sets. Net weights, such as fixed feature initialization or extractors, help to solve most of the discussed problems. Training very deep networks is expensive. It takes weeks to train the most complex models on hundreds of expensive GPU-powered machines. The learning method/hyper parameter of deep learning is a black art, and there is not much theory for us to guide. In transfer learning, the data set is trained by using the underlying network and the underlying problem, and then reuse the learned attributes or transfer them to the second target network to train the task and target data set. If the characteristics are general, meaning suitable for basic tasks and objective tasks, not specific to the main task. **Flow of Algorithm**: The flow of the algorithm and Gesture reorganization is shown in the Fig. 7. **Layer 1**: The Gaussian blur filter is applied and some threshold value will be set to the image which is captured by Open CV to obtain the processed image after feature extraction. This processed image is passed to the CNN model for prediction and if a letter is detected for more than 50 frames then the letter is printed and taken into consideration for forming the word. Space between the words is considered using the blank symbol.

Layer 2: This layer recognizes different fonts, displays similar results during recognition. It also tries to classify collections using a classifier.

Flowchart:

The overall design flow of the system is represented in the form of flowchart shown in Fig. 8. The flowchart shows the complete architecture of the system. The device is split into phases—schooling section and checking out section. In the training period, the gestures will be captured by the camera using implicit camera. The captured

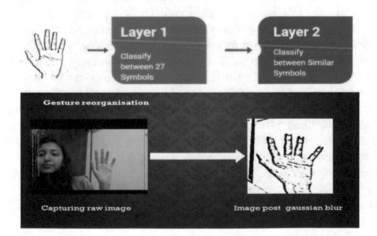

Fig. 7 Flow of algorithm and gesture reorganization

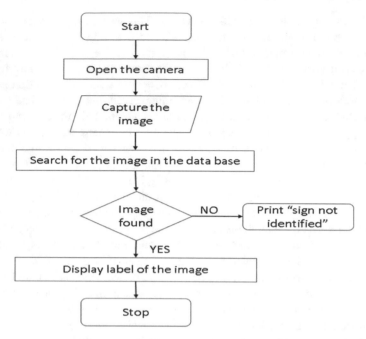

Fig. 8 Flow chart of the designed algorithm

images will be identified by using the image processing techniques such as image enhancement and segmentation, process of color filtering, skin segmentation and reduction of noise techniques. The images are further processed to extract the vital features using count value. They will be reserved in a file folder and then compared with the test images. In test period, the user makes gestures in front of the camera, the gestures are captured as image by the camera using virtual mouse. Images are further processed and their features will be extracted. The test images will be compared with the stored images. If the gestures are recognized, the output is produced in the form of text. Otherwise an error message is generated. **Training and Testing**: The training data convert input images (RGB) into grayscale and apply Gaussian blur to remove unnecessary noise. Adaptive threshold is applied to extract the hand from the background and resize the images to 64X64. After applying all the above operations, the pre-processed input image is loaded into the model for training and testing. The prediction layer estimates how likely the image will fall under one of the classes. So that the output is normalized between 0 and 1 and such that the sum of each value in each class summates to be 1. This is achieved using the soft max function.

The result of the predicted level will initially be a bit far from the actual value. In this way the network will be trained with the labeled data.

The images related to the testing have been shown in the Figs. 9, 10 and 11

Fig. 9 Training data from 0 to Z

Fig. 10 RGB testing images from 0 to O

Fig. 11 RGB testing images from P to Z

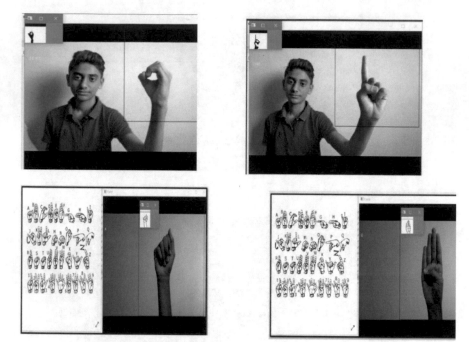

Fig. 12 Demonstrating the various letters and numbers

5 Results and Discussion

The discussion about the obtained results is as follows. Figure 12 shows the demonstration results showcasing various letters 7 numbers. The prediction window has two frames, one is the actual frame of 300 × 300 in dimension to capture the user input and another frame is of 64 × 64 which is a Gaussian filter window. The inputs of the user are captured and converted to gray images by applying Gaussian filter and these images are used for prediction. As shown in the pictures, the user's hand textures, background colors and other related parameters are changing and the prediction model is able to recognize any changes at the user's end and provide us the proper prediction results.

6 Conclusion

This paper finally concludes about the implementation of an automatic sign recognition system in real-time through migration learning. Implementation using basic methods is more effective than complex methods. The time constraints and difficulties involved in creating records from scratch is present. Until then, it is best to have a data set that can be used. In our live demonstration, some letters are more difficult to classify, such as "a" and "i", because they have a small difference ("I" puts the little finger up). The classification effect is very good. With pictures, there are many possibilities for future work. Possible extensions of the project include extending the gesture recognition system to all ASL letters and other non-letter gestures. The scope of the project can also be extended to various other applications, such as the use of gestures to control robot navigation.

References

1. M.S. Mahamud, M.S.R. Zishan, Watch IT: an assistive device for deaf and hearing impaired. In *2017 4th International Conference on Advances in Electrical Engineering (ICAEE)* (2017). https://doi.org/10.1109/icaee.2017.8255418
2. A.S. Dhanjal, W. Singh, Tools and techniques of assistive technology for hearing impaired people. In: *2019 International Confer- ence on Machine Learning, Big Data, Cloud and Parallel Computing (COMITCon)* (2019). https://doi.org/10.1109/comitcon.2019.886245.4
3. H. Ren, M. Meng, X. Chen, Wireless assistive sensor networks for the deaf. In: *2006 IEEE/RSJ International Conference on Intelligent Robots and Systems* (2006). https://doi.org/10.1109/iros.2006.282354
4. M. Wald, "Captioning for deaf and hard of hearing people by editing automatic speech recognition in real time", In: *Proceedings of 10th International Conference on Computers Helping People with Special Needs ICCHP*, LNCS 4061, pp. 683–690 (2006)
5. S. Ahmed, "Electronic speaking system for speech impaired people: Speak up", in *2015 International Conference on Electrical Engineering and Information Communication Technology(ICEEICT)*, pp. 1–4 (2015)
6. Mohammed Waleed Kalous, Machine recognition of Auslan signs using Power Gloves: towards large-lexicon recognition of sign language
7. M.M. Zaki, S.I. Shaheen, Sign language recognition using a combination of new vision based features. Patt. Recogn. Lett. **32**(4), 572–577 (2011)
8. P. Nikita et al., Active learning in electronic measurements and instrumentation course through hands-on. J. Eng. Educ. Transform., [S.l.], Jan. 2016. ISSN 2394–1707. Available at: http://www.journaleet.in/index.php/jeet/article/view/85543. https://doi.org/10.16920/jeet/2016/v0i0/85543
9. B.H. Shraddha, et al., Model based learning of linear integrated circuit course: a practical approach. J. Eng. Educ. Transform., [S.l.], Jan. 2016. ISSN 2394–1707. Available at: http://journaleet.in/index.php/jeet/article/view/85549. https://doi.org/10.16920/jeet/2016/v0i0/85549

Internet of Things Security: A Blockchain Perspective

Mohammad Luqman and **Arman Rasool Faridi**

Abstract Since the arrival of computers in 1950s, technology has leaped over many generations at a rapid pace. Computers have found their use in every human activity, namely scientific experiments, industries, transportation, medicine, healthcare, home management, etc. Internet of things (IoT) is one such computing paradigm that aims to assist and automate various trivial and non-trivial aspects of these activities. IoT generates a huge amount of data at a fast rate. The data can be trivial, personal, or sensitive to someone or an organization. Hence, securing IoT data and its aggregation is of utmost importance. The paper gives a brief introduction of IoT in terms of its security aspects to understand the various facets of malicious attacks possible on it. Then, how blockchain, an ever-popular security suite, can be used to provide security to IoT in its various integrations is presented. Finally, some future research issues are discussed towards blockchain-IoT future.

Keywords Internet of things · Blockchain · Security · Internet of vehicles · Internet of medical things · Consensus algorithms

1 Introduction

Today's modern society is a data driven society. Personalized services are facilitating the ease of life in people's life round the clock. People commuting to home wants the shortest path with the least traffic in their route planning system; healthcare workers require accurate and timely reports of a patient body conditions or homemaker could use their smartphone to control AC or other devices in a smart home. This fine-grained insertion of computers in human life was made possible due to two major reasons that are the small form factor of microcontrollers and universal connectivity of embedded devices to other unassociated devices with the help of Internet. Internet

M. Luqman (✉) · A. R. Faridi
Aligarh Muslim University, Aligarh, Uttar Pradesh 202002, India
e-mail: luqman.geeky@gmail.com

A. R. Faridi
e-mail: ar.faridi.cs@amu.ac.in

of things (IoT) is one such technology which is driving our data driven society. It is a universal network of things (smart objects) where things can get instructions from either human, from other things or from prebuilt commands.

Before we discuss various ways through which blockchain can secure IoT, a brief introduction of IoT and blockchain is needed to understand their significance in real-world applications. This paper aims to provide readers with the following details:

1. Introduction of IoT and its security requirements.
2. Acute analysis of blockchain and its functioning.
3. Discuss how various IoT networks are vulnerable to internal/external attacks.
4. Discuss the roles in which a blockchain can be used to provide security services to IoT and its similar network.

2 Internet of Things

IoT enables the soft touch of computers with the human life. The data collected are not stored by the IoT devices by themselves but instead are send to fog or cloud computing network where the data are stored and statistically analyzed for behaviors, routines, or patterns to enhance the quality of service provided to its user. Each device in an IoT network is directly connected to Internet thus has a unique address to identify itself over the network. This allows user's comfort to access their IoT device anywhere and anytime. IoT device provides and supports lots of applications in different areas such as transportation, industries, battlefields, healthcare, smart transportation systems, smart cities, smart traffic systems, e-governance, smart retails, smart farming, e-gaming, smart home, smart equipment, and automated security and surveillance systems. A survey by International Data Corporation (IDC) [1] indicates that more than 41.6 billion devices will operate by 2025 which will generate approx. 79.4 zettabytes of data. A survey by Kagan [2] estimates that 28% homes in USA will be smart by 2021 compared to 12.5% in 2016. Business insider intelligence [3] predicts 12 million sensors will be used in agriculture by 2023. Forrester [4] predicts that more than 250 million smart home devices will be installed in US homes by 2021.

IoT device being, generally, humancentric generates and transmits data which can be private, secretive, or possess critical value that create security and privacy concerns. Hence, securing the IoT data generated and transmitted as well as need of a secure centralized storage via fog/cloud computing to permanently store it are critically important. IoT security majorly consists of three security layers: hardware security, network security, and data security. **Hardware security** deals with securing the IoT device from tampering, destroying, depackaging, side-channel attacks, and other attacks to sensors and chips. **Network security** concerns with preventing crypt-analysis of embedded security cipher and the secure data transfer from transport layer. **Data security** concerns with secure execution of OS, application, and software on the IoT device, to provide secure support and middleware services to other

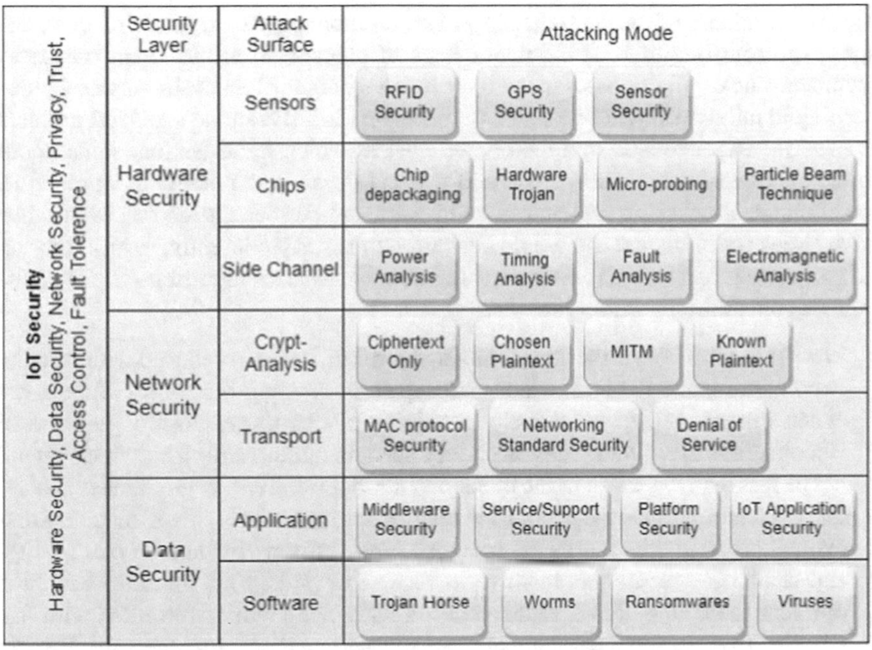

	Security Layer	Attack Surface	Attacking Mode			
IoT security Hardware Security, Data Security, Network Security, Privacy, Trust, Access Control, Fault Tolerence	Hardware Security	Sensors	RFID Security	GPS Security	Sensor Security	
		Chips	Chip depackaging	Hardware Trojan	Micro-probing	Particle Beam Technique
		Side Channel	Power Analysis	Timing Analysis	Fault Analysis	Electromagnetic Analysis
	Network Security	Crypt-Analysis	Ciphertext Only	Chosen Plaintext	MITM	Known Plaintext
		Transport	MAC protocol Security	Networking Standard Security	Denial of Service	
	Data Security	Application	Middleware Security	Service/Support Security	Platform Security	IoT Application Security
		Software	Trojan Horse	Worms	Ransomwares	Viruses

Fig. 1 Security overview of IoT

devices on the network and preventing execution of virus or similar rogue application. These attacks on the three layers of IoT are shown in Fig. 1 above. Blockchain is an upcoming solution which can guarantee security and privacy of IoT.

3 Blockchain

Blockchain is a digital ledger which is distributed, transparent, incorruptible, and equally accessible to its participants. It can store transactions which are immutable and verifiable. Blockchain network allows applications to store transactions in secure and peer-to-peer (P2P) manner. Blockchain forms the foundation of Bitcoin which is the first known cryptocurrency. Although blockchain was formally introduced with Bitcoin in a paper published by Satoshi Nakamoto [5] in 2008, but earlier attempts were made as well. David Chaum [6] in his dissertation proposed a protocol similar to blockchain in 1982. Stuart Haber and W. Scott Stornetta [7] in 1991 introduced a cryptographic system to securely timestamp documents. Stuart Haber, W. Scott Stornetta, and Dave Bayer [8] in 1992 improved upon the previous work on timestamping documents by using Merkle trees, hashcash [9] was formally introduced in 2002 paper as CPU cost function which was to be used in PoW. Blockchain benefits from the drawbacks of centralized system such as no single point of failure, amount of trustworthiness of the centralized system, no monopoly on transaction cost, and speed or security. Some of the major uses of blockchain are in smart contracts, cryptocurrencies, digital assets, and distributed storage. Blockchain was developed keeping in mind the Byzantine's general problem [10]. Byzantine's general problem is a unique characteristic of a distributed system which indicates that some nodes in the distributed system may fault, and it is not possible to reliably determined its truthfulness. Blockchain guarantees valid state and permanent records even when adversaries try to disrupt the whole system, by properly consenting every transaction with each participants involved using some consensus algorithm. Some of the famous consensus algorithms are

- *Proof of work (PoW)* is the most famous algorithm. It was introduced with Bitcoin. People, group, or network which participate in creating blocks for Bitcoin are called miners. Miners participate in mining blocks to get rewards in form of Bitcoin. PoW algorithm relies on energy spent in mining any block which should not be less than the worth of the block produced. PoW trivially guarantees Byzantine's general problem. Bitcoin used hashcash to create a highly difficult mathematical problem which uses the previous block information among other things and challenges the miners to solve to receive rewards. First miner to solve the problem can add the block to the network and he/she will be rewarded with one Bitcoin. To prevent rapid mining of blocks by adding more compute units, Bitcoin network will automatically increase the difficulty of mining the new blocks by increasing the mining difficulty of the whole network.
- *Proof of stake (PoS)* is another famous consensus algorithm which is currently adopted by Solana, Cardano, Nxt etc., cryptocurrencies. PoS is deemed as alternative to the drawbacks of the PoW. The first cryptocurrency to use PoS was PPCoin [11]. PoS uses age of coin to determine the stake rights of a participant. Higher the age of a coin, more rights are given to its owner as well as some rewards. PoS encourages its user to hold the coins for longer time. Using coin age as important parameter in creating consensus, resource wastage is automatically reduced. Attackers trying to attack network needs to hold sufficient number of coins as well as coin age to successfully attack it.
- *Practical Byzantine fault tolerance (PBFT)* [12] is an improved and efficient version of Byzantine fault tolerance algorithm. Byzantine fault tolerance algorithm was used in distributed computing which provides better results but require exponential time complexity. In 1999, when PBFT was introduced, then the wider adoption of algorithm was started. PBFT decreases algorithmic complexity from exponential to polynomial which was a major improvement. It involves five step to achieve consensus that are request, pre-prepare, prepare, commit, reply.
- *Delegated Proof of Stake (DPoS)* [13] is an improved version of PoS, which was developed by Daniel Larimer in the year 2014. Bitshares is a cryptocurrency that uses DPoS. It utilizes less resources to operate the network and is more efficient compared to PoS and PoW algorithms. This algorithm selects witness based on the stakes hold by it for every node. Accounting rights are given to those witness which have participated in the network and are in top N position. The number N is

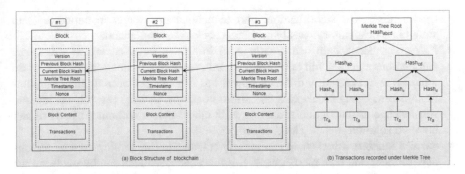

Fig. 2 Block structure of a blockchain

determined by the voting stakeholders when at least 50% of the voting stakeholders believe there is enough decentralization in the network.

The basic unit of a blockchain is a block which consists of version number, the previous hash value, current hash value, Merkle tree root, timestamp, and a nonce. Merkle tree or hash tree is a tree data structure, generally a binary tree, whose non-leaf nodes are hashes of its child nodes label while leaf nodes are hashes of data blocks. Nonce is a random value whose requirement is, it should be lower than or equal to the hash target set by the network. Figure 2 shows the basic structure of a block.

4 Applications of Blockchain in IoT and Similar Networks

4.1 IoT with Blockchain

IoT devices regularly perform trivial tasks on behalf of is owner, sense data, and transmit it to store in the cloud. But IoT devices are incapable to properly manage data or provide data filtration and reporting service. Some of the benefits which will come by integrating blockchain with IoT [14] are as follows:

- *Auditability*: Every transaction such as ordering, payments, delivery receipt, etc., can be stored in blockchain which can be used as proofs of service. Blockchain permanently stores records which is immutable and incorruptible and hence can be audited later for obtaining proofs.
- *Reliability*: Blockchain stores data in immutable and distributed form which guarantees against malicious data tampering or data loss. Moreover, blockchain can enable sensor data traceability and accountability which improves overall reliability of the network.
- *Identity*: A single blockchain can easily provide identity service in a IoT network [15]. It can easily identify devices by the data transmitted by them and provide

trusted, distrusted authorization, and authentication system in network. Data stored in blockchain can easily be traced back to its source node.

- *Autonomy*: Blockchain can greatly help in *hardware as a service* and smart autonomous assets. It will optimize the hardware usage in situations such as when a hardware is constantly drawing power but it is not performing any computational work. Using blockchain and smart contracts, devices can self-decide and self-manage its interaction and usage with other devices without the need of any backend solution such as cloud or fog computing.
- *Service Markets*: Blockchain can greatly enhance in the development of various new services and data marketplace. Services and payments can be made by the device itself securely even in trustless environment by using blockchain.
- *Secure code update/deployment*: Blockchain can provide secure and immutable data storage which can be used to provide regular code updates or deployment securely. Manufactures can reliably send updates and be ensured that the code update process will be successfully processed.

4.2 Internet of Vehicles (IoV) with Blockchain

IoT engulfs existing vehicular ad hoc network (VANET) into the new concept of Internet of vehicles (IoV) [16] in which nodes transmit data to everyone in the network, i.e., vehicle to vehicle, vehicle to traffic system, vehicle to road, vehicle to city, etc. Sensors embedded in vehicles generate critical information which is processed to extract useful information such as fuel efficiency, maintenance prediction, smart braking and speed control, and smart navigation. As a result, IoV generates lots of information which can cause unauthorized access, DoS attack, information and transmission security, etc. Blockchain can be used to provide following features to IoV:

- *Traffic safety and efficiency*: Blockchain can reliably transmit real-time vehicular information to its peer as well as to the centralized storage using cryptographic methods [17].
- *Micro-payment gateway service*: Using blockchain and smart contracts, payments of e-toll, e-parking, taxes, insurances, etc., can be made autonomously. Blockchain can be used to reliably made these transactions and hence allows fast and efficient IoV operation [18].
- *Vehicle insurance* and *audits*: IoV can collect car usage information and driver's driving tendencies to calculate premium for insurance. Blockchain can immutably and efficiently store these data to provide reliable audit results when queried by the insurance company [19].
- *Improved security*: Blockchain provides immutable and distributed storage. It can be used to reliably provide code updates, secure sensor data transmission, tolerance to data theft and hacking which results in efficient and resilient system [20].

4.3 Internet of Medical Things (IoMT) with Blockchain

IoMT [21] plays vital role now in healthcare. Sensors gives real-time reports of various body conditions as well predications, analysis, and reporting of medical data for future use. Medical data are very sensitive and private for any individual. Proper handling and transmission of medical data are required so that any data leakage or hacking can be prevented. Incorrect data or missing information can prove lethal for patients. So, IoT data need to be securely stored, managed, and disseminated. Blockchain can be used to secure IoMT and its data. Below are some benefits of using blockchain with IoMT:

- *Medical data management*: Smart contracts can restrict data access/share permissions along entities like doctor, hospital, research organizations, etc.
- *Medical data storage*: Blockchain can be used to provide secure data storage to store monitoring data of patient's conditions, activities or to monitor the development of a chronic disease over time. It will also enhance availability of data from malicious attacks.
- *Enable P2P applications*: Though healthcare data are generally stored in centralized databases, blockchain can guarantee secure decentralized storage service enabling P2P applications and systems that will improve the overall system.
- *Ease of data transformation*: Currently, data transformations are done by third parties in hidden and costly manner. Invested parties cannot know the process involved in data transformation which reduces safety, cost involved, regulations, and policies compliance. Blockchain with smart contracts can enable data transformation in transparent and cost-effective manner.

4.4 Internet of Battlefield Things (IoBT) with Blockchain

IoBT [22] is the amalgam of IoT with military applications, war machines, or human war-fighters. Future wars will be fought by machines on behalf of their human operators. Planning war strategies, extracting enemy secret information, enemy key assets surveillance, optimized soldier deployment, precise defensive planning, and performing accurate surgical strikes on enemy troops are some of the applications where IoBT will be very useful for militaries around the globe. Some of the benefits of using blockchain in securing IoBT are as follows:

- *Secrecy* and *authentication*: Blockchain stores data with keeping attributes like data integrity, data confidentiality, and data authentication. Once a command is issued by army leadership, it will not be possible to alter it by any external entities or it is not possible to be accessed by unauthorized persons and is only visible to properly authenticated parties.

- *Decentralization of data security*: Since data will be stored in a decentralized nature, it will be very difficult for enemies to attack or modify the data in comparison with centralized systems which will help in clear and coordinated attacks.

4.5 Green IoT with Blockchain

Green IoT [23] aims to drive toward eco-friendly sustainable development of IoT and its assisting technologies. IoT is exploding in its number due to affordable access to high-speed Internet and cheap production of microcontrollers. Current explosion of IoT devices is neither sustainable nor environment friendly. Hence, green IoT focusses on the reduction of energy consumption by IoT devices, reducing carbon footprint of IoT devices and using renewable energy sources in place of fossil fuels in powering IoT devices.

- *Green energy*: IoT can be used in smart grids to conserve energy and reduce wastage of energy [24]. Blockchain can enable secure transmission of critical information regarding power generation, power transmission and distribution, and power consumptions. Energy trading, electric vehicles, microgrid operations, and cyber-physical security are some areas of smart grids which can be hugely benefitted by the combination of blockchain with IoT.
- *Green Computing*: Cloud computing caters to our large demands of computing power and storage system by individuals or corporations [25]. But power consumptions of cloud computing or other similar types of computing are also creating major issue toward green goals. Green computing aims to introduce eco-friendliness along with sustainability to different computing types by efficiently using computing resources, minimizing energy uses, and reducing carbon emissions.
- *Green finance and e-governance*: Green finances enable efficient data availability, increased data sharing, fast access, and reduces cost to store or transmit data which overall optimize financial strategies that are built upon these data [26]. Cryptocurrencies and smart contracts can be used as a greener alternative to current conventional approach. Cryptocurrencies can exchange currencies or do international money transfer in real-time with no middlemen involved which saves additional energy, cost, and time. Smart contracts will reduce wastage by automating purchase orders, order monitoring or service deliveries, etc. E-governance can highly benefit from blockchain-IoT integration by eliminating excess paperwork, reduced public service delivery time, and minimizing manual interventions.

4.6 Future Research Issues

IoT has a huge role to play in future whether in civilian issues (Internet of vehicles, Internet of energy, etc.) or military issues (Internet of drones, Internet of battlefield things, etc.). IoT will continue to provide interactive smart services but the research enabling IoT lacks properly secured data communication, efficient data aggregation, access controls, etc. Thus, it is the need of the hour to increase the research focus of researchers on securing IoT. Some of the future research issues that should be worked upon immediately are as follows:

- *Lack of datasets*: There is a general lack of IoT dataset. IoT devices produce heterogeneous data which are often controlled by their device manufactures. Open public datasets are required to further analyze and optimize IoT device activities.
- *Merging artificial intelligence (AI) with IoT and blockchain*: Artificial intelligence (AI) has opened new avenues in automating services to consumers. But research in the field of AI with blockchain and IoT are very lacking. AI can easily detect fraudulent activities, prevent hacking and tampering activities, make sound decisions on behalf of its users, provide automation services, predict possible disease from patient history [27], etc., into the IoT network and blockchain.
- *Prediction of IoT device security*: Hospitals, assisted driving, battlefield surveillance, etc., are just some areas where physical security of IoT device is highly critical to the human lives. IoT devices can be destroyed whether by malicious intent or due to constant usage over time. Either case requires that security of IoT device can be made well in advance to prevent an unknown accident/loss.
- *Optimization of blockchains and blockchain-based platforms*: Lots of work has been done to improve the efficiency of blockchain but still enterprises do not wish to run their own blockchains due to high-energy costs and operational cost. For wide adoption of blockchain by enterprises and to create an optimized solution for IoT, research is further needed in optimizing blockchains in terms of energy efficiency, computational needs, and memory requirements.
- *Efficient* and *lightweight blockchain-based security solutions*: Blockchains are too computationally heavy to be run by the IoT device themselves. Hence, IoT devices cannot secure themselves from tampering, abusing, using anti-forensic techniques, etc., from a malicious third-party by using blockchains by themselves. Hence, blockchain solutions that can help in determining the security status and the information leakage during hacking process to investigators are needed to catch the culprit involved and prevent such activities in future.
- *Scalability* and *cost-efficiency*: Blockchain implementation should be scalable and cost-effective for huge IoT networks. IoT networks are dynamic and rapidly changing. Consensus algorithms should also be elastic enough to accept rapid changes to IoT network. Some of the current consensus algorithms are scalable compared to others but they lack in terms of security provided.
- *Uniform* and *interoperable security system/protocols*: IoT devices are heterogeneous in nature which causes issues in standards development and interoperability. A defined set of protocols or rules should be developed in conjunction

with blockchain which should declare a strict framework that dictates software and hardware compliance. Framework should be multi-layered like OSI. Layer-wise classification will define roles to specific layers and determines proper security level and transmission protocols before starting any data transmission within devices.

5 Conclusion

IoT and blockchain have highly steered the modern world toward smart and data driven future. Gone are the days when human potentials are destroyed in performing trivial, repetitive tasks. IoT devices are characterized by their ability to perform redundant tasks easily, and for intelligent tasks, it automatically senses some data using sensors and performs the task based on the sensed data. But IoT device cannot guarantee the security of sensed data or its transmission to centralized storage or other nodes. Blockchain, a distributed ledger with cryptography at its helm aims to provide security services to IoT. Blockchain with IoT can guarantee security, auditability, and incorruptibility while performing autonomous tasks. The IoT, blockchain, and AI combined will become the building blocks of the futuristic smart cities, smart healthcare, smart transportation, etc.

References

1. "The Growth in Connected IoT Devices Is Expected to Generate 79.4ZB of Data in 2025, According to a New IDC Forecast."
2. "Smart Homes In The U.S. Becoming More Common, But Still Face Challenges | S&P Global Market Intelligence."
3. "Smart Farming in 2020: IoT Sensors & Precision Agriculture - Business Insider."
4. "Future Of Retail 2020: Retailers Test Consumer-Facing And Operational IoT Solutions."
5. S. Nakamoto, "Bitcoin: A Peer-to-Peer Electronic Cash System."
6. D. Lee, "Computer Systems Established, Maintained and Trusted by Mutually Suspicious Groups," (1982)
7. S. Haber, W.S. Stornetta, How to time-stamp a digital document. J. Cryptol. 3(2), 99–111 (1991)
8. D. Bayer, S. Haber, W. S. Stornetta, "Improving the efficiency and reliability of digital time-stamping," in Sequences II, Springer New York, pp. 329–334 (1993)
9. A. Back, "Hashcash-A Denial of Service Counter-Measure," 2002.
10. L. Lamport, R. Shostak, M. Pease, The Byzantine generals problem. ACM Trans. Program. Lang. Syst. (TOPLAS) 4, 382–401 (1982)
11. S. King, S. Nadal, "PPCoin: Peer-to-Peer Crypto-Currency with Proof-of-Stake," (2012)
12. M. Castro, B. Liskov, "Practical Byzantine Fault Tolerance," (1999)
13. "Delegated proof of stake - Bitcoin Wiki."
14. M.A. Khan, K. Salah, IoT security: review, blockchain solutions, and open challenges. Futur. Gener. Comput. Syst. 82, 395–411 (2018)
15. X. Zhu, Y. Badr, "Identity management systems for the internet of things: a survey towards blockchain solutions," Sensors (Basel, Switzerland) 18(12) (2018)

16. L. Zhang et al., Blockchain based secure data sharing system for internet of vehicles: a position paper. Vehicular Commun. **16**, 85–93 (Apr. 2019)
17. R.A. Sanchez, Acosta Rodriguez, W.M. Prieto, "Control of vehicular traffic through the IoV," Contemp. Eng. Sci., **10**(33), 1643–1650 (2017)
18. A. Ensor, S. Schefer-Wenzl, I. Miladinovic, "Blockchains for IoT payments: a survey," (2019)
19. R. Jabbar, M. Kharbeche, K. Al-Khalifa, M. Krichen, A.K. Barkaoui, Blockchain for the internet of vehicles: a decentralized IoT solution for vehicles communication using ethereum. Sensors (Switzerland) **20**(14), 1–27 (2020)
20. Y.R. Chen, J.R. Sha, Z.H. Zhou, "IOV privacy protection system based on double-layered chains," Wireless Commun. Mobile Comput., **2019** (2019)
21. J.M. Roman-Belmonte, H. de la Corte-Rodriguez, E.C. Rodriguez-Merchan, "How blockchain technology can change medicine," *Postgraduate Medicine*, vol. 130(4). Taylor and Francis Inc., pp. 420–427 (2018)
22. A. Kott, A. Swami, B.J. West, The internet of battle things. Computer **49**(12), 70–75 (2016)
23. F.K. Shaikh, S. Zeadally, E. Exposito, Enabling technologies for green internet of things. IEEE Syst. J. **11**(2), 983–994 (2017)
24. T. Alladi, V. Chamola, J.J.P.C. Rodrigues, S.A. Kozlov, Blockchain in smart grids: a review on different use cases. Sensors (Switzerland) **19**(22), 1–25 (2019)
25. P.K. Sharma, M.Y. Chen, J.H. Park, A software defined fog node based distributed blockchain cloud architecture for IoT. IEEE Access **6**, 115–124 (2018)
26. K. Anwar, J. Siddiqui, S. Saquib Sohail, "Machine learning techniques for book recommendation: an overview," SSRN Electron. J., (2019)
27. M. Atif, J. Siddiqui, F. Talib, "An overview of diabetes mellitus prediction through machine learning approaches," in *2019 6th International Conference on Computing for Sustainable Global Development (INDIACom)*, pp. 1145–1150 (2019)

Fuzzy C-Means Clustering of Network for Multi Mobile Agent Itinerary Planning

Nidhi and Shuchita Upadhyaya

Abstract Mobile agent (MA) works potentially efficiently in reducing network bandwidth consumption for distributed computing. In an MA-based system, a small-sized processing code is transmitted through the network rather than the raw data. Using strong migration capability, MA's processing code is being executed at each targeted node. Subsequently, useful and significant information is delivered to the intended user. Despite its benefit, MA has its issues also. Finding the appropriate number of MAs to be dispatched, the set of targeted nodes and their sequence of migration are the major issues related to multi-mobile agent itinerary planning (MIP). This paper tries to find a suitable number of MAs by considering the size of the data payload of a single MA and the load to be carried out by the MAs from the whole network. Moreover, this paper gives an idea of partitioning the given network into load balanced and nonoverlapping clusters using a fuzzy c-means clustering algorithm.

Keywords Clustering · Itinerary planning · Mobile agent

1 Introduction

A mobile agent is an autonomous and intelligent application program that is used to perform any data processing task on behalf of a remote user. The processing element (PE) can deploy one or more mobile agents to complete the assigned task. MA migrates to each targeted node locally and processes the data to gain some significant information [1–4]. After processing the data, aggregation of processed data will be performed. Thus, only significant accumulated information is transmitted to PE. On behalf of this retrieved information, a good decision can be made by the remote user. In other words, instead of transmission of a large amount of raw data; MA (a small-sized source code) can be transferred to each targeted node to process the data toward the completion of the assigned task [4–7]. This progressively processed and aggregated data is carried out by the MA in its data payload. Further, MA returns to

---

Nidhi (✉) · S. Upadhyaya
Department of Computer Science and Applications, Kurukshetra University, Kurukshetra, India
e-mail: nidhikashyap@kuk.ac.in

© The Author(s), under exclusive license to Springer Nature Singapore Pte Ltd. 2023
Y.-D. Zhang et al. (eds.), *Smart Trends in Computing and Communications*, Lecture Notes in Networks and Systems 396, https://doi.org/10.1007/978-981-16-9967-2_55

deliver this retrieved information to the PE to unload itself. Hence, only small-sized source code and retrieved significant information are getting transmitted through the network. Therefore, it will save large consumption of network bandwidth.

Besides, MA has one more exceptional characteristic of strong mobility that it can work even after disconnection also. During processing, it does not need any network connection. After processing when it will get connected it will be migrated to the next node and resume itself. In conclusion, MA needs an Internet connection only at the time of migration. In contrast, it does not need an Internet connection during the processing of data.

Moreover, when a single MA becomes insufficient for the given network then multiple mobile agents can be dispatched by the PE. Each MA has to migrate to the targeted set of nodes toward the completion of its task. Their route of migration to the assigned set of nodes is called itinerary planning. Whenever a single MA has been deployed to perform the task, the designed itinerary is called single mobile agent itinerary planning (SIP) [1, 8, 9]. Large delay and reliability are the most prominent issues related to SIP. In contrast, whenever more than one MA is dispatched, then a planned itinerary is called multi-mobile agent itinerary planning (MIP), in which, all mobile agents have their own set of nodes, and relative to each set they have separate itinerary planning. MIP came into existence to overcome the aforementioned issues of SIP. Despite its benefits, MIP leads to several issues also. Respective research has been done to resolve these specific issues. This paper suggests an algorithm to partition the network into k number of disjoint domains. Moreover, by considering the threshold value of a single MA, the appropriate number of MAs for the respective domain can be determined.

2 Related Work

Several algorithms have been proposed by the researchers to cluster the network for multi-mobile agent itinerary planning. They dealt with this identified problem in their way by keeping mobile agent routing into consideration. Kuila [10–13] suggested algorithms are using k-means and x-means clustering algorithms. These algorithms are very useful for large-scale problems. After the choice of cluster heads (CHs), nodes are associated with the nearest CH. Further, a single MA is deployed to each cluster to perform the assigned task.

Mpitziopoulos [4, 6, 9, 14, 15] algorithms use a tree-based approach to distribute their network into clusters. A dedicated single MA is responsible for each stemmed branch. Tree-based clustering is not useful for time-critical problems. Chou [8, 16–18] algorithms use Genetic algorithm (GA)-based evolutionary technique. Due to the initialization of the population on a random basis, GA-based MIP is not useful for time-based applications. [9] Second Near Optimal Itinerary Design (SNOID), [2] Tree-Based Itinerary Design (TBID) distributes the network into concentric ellip- tical or circular zones. However, not all sensor nodes need to be scattered in any geographical shape.

Chen [1] Central Location-based MIP (CL-MIP), [3] Energy Efficient MIP (EMIP) partitions their network by considering visiting center location (VCL) as the center of, clusters to be made. VCL can be taken any node which is situated at the center of the dense area. Source Grouping-based MIP (SG-MIP), [5] Directional Source Grouping-based MIP (DSG-MIP), [19] Angle Gap-based MIP (AG-MIP) distributes their network into directional sector zones by considering an angle θ. However, these algorithms are limited in determining the angle θ.

A lot of research has been dealt with the issue of partitioning the network through heuristic approaches, optimization algorithms, or evolutionary techniques. Most weaknesses and the robustness of aforesaid algorithms have been discussed. To get the suitable partitioning of the network into domains, soft computing is the best technique. Using a fuzzy C-means clustering algorithm, an individual sensor node can be associated with the appropriate domain. This algorithm states that each node has some degree of belongingness to each domain center (DC). But the node will be associated with whom it has a higher degree of membership. The proposed algorithm uses this technique to make clustering into the number of disjoint and load balanced domains by keeping the expected estimated assigned load in advance into consideration. Through the suggested algorithm, a node is associated with the DC with maximum membership value. In addition, if a single node has the same membership value with more DCs, then it will be connected with the DC having the least expected estimated assigned load so that each domain would get load balanced.

3 Proposed Method

In the proposed clustering algorithm, a network having n number of nodes is to be partitioned into k number of domains (clusters). Each domain has a single domain center (cluster head). For this, suitable nodes have to be chosen to represent domain centers (DCs). The remaining nodes of the network are to be associated with appropriate DC to form disjoint domains.

While clustering, cluster overlapping and load balancing are the most important issues on which research can be carried out. In the proposed algorithm both issues have been resolved. In crisp or traditional partitioning technologies, each node has the characteristic value of either the node belonging to any particular domain or not. However, using fuzzy logic, the degree of belongingness of each node relative to each domain center can be determined. This degree of belongingness is called degree of membership (μ). Afterward, the ith node is assigned to jth DC by considering, $\mu_{i,j} \in [0, 1]$, which has been described through Eq. 1. Where $\mu_{i,j}$ can be determined through Eq. 2. This value is inversely proportional to the distance value. If the distance is small, then it tends to greater membership value and vice versa.

$$\mu i, j = \left[\begin{array}{ll} 1 & \text{if } i = j \\ \dfrac{d_{i,j}^{\frac{-2}{m-1}}}{\sum_{l=1}^{k} d_{1,j}^{\frac{-2}{m-1}}} & \text{Otherwise} \end{array} \right] \forall i \in N, l \neq j \tag{1}$$

$$\mu i, j = \frac{d_{i,j}^{\frac{-2}{m-1}}}{\sum_{l=1}^{k} d_{i,j}^{\frac{-2}{m-1}}} \qquad (2)$$

where k is the number of domains in which the whole network is to be partitioned.

'm' is the weighing exponent which controls the fuzziness or crispness of the clustering. This 'm' decides the boundaries of the domain as either softer or harder.

Equation 3, specifies that $\mu_{i,j}$ of each source node corresponding to every DC is always 1. It ensures that each source node is assigned to exactly one domain center.

$$\sum_{i=1}^{k} \mu_{i,j} = 1, \ \forall j \in \{1, 2, 3, \ldots, n\} \qquad (3)$$

Equation 4; every node has some membership value w. r. t. each domain. In other words, it assures that each domain has some expected assigned load relative to source nodes in the network, otherwise it needs to be dropped out.

$$\sum_{i=1}^{n} \mu_{i,j} > 0, \ \forall j \in \{1, 2, 3 \ldots, k\} \qquad (4)$$

A brief description of the clustering algorithm as applied to the network for determining candidate DCs and assignment of the rest of the nodes to appropriate DC is given below.

3.1 Problem Statement

Given:

1. $N = \{N i \in i = 1, 2, \ldots, n\}$—Indexed set of n nodes.
2. $D = \{d_{i,j}, i = 1, 2, \ldots, n; \ j = 1, 2, \ldots, n; i \neq j\}$—Table 1: distance table, where d_{ij} is the Euclidean distance between node I and node j and is specified as integer.
3. Maximum number of domains is defined as k where $k < n$.

Determine:

Selection of k nodes to represent as a set DC (candidate domain centers) and then partitioning the network into k equals loaded domains with each domain having a single DC (element of set DC).

To be determined:

Exhaustive partitions of the set of nodes (N_i) into c number of domains.

Table 1 Euclidean distance matrix for each node

$D_{i,j}$	1	2	3	4	5	6	7	8	9	10	11	12	13	14	15
1	0	3	6	3	2	8	8	6	8	11	11	12	12	10	12
2	3	0	3	2	3	7	8	6	8	12	12	12	12	10	12
3	6	3	0	3	6	4	6	6	7	10	10	10	10	8	10
4	3	2	3	0	3	5	6	4	6	10	10	10	9	8	10
5	2	3	6	3	0	8	6	4	6	10	10	10	10	8	10
6	8	7	4	5	8	0	5	8	9	5	5	5	7	9	11
7	8	8	6	6	6	5	0	3	4	6	4	4	4	4	6
8	6	6	6	4	4	8	3	0	2	9	7	6	6	4	6
9	8	8	7	6	6	9	4	2	0	10	8	6	4	2	4
10	11	12	10	10	10	5	6	9	10	0	2	4	8	10	12
11	11	12	10	10	10	5	4	7	8	2	0	2	6	8	10
12	12	12	10	10	10	5	4	6	6	4	2	0	4	6	8
13	12	12	10	9	10	7	4	6	4	8	6	4	0	2	4
14	10	10	8	8	8	9	4	4	2	10	8	6	?	0	2
15	12	12	10	10	10	11	6	6	4	12	10	8	4	2	0

FCM Clustering Algorithm

Some abbreviations which are used in the given algorithm:

N_i—Set of nodes.
n_i—element of set N_i.
DC— Set of domain centers.
dc_j—element of set DC_j.

1. Given: Network with n number of nodes. $N_i = \{1, 2, 3, ...n\}$;
2. Determine membership values of each sensor node *w. r. t.* rest $n-1$ nodes;

   /* Step 3 is evaluated to choose k number of candidate domain centers.
3. Choose top k number of nodes from the original set (N_i) of nodes such that $\sum_{j=1}^{n} \mu_{i,j} : i \neq j$; is maximum; denoted by set DC_j, where $j = 1$ to k, set $N_i = N_i - DC_j$;

   /* Steps 4 and 5 will be evaluated to associate each node to the relative candidate domain center.
4. Determine membership value of each node n_i relative to each dc_j, i.e., $\mu_{i,j}$: $\forall i \in N_i, \forall j \in DC_j$;
5. Associate each $i \in N_i$, to relative $j \in DC_j$ such that n_i will be associated to dc_j with whom it has maximum membership value. If for any $i \in N_i, \exists j \in DC_j$: $\mu_{i,j}$ is equal then n_i will be associated to dc_j, the whose expected estimated load is least;

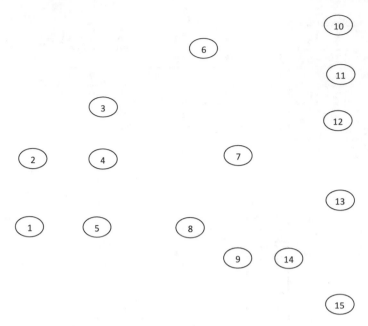

Fig. 1 Illustration network of fifteen nodes

6. Afterward, if there exists any dc_j, whose load to be carried out is less than the threshold value.[1] Then this node, as well the associated nodes need to be connected to some other appropriate dc_j.

4 Instance Formulation of the Network with FCM Clustering Algorithm

For instance, a network of 15 nodes has been depicted in Fig. 1. Furthermore, distance of each node relative to other node is given by Table 1.

The motive of this algorithm is to make domains around DCs. Furthermore, these DCs should be situated at the center of the densest places in the network. From the below, DCs using Eq. 2 can be determined. Uppermost k number of nodes, where $k = 4$, nodes having maximum $\sum_{j=1}^{n} \mu_{i,j} \forall i \in N_i$.

In particular, using above Table 2 selected candidate DCs are DC_4, DC_7, DC_8 and DC_9 with their respective $\mu_{i,j}$ are 1.14, 1.11, 1.12and 1.12. In below Table 3, membership value of each node w. r. t. each candidate DC, using Eq. 2 has been determined.

---

[1] Threshold value of a MA is the minimum number of nodes for which a single MA may be deployed. For the illustrated network, this threshold value has to be taken 2 nodes.

Table 2 Membership values of each node w. r. t. other nodes in the network

$D_{i,j}$	1	2	3	4	5	6	7	8	9	10	11	12	13	14	15
1	0	0.14	0.07	0.14	0.21	0.05	0.05	0.07	0.05	0.04	0.04	0.03	0.03	0.04	0.03
2	0.13	0	0.13	0.19	0.13	0.06	0.05	0.06	0.05	0.03	0.03	0.03	0.03	0.04	0.03
3	0.07	0.14	0	0.14	0.07	0.11	0.07	0.07	0.06	0.04	0.04	0.04	0.04	0.05	0.04
4	0.11	0.17	0.11	0	0.11	0.07	0.06	0.09	0.06	0.03	0.03	0.03	0.04	0.04	0.03
5	0.19	0.13	0.06	0.13	0	0.05	0.06	0.09	0.06	0.04	0.04	0.04	0.04	0.05	0.04
6	0.06	0.06	0.11	0.09	0.06	0	0.09	0.06	0.05	0.09	0.09	0.09	0.06	0.05	0.04
7	0.04	0.04	0.06	0.06	0.06	0.07	0	0.12	0.09	0.06	0.09	0.09	0.09	0.09	0.06
8	0.06	0.06	0.06	0.08	0.08	0.04	0.11	0	0.17	0.04	0.05	0.06	0.06	0.08	0.06
9	0.04	0.04	0.05	0.06	0.06	0.04	0.08	0.17	0	0.03	0.04	0.06	0.08	0.17	0.08
10	0.04	0.04	0.05	0.05	0.05	0.09	0.08	0.05	0.05	0	0.24	0.12	0.06	0.05	0.04
11	0.04	0.03	0.04	0.04	0.04	0.08	0.1	0.06	0.05	0.19	0	0.19	0.06	0.05	0.04
12	0.03	0.03	0.04	0.04	0.04	0.08	0.1	0.07	0.07	0.1	0.2	0	0.1	0.07	0.05
13	0.03	0.03	0.04	0.04	0.04	0.06	0.1	0.06	0.1	0.05	0.06	0.1	0	0.19	0.1
14	0.03	0.03	0.04	0.04	0.04	0.04	0.08	0.08	0.16	0.03	0.04	0.05	0.16	0	0.16
15	0.04	0.04	0.05	0.05	0.05	0.04	0.08	0.08	0.11	0.04	0.05	0.06	0.11	0.23	0

Table 3 Membership values of each node with respect to the domain centers

Node N_i	Membership value w.r.t. DC_4	Membership value w.r.t. DC_7	Membership value w.r.t. DC_8	Membership value w.r.t. DC_9
1	0.4	0.2	0.2	0.2
2	0.5	0.1	0.2	0.1
3	0.4	0.2	0.2	0.2
5	0.4	0.2	0.3	0.2
6	0.3	0.3	0.2	0.2
10	0.2	0.3	0.2	0.2
11	0.2	0.4	0.2	0.2
12	0.1	0.4	0.2	0.2
13	0.1	0.3	0.2	0.3
14	0.1	0.2	0.2	0.4
15	0.1	0.2	0.2	0.4
$\sum \mu_{i,j}$	3.0	2.9	2.5	2.6

The expected assigned load for DC_4, DC_7, DC_8 and DC_9 are 3.0, 2.9, 2.5 and 2.6, respectively. Then, nodes have been associated with appropriate DC. The node has been connected to the DC with whom it has maximum membership value. Further, if any node had the same membership value with 2 or more DCs then the node has been connected to the least loaded DC. In the below Table 4, the association of all

Table 4 Assignment of successive nodes to the domain centers

Assignment of successive nodes to the domain centers

Nodes with possible DCs	DC_4 (assigned load (3), associated nodes)	DC_7 (assigned load (2.9), associated nodes)	DC_8 (assigned load (2.5), associated nodes)	DC_9 (assigned load (2.6), associated nodes)
1	3, {1}	2.7,{}	2.2, {}	2.5, {}
2	3, {1, 2}	2.6, {}	2.1, {}	2.3, {}
3	3, {1, 2, 3}	2.4, {}	1.9, {}	2.1, {}
5	3, {1, 2, 3, 5}	2.2, {}	1.6, {}	2, {}
6	2.7, {1, 2, 3, 5}	2.2, {6}	1.4, {}	1.8, {}
10	2.5, {1, 2, 3, 5}	2.2, {6, 10}	1.2, {}	1.6, {}
11	2.3, {1, 2, 3, 5}	2.2, {6, 10, 11}	0.9, {}	1.4, {}
12	2.2, {1, 2, 3, 5}	2.2, {6, 10, 11, 12}	0.7, {}	1.1, {}
13	2.0, {1, 2, 3, 5}	1.9, {6, 10, 11, 12}	0.5, {}	1.1, {13}
14	1.9, {1, 2, 3, 5}	1.7, {6, 10, 11, 12}	0.2, {}	1.1, {13, 14}
15	1.8, {1, 2, 3, 5}	1.4, {6, 10, 11, 12}	0, {}	1.1, {13, 14, 15}

nodes has been depicted. There is a single candidate domain center, DC_8 having an estimated expected load of 2.5 but none of the nodes has been connected to it. The load carried out by this node is zero, which is less than the threshold value. In particular, DC_8 needs to be dropped out as candidate DC. Also, it needs to be connected to some other DCs. Membership value of N_8 w. r. t. DC_4, DC_7 and DC_9 can be determined again using the Eq. 2. Besides, its respective membership values are 0.2, 0.3 and 0.5. Hence, N_8 will be connected to DC_9.

Finally, set of nodes assigned to DC_4, DC_7 and DC_9 are {1, 2, 3, 5}, {6, 10, 11, 12} and {8, 13, 14, 15} with their respective loads 1.8, 1.4 and 1.6. Hence, all nodes are mostly equally loaded containing same number of nodes. Therefore, network of 15 nodes has been partitioned into $c = 3$, number of disjoint domains {1, 2, 3, 4, 5}, {6, 7, 10, 11, 12} and {8, 9, 13, 14, 15} which are significantly load balanced.

5   Conclusion

This paper has presented an algorithm for partitioning the network into a required number of domains for multiple mobile agent itinerary planning. This paper presented the idea that the domain whose data is to be carried out by the MA is less than the threshold payload required to drop out. Therefore, by considering the threshold value adaptively for a single mobile agent, the decision of deploying an appropriate number of mobile agents for each domain can be taken. Using fuzzy c-means clustering each node is associated with the domain for which it has the largest membership value and is assigned to exactly one domain. Thus, each domain becomes non-overlapped. Hence, MA visits each targeted node once. Therefore, it incurs less network bandwidth and reduces energy consumption. Also, it tries to balance each domain. Consequently, the proposed algorithm is a disjoint and balanced clustering algorithm so that an appropriate number of MAs can be deployed to an MA-based system.

References

1. M Chen, S Gonzalez, Y Zhang, V Leung, "Multi-agent itinerary planning for wireless sensor networks." In: International Conference on Heterogeneous Networking for Quality, Reliability, Security and Robustness, (Berlin, Heidelberg, 2009)
2. D Gavalas, C Konstantopoulos, B Mamalis, G Pantziou, "New techniques for incremental data fusion in distributed sensor networks," in *Proceedings of the 11th Panhellenic Conference on Informatics (PCI'2007)* (2007)
3. J Wang, Y Zhang, Z Cheng, X Zhu, "EMIP: energy-efficient itinerary planning for multiple mobile agents in wireless sensor network." Telecommun. Syst. **62**(1), 93–100 (2016)
4. A Mpitziopoulos, D Gavalas, C Konstantopoulos, "Deriving efficient mobile agent routes in wireless sensor networks with NOID algorithm," in *IEEE 18th International Symposium on Personal, Indoor and Mobile Radio Communications* (2007)
5. M Chen, S Gonzalez-Valenzuela, VCM Leung, "Directional source grouping for multi-agent itinerary planning in wireless sensor networks," in *International Conference on Information and Communication Technology Convergence (ICTC)* (2012)
6. D Gavalas, IE Venetis, C Konstantopoulos, "Energy-efficient multiple itinerary planning for mobile agents-based data aggregation in WSNs," Telecommun. Syst., **63**(4), 531–545 (2016)
7. K Lingaraj, RV Biradar, VC Patil, "OMMIP: an optimized multiple mobile agents itinerary planning for wireless sensor networks." J. Inf. Optim. Sci., **38**(6), 1067–1076 (2017)
8. YC Chou, M Nakajima, "A clonal selection algorithm for energy-efficient mobile agent itinerary planning in wireless sensor networks." Mobile Netw. Appl., **23**(5), 1233–1246 (2018)
9. C Konstantopoulos, A Mpitziopoulos, D Gavalas, "Effective determination of mobile agent itineraries for data aggregation on sensor networks." In *IEEE Transactions on Knowledge and Data Engineering* (2009)
10. P Kuila, SK Gupta, PK Jana, "A novel evolutionary approach for load balanced clustering problem for wireless sensor networks." Swarm Evol. Comput., **12**, 48–56 (2013)
11. I Aloui, O Kazar, L Kahloul, "A new itinerary planning approach among multiple mobile agents in wireless sensor networks (WSN) to reduce energy consumption." Int. J. Commun. Netw. Inf. Secur. (IJCNIS) **7**(2), 116 (2015)
12. I Aloui, O Kazar, L Kahloul, A Aissaoui, "A new" data size" based algorithm for itinerary planning among mobile agents in wireless sensor networks." In *Proceedings of the International Conference on Big Data and Advanced Wireless Technologies* (2016)
13. HQ Qadori, ZA Zulkarnain, ZM Hanapi, S Subramaniam, "A spawn mobile agent itinerary planning approach for energy-efficient data gathering in wireless sensor networks." Sensors **17**(6), 1280–1296 (2017)
14. B Liu, J Cao, J Yin, W Yu, B Liu, X Fu, "Disjoint multi mobile agent itinerary planning for big data analytics." J. Wireless Commun. Netw. **1**, 1–12 (2016)
15. M Chen, W Cai, S Gonzalez, V Leung, "Balanced itinerary planning for multiple mobile agents in wireless sensor networks." In *International Conference on Ad Hoc Networks*. (Berlin, Heidelberg, 2010)
16. Q Wu, NSV Rao, J Barhen, SS Iyengar, "On computing mobile agent routes for data fusion in distributed sensor networks." In *IEEE Transactions on Knowledge and Data Engineering* (2004)

17. R Rajagopalan, CK Mohan, P Varshney, "Multi-objective mobile agent routing in wireless sensor networks." In *IEEE Congress on Evolutionary Computation* (2005)
18. W Cai, M Chen, T Hara, L Shu, T Kwon, "A genetic algorithm approach to multi-agent itinerary planning in wireless sensor networks." Mobile Netw. Appl. **16**(6), 782–793 (2011)
19. W Cai, M Chen, X Wang, "Angle gap (AG) based grouping algorithm for multi-mobile agents itinerary planning in wireless sensor networks." (2009)
20. JV De Oliveira, W Pedrycz, Advances in Fuzzy Clustering and its Applications. (Chichester, Wiley, 2007)
21. Pratyay Kuila, Prasanta K. Jana, Clustering and Routing Algorithms For Wireless Sensor Networks: Energy Efficient Approaches. (London: Taylor and Francis Group, 2018)
22. Bo Yuan George, J. Klir, Fuzzy Sets and Fuzzy Logic Theory and Applications (New Delhi, Prentice Hall of India, 2005)

Acceptance of Eco-Friendly Substitute of White Skim Coat (Wall Putty) in Mumbai

Mandar Anil Chitre, Shivoham Singh, and Manuj Joshi

Abstract Wall putty available in the market is white in colour and made from perishable raw materials. It is possible to have a substitute product, black in colour and made from using waste powders from quarry and construction. Replacement of the current white putty with this substitute black putty is technically possible. Our research is on the acceptance of this product. On conducting a survey in Mumbai and surrounding suburbs, we asked questions on the awareness, acceptance of self and perceived acceptance of the industry. The results are compiled and analyze. The resistance shown by the various influencers of the industry is significantly on the higher side. However, as researchers we feel that there is certainly a possibility to make this change and develop the market in the direction of usage of black putty which is actually an eco-friendly product using waste materials. Another important benefit of using this product is that it actually reduces pollution. The waste product used as raw material is very fine, and these particles create air pollution. We have also studied and worked out the cost of this substitute product which is actually cheaper than the current white putty. The research has further scope to be carried out in other areas of the country and interacting with more population.

Keywords Building materials · Wall putty · Waste management · Cement mortars · Wall finishes

1 Introduction

As defined by Wikipedia; putty is a material with high plasticity, similar in texture to clay or dough, typically used in domestic construction and repair as a sealant or filler. Although some types of putty (typically those using linseed oil) slowly polymerize and become stiff, many putties can be reworked indefinitely, in contrast to other

---

M. A. Chitre · S. Singh · M. Joshi (✉)
Pacific Academy of Higher Education and Research Society, Udaipur, Rajasthan, India
e-mail: manujjoshi@gmail.com

M. A. Chitre
e-mail: mandar.chitre@batonconsultants.com

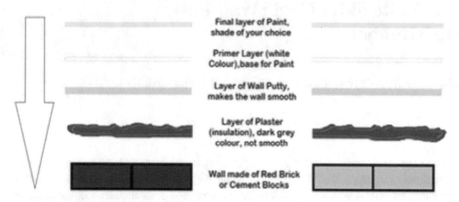

Fig. 1 Layers of a wall as per SCP

types of filler which typically set solid relatively rapidly. It is a material which is used regularly in construction of walls, ceilings, and other surfaces. After plastering, the wall is not smooth finished or not on "zero-zero" level. If painting is done directly after plaster, it will have two effects.

1. The finish of the paint will be very poor because of the undulations in the wall.
2. The consumption of paint and primer will be very high and un-economical for construction.

Wall putty is the material which solves this problem. This fine paste fills up the voids and makes the wall as smooth as a mirror. Thus, the primer + paint consumption is reduced, and a good surface radiates an excellent finish.

As illustrated in Fig. 1, the wall has layers. Each layer has its own significance, and the wall putty is applied after the plaster and before the primer.

The wall putty is a very important material in construction. It is used on both internal and external walls. This paste can also be used as a filler material in those applications which do not demand water proofing or strength. This is because the wall putty is merely a filler paste. It is not a water proofing material which can hold water from passing or prevent it from penetration. The putty itself does not have a strong compressive strength and therefore will not last if applied to accept any load which is compressive, flexural, or shear. Thus, to conclude, we will say that the wall putty is purely a filler paste which is used to make the wall "zero-zero" removing undulations and make it ready to accept the next layers.

Colour of Wall Putty:

In India, since the last two decades, wall putty has been white in colour. This has created a precedence that putty should always be white. In all other tropical countries where putty is used, (the name is skim coat) the colour is dark.

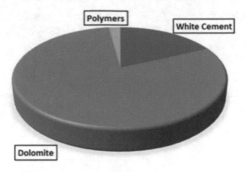

Fig. 2 Wall putty—raw materials

Table 1 Binder, filler, and enhancer

The binder: cement	The filler: dolomite	The enhancer: polymers
Is responsible for binding all material. It reacts with water and releases heat of hydration. The effect is that it becomes hard after drying	It gives the body to the product. This is an inert material which does not react and provides the mass required	These are chemicals which enhance the properties of the product, like adhesion, workability, and water retention

The Wall Putty Raw Materials:

There are three main raw materials of wall putty. Each component has its own role to play (Fig. 2 and Table 1).

White Cement—As defined by Wikipedia, a cement is a binder, a substance used for construction that sets, hardens, and adheres to other materials to bind them together.

The cement used for decorative purpose is white. The normal cement which is available and used in bulk is grey in colour. Cement is made from limestone. White cement availability is limited, because it is restricted to only those mines which produce limestone which is free from Fe_2O_3 and a few other minerals which impart the dark shade. A few mines in India have limestone which is suitable for producing white cement. Thus, white cement is limited, expensive, and to be used for decorative purpose.

Dolomite— It is a mineral available in abundance across regions of Rajasthan, MP, Chattisgarh, Maharashtra, AP, and Tamil Nadu. This yellowish rock becomes white in colour when it is fine grinded to a size of 300 mesh or 50 microns or 10–3 mm. This powder can be thoroughly mixed with white cement to achieve homogeneity of over 99%.

Polymers—Chemicals such as MHEC and RDP are added in very small proportions. These are the enhancers which react with water and cement to produce properties which help the product perform. Adhesion, workability, water retention, etc.,

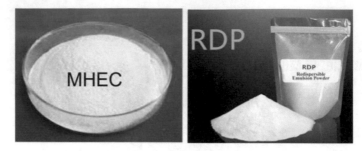

Fig. 3 MHEC and RDP

are very important properties for the wall putty, and these polymers enable them. They are white in colour (Fig. 3).

2 Background

In context to our paper, we have considered six papers/thesis/articles, which were reviewed and the summary of it is as follows.

The first review is of researcher Kantharia Mohan, who has done a thesis on usage of polymer and nanomaterials in cement mortars. The researcher states the various admixtures which are added to the mortars to enhance the required properties. He also gives importance of durability and strength of the mortars. Our product is also a mortar and usage of such admixtures are considered for the same. Our research includes usage of waste materials and creates a substitute. This review was useful to us on the admixture and properties usage [1].

The second review conducted is on the usage of clay tile waste as a partial replacement of cement in the masonry mortar. Researcher Jiji Antony gives importance to usage of waste materials and its feasibility. This is a good review for us as we are also using waste materials [2].

The third review is on the use of lightweight filler materials. The objective is to improve the insulation properties. The researchers have shown the effects of the dosage of different components to optimize the performance of the final product [3].

The fourth review is about usage of activated recycled powder from construction and demolition waste. The research done by four researchers talks about use of hybrid recycled powder (HRP) collected from construction and demolition waste (C&DW). This has partial cement which is un-hydrated or not reacted. This is an important reference for us because we are also recommending usage of waste material which is inert and not reactive. This paper uses waste which is partially active [4].

The fifth review is on the new product as exterior wall putty. The product that we are proposing is for both internal as well as external. Researchers have highlighted the requirements of the exterior putty which are more and need to be tougher than the internal putty. Usage of certain pozzolanic materials enhance the properties

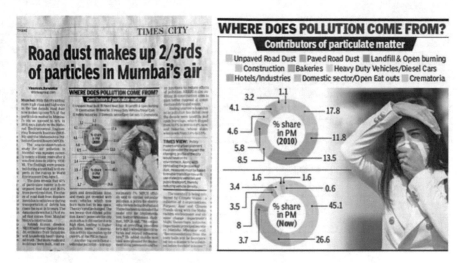

Fig. 4 2/3rd particles of Mumbai's pollutant are road dust and construction [7]

and achieve better results in workability, water resistance, high-bonding strength, and dynamic crack resistance. Article from industry magazine Drycotec Diaries published in January 2020, an expert of the industry Mr. Anantha Krishnan Subraveti mentions that colour of the putty does not matter and he does not consider it a customer complaint. This reference is important to us because we are proposing a black colour putty [5, 6].

Research Gap: No such study is conducted to identify the market potential.

Problem: About the current scenario in Mumbai (Fig. 4).

Pollution is rampantly increasing in Mumbai. Figure 3 by Times of India shows that 2/3rd particles of Mumbai's pollutant is road dust and construction. Manufactured sand is the major culprit. Construction quarries which produce rocks and sand generate a lot of dust. This dust is a fine powder of less than 50 microns and consists of inert rock. Its usage in construction is very limited because it contains deleterious material which affects the strength of the concrete and is harmful to the structural loading.

Consumption of White Putty:

Mumbai consumes approximately 20–25,000 metric tonnes of wall putty every month. This product is produced by mining of dolomite and white cement. Then, mixing all the components together after accurately weighing each one.

The Solution:

It is easily possible that we use this quarry dust (rock particles below 50 microns) in the wall putty. However, this wall putty will certainly not be white in colour, because a large component is this filler which will be black or grey in colour.

Table 2 Black putty is cheaper than the current white putty

White putty 40 kg bag—raw materials

Material description	Function	Unit cost / kgs	Quantity (kgs)	Total cost
White cement	Binder	22	8	176
Dolomite	Filler	3	32	96
Chemicals	Enhancer	300	0.3	90
Total raw material cost				362.00

Black putty 40 kg bag—raw materials

Material description	Function	Unit cost/kgs	Quantity (kgs)	Total cost
Grey cement	Binder	6.5	8	52
Quarry dust	Filler	0.3	32	9.6
Chemicals	Enhancer	300	0.3	90
Total raw material cost				151.60

Commercial Solution:

The product we have designed has the following combination and is a proven substitute of current wall putty which is white in colour.

Table 2 indicates that the black putty is cheaper than the current white putty. Manufacturers can produce 2.3 kgs of this dark grey putty in the price of 1 kg of white putty. This will help in reducing a small part of the cost of construction and eventually help the final buyer. We have tried and tested this product and are perfectly replaceable with the white putty.

Market References:

The products which are not white are currently available in the market and we wish to provide two such examples as below (Fig. 5).

One such, e.g. reference provided is of Dalmia magic putty. A product which is not white but slightly light red or pink in colour. This product is actually still struggling to perform in the market (Fig. 6).

Fig. 5 Dalmia magic putty colour light red or pink

Fig. 6 GoMix wall putty colour grey

The market has a few products in the grey colour zone but they are used in the semi-filler applications to cover up the errors of RCC or any other reason. Not on mass scale.

3 Methodology

The main objective of our research is to find the acceptance of eco-friendly product of black putty in Mumbai market.

To find if the market will accept this product, we decide to conduct the market research in Mumbai. This research would reveal if the market is in a position to accept this product or not.

Process:

Area sampling is done by us. The sample details are as in Fig. 7.

Data Collection—Interviews (scheduled).

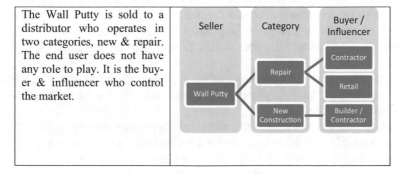

Fig. 7 Sample details

Considering the market of Mumbai, we have selected a list of 30 retailers and applicators (painters) who do business in the Mumbai and surrounding areas. Different areas were selected for considering the retailers for the sample.

The following questions were asked:

Q1. Are you aware that the wall putty is not white and is dark grey when used outside India in markets like Dubai and Singapore?
Q2. Are you aware that this product is not white and has the same features, quality, and functioning as the white putty?
Q3. If this product is offered to you, will you keep in your retail outlet and sell it to your customers?
Q4. Do you think the other retailers in the market will accept and start selling this product?
Q5. What according to you will be the reasons of success or failure of this product?

The short interview captured the reactions of the sample lot. The questions selected are direct to the point and adequate to gain the objective of this research. This paper is limited to studying the acceptance and reasons for the same before proceeding to the next level.

The territory distribution for collection of data was as below, coverage of the city is complete and the response generated would truly resemble the opinion of all the influencers of the city (Fig. 8).

Limitations:

Due to the pandemic situation and lockdowns, all interview work has been done over phone, and the size of sample has been limited to 30.

4   Results

The data compiled and analyzed give the following results (Fig. 9):

The awareness levels in the sample size were negative. Overall, they are not aware about such a product which exist in other markets (Fig. 10).

Since there is no awareness of the product. When explained most of them were not sure if it would work. Therefore, the confidence level of this product seems to be weak (Fig. 11).

In regards to their own acceptance of the product. Majority was on the negative side. Therefore, we can surely imply that the product will not be easily acceptable to the influencers (Fig. 12).

A strong perception in the selected sample about others not accepting this product. The confidence level felt by us as interviewers indicate that the problem is big. The product will definitely struggle to get an acceptance in the market (Fig. 13).

Maximum respondent said that it is the mindset which would be the major reason for rejection. The level of confidence displayed by them was very strong.

Locations	
Ambernath	Khopoli
Ambernath	Kurla
Andheri	Manpada
Andheri	Mira Road
Bhandup	Nalasopara
Byculla	Panvel
Chembur	Sanpada
Digha	Santa Cruz
Dombivali	Sion
Ghatkoper	Palghar
Ghatla	Thane
Jogeshwari	Turbhe
Kalyan	Ulhasnagar
Kausa	Vasai
Kharghar	Wadala

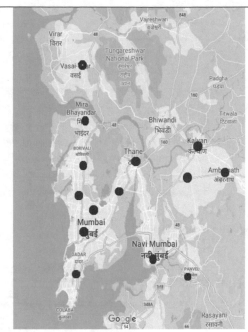

Fig. 8 Territory distribution

Fig. 9 Product awareness

5 Conclusion and Future Scope

It was really surprising to understand that the reason for rejection was not the colour. They all accepted that it is the mindset which would need to change. The respondents

Fig. 10 Product benefits

Fig. 11 Product acceptance—self

agree that the colour would be an issue but they also understand that primer will be able to cover it. This particular observation makes us believe that there is scope for this product to get established in the market and it has a future.

To Conclude:

- Acceptance of this product in the market is possible but challenging
- Work will have to be done by the manufacturer wants to launch this product
- Focus on education and concept selling will enable the change

Scope for Future Work:

- This research was done in Mumbai and suburbs
- Other parts of the country can be targeted for future research

Fig. 12 Product acceptance—perception of others

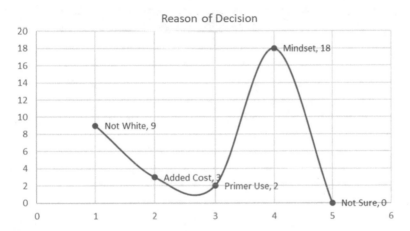

Fig. 13 Product acceptance—reason of decision

- The researcher could get a different report in the rural areas where price sensitivity is higher.
- The product could be introduced in rural areas and then gradually moved for acceptance in semi-urban and urban markets.

References

1. M. Kantharia, *Strength Investigation of Cement Mortar Using Polymer And Nano Materials*, Amity University M.P. (2020), http://hdl.handle.net/10603/309937
2. J. Antony, Investigation of clay tile waste as partial replacement of cement in masonry mortar and building blocks. Cochin University of Science and Technology (2019), http://hdl.handle.net/10603/309913
3. M. Kupinski, K. Stobieniecka, K. Skowera, Influence of Lightweight Fillers on the performance of cement based skim coat. Struct. Environ. **12**, 5–11 (2020). https://doi.org/10.30540/sae-2020-001
4. T. Meng, Y. Hong, K. Ying, Z. Wang, Comparison of technical properties of cement pastes with different activated recycled powder from construction and demolition waste (2021), https://doi.org/10.1016/j.cemconcomp.2021.104065
5. G.L. Zhang, L.W. Mo, J.Y. Sun, J.B. Chen, J.Z. Liu, Preparation and experimental study on a new type of exterior wall putty. Adv. Mater. Res. **671–674**, 1914–1917 (2013). https://doi.org/10.4028/www.scientific.net/amr.671-674.1914Researchers
6. A.K. Subraveti, *Issues and Possible Reasons in the Application of Wall Putty, Drycotec Diaries* (2020), www.drycotec.com/magazines
7. https://timesofindia.indiatimes.com/city/mumbai/road-dust-makes-up-2/3rds-of-particles-in-mumbais-air/articleshow/81300796.cmsvisited on 07 Jun 2021

Economics of Immutability Preserving Streaming Healthcare Data Storage Using Aggregation in Blockchain Technology

Sachin Gupta and Babita Yadav

Abstract The contemporary technology landscape is big data oriented and driven by analytics. The paradigm shift fueled by rapid data generation and higher storage capabilities has resulted in a massive transition across data storage technologies. Traditional database management systems being used in record-based transactional applications are paving the way for new rapid engagement-based data stores leading the newer applications. This transition is a result of a sharp growth trajectory witnessed across many parallel technology landscapes including cloud for storage and processing, Internet of things (IoT) for rapid data generation and transmission, support for unstructured, semi-structured and structured data by social media platforms, and polyglot persistence supported by NoSQL, to name a few. The emerging applications thrive on insights acquired by data analytics, and immutability of data storage is expected to be one of the key factors in growth of forward leaning enterprises. There has been a constant rise in application development requiring a full history of transactions, which is both trustworthy and traceable. Such applications mandate the append-only nature of the data storage to support analytics and trust. Immutability is also the underlying premise of popular cloud native storage systems. This paper explores blockchain technology as a solution for privacy and disclosure compliance in stored healthcare data. Blockchain as a data structure is primed to store only small amounts of data to maintain the properties of immutability, tamper proofing, security, and transparency in applications and is not suitable for storing big data. However, healthcare blockchain implementations have several possibilities for storage management, and the paper presents a comparative analysis of the aggregation-based storage with IoT-based streaming e-healthcare application data for the use case. The in-depth analysis for the potential blockchain storage in the paper includes the cost factors, in addition to immutability and privacy preservation, and the results show that costs may be saved up to an order of magnitude 300 with aggregation.

S. Gupta (✉) · B. Yadav
SoET, MVN University, Aurangabad, India
e-mail: sachin.gupta@mvn.edu.in

B. Yadav
e-mail: babita.yadav@mvn.edu.in

© The Author(s), under exclusive license to Springer Nature Singapore Pte Ltd. 2023   611
Y.-D. Zhang et al. (eds.), *Smart Trends in Computing and Communications*, Lecture Notes in Networks and Systems 396, https://doi.org/10.1007/978-981-16-9967-2_57

Keywords Healthcare · Compliance data · Immutability · Privacy · Storage economics · Blockchain

1 Introduction

There has been an exponential growth of social and economic activities in online mode. Not only has the routine average consumption of data per user shot up in the past decade, similar amounts of content and data are also being created per user as a by-product. While most user generated data and content can be considered "harmless" to share, there exist several domains of individual data that unquestionably classify as private data. Trust as a commodity is becoming increasingly difficult to maintain with social media rapidly pushing us in the era of data exploitation. Governments are waking up to this reality, and we have witnessed some landmark data privacy legislation happening across the globe in the recent decades. As per the UNCTAD data [1] analyzed for 194 countries, it has been reported that 128 countries have created stringent laws for data and privacy protection of their citizens. Non adherence of the data privacy laws is punishable with heavy fines [2].

The rules of engagement with databases are changing as the technology spectrum is shifting from pure storage to analytical engines. The long drawn "AI winter" [3] ended almost a decade back with a symbiotic increase in data volumes as well as the computation capacity. The new generation data stores are thus not designed for storage alone, they are preferred to be analytical to handle a variety of big data including social media posts, Internet of things (IoT), and immutability is desirable in synchronization the velocity requirements and with the cloud native nature of applications. Analyzing the data for business intelligence and useful patterns requires high fidelity data, with a wide time range of data generation, and the previously used mutable data storage with data overwriting is no longer a good choice in many scenarios. The stock market trends data for instance requires the best of big data storage [4, 5] and immutability for analytics. It would be similarly wise to store the disaster and its management datasets on immutable storage for in-depth analytics and prediction [6].

The recent developments in blockchain have made it possible to ensure the desirable traits of both immutability as well as verifiable data security compliance as discussed in the above sections. This paper presents the economics of immutability preservation by using on-chain blockchain storage and a comparative cost analysis in case of off-chain storage methods with immutable storage and protected via the blockchain technology. The resting heart rate data with 9000 data points from the personal fitbit app health profile [7] of the author have been used for the paper. The data are IoT-based with analytics desired, personal healthcare data governed by data privacy laws, and are a good representative use case for assessing the storage cost impact of using blockchain technology on immutable and GDPR privacy preserving off-chain and on-chain storage methods.

The remainder of the paper has been structured as follows. Section 2 discusses the research and development in the domains of immutability and private data disclosure compliance through a brief literature survey, while Sect. 3 focuses on explanation of the methodology used in the paper for analyzing the storage costs with immutability on blockchain. In Sect. 4, we have presented the comparative cost analysis on on-chain solutions with and without aggregation as the results from Sect. 3, and a concluding summary of the paper along with future research directions has been presented in Sect. 5.

2   Related Work

This paper deals with both privacy preservation laws and immutability of storage for trust-related issues and the proposed solution being blockchain, it is imperative to understand the research landscape of all these topics. The topics have been discussed by various researchers across several applications, but we did not come across any research work unifying these topics comprehensively in the same paper. The available literature across all the related topics is reviewed and presented ahead with topic wise associations as they appear.

2.1   *Immutability of Storage*

The advancements in cloud-based storage brought the age long myth about immutability being propagated by vendors of storage devices to the fore. The term was being used incorrectly for the difficulty in deletion or modification of a storage-based file by a user. The precise immutability definition as per storage and compliance requires that the data termed as immutable shall not be modified or deleted for contract-bound duration of time by neither men nor software [18]. Immutability is the key USP of a blockchain-based storage [19], and all applications focusing on immutability of data are potential candidates for a shift to blockchain in some form or the other. It is imperative to note that immutability is not a silver bullet. The organizations have to consider what happens when a user actually wants to get his personal medical records deleted. Or the impossible situation of keeping every piece of personal medical data due to it being immutable even after a person dies. Similarly, there might be a requirement to maintain the compliance and regulation data as mutable to allow for possible customer relocation leading to address change or likewise. A split mutable/immutable approach as proposed in some cloud-based solutions like Fluree [20] may help maintain data with different requirements across different media. The pros and cons of on-chain and off-chain storage in blockchains have been comprehensively covered in [32], but we consider only the on-chain storage as a special case in this paper.

2.2   Healthcare Data Privacy Laws

Trust in healthcare analytics has been defined [8] as "a dynamic and always emerging attribute of relationships among people that evolves with time." Bulk of healthcare data is originating from IoT-based medical sensors especially in remote monitoring and telemedicine, and several researchers have proposed trust solutions for such applications. Privacy preservation using a blockchain including a forensic framework for IoT discussed in [21], and a trust framework for collaborative IoT [22] gave a general approach to trust establishment using blockchain. Trust in a healthcare blockchain with a design has also been discussed in [23]. Personal data processing using blockchain to maintain privacy concerns has been adequately covered in [24], while the specific discussion on general data protection regulation (GDPR) [29] compliance has been discussed in [25]. An implementation for a blockchain-based data-sharing platform with GDPR compliance is available in [26] while a Health Insurance Portability and Accountability Act (HIPAA) [28] compliant off-chain data store has been implemented in [27].

2.3   Blockchain in Healthcare

Ever since its origin with the Bitcoin cryptocurrency envisioned as a financial application in the now virally sensational breakthrough research paper [9] of Satoshi Nakamoto in 2008, blockchain has been making waves in every technology domain due to its distributed and immutable ledger property. Estonia is among the early adopters of blockchain for storing healthcare data [10] in the form of medical records of patients. Since then, there has been a significant interest of the healthcare community in this technology. The electronic medical records (EMRs) for the patients with blockchain as a consensual access control mechanism have been discussed by MedRec [11] and HDG [12] while blockchain as an immutable storage mechanism has been discussed in Medibchain [13]. There has been a significant debate on using permissioned vs permissionless blockchains in healthcare owing to the sensitive nature of the patients' medical records and has been discussed in [14, 15]. Almost the entire literature survey focuses on the inability of blockchain to handle copious amounts of data [14] and suggest off-chain solutions but blockchain scalability with large documents still remains a concern [16, 17].

3 Methodology

3.1 Data Description

The potential blockchain use cases for data storage based on decentralized trust relationships can be attributed to applications that work with information transfer and compliance. The applicability of blockchain in this age of internet is universal, and the best representative industries include supply chain management, healthcare, and insurance to name a few. In this paper, the author has used personal fitness data from a popular fitbit application for representative calculation for storage in streaming data on a blockchain. Healthcare as an industry has been specifically chosen here to cover data having both modification and exchange requirements, while working on crucial patient integrity maintenance and law mandated compliance management. We have chosen HIPAA from among the universal laws for representing the legal use case requirements for this paper, as GDPR is a generalized law but for prevention of the misuse or disclosure of the patients' healthcare data is thoroughly covered in HIPAA.

The data include 9000 data points from an IoT wearable fitbit versa device generating sensor-based data for resting heart rate of a user. The data may have multiple stakeholders including the patient, care provider, physician, and the insurance company and is a suitably complex representation of healthcare data under HIPAA regulation. The healthcare data from millions of fitbit devices being used by the customers are currently being stored on company servers with a potential need of load balancing [30].

For the purpose of illustration, we have assumed that the storage system being used by the company is a legacy system with a mutable overwriting-based relational database. The basic problem associated with such systems in case of compliance-based applications is the misinformation due to lack of historical data. With an immutable distributed ledger, it would be easier to provide data ownership to the concerned individual along with traceability of all the previous medical history including diagnosis, consultations, and treatments along with medicine records. The patient's confidence in such a system shall be high on trust and the data thus generated can be used effectively in individual and community AI and ML applications for drug discovery [31] and treatment-related studies. A comparative storage cost analysis using an on-chain blockchain storage with application of aggregation has been done on the acquired fitbit data in the following sections.

3.2 On-Chain Storage Costs

The understanding of storage costs on the blockchain can be considered for public vs private/hybrid global blockchain systems in use being Bitcoin and Ethereum, the latter has been used for cost calculation in our paper. The basis of calculations comes from the Ethereum yellow paper [33] where it is mentioned that the basic unit of payment in the Ethereum network is gas. Each transaction, whether transfer or execution of smart contract, is made with gas which fuels the network. It is basically a fee paid to the miners for maintaining the transaction computations going on.

Any transaction cost may be represented with a simple to calculate formula given by Eq. 1 as shown below, where gas price is the amount of ETH paid per unit of gas and fluctuates on the basis of what the transaction drivers *intend to* pay the miner for this transaction confirmation, and the gas limit is fairly standardized on type of operation. The service is premium, if you pay more gas fees, transaction confirms faster.

$$\text{gas price} * \text{gas limit} = \text{total cost} \tag{1}$$

The gas is measured in wei, which is the smallest unit of ETH. Typically measured in gwei, or gigawei where 1,000,000,000 Gwei is equal to 1 ETH. From the yellow paper, it is also given that each transaction consumes at least 21,000 gas. Similarly, the cost of storing a 256 bit word to storage has been given a cost of 20,000 gas. We have used the Eth Gas Station [34] for the cost of saving the resting heart rate data. The Fitabase [35] research library adds some unique perspective to resting heart data storage with possibilities of heart rate data being generated at variable sampling with readings every 15 s. We assume a single 256 bit word to store the resting heart rate data in each case.

We have calculated the storage costs on 9000 points, and the cost calculations have been shown under different categories of aggregation as per below mentioned cases:

- Case 1 is with no aggregation, with one point stored per transaction as in streaming data.
- Case 2 considers a grouping aggregation with 300 points stored together during a single transaction
- Case 3 considers a temporal aggregation on the device whereby the aggregated data per 15 min is sent to be stored on the blockchain.
- Case 4 considers a daily aggregation of the streaming healthcare data with all points across active 14 h with 3360 data points were aggregated together and stored during one transaction leading to 9000 points being generated in 2.68 days and stored on immutable blockchain storage, while the device keeps mutable storage records of up to one week.

In each case, as per the fitbit data, each data point is read by the sensor every 15 s, which can be considered on a minute basis (4 points) or an hourly basis (240 points) while the device is being actively worn. We also assume an 8 h sleep cycle of the user and maximum 14 h data availability in a day.

4   Results

The results from the study have been shared in Table 1. The table has been split into three subsections where the first subsection considers the average gas cost to be 73 Gwei (as per the ETH Gas Station data available on 11th October 2021 at 5 PM IST), and the second and third subsections record the data values with maximum cost at (105 Gwei) and minimum cost (69 Gwei).

The comparative cost analysis with the four case scenarios has been plotted in Fig. 1 which shows the cost of storage across all cases across the average, high, and low gas costs.

5   Conclusions and Future Research

Healthcare data may come in various forms, shapes, and sizes. The streaming IoT-based healthcare data are not amenable to be stored on a blockchain owing to heavy expenses of generating a single transaction per data point but with a clever aggregation mechanism as shown in the paper, a partially mutable storage capable of holding a few days data on the IoT sensor device and a partial immutable storage on the blockchain can be used to reduce the costs by approximately 300% as shown in the paper. Paying the highest gas costs invariably speeds up the mean time to confirm the transaction on blockchain in seconds, but the speed/cost tradeoff between the average and the lowest cost is not as significant and thus its always advisable to pay the average gas fee on Ethereum rather the lowest which may make the transaction wait for very long periods.

It would be interesting to have similar calculations as a future work for storage costs in the case of off-chain data storage methods of blockchain in the presence of gas costs associated with execution of smart contracts and to analyze whether the costs of blockchain storage may be offset by storing the data off-chain as the hash and smart contracts are executed on-chain.

Table 1 Ethereum on-chain storage costs

Data stored	Gas used	MTTC (Sec)	Tx fees (ETH)	TX fees (Dollars)	Total cost (9000 points)	Remarks
1	41,000	624	0.002993	$10.31	$92,798.01	Streaming 9000 points storage
300	6,021,000	2270	0.439533	$1,514.19	$45,425.74	300 data points stored with each transaction
60	41,000	624	0.002993	$10.31	$1,546.63	Data aggregated every 15 min
3360	41,000	624	0.002993	$10.31	$27.63	Daily storage (3360 points aggregated over 14 h)
Tx costs @105 Gwei (highest)						
1	41,000	28	0.004305	$14.83	$133,476.48	Streaming 9000 points storage
300	6,021,000	1413	0.632205	$2,177.95	$65,338.39	300 data points stored with each transaction
60	41,000	28	0.004305	$14.83	$2,224.61	Data aggregated every 15 min
3360	41,000	28	0.004305	$14.83	$39.75	Daily storage (aggregated over 14 h)
Tx costs @69 Gwei (lowest)						
1	41,000	1075	0.002829	$9.75	$87,713.19	Streaming 9000 points storage
300	6,021,000	22,464	0.415449	$1,431.22	$42,936.65	300 data points stored with each transaction
60	41,000	1075	0.002829	$9.75	$1,461.89	Data aggregated every 15 min
3360	41,000	1075	0.002829	$9.75	$26.12	Daily storage (aggregated over 14 h)

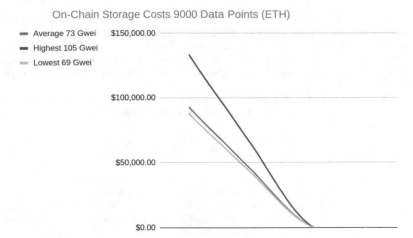

Fig. 1 Storage costs for 9000 data points in Eth with variable gas fees

References

1. "Data Protection and Privacy Legislation Worldwide," *UNCTAD*. [Online]. Available: https://unctad.org/page/data-protection-and-privacy-legislation-worldwide. Accessed 07 Oct 2021
2. P. Yifat Perry, "Meeting Data Compliance with a Wave of New Privacy Regulations: GDPR, CCPA, PIPEDA, POPI, LGPD, HIPAA, PCI-DSS, and More", Cloud.netapp.com (2021). [Online]. Available:https://cloud.netapp.com/blog/data-compliance-regulations-hipaa-gdpr-and-pci-dss. Accessed 07 Oct 2021
3. "AI winter - Wikipedia", En.wikipedia.org (2021). [Online]. Available: https://en.wikipedia.org/wiki/AI_winter. Accessed 08 Oct 2021
4. S. Kalra, S. Gupta, J.S. Prasad, predicting trends of stock market using SVM: a big data analytics approach. In: *Data Science and Analytics. REDSET 2019. Communications in Computer and Information Science* ed. by U. Batra, N. Roy, B. Panda, vol. 1229 (Springer, Singapore, 2020). https://doi.org/10.1007/978-981-15-5827-6_4
5. S. Kalra, Dr. S. Gupta, "Performance evaluation of machine learning classifiers for stock market prediction in big data environment". Published—J. Mech. Continua Math. Sci. Web Sci., **14**(5), (2019). https://doi.org/10.26782/jmcms.2019.10.00022
6. S. Gupta, B. Gupta, Performance modeling and evaluation of transportation systems using analytical recursive decomposition algorithm for cyclone mitigation. J. Inf. Optim. Sci. **40**(5), 1131–1141 (2019). https://doi.org/10.1080/02522667.2019.1638003
7. "Fitbit Official Site for Activity Trackers and More", Fitbit.com, 2021. [Online]. Available: https://www.fitbit.com/global/in/home. Accessed 08 Oct 2021
8. P. Holub, F. Kohlmayer, F. Prasser, M.T. Mayrhofer, I. Schlünder, G.M. Martin et al., Enhancing reuse of data and biological material in medical research: from FAIR to FAIR-health. Biopreserv Biobank. **16**, 97–105 (2018)
9. S. Nakamoto, "Bitcoin: A Peer-to-Peer Electronic Cash System" (2008). [Online]. Available: https://bitcoin.org/bitcoin.pdf. Accessed 08 Oct 2021
10. T. Heston, A case study in blockchain healthcare innovation. Int J Curr Res. **9**, 60587–60588 (2017)
11. A. Ekblaw, A. Azaria, J.D. Halamka, A. Lippman, "A case study for blockchain in healthcare:"MedRec" prototype for electronic health records and medical research data", 13 (2016)

12. X. Yue, H. Wang, D. Jin, M. Li, W. Jiang, Healthcare data gateways: found healthcare intelligence on blockchain with novel privacy risk control. J Med Syst. **40**, 1–8 (2016)

13. A. Al Omar, M.S. Rahman, A. Basu, S. Kiyomoto, "Medibchain: a blockchain based privacy preserving platform for healthcare data", In: *Proceedings of the International Conference on Security, Privacy and Anonymity in Computation, Communication and Storage* (Guangzhou, China, Springer, 2017)

14. O. Choudhury, H. Sarker, N. Rudolph, M. Foreman, N. Fay, M. Dhuliawala et al., Enforcing human subject regulations using blockchain and smart contracts. Blockchain in Healthcare Today. **1**, 1–14 (2018)

15. A. Dubovitskaya, Z. Xu, S. Ryu, M. Schumacher, F. Wang, "Secure and trustable electronic medical records sharing using blockchain" (Washington, D.C., United States, American Medical Informatics Association, 2017)

16. Q. Xia, E. Sifah, A. Smahi, S. Amofa, X. Zhang, BBDS: blockchain based data sharing for electronic medical records in cloud environments. Information **8**, 44 (2017)

17. P. Mamoshina, L. Ojomoko, Y. Yanovich, A. Ostrovski, A. Botezatu, P. Prikhodko et al., Converging blockchain and next-generation artificial intelligence technologies to decentralize and accelerate biomedical research and healthcare. Oncotarget **9**, 5665–5690 (2018)

18. B. Tolson, Litigation Hold and Data Immutability (2021). [online] Archive360.com. Available at: https://www.archive360.com/blog/litigation-hold-and-data-immutability-why-a-litigation-hold-does-not-meet-the-regulatory-definition-of-immutability. Accessed 9 Oct 2021

19. Network World, How data storage will shift to blockchain (2021). [online] Available at: https://www.networkworld.com/article/3390722/how-data-storage-will-shift-to-blockchain.html. Accessed 9 Oct 2021

20. Fluree, Fluree | The Web3 Data Platform Fluree (2021). [online] Available at: https://flur.ee/. Accessed 9 Oct 2021

21. D.-P. Le, H. Meng, L. Su, S. L. Yeo, V. Thing, "BIFF: a blockchain-based iot forensics framework with identity privacy," in *Proc. TENCON IEEE Region Conference*, pp. 2372–2377 (2018)

22. B. Tang, H. Kang, J. Fan, Q. Li, R. Sandhu, "IoT passport: a blockchain-based trust framework for collaborative Internet-of-Things," in *Proc. 24th ACM Symposium Access Control Models Technology*, pp. 83–92 (2019)

23. A. Dwivedi, G. Srivastava, S. Dhar, R. Singh, A decentralized privacy-preserving healthcare blockchain for IoT. Sensors **19**(2), 326 (2019)

24. K. Rantos, G. Drosatos, K. Demertzis, C. Ilioudis, A. Papanikolaou, "Blockchain-based consents management for personal data processing in the IoT ecosystem," in *Proceedings 15th International Joint Conference e-Bus Telecommunications*, pp. 572–577 (2018)

25. M. Rhahla, T. Abdellatif, R. Attia, W. Berrayana, "A GDPR controller for IoT systems: application to e-Health," in *Proceedings IEEE 28th International Conference Enabling Technology Infrastructure Collaborative Enterprises (WET-ICE)*, pp. 170–173 (2019)

26. B. Faber, G.C. Michelet, N. Weidmann, R.R. Mukkamala, R. Vatrapu, BPDIMS: a blockchain-based personal data and identity management system. In *Proceedings of the 52nd Hawaii International Conference on System Sciences* (Honolulu, Hawaii, United States, Hawaii International Conference on System Sciences (HICSS), 2019)

27. K. Griggs, O. Ossipova, C. Kohlios, A. Baccarini, E. Howson, T. Hayajneh, Healthcare blockchain system using smart contracts for secure automated remote patient monitoring. J. Med. Syst. **42**, 1–7 (2018)

28. UNITED STATES, The Health Insurance Portability and Accountability Act (HIPAA). [Washington, D.C.], U.S. Dept. of Labor, Employee Benefits Security Administration (2004). http://purl.fdlp.gov/GPO/gpo10291

29. EU General Data Protection Regulation (GDPR): Regulation (EU) 2016/679 of the European Parliament and of the Council of 27 April 2016 on the protection of natural persons with regard to the processing of personal data and on the free movement of such data, and repealing Directive 95/46/EC (General Data Protection Regulation), OJ 2016 L 119/1

30. M. Yadav, S. Gupta, Hybrid meta-heuristic VM load balancing optimization approach. J. Inf. Optim. Sci. **41**(2), 577–586 (2020). https://doi.org/10.1080/02522667.2020.1733190
31. J. Vamathevan, D. Clark, P. Czodrowski et al., Applications of machine learning in drug discovery and development. Nat. Rev. Drug. Discov. **18**, 463–477 (2019). https://doi.org/10.1038/s41573-019-0024-5
32. X. Xu et al., "A Taxonomy of Blockchain-Based Systems for Architecture Design", In *Proceedings of 2017 IEEE International Conference on Software Architecture, ICSA 2017*, pp. 243–252 (2017)
33. G. Wood, Ethereum: a secure decentralised generalised transaction ledger. Ethereum Project Yellow Paper **151**(2014), 1–32 (2014)
34. Eth Gas Station [Online] Available: https://legacy.ethgasstation.info/calculatorTxV.php. Accessed 9 Oct 2021
35. "Fitbit Research Library", Fitabase.com. [Online]. Available: https://www.fitabase.com/research-library/. Accessed 9 Oct 2021

A Comparative Cost Analysis of Organizational Network Security Test Lab Setup on Cloud Versus Dedicated Virtual Machine

Sachin Gupta, Bhoomi Gupta, and Atul Rana

Abstract The global network infrastructure spectrum is witnessing its fastest growth since the last decade with concurrent rise in cloud computing, Internet of things (IoT), and edge computing. There has been a multitude of heterogeneous networking devices spanning different configurations and using a variety of access methods. A parallel evolution of the network infrastructure security is happening with increasing attempts to exploit the security vulnerabilities in mission critical cyber-assets of organizations. Several organizations invest heavily in security research using lengthy and cryptic mathematical models while ignoring the practical network implementation situation and focus only on the monetary implications of the attack and defense. Attack tree has evolved as a convenient and cost effective way of plotting the network in which an attack may take place and can also help organizations understand the way it can be defended. Attack trees combined with the MITRE ATT&CK framework are widely used for crown jewels risk assessment globally. However, the major challenge for information security experts using the attack tree methodology lies in manually creating the attack tree and plotting all the crown jewels and perimeter network so that it can be defended from attackers. We propose a test lab setup for simulation and attack tree generation, which can be used in conjunction with the MITRE ATT&CK framework and allow us to create and assess various attack scenarios while providing flexibility in subnet configuration and movement, addition or removal of networking devices. The lab can be cloud hosted with a popular cloud hosting on Microsoft Azure or may be created on a VM within a dedicated high-resource machine to be used as a portable testbed. The results indicate that both services have their own pros and cons based on the hours of usage, and the dedicated resource VM testbed may

S. Gupta (✉) · A. Rana
SoET, MVN University, Haryana, India
e-mail: sachin.gupta@mvn.edu.in

B. Gupta
MAIT, GGSIPU, Delhi, India
e-mail: bhoomigupta@mait.ac.in

A. Rana
Research Scholar, MVN University, Haryana, India

© The Author(s), under exclusive license to Springer Nature Singapore Pte Ltd. 2023   623
Y.-D. Zhang et al. (eds.), *Smart Trends in Computing and Communications*, Lecture Notes in Networks and Systems 396, https://doi.org/10.1007/978-981-16-9967-2_58

perform better in a low-risk potential small network while the cloud-based approach is useful for the scalable organizations with high-threat potential.

Keywords MITRE framework · Attack tree · Test lab · Network security

1 Introduction

The term cybercrime, cybercriminals, and cyberattack are not new for us in today's world, but with increased use of remote working, automation, and increased device connectivity, the term cyberattack has reached new horizons of all time high.

Businesses have gone online and organizations trying to provide all the information as readily available to its customers in a self-serving manner has also allowed cybercriminals to infiltrate the organization even more than ever. If we look at the last 5 year data of cybercrimes, it has been on a constant growth trajectory and has also triggered safety costs as per a report published by cybersecurity ventures [1] which shows that spending on cybersecurity all around the globe was even more than $ 1 trillion (2017–2021). The attack surface of cybercrime is so high that it seems very difficult to understand how we can even protect our identity and assets. Cybercrime is causing unpreceded damage to both industry as well as the personal lives of individuals. It has also been observed that by end of 2016, a business becoming a victim of a cyberattack was 21 s but this has been reduced to 14 s by year 2019 and is also expected to reduce further to 11 s in 2021. Phishing, ransomware, and malware are few of the fastest growing cybercrime mechanisms. This has caused extremely damaging network infections and big downtime with a lot of reputation damage. If we look at the data of damage caused only by ransomware which was $5 billion in 2017 which is already 15 times higher if compared to year 2015 have also increased further to $11.5 billion in 2019 and expected to grow $20 billion in 2021 as per the report published by Herjavec group [1].

Over a decade, cloud computing opened a new horizon for businesses and has generated exciting opportunities for organization in terms of flexibility, cost, high availability and scalability with reduced costs. Many companies have made a move from traditional data centers to cloud environments to help them get better returns with lower costs. The on-demand access to network resources has made it very scalable. According to a recent Gartner press report, [14] published the demand of cloud computing has been rising every year and it is expected to be reaching up to $308 billion by end of year 2021 which is far higher when compared to the previous years. With the increased use of cloud computing, it brings new risks also with it which are even higher than that of traditional data center environments.

Security operation centers in organizations find it difficult nowadays to keep pace with the volume of cyberattacks and volume of events received every day. Most of the security tools in the market focus on alerting when an event has happened rather than making a predictive guess and showing the insight about the weakness and strengths within the organizations. Another survey shows that most of the organizations are finding it difficult to understand the network as it is huge and that is one of the biggest reasons for defending it against adversaries.

Attack tree (which was first introduced by Bruce Schneier in 1999) is a convenient way of plotting the network in which an attack may take place and can also help organizations understand the way it can be defended. There are lots of research papers published on attack trees which show lots of academic ways with the help of mathematical manners to understand metrics related to probability, cost, and impact, but this research paper will focus more on how we can create an attack tree and how organizations can be benefited from such an attack tree.

Attack trees are diagrams that represent assets of the organization that can be a target to an attacker, it also provides a mechanism to not only model threats against organization but also helps in identifying a countermeasure for that threat. It is similar to a red team blue team exercise in one attack tree. Security analysts can define a way of accessing and breaching the defense while other teams can suggest a countermeasure against that attack, i.e., placing a detective or preventive tool against it. It provides graphical methods of how a security of an organization looks like and represents possible attacks against those security measures. Attack tree may consist of a few root nodes which represent a final target of an attacker but also several child nodes that help an attacker reach its target. There arc also few connectivity's that help an attacker to reach to the next child node, and these are attack methods.

Attack tree can become a very big and complex representation based on organization size and also depends on some specific attack. Attack tree can also prove beneficial in defining the strategy on how to protect assets from an attack and understand the scenario to mitigate the threat to the root node. In practical situations, this may not always be possible to eliminate the threat with 100 percent accuracy but we can always make it more difficult for an attacker to reach the root node. We can try to increase the child nodes between an attacker and the root node but this comes with a cost and that can be a cost of placing a new security tool or hiring a new security analyst to administer and monitor the tool.

2 Related Work

Attack trees were introduced for the first time in 1999 by Bruce Schneier [2] to understand the attacker, their goals, likelihood of an attack, and the need of threat modeling. The concept of root node, leaf node, and also Boolean nodes (AND, OR) is also introduced. The concept of attack tree [9, 10] and attack graph is used to identify threats based on vulnerabilities in the organization with possible impact that they can cause. The attack tree and graph were very good in explaining "what if" situation of an attack while they lacked in the graphical virtualization as it is very complex to build and maintain for long run which has been explained by Steven Noel [13] managing attack graph complexity paper and also precision on manipulation of the attack tree construction by study of transformation and attack tree [11, 12] with attributes and projection of the tree.

There was another paper published in 2005 [3] which reached to another side of attackers and this time it is insider attackers within the organization and the paper proposed a framework to detect malicious activities from an authorized insider within the organization. This concept was to generate an alarm when it assumes that a malicious activity is about to trigger which was a shortcoming with the intrusion detection systems which work on the signature bases and trigger when some malicious operation has already been performed. The concept was to use an additional framework on top of the intrusion detection system and build an attack prediction system.

In a research paper published in 2006 [4], it was researched to provide foundation to the attack tree which was done to answer a few questions that were not provided previously. This research paper described semantics in terms of attack, attack suits, valuation, and projection in a manner that is formal in nature and also provided algebraic conditions that provide better results. This paper also looked at the fault that were proposed in the previous paper [5] where author preferred attack net which is a formalization of threat analysis based on petri net more than the attack tree and lacking in semantics formalization which was later improved by another research paper published [6] that successfully applies attacks trees to multi-model Internet attacks. The paper proposed a new approach and it proposed a new word called bundle which is a connection from a node to multiset node, and this bundle should be executed to execute an attack which is different from Schneier's proposed attack tree.

In a research paper published in 2007 [7] which highlighted the limitation of attack trees being time consuming to generate for a complex system and proposed a recommendation that automation can help in this shortcoming. The paper proposed combining attack libraries with analyst provided specifications can make the automation easier. This paper also talks about a protection tree which provides a methodical approach of mitigating weakness highlighted by the attack tree. The few important words like probability, cost, and impact also were used, and metrics were created to draw a conclusion but according to this paper, it is a bit hard to guess.

In [13], Barbara Kordy (2010) has proposed a new tree structure named as attack defense tree which uses an interactive approach that supports semantic approach that helps in prompt response in an event of attack. A new research paper which was published in 2015 [8] proposed a new idea of the first foundation of the SAND attack tree which is the extension of an attack tree. Here, in this approach, a new attack tree is defined with a new sequential operator and then a new graph which is a series–parallel graph has been defined on which the semantics of the new attack tree is based.

In a research paper, [8] published by Jhawar, Kordy, Mauw, Radomirović, and Trujillo-Rasua, 2015 first formal foundation of SAND tree was proposed which was an extension on attack tree by addition of sequential operator SAND, which helps in modeling of ordered events. This paper is a further extension to multiset semantics to SAND attack trees with series–parallel graphs, and it helps in interpreting the tree using both commutative and sequential conjunctive refinement.

3　Methodology

Attack trees are a very interesting and innovative way to understand the current status of an organization with regards to its defense mechanism as well as its exposure and coverage level for an attacker. Attack trees can also be used based on application or as a whole for an organization, and they are a good way to document and realize the threat possessed by an organization. There are, however, a few common questions related to the understanding of an attack tree with reference to an organization, and they include

- What is the best way to model an attack tree?
- How can we attach a framework with regards to tactics and techniques used by attackers?
- How can attack trees help in understanding the risk for an organization?
- Are attack trees a solution going forward with regards to complexity and exponential growth?

In this paper, we have created the template for a test bed to model the attack tree of a virtual organization and to assess it with MITRE ATT&CK framework which is a well-known attack framework based on tactics and techniques based on real-world observations. The configuration thus created has been used for comparative cost analysis of the attack tree modeling using the physical network, a cloud-based approach, or a VM-based setup on a dedicated high-resource machine.

3.1　Cloud-Based Testbed

We have used the Azure cloud environment to estimate the costs for setting up the lab [15]. This will require a subscription to Azure environment and few resource groups within the subscription to place resources according to the device type, e.g., all servers in one resource group with balancing [20]. Also we can use Microsoft security tools to create a security operation center in this test environment. The main aim of creation of this lab is to simulate the same kind of environment that is used in the organizations and then create an active attack defense environment for testing the attack tree.

The estimated costs for a security-based network in Azure are shown in Table 1.

3.2　Dedicated Machine VM-Based Testbed

The dedicated machine with a testbed virtual network can be created on a machine with following hardware configuration:

Table 1 Estimated cost for an Azure-based testbed for security

Microsoft azure estimate		
Service type	Description	Estimated monthly cost ($)
Storage accounts	1 K GB with 10 K operations	21.84
Virtual machine scale sets	Licensed 3.5Ghz 730 h win machines	102.20
Azure DNS	Public domain name service	0.00
Network watcher	1 gigabyte logs	5.80
Azure firewall	Single firewall 730 h	912.50
Azure bastion	5 gigabyte 730 h monthly usage	138.70
Azure front door	1000,000 custom requests	138.00
Azure firewall manager	Free with firewall	0.00
StorSimple	Storage size dependent	0.00
Azure active directory external identities	Active directory service	0.00
Azure sentinel	Daily logs up to 100 numbers	8880.00
VPN gateway	128 tunnels p2s	0.00
Azure monitor	Log analytics: 0 GB daily logs ingested, 1 month data retention; Application Insights: 0 GB daily logs ingested, 3 months data retention, 0 multi-step Web tests; 1 VM monitored X 1 metric monitored per VM, 1 log alert at 5 min frequency, 0 additional events, 0 additional emails, 0 additional push notifications, 0 additional Web hooks (in millions)	1.60
Support	**Support**	0.00
	Total	**10,200.64**
	Annualized	**1,20,000.00 (Approx)**

Motherboard supporting 6 GB/s or above, 8 core processor, 256 GB RAM, 2 TB SSD, and 16 GB HDD or above. This shall allow for setting up a virtual network with high availability and optimized as per the plan indicated below:

- Install a BASE machine and create a "full clone" for each machine (disk consuming), i.e., around 20 GB per machine + the applications space.
- Install a BASE machine and create a "linked clone."

The second mentioned technique will use a base 20 GB image used by all the VM's, i.e., every further machine saves 20 GB of disk space each in this technique.

For e.g., base machine = 20 GB.

Machine 1 = 0 GB of linked machine + VRAM of its own + application data.

Table 2 Estimated cost for a dedicated VM-based testbed for security

Dedicated machine estimate		
Part	Description	Estimated cost ($)
Motherboard	MSI MAG B460M 6 Gb/s Intel Motherboard	190
Processor	Intel Core i9-11900KF 8-Core 3.5 GHz	600
RAM	256 GB (8 × 32 GB) DDR4	1660
SSD	2 TB 3,500 MB/s SSD	250
HDD	16 TB HDD 6 Gb/s 512e/4Kn	339
Firewall	NETGATE 5100 PFSENSE + SECURITY GATEWAY [17]	699
Squid	Squid for proxy	0
Server	Windows server 2012 (DNS, DHCP)	429
Client machines	Win 7, 8, 10	300 (per machine) 3000 (10 machines)
Client machines	Kali Linux, Ubuntu	0
AD	Windows server 2016 (active directory)	1069
IDS	Security onion for security monitoring tool	0
Honeypot	Tpot for honeypot services [18]	0
Log	Sysmon log system activity to the Windows event log	0
SIEM	Splunk/Sentinel/Graylog for SIEM solution	0 or 2000 (Annualized)
	Total	**9537**

All machines using this technique access the same portion of the HDD simultaneously. We should not use this technique if we are running an HDD which can lead to bad sectors. An SSD is highly recommended for such access. RAM disk out performs, we need to allocate physical RAM for it though.

The cost estimates available as on date using Newegg.com [16] and individual vendors software for the recommended hardware configuration are listed in Table 2:

The software can be run on VMware 16 Workstation Pro, available at [19] for $199. Adding miscellaneous costs, we may approximate the total system cost at a $10,000 onetime investment.

4 Results

The comparative analysis of the details presented in Sect. 3 cannot be considered on the presented costs alone. The hidden cost factors in both scenarios exist and are summarized in Table 3 for an easy reference. Keeping the target variable fixed at simulation of the attack trees on the testbed machines, the comparable variable parameters have been discussed.

Table 3 Comparative analysis of testbeds

Type of testbed	Scalability	Human monitor expertise	Cost of setup	Dynamic configuration	Ease of experiment	Annualized charges ($)
Cloud (Azure) testbed	High	Low (built-in analytics)	Fixed + Pay as you use	Easy	Low	1,20,000
Dedicated VM testbed	Low	High (analysis needed)	Fixed one time cost	Difficult	High	10,000*

* The prices used may vary as per market

If we keep on increasing the size of the VM testbed, the comparative cost difference keeps decreasing as the licensing costs go up. The number of machines vs licensing costs has been depicted in Fig. 1, but the scalability is an issue here as explained below:

The 256 GB RAM limitation imposes an upper limit on the number of VM client machines even with cloning methods. The cost dynamics for the testbed considering a 2 GB clone machine size and a 256 GB RAM availability shall reach $30,000 for licensing with a $300 individual license for a maximum of 100 machines occupying 200 GB RAM leaving 56 GB operational to run servers, network switches, and host operating system utilities.

Each component in Table 3 is explained below for an easy reference:

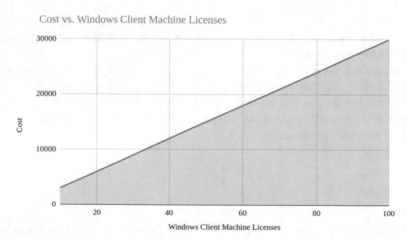

Fig. 1 Cost variation for dedicated VM with increasing users

- Scalability: The cloud-based testbed shall provide an easy scalability across network configuration, while the dedicated machine is limited by the RAM capacity and SSD size.
- Monitor Expertise: The testing person has to be highly skilled across several technology domains including virtualization and networking besides security in case of a dedicated VM testbed, while the built-in analytics dashboards with Azure make the task easier in case of cloud-based testbeds.
- The cost of setup for Azure is recurring and billed monthly while the dedicated VM cost is onetime investment.
- The services running on Azure cloud testbed can be dynamically integrated with new requirements while it shall remain difficult to change the configuration of a running dedicated VM testbed.
- The ease of experiment in case of a cloud-based testbed shall remain limited to the permissible frequency of use cases provisioned by the tier while an unlimited number of trials can be handled on the dedicated VM testbed.
- The cloud-based model is safe from disasters as discussed in [21].

5 Future Work and Conclusions

The comparative cost analysis indicates that establishing a testbed for up to 100 client machines is still a lot cheaper than the annualized cost of a security testbed established on the cloud with basic free tier usage of most components. A cloud-based VM setup with a bare metal cloud service provider and custom software can lower the costs in comparison with both approaches and needs to be studied further. It would also be possible to research the risks attached to an organization by setting scenario-based experiments on the testbeds using the attack tree which may help us model the risk associated and corresponding mitigation strategies. Gamification of network security assessment through the testbeds is another open area of research.

References

1. Herjavecgroup.com Official Annual Cyber Crime Report 2018. [online] Available at: https://www.herjavecgroup.com/wp-content/uploads/2018/12/CV-HG-2019-Official-Annual-Cybercrime-Report.pdf. Accessed 10 Oct 2021
2. B. Schneier, "Attack trees: modeling security threats," Dr. Dobb's Journal, December (1999)
3. I. Ray, N. Poolsapassit, Using attack trees to identify malicious attacks from authorized insiders. In Proceedings of the 10th European Conference on Research in Computer Security (ESORICS'05) (Springer, Berlin, Heidelberg, 2005), pp. 231–246. https://doi.org/10.1007/11555827_14
4. S. Mauw, M. Oostdijk, Foundations of attack trees. Lecture Notes Comput. Sci. **3935**, 186–198 (2006). https://doi.org/10.1007/11734727_17
5. J. Stefan, M. Schumacher, Collaborative attack modeling. In Proc. SAC 2002 (ACM, 2002), pp. 253–259

6. T. Tidwell, R. Larson, K. Fitch, J. Hale, Modeling internet attacks, in Proceedings of the 2001 IEEE Workshop on Information Assurance and Security (2001)
7. K. S. Edge, G.C. Dalton II, R.A., Raines, R.F., Mills, *"Using Attack and Protection Trees to Analyze Threats and Defenses to Homeland Security"* MILCOM 2007 (2007), pp. 1–7
8. R. Jhawar, B. Kordy, S. Mauw, S. Radomirovic, R. Trujillo-Rasua, Attack trees with sequential conjunction. IFIP Adv. Inf. Commun. Technol. **455** (2015). https://doi.org/10.1007/978-3-319-18467-8_23
9. S. Noel, S. Jajodia, "Managing attack graph complexity through visual hierarchical aggregation", in Proceedings of the 2004 ACM workshop on Visualization and data mining for computer security (New York, USA, 2004), pp. 109–118
10. O. Sheyner, Scenario Graphs and Attack Graphs. Ph.D. thesis, Carnegie Mellon University (2004)
11. S. Noel, S. Jajodia, "Managing attack graph complexity through visual hierarchical aggregation", in Proceedings of the workshop on Visualization and data mining for computer security (New York, USA, 2004), pp. 109–118
12. S. Mauw, M. Oostdijk, Foundations of attack trees, in *ICISC 2005*. ed. by D.H. Won, S. Kim. LNCS 3935. (Springer, Heidelberg, 2005), pp. 186–198
13. B. Kordy, S. Mauw, S. Radomirovic, P. Schweitzer, "Foundations of Attack–Defense Trees," In: LNCS (Springer, Heidelberg, 2010). Available at http://satoss.uni.lu/members/barbara/papers/adt.pdf
14. Gartner, Gartner Forecasts Worldwide Public Cloud End-User Spending to Grow 18% in 2021 (2020). [online] Available at: https://www.gartner.com/en/newsroom/press-releases/2020-11-17-gartner-forecasts-worldwide-public-cloud-end-user-spending-to-grow-18-percent-in-2021. Accessed 16 Oct 2021]
15. All prices shown are in US Dollar ($). This is a summary estimate, not a quote. For up to date pricing information please visit https://azure.microsoft.com/pricing/calculator/. This estimate was created at 10/16/2021 6:09:58 AM UTC
16. Newegg.com. 2021. Newegg—Shopping Upgraded. [online] Available at: https://www.newegg.com/. Accessed 16 Oct 2021
17. Netgate. 2021. Netgate 5100 pfSense+ Security Gateway. [online] Available at: <https://shop.netgate.com/products/5100-pfsense> [Accessed 16 October 2021].
18. GitHub, GitHub—telekom-security/tpotce: T-Pot - The All In One Honeypot Platform (2021). [online] Available at: <https://github.com/telekom-security/tpotce>. Accessed 16 Oct 2021
19. V. Pro, VMware Workstation 16 Pro. [online] Store-us.vmware.com (2021). Available at: https://store-us.vmware.com/vmware-workstation-16-pro-5424176500.html. Accessed 16 Oct 2021
20. M. Yadav, S. Gupta, Hybrid meta-heuristic VM load balancing optimization approach. J. Inf. Optim. Sci. **41**(2), 577–586 (2020). https://doi.org/10.1080/02522667.2020.1733190
21. S. Gupta, B. Gupta, Performance modeling and evaluation of transportation systems using analytical recursive decomposition algorithm for cyclone mitigation. J. Inf. Optim. Sci. **40**(5), 1131–1141 (2019). https://doi.org/10.1080/02522667.2019.1638003

Insights into the Black Box Machine Learning Models Through Explainability and Interpretability

Sachin Gupta and Bhoomi Gupta

Abstract Artificial intelligence (AI) and machine learning (ML) technologies are considered to be the Holy Grail for the researchers across the world. The applications of AI and ML are proving disruptive across the global technological spectrum, and there is practically no area which has been left untouched by these technologies right from computer science to manufacturing, healthcare, insurance, credit ratings, cybersecurity, and many more. It would not be an exaggeration to say that it is the next big thing after the advent of the Internet and potentially holds a similar impact in touching the lives of human beings. Whilst most researchers using machine learning in research across diverse domains do not need to look beyond the model abstraction for their work, the need for understanding what is happening beneath the surface is sometimes necessary. This becomes especially important in the cases where the predictions are too good to be apparently true, and the researcher running the model is not sure about its validity as the logic for prediction is obscure. The process of feature engineering brings in more accuracy to predictions, but in the absence of intuitive background information regarding the features, the task gets more challenging. The scientific reasoning has been driven by logic through ages, and the scientist community remains sceptical of the results unless they can extract useful insights from the black box ML models. The paper applies five popular explainability algorithms being used by the research community to demystify the abstract nature of ML black box models and compare the relative clarity of the insights being provided individually by each from a practitioner's perspective using the publicly available UCI wine quality dataset.

Keywords Machine learning · Black box models · Feature engineering · Interpretability · Explainability

S. Gupta (✉)
MVN University, Aurangabad, India
e-mail: sachin.gupta@mvn.edu.in

B. Gupta
Maharaja Agrasen Institute of Technology, Delhi, India
e-mail: bhoomigupta@mait.ac.in

1 Introduction

Data are the new oil, and a similar analogy can be thought of as the advances in machine learning are lubricating the fast paced wheels of innovation and technology. There are a multitude of ML models being used by the scientific community, and with the explosion of data and ever increasing computational power, the accuracy of the models is also getting near perfect. This remarkable increase in accuracy is certainly a desirable attribute of ML models, but it is not the only most important thing [1]. There has been a long standing debate between those advocating the computational efficiency of neural network-based deep learning at the cost of human interpretability versus the researchers who favour the traceability of regression methods for an enhanced experience of explainable artificial intelligence (XAI) [2, 3]. The famous explainable machine learning challenge of 2018 which was co-hosted by FICO, Google, Berkeley, Oxford, UC Irvine, MIT, and Imperial invited the participants to write a black box ML model and explain its working but the winning team flipped the challenge with writing a solution that was a globally interpretable white box and worked with the same accuracy [4]. The challenge has since then brought forward a tremendous amount of research in explainability of the machine learning models. Whilst the black box models still hold a position of dominance in the decision-making process of the ML world for low-stake decisions involving no risk to human lives, it is no longer the only option. With increasing penetration of ML in life-saving applications including critical healthcare, robotic surgeries, and the pandemic predictions, it is becoming imperative to have insights into "how" the prediction is coming through and "what" portions of data it is using for the same [5].

2 Related Work

The earliest use of interpretability, though not exactly in the same terms as explainability, is due Miller [6], who defined interpretability by correlating the cause of decision-making to the degree of human understanding. In the specific context of machine learning, Kim et al. [7] described interpretability in the terms of the ease of human reasoning and traceability of any prediction made by a model, and the degree of consistency with which the reasoning could be made. The contextual definition evolved in recent times with a paper by Kim and Doshi-Velez redefining interpretability in terms of "explainability" in understandable ways to humans [8]. The model explainability as a trust issue has also been explored by Ribeiro et al. for classifier prediction [9], whilst the new European General Data Protection Regulation [10] guidelines make explainability of algorithmic decision-making mandatory as a fundamental right in the context of data. A detailed survey of the black box models being used in ML can be attributed to recent work of Guidotti et al. [11] which makes a general classification of the explanation models into three distinct categories based on the granularity of explainability. The coarse grain methods just providing an approximate explanation of the model as a black box are classified as global [12, 13] whereas the local methods [9, 14, 15] allow better insights into the mechanism by providing outcome specific explanations over instance specific datasets. Whilst the global methods are sometimes as obscure as a black box, at times, they provide useful insights in the working of a model consistently across different datasets; on the contrary, the local models give detailed insights but the insights obtained are highly variable across different datasets and are thus unreliable at times. In the next section, we discuss noteworthy model-agnostic global and local ML explainability models including permutation importance [16], partial dependence plots [17], individual condition execution [18], local surrogate [9] and SHAP [19].

3 Methodology

The famous white wine quality dataset from UCI machine learning repository [20] has been used as a means to procure multivariate data of a sufficient quantity, and with 12 attributes and 4898 rows, the dataset is suitable for making a fair assessment of the explainability models. The dataset was also chosen for the desirable properties with no null values and having only numeric data types including 11 float type of variables/1 integer for the ease of study. The features in the dataset used are described in [20] and the chosen target variable is "quality" as recorded by the sommeliers. After cleaning the dataset, dropping the target attribute and applying the basic exploratory data analysis to the remaining 11, the following insights were discovered from the histogram. No feature engineering has been applied for the study. Only sample plots shown in Fig. 1.

From the plots, it is clearly apparent that some features need to be engineered to compensate for the skewed data. The volatile acidity as seen from Fig. 1 is a potential feature engineering candidate for instance. In order to assess model-agnostic interpretability methods, we have not used any feature engineering at this stage as well.

To further understand the correlation between different attributes, and especially with the target attribute of quality, a heatmap was also plotted as per Fig. 2, which gave additional insights into the relative importance of the attributes. The correlation of quality was observed having the highest degree of positive correlation with alcohol content with 0.44, whilst a high-negative correlation is seen with the density of the wine at −0.31. The other attributes can also be seen in context and needs a good domain knowledge to interpret other than the usual taste palette wine parameters of acidity and alcohol.

Having explored the dataset through basic exploratory analysis methods, we now split our dataset in the ratio of 80:20 for training and testing the machine learning model, in order to apply the post-hoc analysis methods. It is interesting to observe that although it would be difficult to predict "why" our ML model is giving certain answers, it is still possible to understand "how" it is coming to those conclusions.

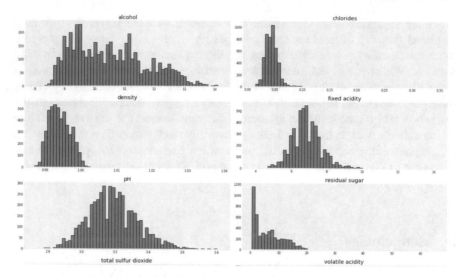

Fig. 1 Basic EDA of the white wine quality dataset

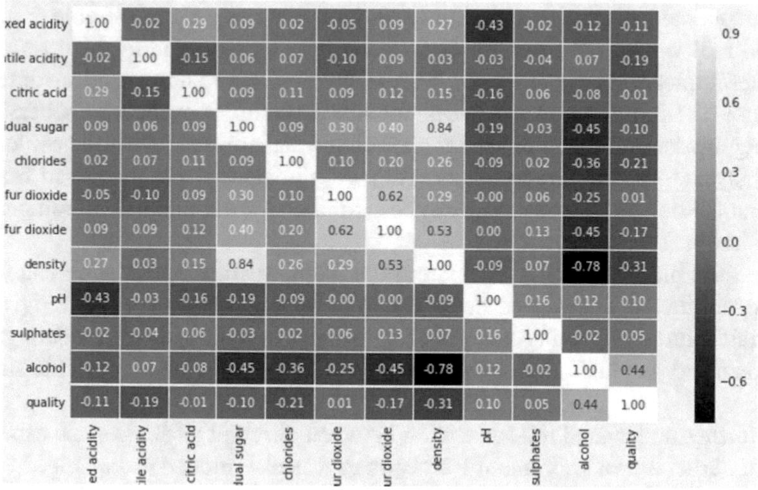

Fig. 2 Attribute correlation heatmap

This interpretability gives us the ability to improve our work through debugging and makes it suitable for better feature engineering and therefore a better human decision-making process. It adds the essential human TRUST component to the otherwise black box model.

3.1 Feature Importance

A baseline understanding of the relative importance of the features was obtained by plotting the "feature importance" function of XGB which allows for an analysis based on feature weights as well as that based on feature coverage in samples. The native feature importance given by both methods has been shown in Fig. 3a, b.

The plots obtained by the two native XGBoost methods to plot global relative importance of features are inconclusive, and this anomaly needs to be investigated even though the feature by sample coverage seems more in synchronization with the

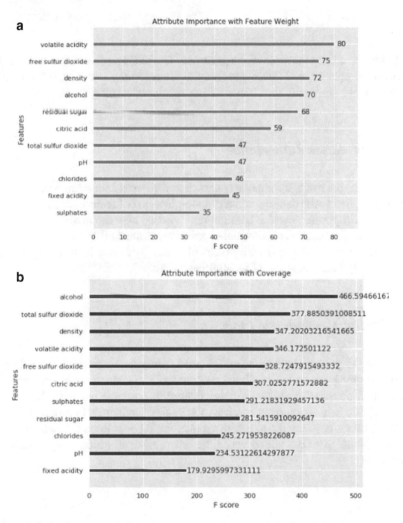

Fig. 3 **a**. Relative importance of the features as per feature weight. **b**. relative importance of the features as per coverage

Fig. 4 Permutation
importance using Eli5 for
relative importance of
features

Weight	Feature
0.0968 ± 0.0058	alcohol
0.0790 ± 0.0106	volatile acidity
0.0370 ± 0.0028	free sulfur dioxide
0.0224 ± 0.0038	residual sugar
0.0214 ± 0.0045	citric acid
0.0147 ± 0.0022	density
0.0145 ± 0.0034	chlorides
0.0130 ± 0.0050	total sulfur dioxide
0.0115 ± 0.0033	sulphates
0.0114 ± 0.0024	pH
0.0108 ± 0.0043	fixed acidity

results obtained from the multivariate heatmap. The inconclusive nature of these plots also leads to opacity, and that is precisely the reason why interpretability methods may be used to demystify the "hows" of our machine learning model.

3.2 Permutation Importance

The "Explain like I'm 5" or popularly known as the permutation importance method Eli5 [16] applied to the dataset produces a weighted list of features as given in Fig. 4. Permutation importance grades the features based on how the model accuracy deteriorates when the feature is not available or its values are shuffled across the column. The method when compared with results obtained by native feature importance is comparatively better in terms of its intuitive nature.

3.3 Partial Dependence Plots

"The partial dependence plot (short PDP or PD plot) shows the marginal effect one or two features have on the predicted outcome of a machine learning model" [17]. The method does not provide a relative feature importance but indicates how our target variable gets affected by the change in value of a feature. Figure 5 shows how sulphates values between 0.5 to 0.6 give a maximum boost to quality which remains relatively stable after 0.6 with a very little but consistently increasing contribution with values.

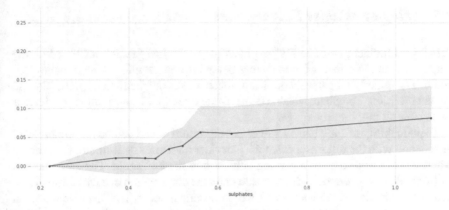

Fig. 5 Partial dependence plot showing the effect of sulphates on quality

3.4 Individual Conditional Expectation (ICE) Plots

The PD plot is built around average behaviour and hides the individual instance effect on the contribution of a feature on the prediction. The ICE plots are instance specific may uncover heterogeneous relationships and give better insights into the average behaviour of the PD plots by plotting a number of instances. In Fig. 6, the ICE plot clearly shows a distinct jump in the quality prediction based on sulphates values between 0.5 and 0.6, and this might be a characteristic to predict a good quality white wine.

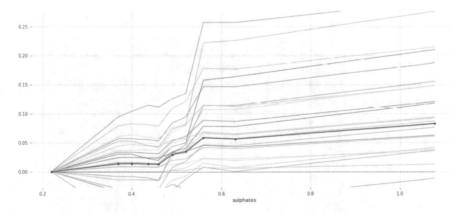

Fig. 6 Independent conditional expectation plot for sulphates contribution to quality

3.5 *SHapley Additive ExPlanations (SHAP)*

SHAP by Lundberg and Lee [19] explains the individual predictions based on the output as a sum of effects of each feature being introduced into a conditional expectation. The underlying game theoretic approach makes the predictions pretty close to human intuition. The Shapley value is the average marginal contribution of a feature value over all possible coalitions.

Explainability of the Shapley values is very intuitive and can be reasonably well understood with the following example. Consider n features or attributes for a model which are allowed to enter a gaming contest in a random order. The game here is called predict to win, and scoring is based upon the contribution of each feature. If we denote the Shapley value with ϕ_{ab}, it would be calculated as the "average marginal contribution" of the feature instance f_{ab} after it is added to the mix of other feature values that have entered the contest before it.

$$\phi_{ab} = \sum_{\text{All Orderings}} \text{val}(\{\text{features.before.}b\} \cup x_{ab}) - \text{val}(\{\text{features.before.}b\}).$$

Shapley values are considerably the best to predict the importance of features based on its calculation, and Fig. 7 shows the feature importance as seen from the Shapley perspective. It is interesting to have a comparison of this technique with the permutation importance, and the authors consider that Shapley is a better metric since permutation-based feature importance computes based on the degradation in model performance whilst Shapley computations account for magnitude of feature attributions.

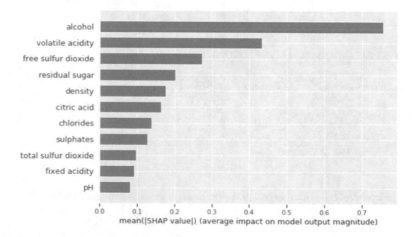

Fig. 7 Shapley values-based feature importance

Fig. 8 High quality wine parameters through shapley

The Shapley values allow for several interactive plots like the one plotted in Fig. 8, which shows the components of a "high" quality (quality >5) wine in terms of the features along with the relative contribution of each feature to the quality score.

4 Results

The first glance at the various outputs obtained through several interpretability models throw surprising variations on the relative feature importance and have been shown in Table 1. The most interesting outcome of this tabulation is a high degree of similarity that exists between the Shapley value predictions and the average of all other methods *minus* the basic heatmap approach. This relationship between results can also be understood in context of the ensemble-based approaches of ML for an easy perspective or analogy.

Table 1 Feature ranking by different models for quality prediction

Feature	Heatmap	XGB feature by weight	XGB feature by coverage	Permutation importance (Eli5)	Average without heatmap	SHAP
Alcohol	1	4	1	1	1	1
pH	2	7	10	10	9	11
Sulphates	3	10	7	9	8	8
Free sulphur dioxide	4	2	5	3	3	3
Citric acid	5	6	6	5	5	6
Residual sugar	6	5	8	4	6	4
Fixed acidity	7	9	11	11	10	10
Total sulphur dioxide	8	7	2	8	5	9
Volatile acidity	9	1	4	2	2	2
Chlorides	10	8	9	7	8	7
Density	11	3	3	6	4	5

Table 2 Comparison of interpretability methods, with special reference to speed of execution

Method	Native feature importance	Eli5 (permutation importance)	Partial dependence plots	LIME	SHAP
Type	Global	global: black box models/local: white box models	Global	Local	Hybrid
Consistency	Inconsistent	Consistent	Consistent	Consistent	Consistent
Speed	Fast	Slow with more features	Fast	Fast	Moderate
Reliability	Low	Moderate	Moderate, increase with ICE	High	High
Model-agnostic	No	Partially	Yes	Yes	Yes

The findings indicate that although none of the individual interpretability models can exclusively guarantee a great insight into the behaviour of the ML model, and there has to be a back and forth approach of going through various methods to uncover the real insights. Table 2 lists the basic differences between models based on the literature, along with the findings of the study on the comparative speed of operation of each calculated as a function of time.

5 Conclusion and Future Work

The ML scientists always try to strike a balancing role between bias and variance for accuracy of the models, and there always exists a trade-off between the two. The study of this paper indicates that a trade-off in interpretability to accuracy also exists if we consider the interpretability of the models. There exists a direct relationship between the accuracy of a model to its complexity, which is a result of the using multiple stacked models or ensemble-based boosting, but the two are inversely proportional to interpretability with each level of a model addition making it increasingly difficult to understand its outputs.

It can also be concluded that any model with a variable number of parameters like XGBoost is harder to understand since the parameters tend to increase with the volume of training data. Parametric models on the other hand can provide easier insights with coefficient values for interpretability.

The most important takeaway from the study is that none of the interpretability models is an absolute authority, and any assumption made by looking at the output of a single interpretability technique can (and most certainly will) go absolutely wrong with its dependency not only on the concerned model, but also with the values available in the raw data. The scope of interpretability and explainability of the ML models in the presence of hyperparameter tuning with complex models remains an interesting research area and is expected to precipitate further research in associated domains.

References

1. N. Burkart, M. F. Huber, "A survey on the explainability of supervised machine learning," *Journal of Artificial Intelligence Research, 70* (2021). https://doi.org/10.1613/JAIR.1.12228
2. A. Datta, M. Fredrikson, K. Leino, K. Lu, S. Sen, Z. Wang, "*Machine Learning Explainability and Robustness,*" (2021). https://doi.org/10.1145/3447548.3470806
3. C. Rudin, J. Radin, Why are we using black box models in ai when we don't need to? a lesson from an explainable ai competition. *Harvard Data Science Review, 1*(2) (2019). https://doi.org/10.1162/99608f92.5a8a3a3d
4. "Explainable Machine Learning Challenge", FICO Community, Available: https://community.fico.com/s/explainable-machine-learning-challenge. [Accessed: 03-Sep-2021]
5. A. Vellido, The importance of interpretability and visualization in machine learning for applications in medicine and health care. *Neural Computing and Applications, 32*(24) (2020). https://doi.org/10.1007/s00521-019-04051-w
6. T. Miller, Explanation in Artificial Intelligence: Insights from the social sciences. Artif. Intell. **267**, 1–38 (2018). https://doi.org/10.1016/j.artint.2018.07.007
7. B. Kim, R. Khanna, O.O. Koyejo, Examples are not enough, learn to criticize! Criticism for interpretability. In *Advances in Neural Information Processing Systems* (MIT Press, Cambridge, MA, USA, 2016), pp. 2280–2288
8. F. Doshi-Velez, B. Kim, *Towards a Rigorous Science of Interpretable Machine Learning*. arXiv 2017, arXiv:1702.08608
9. M.T. Ribeiro, S. Singh, C. Guestrin, *"Why Should I Trust You?": Explaining the Predictions of Any Classifier*. In Proceedings of the 22nd ACM SIGKDD International Conference on Knowledge Discovery and Data Mining, (San Francisco, CA, USA, 2016), pp. 1135–1144
10. European Commission. General Data Protection Regulation (2016). Available online: https://eur-lex.europa.eu/legal-content/EN/TXT/PDF/?uri=CELEX:32016R0679. Accessed 03 Sep 2021
11. R. Guidotti, A. Monreale, S. Ruggieri, F. Turini, F. Giannotti, D. Pedreschi, A survey of methods for explaining black box models. ACM Comput. Surv. **51**(5), 1–42 (2018)
12. M. W. Craven, J. W. Shavlik, Extracting tree-structured representations of trained networks. In Neural Information Processing Systems (Cambridge, MA: MIT Press, 1995), pp. 24–30
13. N. Frosst, G. E. Hinton, Distilling a neural network into a soft decision tree. In Proceedings of the First International Workshop on Comprehensibility and Explanation in AI and ML 2017 Colocated with 16th International Conference of the Italian Association for Arti-ficial Intelligence (AI*IA 2017). CEUR Workshop Proceedings, vol 2071 (2017)
14. B. Kim, C. Rudin, J. Shah, The Bayesian case model: a generative approach for case-based reasoning and prototype classification. In Proceedings of the 27th International Conference on Neural Information Processing Systems, vol 2 (Cambridge, MA, MIT Press, 2014), pp. 1952–1960
15. M. T. Ribeiro, S. Singh, C. Guestrin, *Anchors: High-precision model-agnostic explanations. InAAAI* (NewOrleans, Louisiana, AAAI Press, 2018), pp. 1527–1535
16. A. Fisher, C. Rudin, F. Dominici, *Model Class Reliance: Variable Importance Measures for any Machine Learning Model Class, from the "Rashomon" Perspective*. arXiv 2018, arXiv:1801.01489
17. J.H. Friedman, Greedy function approximation: A gradient boosting machine. Ann. Stat. **29**, 1189–1232 (2001)

18. A. Goldstein, A. Kapelner, J. Bleich, E. Pitkin, Peeking inside the black box: Visualizing statistical learning with plots of individual conditional expectation. J. Comput. Gr. Stat. **24**, 44–65 (2015)
19. S.M. Lundberg, S.I. Lee, A unified approach to interpreting model predictions. In *Advances in Neural Information Processing Systems* (MIT Press, Cambridge, MA, USA, 2017), pp. 4765–4774
20. P. Cortez, A. Cerdeira, F. Almeida, T. Matos, J. Reis, Modeling wine preferences by data mining from physicochemical properties. Decision Support Syst., Elsevier **47**(4), 547–553 (2009)

An Improved AODV Routing Algorithm for Detection of Wormhole and Sybil Attacks in MANET

Shreeya Mishra⓪, Umesh Kumar⓪, and Komal Mehta Bhagat⓪

Abstract Wireless sensor network is a promising technology in the current scenario due to its wider area of research. With the advancement of technology, communication in mobile ad hoc networks has become easier and efficient with some vulnerability. Secured system communication is required, but with the evolution and advancements in the technology, threats also increase. During communication, various kinds of attacks may occur in the mobile ad hoc network. In this paper, wormhole attack in AODV protocol is discussed. Different parameters affect the working of protocols; here parameters end-to-end delay and PDR ratio are discussed with the main aim to detect the wormhole attack and provide an efficient solution to minimize the risk of attack.

Keywords WSN · MANET · Wormhole attack · Sybil attack · End-to-end delay

1 Introduction

A wireless sensor network consists of a large number of nodes that gather information from the environment and transmit it to the base station via wireless signals [1]. Due to unstable environment and unattended deployment, WSN is oversensitive in failures [2]. In WSN, node movement is a centralized type. In a wireless network, mobile ad hoc networks (MANET) is a self-configured and small-range instant network, where wireless devices directly communicate with each other [3]. MANET network is designed anywhere and anytime in such a way to support the mobility of users in the network. Such networks serve blocking [4]; therefore, the performance depends

S. Mishra (✉) · U. Kumar
Govt. Women Engineering College, Ajmer, India
e-mail: shreeyamishra9@gmail.com

U. Kumar
e-mail: umesh@gweca.ac.in

K. M. Bhagat
Bhagwan Parshuram Institute of Technology, Delhi, India
e-mail: komalbhagat@bpitindia.com

© The Author(s), under exclusive license to Springer Nature Singapore Pte Ltd. 2023   645
Y.-D. Zhang et al. (eds.), *Smart Trends in Computing and Communications*, Lecture Notes
in Networks and Systems 396, https://doi.org/10.1007/978-981-16-9967-2_60

on the stability of the network architecture [5]. MANET is a decentralized type of WSN. WSN interacts with the environment, while MANET is closure to human interaction. MANET has security and routing problems because of its complexity in nodes establishment, and also nodes need to be adaptable in network topologies as an ad hoc network changes their topology quickly and unpredictably [6, 7].

AODV in MANET, where a device sends a packet to another device for communication, first looks up a route to the destination in its routing table if this exists in the data packet; otherwise, it starts the route discovery process. In this case, the device will send an RREQ message in broadcast mode. The RREQ message has the source address, a destination address, source sequence number, broadcast identifier, and the most recent sequence number between the source device and the destination device. The device will be waiting for a reset route message (RREP) message to be received. If an RREP message is not received after a fixed time, the device will wait for the RREP message to be received. The site will try again a finite number of times. If it fails again, the device concluded that there is no route to that destination. When a device receives an RREQ message for a destination, one of the following actions may occur: The device that received the RREQ message has no information about the error, so it resends the message to its neighboring devices via broadcast. It also temporarily stores the route back to the destination device [8]. The device that receives the message is the destination device or has route information to destination t. In either case, it sends an RREP message back to the source device via the route by which the device received the first RREQ message. All devices maintain a routing table with the information of each destination of interest, created from the received RREQ messages, and remain active if used by any active neighboring device. Therefore, author [9] proposed a new protocol that is WAODV (waiting AODV) in which the most stable path between source and destination is chosen by using hello packets to the intermediate node to find the most stable neighbor and thus path. TAODV [10] is the extension form of AODV. It uses trust values and thus sends three messages trust request message (TREQ), trust reply message (TREP), and trust warning message (TWARN).

In a wireless sensor network, different types of attacks can be seen. Attacks are classified as active, passive, internal or external, different protocol layers, cryptography, or non-cryptography related [11]. Active Attacks create a hole in following security issues in protocols, sensor node destruction, and data alteration, which degrades the performance of the WSN protocol ([12], [13]). Gray hole attack and black hole attack are like each other. In the gray hole with a certain probability, it drops the packets, due to this it is difficult to detect and also causes network congestion [14].

Sybil attack: In a network, present malicious nodes create multiple identities to the other node. This type of attack is common in MANET [15].

Wormhole attack: The most threatened type of attack seen in the MANET is a wormhole attack. It takes place in the form of forwarding the packets in between the malicious nodes [16]. In wormhole attack work, two malicious nodes are present: One malicious node gets the packet from the source and forwards to next node which seems to be malicious node, and it further forwards to the destination [17].

2 Literature Survey

MANET has been analyzed by the authors for the detection of attacks and protocols that have been used. AODV protocol in MANETs is designed to choose the least congested path between source and destination, but choosing the least path is not enough as it does not guarantee a stable path. Routes are only maintained when they are active that is they are of use. During routing, security threats are present. In [18], author detects the black hole attack in AODV protocol and compares the parameters like throughput and PDR with traditional AODV. Authors [19] have discussed different detection and prevention methods of wormhole attack along with merits and demerits. Authors [20] suggest a method for wormhole attack called (DELPHI) delay per hop indication. This method calculates the mean delay per path of every path by using the multipath approach. In ad hoc on-demand multipath distance vector (AOMDV), an extension of AODV is proposed by the authors [21]. In this protocol, the RTT method is used in which the RTT value from source to destination is calculated for each route.

Authors [22] proposed a secure tracking of node encounters (SECTOR). In this protocol, based on the speed of transmitted data, distance between two nodes is calculated. In [23], authors suggest a method to detect wormhole attacks by applying routing protocol IPv6 for low-power and lossy networks. To detect wormhole attack on the mechanism on packet leashes is suggested by the authors [24]. This method considers geographical-based protocol which relies on the current location and transmission time of the packet and in temporal leashes clock is tightly synchronized and it is not dependent on the GPS information. Authors [25] proposed a mechanism based on directional antennas that prevent from wormhole attack. In this method after hearing a signal from a neighbor with the assumption that the antennas on all the nodes are aligned, the neighbor list is built securely. Distance consistency-based secure localization scheme is proposed by the authors [26] to detect wormhole attack. To avoid Sybil attack, author [27] collects the information of routes by using an intelligence algorithm. In this algorithm, a Sybil attack is detected when there is an energy change in an active network.

3 Proposed Work

The most common attack in MANET is a wormhole attack in a network, and the malicious node between source and destination creates s tunnel and affects the transmission of the packets. The proposed system is the modified version of AODV. An exploration of the possible control measures to reduce the risk of exploitation of these vulnerabilities by wormhole attacks is presented. Flowchart is given in Fig. 1

The simulation has been done in NS2 simulator tool. The tool helps to study the dynamic nature of communication in networks. It helps to design a network protocol by providing a comprehensive environment; under specific conditions, it creates

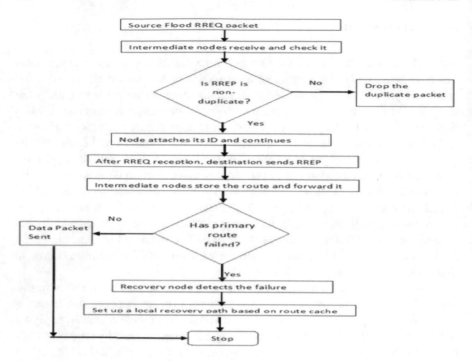

Fig. 1 Flowchart of proposed algorithm

virtual scenario and helps to analyzing their performance. Here in the network, 50 nodes are taken, duration for simulation was 90 s, and other parameters are listed in Table 1. In this subsection, simulations results show the effect on data communication when the number of attackers is increased in the network. Figure 1 shows the working of proposed work.

Table 2 shows the effect of an increase in the number of attackers on end-to-end delay when compared to traditional AODV, WAODV, TAODV, and proposed routing protocols in MANETs.

Table 1 Simulation parameters

Parameters	Value
Simulator	NS2
Simulation time	90 s
Topology	800 m X, 800 m Y
No. of nodes	25,50,75,100,125
Max segment size	512 bytes
Traffic type	FTP
Routing protocol	AODV, WAODV, TAODV

Table 2 Average end-to-end delay with number of attackers in wormhole attack

Nodes	Proposed work	AODV	WAODV	TAODV
25	9.18	26.52	72.56	11.23
50	12.45	21.18	82.43	14.31
75	11.34	33.45	78.34	13.52
100	11.21	36.34	74.05	12.23
125	10.52	29.21	69.36	11.44

Increasing the number of attackers in a network has an impact on data communication, as we will see in this section's simulation findings. To illustrate this, Fig. 2 compares the end-to-end latency for the standard AODV, WAODV, TAODV, and proposed MANET routing protocols to those for an increase in the number of attackers.

Increasing the number of attackers in the network results in an increase in the average packet delivery ratios (PDR) of all the compared routing methods (Fig. 3). In other words, as the number of attackers on the network grows, so does their likelihood of joining a particular assault path. As a result, attackers are dropping more packets throughout the transmission process. When using the TAODV protocol, the attacker

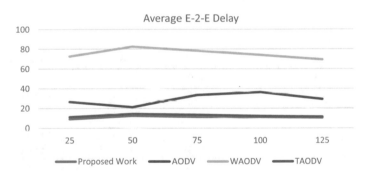

Fig. 2 End-to-end delays with increase in number of attackers in the network

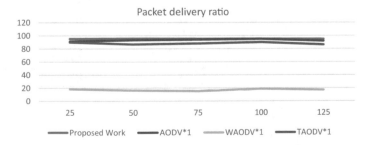

Fig. 3 Average PDR with increase in number of attackers in the network

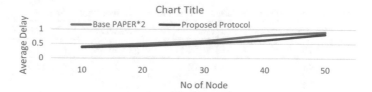

Fig. 4 Average end-to-end delays performance in Sybil attack

utilizes false information during the route discovery phase to ensure that he will be a part of every route found during the route discovery process, but the proposed approach loses the least amount of data when using the WAODV attack protocol.

Figure 4 depicts the impact of increasing the number of Sybil attack nodes in a MANET's protocols. The end-to-end latency of the network increases the delay of the packet as the number of attackers grows. EED is still higher than the proposed procedures in AODV because more attackers reach their objective and create a greater average delay (Table 4).

According to Fig. 5 for AODV and suggested work routing protocols, increased network and the PDR of the compared routing methods show that the PDR of the proposed protocol is much higher than that of the AODV routing protocols (Table 5 and Fig. 5).

Table 3 Packet delivery ratio

Nodes	Proposed work	AODV	WAODV	TAODV
25	9.18	26.52	72.56	11.23
50	12.45	21.18	82.43	14.31
75	11.34	33.45	78.34	13.52
100	11.21	36.34	74.05	12.23
125	10.52	29.21	69.36	11.44

Table 4 Average end-to-end delays performance in Sybil attack

Nodes	Base paper [38]	Proposed protocol
10	0.4	0.38
20	0.5	0.43
30	0.6	0.53
40	0.8	0.63
50	0.9	0.83

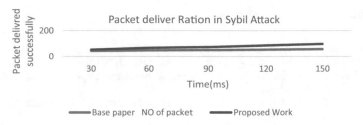

Fig. 5 Effect of packet-on-packet delivery ratio in Sybil attack

Table 5 Effect of packet-on-packet delivery ratio in Sybil attack

Time	No of packets Base paper [27]	Proposed work
30	43	52
60	47	67
90	50	73
120	53	86
150	57	97

4 Conclusion

MANETs are spontaneously deployed mobile networks that do not require any preexisting infrastructure over a geographically restricted area. Without the need for centralized management or set network architecture, nodes are often both autonomous and self-organized. MANETs are prone to a specific routing error known as a wormhole attack and Sybil attack because of their decentralized structure. Malicious nodes tunnel traffic from their locations to the target node through an attack called a wormhole and malicious which creates multiples fake IDs called Sybil. A fake route with a lower hop count is created as a result of wormhole assaults. This false route allows bad nodes to deliver or drop packets depending on how they feel about the originating node. The purpose of this piece is to put an end to these kinds of attacks. When it comes to defending against wormhole attacks, it investigates the usage of an artificial immune system (AIS). The suggested method detects and bypasses wormhole nodes quickly and without degrading the network's overall performance. We compare the suggested technique to alternative solutions in terms of packet delivery ratio and end-to-end latency to see how well it performs. In a simulation, the proposed technique outperforms by 8% in the term of packet delivery ratio by previous schemes in preventing a wormhole attack and Sybil attack.

References

1. J. Tian, et al., "Study on wireless sensor networks." 2010 International Conference on Intelligent System Design and Engineering Application, vol. 2 (IEEE, 2010)
2. N.N. Datta, K. Gopinath, A survey of routing algorithms for wireless sensor networks. J. Indian Inst. Sci. **86**(6), 569 (2006)
3. T.H. Hadi, MANET and WSN: what makes them different? Int. J. Comput. Netw. Wirel. Commun. (IJCNWC) **7**(6), 23–26 (2017)
4. Dr. S. S. Aarti, "Tyagi, "Study of manet: Characteristics, challenges, application and security attacks". Int. J. Adv. Res. Comput. Sci. Softw. Eng. **3**(5), 252–257 (2013)
5. S. A. K. Al-Omari, P. Sumari, "An overview of mobile ad hoc networks for the existing protocols and applications." arXiv preprint arXiv:1003.3565 (2010)
6. A.O. Bang, P.L. Ramteke, MANET: history, challenges and applications. Int. J. Appl. Innovation Eng. Manage. (IJAIEM) **2**(9), 249–251 (2013)
7. G. Wang, G. Wang, An energy-Aware geographic routing protocol for mobile ad hoc networks. Int. J. Softw. Inf. **4**(2), 183–196 (2010)
8. P.K. Maurya et al., An overview of AODV routing protocol. Int. J. Modern Eng. Res. (IJMER) **2**(3), 728–732 (2012)
9. T. Jamal, S.A. Butt, Malicious node analysis in MANETS. Int. J. Inf. Technol. **11**(4), 859–867 (2019)
10. X. Li, M.R. Lyu, J. Liu, "A trust model based routing protocol for secure ad hoc networks." 2004 IEEE Aerospace Conference Proceedings (IEEE Cat. No. 04TH8720) vol 2 (IEEE, 2004)
11. R. Hunt, "Network security: the principles of threats, attacks and intrusions, part1 and part 2." (2004)
12. M. Karthigha, L. Latha, K. Sripriyan, "A comprehensive survey of routing attacks in wireless mobile ad hoc networks." 2020 International Conference on Inventive Computation Technologies (ICICT) (IEEE, 2020)
13. L.K. Bysani, A. K. Turuk, "A survey on selective forwarding attack in wireless sensor networks." 2011 International Conference on Devices and Communications (ICDeCom) (IEEE, 2011)
14. M. Sathish et al., "Detection of single and collaborative black hole attack in MANET." 2016 International Conference on Wireless Communications, Signal Processing and Networking (WiSPNET) (IEEE, 2016)
15. SHIATS, Allahabad. "A survey on comparison of secure routing protocols in wireless sensor networks." Int. J. **5**(3) (2016)
16. N. Choudhary, L. Tharani, "A survey of routing attacks in mobile ad hoc network." IEEE Security in Wireless Mobile AD Hoc and Sensor Networks, pp. 85–91 (2007)
17. B. Kannhavong et al., A survey of routing attacks in mobile ad hoc networks. IEEE Wireless Commun. **14**(5), 85–91 (2007)
18. A. Sharma et al., Prevention of black hole attack in AODV routing algorithm of MANET using trust based computing. Int. J. Comput. Sci. Inf. Technol. **5**(4), 5021–5025 (2014)
19. M.K. Verma, R.K. Dwivedi, "A survey on wormhole attack detection and prevention techniques in wireless sensor networks." 2020 International Conference on Electrical and Electronics Engineering (ICE3) (IEEE, 2020)
20. H.S. Chiu, K.S. Lui, "DelPHI: wormhole detection mechanism for ad hoc wireless networks." 2006 1st international symposium on Wireless pervasive computing (IEEE, 2006)
21. P. Amish, V.B. Vaghela, Detection and prevention of wormhole attack in wireless sensor network using AOMDV protocol. Procedia Comput. Sci. **79**, 700–707 (2016)
22. S. Čapkun, L. Buttyán, J.P. Hubaux, "SECTOR: secure tracking of node encounters in multi-hop wireless networks." Proceedings of the 1st ACM workshop on Security of ad hoc and sensor networks (2003)
23. G.-H. Lai, Detection of wormhole attacks on IPv6 mobility-based wireless sensor network. EURASIP J. Wirel. Commun. Netw. **2016**(1), 1–11 (2016)

24. Y.-C. Hu, A. Perrig, D.B. Johnson, Wormhole attacks in wireless networks. IEEE J. Sel. Areas Commun. **24**(2), 370–380 (2006)
25. L. Hu, D. Evans, "Using directional antennas to prevent wormhole attacks." NDSS. **4** (2004)
26. H. Chen et al., A secure localization approach against wormhole attacks using distance consistency. EURASIP J. Wireless Commun. Netw. **8**, 2010 (2010)
27. W. Jlassi et al., "A novel dynamic authentication method for sybil attack detection and prevention in wireless sensor networks." AINA (1) (2021)

An Ensemble Model (Simple Average) for Malaria Cases in North India

Kumar Shashvat, Arshpreet Kaur, Ranjan, and Vartika

Abstract Malaria is an infectious disease borne due to mosquitoes that attacks humans and other animals' bodies. Malaria is a part of the plasmodium group caused by single-celled microorganisms. This study proposes the use of ensemble model using the three regression algorithms that are linear regression, support vector machine (SVM), and auto-Arima techniques and comparing their results. Predictions of plasmodium virus cases are made with the use of linear regression, support vector machine, and auto-Arima algorithms. The accuracy of prediction is measured by calculating the explained variance score, mean squared error rate, and root mean squared error rate. Our aim is to get better prediction results compared to the individual algorithms by combining the results of these individual models. The proposed work determines the accuracy of linear regression, support vector machine, and auto-Arima and ensembles together to find the trend of prediction using simple Average. A comparison of performance among the three regression techniques indicated the SVM model performs the best and has small RMSE and MAE values. But, by introducing the technique of ensemble modeling using simple average, combining the prediction of these three algorithms results in the lowest RMSE and MAE values.

Keywords Malaria · Support vector machine · Linear regression

1 Introduction

For centuries, malaria has become a problematic disease in India. Malaria is a mosquito-borne disease that causes infection and affects both humans and animals [1–6]. It took ten to fifteen days to show the symptoms after an infected mosquito bit the human body. It should be treated properly; otherwise, the effects of the malaria will

K. Shashvat (✉) · A. Kaur · Ranjan · Vartika
Department of Computer Science and Engineering (ACED), Alliance University, Bangalore, Karnataka, India
e-mail: shashvat.sharma@alliance.edu.in

A. Kaur
e-mail: arshpreet.kaur@alliance.edu.in; arshpreet@nitdelhi.ac.in

© The Author(s), under exclusive license to Springer Nature Singapore Pte Ltd. 2023   655
Y.-D. Zhang et al. (eds.), *Smart Trends in Computing and Communications*, Lecture Notes in Networks and Systems 396, https://doi.org/10.1007/978-981-16-9967-2_61

recur months later. Malaria is a part of the plasmodium group caused by single-celled microorganisms [7–14].

The diseases can be eliminated by using bugs spray and mosquito nets to prevent them from biting humans. It can also be controlled by taking measures by spraying insecticides and draining collected water and having a proper drainage system [2, 15–20]. In the 1930s, there was no single person in the country who was not infected by the disease. Due to malaria, the loss in economy in man-days was at Rs.10,000 million per year in 1935 [20–22].

The disease has majorly affected the tropical and subtropical region of India. Malaria is majorly associated with poverty and due to this economic development have negative effects. Malaria is categorized into severe or uncomplicated by World Health Organization (WHO) [22–24]. By the year 2020, there is only one vaccine that reduces the chances of malaria by 40% in children. From 2010 to 2014 that rate of malaria decreased, but from 2015 to 2017, the cases of malaria reached 231 million. Malaria is a disease associated not only with poverty, but the affirmation has suggested that it has a major hindrance to economic development [25, 26]. The Malaria Eradication Research Agenda (malERA) drive was an instructive process to rectify which parts of research and development (R&D) must be marked for worldwide elimination of malaria. India is considered traditionally native for both plasmodium vivax and plasmodium *falciparum* malaria and has a past of successes [27, 28]. Over 75% cases of malaria come from India of the total cases in Southeast Asia. In 2016, over 50% of India (698 million) was in danger of getting infected by malaria as mentioned by the World Malaria Report, which also stated that India reckon 6% of total malaria cases, 6% deaths, and 51% of the total malaria cases. In 2017, India ranked 46th with a death ratio of 1.97 per 100,000 individuals. The health of the disease is a complex issue in India developing from the geo-ecological diversity of the country [26, 27]. 80% of cases in India is caused by high indigenous areas. 95% of the population of the country lives in areas where the chances of getting infected is high [27–29].

This paper uses various techniques like support vector regression, linear regression, and Arima models to come up with a comparative analysis and see the effect and trend of the study. To improve the accuracy of our model, we combine the results from these techniques using a simple ensemble model. The aim of this paper is to analyze the prediction modeling results by overcoming the prediction error of different models. The performance of the model is evaluated in terms of mean average error, mean square error, explained variance score, and root mean square error. Based on the error rates and scores, the proposed model performs better than any of the individual techniques.

Month	2013	2014	2015	2016	2017
1	1	1	0	0	2
2	0	1	0	0	1
3	0	2	1	2	0
4	0	2	2	4	6
5	5	6	0	5	14
6	19	10	9	27	19
7	28	15	8	27	29
8	27	26	33	54	40
9	17	28	37	23	30
10	9	5	19	6	7
11	2	5	5	2	3
12	4	3	4	0	0

Fig. 1 Malaria dataset

2 Materials and Methods

We undertook detailed experimental and evaluator steps to test the effectiveness of the proposed model. For our experiment, we use a malaria dataset. The dataset is having raw data and stationary data.

2.1 Dataset

The dataset holds the monthly data from January to December for five years that is from 2013 to 2017 and the respective number of malaria positive. This dataset contains raw and stationary data with a total number of 60 inputs (Fig. 1).

The dataset is divided such that 85% of the dataset is used to train the model, and 15% is used to test the dataset to get the performance of the model. The dataset contains two columns month and PV where PV denotes the no. of cases in each month from year 2013 to 2017.

2.2 Model Architecture

The flowchart in Fig. 2 represents the whole architecture of the model which consists of the use of three regression algorithms that is linear regression, support vector machine, and auto-Arima model. The model uses the dataset of malaria which contains the total number of positive cases of malaria from the year 2013 to 2017 for each month that is January to December. The proposed model contains the use of linear regression, support vector machine, and auto-Arima algorithms. The accuracy of prediction is measured through calculating the explained variance score, root mean squared error, and mean squared error. After applying all these models, we took the average of the prediction from all the three regression models and ensemble them together to get the better prediction and less RMSE value. Ensemble modeling is a technique which combines the results of different models to improve the overall

Fig. 2 Model architecture

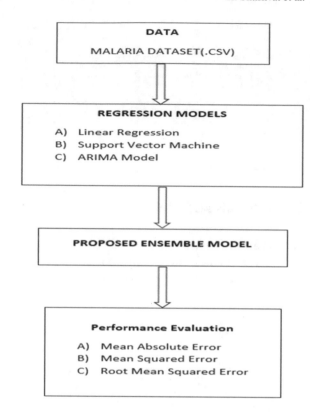

performance. The performance of our model is evaluated in terms of mean average error, mean square error, root mean square error, and explained variance score.

2.3 Models Applied

(A) **Linear Regression**: The term linearity in linear regression refers to the relationship which is linear between two variables. Linear regression is used to determine the value of a dependent variable (y) based on a given independent variable (x). So, regression techniques are used to find linear relation between independent and dependent variables. Figure 3 represents the independent variable (x) on the x-axis and dependent variable (y) on the y-axis, which gives a straight linear line relationship between x and y that best fits the data points. The equation

$$y = a + b * x,$$

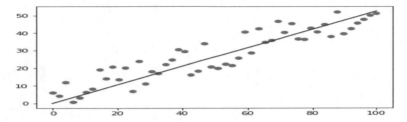

Fig. 3 Linear regression plot from Google Image

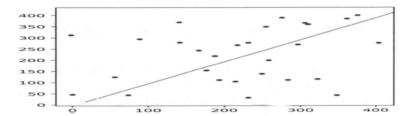

Fig. 4 Support vector machine plot from Google Image

where y is the dependent variable, x is the independent variable, and b is the slope of the line that intercept the y-axis.

(B) **Support Vector Regression**: Support vector machine, commonly known as SVM, is a supervised machine learning algorithm that is helpful for both regression problems and classification problems. Support vector machine aims to find a hyperplane in N-dimension space to classify the data points. There can be a number of hyperplanes that can be chosen to separate two classes of data points. The aim of support vector machine is to find a plane which contains the maximum number of margins. The data points should have maximum distance from the margin to provide some future data points correctly. Hyperplanes are decision boundaries that help classify the data points (Fig. 4).

(C) **Arima Model**: Autoregressive integrated moving average abbreviated as ARIMA is a model that defines a time series which is based on the past experiences that is it predicts the new values based on what it learns from previous time series values. The Arima model is used to forecast future values. The Arima model describes the correlation between data points. An Arima model is described by three terms: p, d, q where p is the order of the autoregressive term, q is the order of the moving average term, and d is the number of differencing required to make the time series stationary (Fig. 5).

(D) **Ensemble Model Using Simple Average**: Ensemble modeling is a technique of combining different types of analytical models and then predicting their results in combined score in regards to maximize the accuracy of predictive analytics. In this method, we take an average of results from all the models

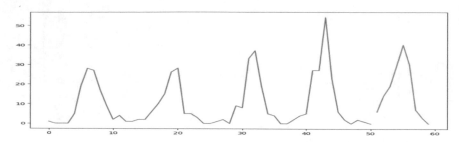

Fig. 5 Arima model plot

and use it to make the final prediction. Algorithm for ensemble learning the above models:

1. Input: Independent variables set (Features) X_i where i is 1, 2, 3…n. Output: Predicting for PV.
2. Using regression on the independent variables, predicting using SVM, LR, and auto-Arima.
3. Take the simple average of the prediction values from each individual model.
4. Use the average of predictions obtained from all three models to compute the final prediction values using the ensemble modeling.
5. Comparison of individual prediction rate with ensemble model error rate.

3 Results and Discussion

Monthly indices of malaria cases in the North India Region from year 2013 to 2017 show an increment of cases from June to September. The cases were higher for these months than the other months in each year.

The model is trained on 85% of the dataset for each model and tested on 15% of data. The correctness of the model is determined by noting the RMSE, MAE, explained variance score's values. Table 1 shows the results of the three regression algorithms that are linear regression, support vector machine, and Arima model.

Table 1 Comparison among prediction models to monthly data of malaria cases in North India region from year 2013 to 2017

Methods	MAE	MSE	RMSE	Explained variance score
Linear regression	5.96	48.16	6.93	0.68
Support vector machine	4.55	36.77	6.06	0.77
Arima model	5.24	52.48	7.24	0.72
Ensemble model using simple averaging	4.06	18.09	4.25	0.89

A comparison of performance among the three regression techniques indicated the SVM model performed the best and has small RMSE and MAE values. But, by introducing the technique of ensemble modeling using simple average, combining the prediction of these three algorithms results in the lowest RMSE and MAE values. In this method, we take an average of predictions from all the models and use it to make the final prediction. Table 1 shows that ensemble model using a simple average performs the best over the given dataset.

Figure 6 shows the plot for the linear regression algorithm and for the given malaria dataset on a monthly basis, and the results for this regression algorithm comes out to be 5.96 MAE, 48.16 MSE, 6.93 RMSE, and variance score is 0.68. The results show that linear regression does not give accurate predictions. Figure 7 shows the plot for the SVM model for which MAE is 4.55, MSE is 36.77, RMSE is 6.06, and variance score is 0.77 which show that the model performs best in all these regression algorithms. Figure 8 shows a plot for the Arima model for which MAE is 5.24, MSE

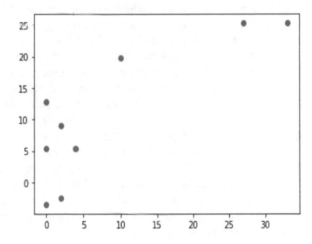

Fig. 6 Prediction for linear regression model

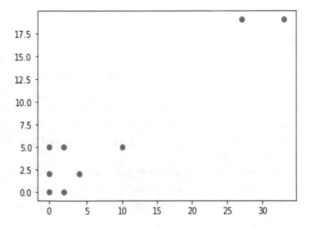

Fig. 7 Prediction for SVM model

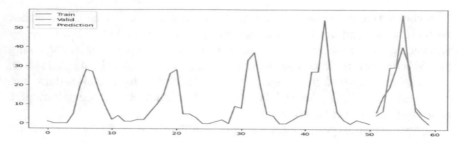

Fig. 8 Prediction for Arima model for malaria dataset where blue curve shows the training set, the yellow curve is the validation set, and the green curve is the prediction set for the given data

is 52.48, RMSE is 7.24, and variance score is 0.72 which perform quite well for the given dataset.

A comparison of performance among the three regression techniques indicated the SVM model performed the best and has small RMSE and MAE values. But, by introducing the technique of ensemble modeling, combining the prediction of these three algorithms results in the lowest RMSE with 4.25, MAE with 4.06, MSE with 18.09, and variance score with 0.89 values. Table 1 shows that ensemble model using a simple average performs the best over the given dataset.

Malaria is one of the most common tropical diseases in Northern parts of India having the most number of cases in the month of June–September due to the monsoon season. Malaria is caused due to the mosquito biting on humans and animals. Poor drainage and sanitation play a major role in the spread of disease.

In the proposed methodology, we contrast the data of malaria cases in the North India region from 2013 to 2017 using the three regression and time series models, and then, we ensembled them together by taking the simple average of their prediction and found that the ensemble model gave the optimal correct predictions and lowest error rates compared to other techniques. It was also noted that malaria cases were highest in the mid of every year.

4   Conclusion

In the proposed study, we contrast the data of malaria cases in the North India region from 2013 to 2017 using the three regression and time series models, and then, we ensemble them together by taking the simple average of their prediction and found that the ensemble model gave the optimal correct predictions and lowest error rates compared to other techniques. It was also noted that malaria cases were highest in the mid of every year.

However, it is to be remembered the merits and limitations of the three regression models for which the model predicts the optimal results. The linear regression makes estimation procedure simple and easy to understand and interpret the linear equation.

Linear regression is prone to overfitting when the data are linearly separable, but in reality, the data are not separable, and the straight line relationship between independent and dependent variables is not true every time. Also, it is prone to outliers. The SVM model provides an optimal solution and allows mapping over the data, but when the dataset becomes large, it requires a large amount of tuning parameters and creates complexity. Arima model is a very popular model and is suitable for all types of time series data. But, if the data are insufficient, then it will not predict accurate results. Future studies should be able to improve these limitations. Also, this model can be used to predict the results for other public health diseases like typhoid. Furthermore, the study should encourage other researchers to carry out different tests on the different time series predictions that explains alertness for the study.

References

1. Malaria, South East Asian Quinine Artesunate, Artesunate versus quinine for treatment of severe falciparum malaria: a randomised trial. The Lancet **366**(9487), 717–725 (2005)
2. M. Poostchi, K. Silamut, R.J. Maude, S. Jaeger, G. Thoma, Image analysis and machine learning for detecting malaria. Transl. Res. **194**, 36–55 (2018)
3. V. Sharma et al., Malaria outbreak prediction model using machine learning. Int. J. Adv. Res. Comput. Eng. Technol. (IJARCET) **4**(12) (2015)
4. P. Kumar Rai, M.S. Nathawat, M. Onagh, Application of multiple linear regression model through GIS and remote sensing for malaria mapping in Varanasi District, India (2014)
5. R. Verma et al., Identification of proteins secreted by malaria parasite into erythrocyte using SVM and PSSM profiles. BMC Bioinform. **9**(1), 201 (2008)
6. Ch. Sudheer, S.K. Sohani, D. Kumar, A. Malik, B.R. Chahar, A.K. Nema, B.K. Panigrahi, R.C. Dhiman, A support vector machine-firefly algorithm based forecasting model to determine malaria transmission. Neurocomputing **129**, 279–288 (2014)
7. S. Suryawanshi, V.V. Dixit, Comparative study of Malaria parasite detection using euclidean distance classifier & SVM. Int. J. Adv. Res. Comput. Eng. Technol. (IJARCET) **2**(11), 2994–2997 (2013)
8. R. Anokye, E. Acheampong, I. Owusu, E. Isaac Obeng, Time series analysis of malaria in Kumasi: using ARIMA models to forecast future incidence. Cogent Soc. Sci. **4**(1), 1461544 (2018)
9. M.I. Musa, Malaria disease distribution in Sudan using time series ARIMA model. Int. J. Public Health Sci. **4**(1), 7–16 (2015)
10. A.E. Jones, A.P. Morse, Application and validation of a seasonal ensemble prediction system using a dynamic malaria model. J. Clim. **23**(15), 4202–4215 (2010)
11. T. Smith, A. Ross, N. Maire, N. Chitnis, A. Studer, D. Hardy, ... M. Tanner, Ensemble modeling of the likely public health impact of a pre-erythrocytic malaria vaccine. PLoS Med **9**(1), e1001157 (2012)
12. K. Shashvat, R. Basu, A.P. Bhondekar, Application of time series methods for dengue cases in North India (Chandigarh). J. Public Health 1–9 (2019)
13. K. Shashvat et al., Comparison of time series models predicting trends in typhoid cases in northern India. Southeast Asian J. Trop. Med. Public Health **50**(2), 347–356 (2019)
14. K. Shashvat et al., A weighted ensemble model for prediction of infectious diseases. Current Pharm. Biotechnol. **20**(8), 674–678 (2019)
15. K. Shashvat, R. Basu, P.A. Bhondekar, A. Kaur, An ensemble model for forecasting infectious diseases in India. Trop. Biomed. **36**(4), 822–832 (2019)

16. O. Sagi, L. Rokach, Ensemble learning: a survey. Wiley Interdisc. Rev. Data Mining Knowl. Discov. **8**(4), e1249 (2018)
17. B. Krawczyk, L.L. Minku, J. Gama, J. Stefanowski, M. Woźniak, Ensemble learning for data stream analysis: a survey. Inf. Fusion **37**, 132–156 (2017)
18. H.M. Gomes, J.P. Barddal, F. Enembreck, A. Bifet, A survey on ensemble learning for data stream classification. ACM Comput. Surv. (CSUR) **50**(2), 1–36 (2017)
19. T. Go, J.H. Kim, H. Byeon, S.J. Lee, Machine learning-based in-line holographic sensing of unstained malaria-infected red blood cells. J. Biophotonics **11**(9), e201800101 (2018)
20. Z. Liang, A. Powell, I. Ersoy, M. Poostchi, K. Silamut, K. Palaniappan, P. Guo, M.A. Hossain, A. Sameer, R.J. Maude, J.X. Huang, CNN-based image analysis for malaria diagnosis. in *2016 IEEE International Conference on Bioinformatics and Biomedicine* (BIBM) (IEEE, 2016) pp. 493–496
21. B.N. Narayanan, R. Ali, R.C. Hardie, Performance analysis of machine learning and deep learning architectures for malaria detection on cell images. in Applications of Machine Learning, vol. 11139 (International Society for Optics and Photonics, 2019), p. 111390W
22. C. Ju, A. Bibaut, M. van der Laan, The relative performance of ensemble methods with deep convolutional neural networks for image classification. J. Appl. Stat. **45**(15), 2800–2818 (2018)
23. T. Strauss, M. Hanselmann, A. Junginger, H. Ulmer, Ensemble methods as a defense to adversarial perturbations against deep neural networks. arXiv preprint arXiv:1709.03423 (2017)
24. E.A. Amrieh, T. Hamtini, I. Aljarah, Mining educational data to predict student's academic performance using ensemble methods. Int. J. Database Theory Appl. **9**(8), 119–136 (2016)
25. P. Bühlmann, Bagging, boosting and ensemble methods. in *Handbook of Computational Statistics* (Springer, Berlin, Heidelberg, 2012), pp. 985–1022
26. T.G. Dietterich, An experimental comparison of three methods for constructing ensembles of decision trees: bagging, boosting, and randomization. Mach. Learn. **40**(2), 139–157 (2000)
27. D. Opitz, R. Maclin, Popular ensemble methods: An empirical study. J. Artif. Intell. Res. **11**, 169–198 (1999)
28. B. Yegnanarayana, *Artificial Neural Networks* (PHI Learning Pvt. Ltd., 2009)
29. A.I. McLeod, W.K. Li, Diagnostic checking ARMA time series models using squared-residual autocorrelations. J. Time Ser. Anal. **4**(4), 269–273 (1983)

IoT-Based Smart City Management Using IAS

Pranali Nitnaware, Supriya Sawwashere, Shrikant V. Sonekar, M. M. Baig, and Snehal Tembhurne

Abstract One of the most important factors in modern information systems of the Indian administrative agencies is the lack of easy access to information through organizations outside of the organization, with the name of the machine-guns in the app. In spite of the abundance of data, and they are locked into proprietary formats, or may not be available due to the lack of an API. In this report, we will focus on the architectural designs that make it easy to share data. Therefore, we focus on a data-based information, see the reference architecture, rather than focusing on the application and communication. Internet of Things (IoT) is a new technology that has emerged in the last couple of years, and it promises to be a continued explosion of data sources. With this technology, it will, in essence, a single "thing" to begin with, the production and transmission of data. For example, each and every lamp post, street lamp, trash bin, power transformer, etc., almost anything that a person can to be created in order to be constantly informed about their disease, their perceptions, and their patterns of use, etc. These data, which, in turn, can be used to lower the level of the business processes of the administration, in order to ensure fast and efficient value-added services. In this report, we will limit ourselves to the architectural design for the integration of new IoT-based systems in a Smart City infrastructure.

Keywords Internet of Things · APIs · Application · Smart City

1 Introduction

It was thought that the combination of data and information that are of critical importance for the development of a new smart app, and it is a core component of the concept of connected data. This means that there is a need to include some additional information about the data, metadata, which the data into a broader context. However, if the digital stream is enriched with additional information such as your device location, device, brand, and the display of the blocks of the intermediate calculations,

P. Nitnaware (✉) · S. Sawwashere · S. V. Sonekar · M. M. Baig · S. Tembhurne
J.D. College of Engineering and Management, Nagpur, Maharashtra, India
e-mail: pnitnaware95@gmail.com

Y.-D. Zhang et al. (eds.), *Smart Trends in Computing and Communications*, Lecture Notes in Networks and Systems 396, https://doi.org/10.1007/978-981-16-9967-2_62

Fig. 1 Main applications of
the IoT

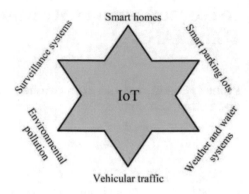

this allows for a richer and more meaningful interpretation of the data, especially for first-time users. Even though it is a technical point of view, it is not difficult to find this information, but in practice, there is no consensus on the standards and/or requirements for the provision of those people. It is a very important aspect of our research is to identify the possible data types (or other systems) that can be used for the Internet of Things. The current implementation of Smart City solutions as is often "isolated," meaning that each application has to provide an end-to-end implementation, which is self-contained and does not work with any other application. In the context of the Internet of Things (IoT) devices that can be integrated on the basis of geographical location and are evaluated with the aid of the analytical system. The sensor of services for the collection of certain data can be used in different types of projects such as monitoring, bikers, cars, and open parking areas. There are many applications in the service sector that is to make use of the Internet of Things, infrastructure, making the operation easier in terms of air pollution and noise, ease of transportation, cars, and video surveillance systems (Fig. 1).

2 Literature Review

2.1 Survey of Existing Methods

Jayti Bhatt, Jignesh Patolia, with the title "Real-Time water quality Monitoring System." In this article, it is stated that in order to ensure access to safe drinking water, the quality of the water needs to be monitored in real time, and an Internet of Things (IoT) is an approach that is based on the monitoring of the quality of the water has been proposed. In this paper, we present a project for the Internet of Things-based water quality monitoring system, which monitors water quality in real time. This system consists of a number of sensors that can measure water quality parameters such as pH, turbidity, conductivity, dissolved oxygen, and temperature. The measured values of the sensors can be processed by a microcontroller, and these values can be remotely transmitted to the main controller, which is a Raspberry Pi with the help of the Zigbee protocol. Finally, the sensor data can be viewed via the app, Web browser, with the help of cloud-based computing.

Michael Lom, Ondrej Pribyl, Miroslav Svitek, under the title of "Industry 4.0", as part of the "Smart Cities." This article focuses on the relationship between the Smart City initiative and the concept of "Industry 4.0." The term "Smart City" has become a phenomenon in recent years, and it is very important, especially since 2008, when the world was hit by the financial crisis. The main reasons for the rise of the Smart City initiative to create a sustainable model of cities, towns, and to the maintenance of the quality of life for those who live in them. The subject of the "Smart City" is not only to be regarded as a technical discipline but in the economic, humanitarian, and legal aspects need to be touched for a while. The concept of Industry 4.0 and the Internet of Things (IoT) is going to be used for the creation of a so-called smart products. The sub-components of the product are in their own minds. Extra intelligence is being used in the manufacturing of the product, and in the post-process, a continuous monitoring of the entire life cycle of the product (intelligent processes. Another important aspect of the Industry 4.0 As a Web of services (iOS), which is true, in particular, to smart transportation and logistics, smart mobility, smart logistics, and energy and Internet access (full), which will determine how natural resources such as electricity, water, olive oil, etc.) correct gebruikt De IoT, iOS, feature, and full, which can be considered as a factor that can lead to a relationship between the Smart City and Industry 4.0—Industry 4.0 can be considered to be a part of the "Smart City."

Zhangwei Sun, Chi Harold Lee, Chatchik Bisdikian, Joel W., Branch, and Bo Yang, whose title is "QOI-Aware energy management in the Internet of Things and Sensory Environments." This article presents an efficient power management system, which provides a useful QOI experience a touch of class-based IoT environments. In contrast to previous efforts, in a transparent manner, and consistent with all of the primary protocols used, and continues to be the energy-efficiency in the long term, and without prejudice to any and reach QOI levels. In particular, the new QOI aware of the concept of the sensor is important for the job" specifically refers to the perceptual features of a touch sensor in the IoT environment, and the QOI requirements, as required by the task. A new concept of "critical transparency" is set up for each task, the selection of sensors for the tasks in the course of time. The power management solution is considered to be dynamic at run-time, to be optimal for the long-term, its traffic statistics, while reducing service time. Finally, to demonstrate the ideas and the algorithms proposed in this paper, a comprehensive sociological study, based on the use of sensor networks for the monitoring of the water levels, is presented, and the effectiveness of the proposed algorithms has been modeled [4].

3 Proposed Algorithm

When the results of a price reduction or assignment, there are multiple clustered machines, more logic programming, in order to access and aggregate the results from each machine, if necessary. In addition, units of work, which must be distributed (correlation) between the machines, so it does not affect the final results? For example, when the identification of the following sentences, a single sentence, it should not be split across multiple nodes, which may lead to recommendations to be split between nodes, and later omitted by the identification of the words in the sentence.

The map and the oppression of the system, and can vary in size, from a single machine to thousands of a cluster of computers, and in some enterprise-class business processes. In the C# programming language provides a set of thread-safe object that can be used quickly and easily to create a map of the reduction, the style, and the programs that are running. The following sections describe some of these objects and provide examples of how to implement a reliable, parallel-card-cutting processes with the help of them.

4 Research Methodology

IoT-end devices and can only respond to a specific vendor's software stack. This will kill the IoT device interactions. In the past, it was possible to create a "dummy" light bulb from a number of different providers; it is now possible to create a "smart" light bulb from just one supplier. This is mainly due to the lack of consistent data, the schema/data model for the IoT devices.

The data from the devices are stored on the system by the supplier in its own format. Therefore, the owner of the center of the city, it is not easy for you to share or re-use the information for any other purpose or to even make them for later use. Visualization/control panel, and device management app, has been closely linked to the devices. For example, it is not going to be easy to develop applications that will be passed to the same IoT devices. An analysis of the use cases of the Indian smart cities (see Appendix A), it appears that some of the sensors and, in particular, cameras, and it can be used in many different areas of application, provided that the tests are well-designed. The current approach to the implementation, it is difficult to do, unless the seller accepts the job.

One of the most important resources in the cities and towns of the data. Easy access to all the data, and will no doubt bring many benefits to the administration, in terms of the definition of "data-driven policy and decision-making." In addition, the access to the data that can drive innovation and new applications of the citizens (Fig. 2).

Since the beginning of the electronic era, many attempts have been made to convince the doctors to enhance the sounds of the heart, in the good faith belief that if the noise level cannot be increased, a greater diagnostic capability can be

Fig. 2 Block diagram

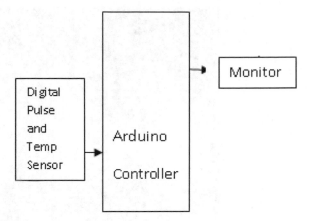

achieved. The rich sounds that can be heard by a physician, by means of a conventional stethoscope to occur at the time of the main valve are closed. In an abnormal heart rate, are there any other comments, it sounds. It had been a whisper is to be heard between the normal voices. The noise is most likely caused by a small opening of the valve, or through the hole in the septum, separates the right and left sides of the heart. Several doctors will be able to hear the same sound, but different interpret. This can lead to an incorrect diagnosis. In addition, it is the exact unit you will be able to reproduce the entire frequency range, the majority of which has been missed by a conventional stethoscope. A tool that was designed in order to use the full range of sound with a precision, and a digital stethoscope to the heart with the aid of appropriate equipment. The extracted signal is then fed to your computer and to detect cardiac abnormalities if any, measurements of physiological parameters, such as heart rate and respiratory rate, are of crucial importance in the medical field. Advances in technology allow for the different measurements for the continuous monitoring of.

Here is an easy way to do it is to measure your respiration with the aid of a displacement of the sensor. This device can be used to monitor your breathing, heart rate, using the appropriate sensor, and heart rate. The response is fast and efficient, as compared with conventional medical devices and equipment. With the help of it, the breath can be measured in a range that can be measured is in the range of 0–999 breaths per minute (Fig. 3).

5 Integrated System

Resource endpoints, providing access to a specific instance or a sub-set of data elements, assets, and will be dealt with by the data warehouses (or databases). For example, you can use these endpoints to request the status of a street lamp or the level of brightness value for the last week, in a particular street lamp. They will handle

Fig. 3 Trans-receiver section

the database queries in the background. It is a type of database that you use to store the resource values are to be specified in the metadata of the source in the catalog, and therefore, the corresponding APIs can be used in applications in order to access the information you need. This is the mechanism you can use to make a bunch for a variety of magazines. Each room is properly configured for the type of the resource. Most of the databases that provide a REST API to interact. Database technology is also likely to be adopted in the future (Figs. 4 and 5).

Fault detection of piezoelectric sensors can only be used for very fast processes under ambient conditions. In fact, there are a lot of applications, which show that the quasi-static measurements of the momentum effect; as an effect, it is located along the neutral axis (y), and the charge will be generated in the direction (x) perpendicular

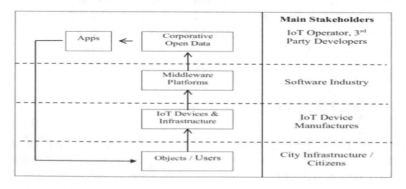

Fig. 4 Working of sections

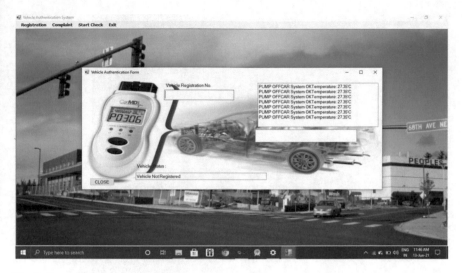

Fig. 5 Generated output

to the line of the armed forces. The amount of energy that depends on the geometrical dimensions of the respective piezoelectric element.

The automatic fuel pump control circuit is composed of a sensor, and a part was built up with the help of the op-amp IC LM324. Op-amps configured here as a comparator. Two of the hard disk drive, the copper wires are inserted in the soil, and to determine whether or not the soil is wet or dry. As the Arduino controller is used to control the entire system, as well as monitoring of the sensor, and if there are more than two of the sensors sense a dry condition, and after, the controller will shut off the engine and turn the engine off if any of the sensors are wet.

6   Patient Monitoring

The LM35 temperature sensor is to help us to get to know the patient's temperature, which has given us the knowledge of the patient's body temperature. The temperature will be, we can judge the condition of the patient, which will help you to decide properly.

A gas pump with the detection, geolocation, which is used in the data-based program that makes use of the supply chain management.

7 Result

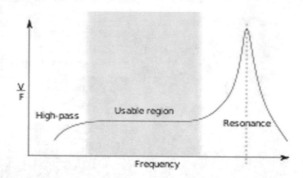

Frequency response of a piezoelectric sensor; output voltage versus applied force.

8 Conclusion

The city should take a more active role to play and will have incentives to make it happen. For example, it is a guarantee of the minimum size, and duration of a service contract, at a lower price, of which the supplier is in very good promotions? In the city, you can detect a variety of prices, for a variety of guaranteed volume levels, and make a decision based on a realistic assessment of what can be done about it. It will also make sure that you use all the different government agencies to coordinate their actions, are much more likely than in the past. In any case, this is a pre-condition for the long-term vision of the Smart City mission.

References

1. V. Albino, U. Berardi, R.M. Dangelico, Smart cities: definitions, dimensions, performance, and initiatives. J. Urban Technol. **21** (2014)
2. L. Arrowsmith, Smart cities: business models, technologies, and existing projects. in *Information Technology Service Research Report* (IHS Technology, 2014)
3. D. Bollier, *How Smart Growth Can Stop Sprawl* (Books, Washington, 1998)
4. A. Bridgwater, Intel Mashery: how to manage an API. *Forbes* (2015)
5. V. Buscher, *Global Innovators: International Case Studies on Smart Cities, BIS Research Paper No. 135* (Ove Arup & Partners Limited, London, 2013)
6. R. Caldwell, Portland, a city of smart growth. The Masthead **54**(1), 29 (2002)
7. A. Caragliu, C. Del Bo, P. Nijkamp, Smart cities in Europe. J. Urban Technol. **2**, 65–82 (2011)
8. E. Castro-Leon, The consumerization in the IT ecosystem. IT Prof. **16**, 20–27 (2014)

9. J. Chambers, Are you ready for the internet of everything? Davos, Switzerland: World Economic Forum, January 15, 2014. https://agenda.weforum.org/2014/01/are-you-ready-for-theinternet-of-everything/. Retrieved 20 Jan 2015

10. H. Chourabi, C. Nam, S. Walker, J.R. Gil-Garcia, S. Mellouli, T.A. Pardo, H.J. Scholl, Understanding smart cities: an integrative framework. in *Proceedings of the 2012 Hawaii International Conference on System Science (HICSS-45)* (2012)

Big Data Analytics and Machine Learning Approach for Smart Agriculture System Using Edge Computing

U. Sakthi⬤, K. Thangaraj⬤, T. Poongothai⬤, and M. K. Kirubakaran⬤

Abstract With the development of the Internet of Things (IoT) and machine learning technology, a smart agriculture environment produces more agricultural land and crop-associated data for knowledge discovery systems. Machine learning decision-making algorithm is applied to discover hidden knowledge patterns from the agricultural data stored in the distributed database. Big data analytics extract useful information from the large, distributed, and complex datasets, which helps the farmer to increase crop yield and quality of the production. The edge computing node collects crop data and land environment data from the agricultural lands using a different kind of IoT sensors. The predicted smart agricultural knowledge pattern can provide needed information to the farmers and other users like an agent, agriculture officers, researchers, and producers to get more profit. Cloud and fog computing provides efficient distributed data storage for big data and execute dynamic operations to predict business intelligence facts to increase production and minimize natural resource utilization. We have compared traditional data mining techniques with the business analytical tool hybrid association rule-based decision tree (HDAT) MapReduce approach for implementing decision tree algorithm to predict and forecast the future needs of the farmer to increase the profit and reduce the resource wastage.

Keywords Precision agriculture · Machine learning · Edge computing

U. Sakthi (✉)
Department of Computer Science and Engineering, Saveetha School of Engineering, SIMATS, Chennai, Tamil Nadu 602105, India
e-mail: sakthiu.sse@saveetha.com

K. Thangaraj
Department of Information Technology, Sona College of Technology, Junction Main Road, Salem, Tamil Nadu 636005, India
e-mail: thangarajkesavan@sonatech.ac.in

T. Poongothai
Department of Computer Science and Engineering, St. Martin's Engineering College, Secunderabad, Telangana 500100, India

M. K. Kirubakaran
Department of Information Technology, St. Joseph's Institute of Technology, Chennai, Tamil Nadu 600119, India

© The Author(s), under exclusive license to Springer Nature Singapore Pte Ltd. 2023
Y.-D. Zhang et al. (eds.), *Smart Trends in Computing and Communications*, Lecture Notes in Networks and Systems 396, https://doi.org/10.1007/978-981-16-9967-2_63

1   Introduction

A computer-based smart agriculture system collects data from the farmer device, IoT sensors, and embedded devices and applies pattern prediction algorithm and deliver required information to the farmers and users. The size of agriculture data collected by Open Government Data (OGD) Platform India, and India Agriculture and Climate Dataset is in the order of gigabytes. The datasets are captured and gathered continuously, so the database size is increased by approximately 92 KB/min in the OGD data repository. The Global Positioning System (GPS) and IoT sensors agriculture dataset consists of crop images, weather images, soil and productivity data, which are stored in a cloud data repository for data analysis. To perform data analysis on large datasets, there is a need for high storage and a powerful computing system to provide efficient and effective data patterns to the users [1]. Big data analytics and cloud computing have been applied to store huge data and perform data analysis in the smart agricultural system [2, 3]. Edge computing is integrated with cloud computing and IoT to provide decisions with low latency time and increase the cloud network bandwidth [4].

Figure 1 shows the smart agricultural model based on IoT and cloud computing technology. Smart farming is the integration of IoT and machine learning process to reduce excessive use of natural resources and reduce the damage of crops. The proposed system provides user-friendly Web-based application to the farmers for acquiring knowledge about the farms and the + environment. Big data are applied everywhere to increase the profit in the business and to analyze the historical data with the following four parameters: volume, velocity, variety, and veracity.

The rest of the paper is organized as follows. Section 2 presents related work in a smart agricultural system using machine learning and edge computing. The architecture and functionalities of the different layers are discussed in Sect. 3. In Sect. 3, the workflow of the proposed system is summarized. Section 4 presents the machine learning algorithm used for data classification. Section 5 describes the

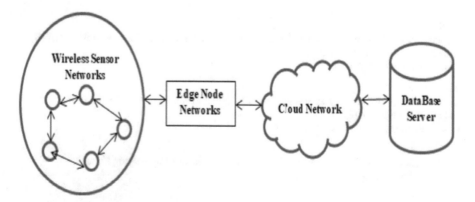

Fig. 1   Smart agricultural decision-making model

experimental setup, execution method, and analyzes the performance of the proposed system. Section 6 concludes the result of the precision agricultural system and future scope.

2 Related Works

Big data analytics and machine learning approach play a significant role in large dataset analysis and efficient decision-making process, which makes the smart agricultural system more profitable, safer, and efficient [5]. Many applications use big data analytics for achieving better profit and make better decisions with less time and cost [6]. Many research works have focused on the knowledge intelligent system for monitoring and controlling the agricultural land using IoT and fog computing with the cloud environment. Edge computing performs data analysis locally without transforming data to the cloud environment, which makes the precision agricultural system provide dynamic knowledge to the farmers and reduce the network latency. The computational load and storage process of the cloud environment are reduced by introducing the edge node in the IoT environment [7]. For example, the farmer can get the knowledge about the humidity of the soil, and he can decide the level of water irrigation to the crops so that the natural resource water can be utilized properly without wastage [8].

Intelligent services are provided to the farmers by applying data classification and prediction algorithms on the large datasets stored in the data repository in the edge node and the decentralized database server [9]. Precision agriculture system includes big data analytics, machine learning classification algorithm, edge computing, and IoT to increase the profit and reduce the wastage. The need for an edge computing nodes in cloud-based environments is analyzed in many research works [10]. The intelligent system does not move all dataset to the cloud server, which transfers partial data to the cloud server, which reduces the network latency and increases the network bandwidth [11]. An effective and efficient smart system is developed to overcome the challenges in the traditional agricultural land management systems. Many researchers and business applications use big data analytics to accomplish great success [12]. A smart agricultural system is a modern application technology to provide, compute, and analyze multisource agricultural data for different management as shown in Fig. 2. It supports water irrigation management, crop management, soil analysis, and fertilizer analysis. The Hadoop distributed file system (HDFS) is used to work with a large amount of data in distributed methods in the cloud network [13]. Google proposed MapReduce programming model for parallel and distributed data analysis approach for large datasets in a cloud environment. The important classification algorithm called decision tree is implemented using the MapReduce processing model to perform data analysis in a distributed manner.

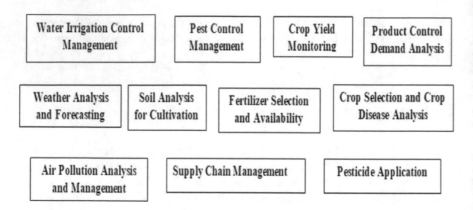

Fig. 2 Functionalities of a smart agricultural system

3 Big Data Analytics in Smart Knowledge-Based Agricultural System

A smart agricultural system includes electronic IoT sensors, edge computing transmission technology, big data analytics, and machine learning. The purpose of this proposed system is to provide required agricultural land data to the farmers and other users without any delay in a precision agricultural environment as shown in Fig. 3. A smart agricultural system consists of four layers, which are the agricultural environment data collection layer, edge computing layer, big data analytics layer, and application layer. All layers are interconnected by the information or data flow. The functionalities provided by each layer are given below.

3.1 Agricultural Environment Data Collection Layer

This layer is responsible for collecting data from sources like the sensors deployed in the agricultural land, Global Positioning System (GPS), and Roadside Unit (RSU) and transfers those data to the next level edge computing layer and cloud server. The data will be in the form of photos, text, images, videos, and pictures.

3.2 Edge Computing Layer

The cloud network connected with the edge computing node via edge gateway using the protocols like Zigbee, Bluetooth, Wi-Fi, NFC, and Ethernet protocol. The edge computing node performs data analysis on the data collected from the data source and sends useful required information to the users.

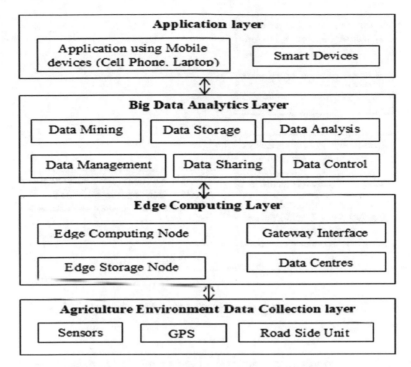

Fig. 3 Layered architecture of the smart agricultural system using big data

3.3 Big Data Analytics Layer

The association rule-based decision tree algorithm is executed to generate knowledge patterns for the farmers. The large volume of data is collected from the data source and stored in the database server for data analysis. The high performance and distributed rule mining algorithm is executed and performs data management.

3.4 Application Layer

The farmer, agent, researcher, scientist, producer, and agent can access the smart agricultural application software using mobile devices connected with the Internet. The user can send a query to the knowledge base server and can receive information about the crop yield, fertilizer availability, pest control, soil details, and water irrigation status.

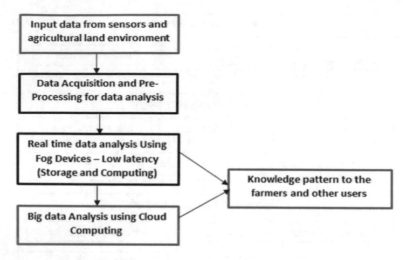

Fig. 4 System framework for big data analytics

4 Machine Learning Approach for the Precision Agriculture

In the big data ecosystem, machine learning technology provides powerful agricultural knowledge prediction algorithm to execute on large datasets. In Fig. 4, we have explained the steps involved in the big data analysis. In this paper, we propose the parallel and distributed association rule mining algorithm and tested with large datasets. The hybrid association rule-based decision tree algorithm is designed to analyze the distributed datasets available in the fog node and cloud computing node. In the supervised machine learning approach, decision trees are widely used prediction algorithm for large datasets. The most widely used supervised algorithm is tree-based algorithm, which gives greater accuracy, easy implementation, and consistency.

5 Performance Analysis

The performance of the proposed has been evaluated by recording the farmer query response time for 6 days. The query processing time is calculated as the sum of user query request time and query response time. Figure 5 shows the various mobile user query response time sent on different days to the cloud environment. QID number represents query identification number, and it is calculated as the time taken by the cloud application service to compute the query and send a response to the farmer user.

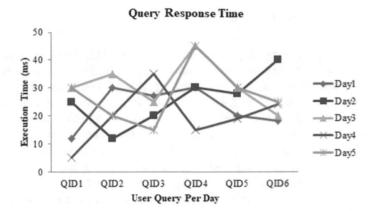

Fig. 5 User query response time

The response can vary based on the type and nature of the query sent by the end user. In Fig. 6, the performance of the various approaches for generating the knowledge from the large datasets is explained. The proposed hybrid association rule-based decision tree-based MapReduce approach gives more accuracy and less error than the other data classification approach. The query response time can vary based on the user request type and nature of the query.

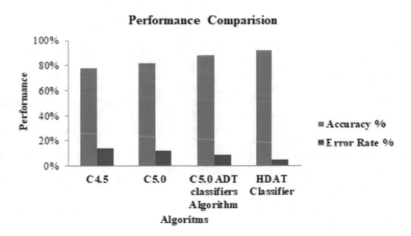

Fig. 6 Performance comparisons of various approaches

6 Conclusion and Future Work

Integration of cloud computing technologies and machine learning approaches swiftly moves traditional agricultural systems to smart agricultural systems. Big data analytics system has been applied in a smart agricultural system to provide knowledge patterns to the farmers to make real-time decisions about farm management. The proposed precision agricultural system is used in the field of science, which leads to maximum crop yield by reducing the usage of resources. The farm management process is optimized with the use of big data and machine learning methods by increasing the crop yield and reducing the resources like fertilizer, water, and pesticide. This system provides the required information to the farmers at the right time and right place. In future, the system can be extended for finding particular crop disease analysis.

References

1. S. Wolfert, L. Ge, C. Verdouw, Big data in smart farming-a review. Agric. Syst. **153**, 69–80 (2017)
2. G.B. Kumar, An encyclopedic overview of "'big data'" analytics. Int. J. Appl. Eng. Res. **10**(3), 5681–5705 (2015)
3. M. Carolan, Publicising food: big data, precision agriculture, and co-experimental techniques of addition. Sociol Ruralis **57**(2), 135–154 (2017)
4. H.R. Zhang, Z. Li, T. Zou, Overview of agriculture big data research. Comput. Sci. **41**(S2), 387–392 (2014)
5. M. Chunqiao, P. Xiaoning, M. Yunlong, Research status and development trend of agriculture big data technology. J. Anhui Agric. Sci. **44**(34), 235–237 (2016)
6. D. Waga, K. Rabah, Environmental conditions' big data management and cloud computing analytics for sustainable agriculture. World J. Comput. Appl. Technol. **2**(3), 73–81 (2014)
7. Z.F. Sun, K.M. Du, F.X. Zheng, Perspectives of research and application of big data on smart agriculture. J. Agric. Sci. Technol. **15**(6), 63–71 (2013)
8. C.S. Nandyala, H.K. Kim, Green IoT agriculture and healthcare application (GAHA). Int. J. Smart Home **10**(4), 289–300 (2016)
9. A. Kamilaris, A. Kartakoullis, F.X. Prenafeta-Boldu, A review on the practice of big data analysis in agriculture. Comput. Electron. Agric. **143**, 23–37 (2017)
10. A.Z. Abbasi, N. Islam, Z.A. Shaikh, A review of wireless sensors and networks' applications in agriculture. Comput. Stand Interface **36**(2), 263–270 (2014)
11. J. Lee, S.H. Kim, S.B. Lee, A study on the necessity and construction plan of the Internet of things platform for smart agriculture. J. Phys. IV **17**(11), 1313–1324 (2014)
12. H. Tian, W. Zheng, H. Li, Application status and developing trend of open field water-saving internet of things technology. Trans. Chin. Soc. Agric. Eng. **32**(21), 1–12 (2016)
13. I. Protopop, A. Shanoyan, Big data and smallholder farmers: big data applications in the agrifood supply chain in developing countries. Int. Food Agribus. Manage. Rev. **19**, 173–190 (2016)

Real-Time Object Detection System with Voice Feedback for the Blind People

Harshal Shah, Meet Amin, Krish Dadwani, Nishant Desai, and Aliasgar Chatiwala

Abstract Lots of people suffer from temporary or permanent disabilities, and one of them is blindness or the visually impaired. According to the World Health Organization (WHO), more than 1 billion people suffer from blindness. Technology may assist blind and visually impaired persons in a number of ways; objects detecting is still a difficult process. There are so many techniques available for object detection. But, the accuracy and efficiency of detection aren't good. Therefore, in this paper, authenticate object detection in real time using the YOLO-v3 and deep learning techniques. The main focus of this paper is to create a smartphone application that is cost-efficient with high accuracy.

Keywords YOLO v3 · Object detection · Machine learning · Text to speech · Deep learning · Image processing · Video processing · Blind people · Computer vision · COCO dataset

1 Introduction

According to the WHO in 2021, around, 1 billion people will suffer from high levels of visual loss. Out of those 1 billion people, 640 million had very low vision, and 39 million was totally blind. Visually impaired persons have complicated challenges while executing things that normal people accept as normal, such as reaching for something, recognizing, and identifying items in their environment. Avoiding impediments and recognizing the things in their environment are a problem for them. The goal of this project is to develop APK deep learning and computer vision technologies that are being used to assist blind persons in overcoming obstacles. The gadget should be able to categorize and identify various things in real time, as well as alert the user of the object class via audio feedback. All of the design and layout considerations are made with visually impaired persons in mind. People's comments and thoughts are considered because they are the ones who will utilize this APK.

H. Shah (✉) · M. Amin · K. Dadwani · N. Desai · A. Chatiwala
Parul University, Vadodara, Gujarat, India
e-mail: harshal.shah@paruluniversity.ac.in

© The Author(s), under exclusive license to Springer Nature Singapore Pte Ltd. 2023
Y.-D. Zhang et al. (eds.), *Smart Trends in Computing and Communications*, Lecture Notes in Networks and Systems 396, https://doi.org/10.1007/978-981-16-9967-2_64

1.1 Deep Learning

Deep learning is a subset of machine learning in artificial intelligence (AI) that uses several layers of neural networks to describe human brain capabilities in data processing and pattern formation for important decision-making. Deep learning is an artificial intelligence (AI) function that comprises networks capable of unlabeled data from an unstructured dataset. Deep learning has grown significantly in the digital world, resulting in an avalanche of data in all kinds of formats and from every corner of the world. Big data can be gathered from a variety of online sources including e-commerce platforms, search engines, social media, and online streaming platforms and others. This huge volume of information is easily accessible and may be shared via cloud computing tools. However, the dataset utilized may well be unstructured and of such a large scale that understanding it, and extracting useful information and patterns might take decades for a human. Deep learning is an unsupervised learning technique that does not require data with specific variables and characteristics. Instead, it uses an iterative approach to get the desired outcome. When dealing with large amounts of data, deep learning typically performs admirably. Deep learning uses neural networks to automatically analyze a large dataset for patterns and tiny correlations. The learnt associations are utilized to interpret new knowledge once the model has been trained using the dataset.

1.2 Computer Vision

Computer vision is an associative subject that analyzes how computers can recognize digital pictures or videos (imagery) at a high level. Although computer vision techniques have been available since the 1960s, there has been a significant improvement in how well the software can explore this type of data due to recent advances in machine learning, as well as leaps forward in data storage technologies, much-improved computing capabilities, and cheap high-quality input devices. Any processing using visual material, such as photos, movies, icons, or pixels in general, is referred to as computer vision. However, there are a few important jobs in computer vision that serve as a framework. A model is trained on a dataset of distinct items in object categorization. The item is identified by the trained model as belonging to one or more of the trained classes. Because the model was trained on a dataset of an object, it should be possible to detect a specific instance of the object in object identification.

1.3 *Object Detection*

Object detection is a computer vision approach for detecting and detecting things in pictures and videos. Object detection may be used to numerical calculations in a scene, determine and monitor their specific location, and effectively label them using this type of identification and localization. Consider the following images: two books, one laptop, and one mobile phone. Object detection helps us to classify the sorts of objects we find while also locating them inside the picture. Bounding boxes are drawn around identified items using object detection. Object detection is strongly connected to other computer vision methods such as image recognition and image segmentation in that it assists in the comprehension and analysis of situations in pictures and video (Fig. 1).

2 Previous Work

There is some previous work already done in the past for helping blind people to help to visualize the surrounding objects with help of object detection techniques. There are some ideas shown below.

- Blind stick navigator
- Smart gloves for the blind
- Talking smart glass blind stick navigator.

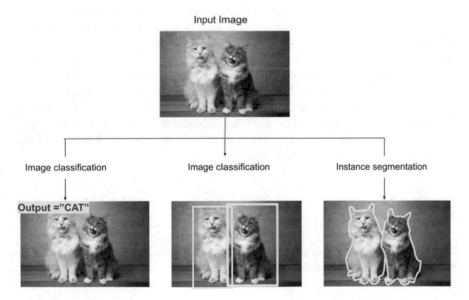

Fig. 1 Object detection

The objective of the present invention is to propose a blind-stick for guiding the blind comprising a location sensor, a micro-controller unit, a temperature sensor, water-level sensor, a heart pulse sensor had a voice module; the present invention relates to the blind stick which includes an electronic sensing mechanism for helping visually impaired people while walking on the road [1]. They have made it smart by having an RF remote which helps to find a stick whenever it is lost.

2.1 Problem and Weaknesses of Current System

(1)   Cost plays an effective role in everything; manufacturing devices include sensors, speakers, and many other tools which increase the cost of the system. Later on, such devices are not affordable for everyone.

(2)   Lifetime support: None of these products gives a lifetime guarantee as some or other day the product needs service. Device can be lost or any issue can come anytime.

(3)   Need to carry the stick or smart-box every time everywhere.

(4)   Low battery can also create an issue at some point.

2.2 Requirements of New System

Blind people or visually impaired people face issues whenever they need to find what they want, or later they need someone's guidance to avoid obstacles or barriers; sometimes, it may happen that guidance is not available; in that case, it becomes difficult for the blind person or their care taker because all of the items and devices mentioned above are expensive and cannot be affordable. To overcome this problem, we are designing an android application that will be available for free for any user. A user should have a smart android phone, and this application will be able to help them in each and every way covering corner things also. This application will work as a guide for blind people, which includes real-time object detection nearby them and informing the user through an assistant and will help them to know the object and the distance between user and object.

3 Objective

The technology uses object detection, and the software will send a message to the user if it finds any possible impediments in the path.

Fig. 2 Object detection code

3.1 Object Detection

For object detection, the tool uses the YOLO-v3 algorithm. For the whole input images, it utilizes a unique neural network. After that the network splits the input images into multiple regions, predicting the bounding regions in the form of boxes with their probability scores (Fig. 2).

3.2 Dataset

Microsoft's Common Object in Context (COCO) dataset [2] is large-scale object detection, segmentation, and captioning dataset. COCO dataset is implemented in this research to train the YOLO model and text reader, which can detect 91 distinct classes. Features of COCO dataset are shown in below.

- 330 K images (>200 K labeled)
- 1.5 million object instances
- 80 object categories
- 91 stuff categories
- 5 captions per image.

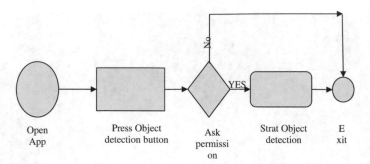

Fig. 3 System work flow

4 Proposed System

The system is built in an android APK that recognizes a variety of objects in real time.

4.1 System Overview

A smartphone is used to capture real-time input data. The APK [3] camera is instantly accessible, and it begins recording the surrounding items, which the application informed them of. To help them learn and navigate their system, the system speaks out about every activity (Fig. 3).

4.2 Implementation

The system is designed to use a combination of technology stacks, which are explained further below. Android Studio is the primary integrated development environment (IDE) built particularly for android application development; it is utilized to develop the app. TensorFlow lite is a lightweight version of TensorFlow that was developed with the mobile operating system and embedded devices in the brain. It gives mobile users a machine learning solution with minimal latency and tiny binary size. TensorFlow includes a collection of fundamental operators that have been optimized for use on mobile devices. Custom operations in models are also supported. The video or image is processed in real time using You Only Look Once (YOLO) [4]. It predicts the observed item in the form of bounding boxes using a single convolutional network that is quicker than the regional convolutional neural network (R-CNN).

Fig. 4 Results

5 Experimental Results and Analysis

The APK [3] can identify and recognize a variety of obstacles that may be encountered when walking, as well as everyday items, and notify the user through assistance. A user must have a smart android phone, and this APK will be able to assist them in any manner possible, even inside and in non-crowded areas (Fig. 4).

6 Conclusion

The suggested solution would be extremely valuable and would improve the user's experience with their android smartphone by offering them assistance. With the user, auto-assistance will play a critical and significant role. Assistance can also be thought of as the user's third eye. Our major goal is to develop an APK [3] that allows blind people or visually impaired people to experience their environment simply by listening to the assistance. As a result, it will aid in the prevention of such mishaps. The mobile devices may be conveniently carried, and the device's camera can be utilized to identify an object in the environment and provide an audio output.

References

1. R.R. Mahmood, M.D. Younus, E.A. Khalaf, N. Thirupathi Rao, Real Time Object Detection for Visually Impaired Person. Annals of R.S.C.B (2021)
2. Microsoft "COCO Dataset". http://cocodataset.org/
3. Android Developer. https://developer.android.com/
4. YOLO weights. https://pjreddie.com/darknet/yolo/

Bird Video Summarization Using Pre-trained CNN Model and Deep Learning

Rachit Adhvaryu, Viral Parekh, and Dipesh Kamdar

Abstract In the past few years, the forms of data have changed drastically from the text formats to images and today, most of the data are available in the video format. With this, there is a huge demand in the techniques that can provide the overall summary of the video. In this paper, we present the summary of birds that are identified from the large datasets using convolutional neural network (CNN). CNN is one of the best image processing and video processing model. The CNN model that we have used for the task is pre-trained AlexNet. The paper clearly proves that the work proposed recognizes the various kinds of birds from the inputted video with accuracy level ranging between 85 and 99%. At last, we provide the overall summary in terms of start time and end time of existence of each bird in the video.

Keywords AlexNet · Convolutional neural network (CNN) · Deep learning (DL) · Machine learning (ML)

1 Introduction

Today, the advancement in the forms of data has changed the complete scenario of research. Earlier, the data was in the form of Text. Later, the data transformed from Text to Images. Today, the data is avalialble in the form of Videos. This has forced the researchers to adopt new technology to generate optimal outputs. Moreover, with the huge datasets in the form of videos, there is an increasing demand in the techniques that can provide the overall summary of the video. Artificial intelligence has played a vital role in changing the thinking power of a humans[1]. The continuous revolution in artificial intelligence forced the humans to adopt its technologies to make their lives more smoother and comfortable.

R. Adhvaryu (✉) · V. Parekh
C. U. Shah University, Surendranagar, Gujarat, India
e-mail: rvadhvaryu@gmail.com

D. Kamdar
V. V. P. Engineering College, Rajkot, Gujarat, India

© The Author(s), under exclusive license to Springer Nature Singapore Pte Ltd. 2023   691
Y.-D. Zhang et al. (eds.), *Smart Trends in Computing and Communications*, Lecture Notes in Networks and Systems 396, https://doi.org/10.1007/978-981-16-9967-2_65

1.1 Artificial Intelligence

Artificial intelligence is also known as machine intelligence; that has capable of interpreting the data in the correct manner, learn that data and put it in execution by flexible adaptation [2]. In short, artificial intelligence commands the machine to try learning the data and generate the desired results. It can easily achievd by developing intelligent systems by learning problem solving techniques of human brains in various situations. The motive of artificial intelligence is to deliver intelligent systems that have an ability of reasoning, learning and problem solving.

1.2 Machine Learning

Machine learning is one of the powerful artificial intelligence application allowing automatic learning of important data from the large sources [3]. The machine learning algorithms create a mathematical model of datasets known as "training dataset", which can be further used in designing and developing intelligent systems. Machine learning also known as Predictive Analytics and it is classified in two types: (a) Supervised Learning, where training dataset contains inputs and its expected outputs determined from inputs and (b) Unsupervised Learning, where training dataset contains only inputs [3].

1.3 Deep Learning

Deep learning differs from machine learning and mainly deals with the algorithms structured from the functioning of neural networks [4]. The execution of deep learning algorithms can be described as problem understanding, dataset identification for detailed analysis, choosing appropriate deep learning algorithm, selecting proper training on large datasets, execution and at last testing it. The major drawback of deep learning is that they learn through interpretations [4] and the deep learning is mainly concerned with datasets.

1.4 Neural Networks

Neural networks identify patterns using some various sets of algorithms. They interpret sensory data through machine observations, clustering or grouping raw inputs [5]. The patterns identified by neural networks are numbers in the form of vectors and all the real-world dataset like images, sound, text or time series are required to be translated to these vector forms. All the operations take place between hidden input and output layers.

1.5 Convolutational Neural Network

A convolutional neural network (CNN) is a deep learning algorithm which expects an image as an input, assigns differetiable learnable weights and biases to different objects known as neurons [6]. For each neuron the output is generated by takeing several inputs, weighing sum over them and then pass it through some functions. CNN requires very low preprocessing as compared to other classification algorithms and methods [6]. CNN has played a vital role in video analytics and is highly recommended where the video processing is required and the ouptuts are expected very quickly as well as accurately. There are lots of convolutional neural network models available that can process the large sets of images and videos in very time span.

2 Related Work

Today, there is a continuos growth in the length and quantity of videos and thus it becomes very difficult for the users to watch each and every video to find the required information from it. Therfore, it is expected that a user must be able to see a summarized video using some kind of tools. Here, summarizing means creating a short story in terms of text, dataset or any format that can provide overall glimpse of the complete video without compromising the main content of the video [7]. Video summarization is one of the best technique to gather all the required informations form the video or its frames [7]. In the following section, we discuss various work on video summarization.

2.1 Audio Visual Video Summarization

This paper proposed to jointly exploit the audio and visual information from the video by developing an Audio Visual Recurrent Network (AVRN) [8]. Moreover, the proposed AVRN is separated into three parts: (1) the two-stream LSTM is used to encode the audio and visual feature sequentially by capturing their temporal dependency. (2) The audio visual fusion LSTM is utilized to fuse the two modalities by exploring the latent consistency between them. (3) The self-attention video encoder is adopted to capture the global dependency in the video [8]. Finally, the fused audio visual information, and the integrated temporal and global dependencies are jointly used in video summary prediction.

2.2 Video Summarization by Learning Relationships Between Action and Scene

This paper proposd a decent deep architecture approach for video summarization in untrimmed videos that parallely recognizes various classes of actions and scenes from the video segemets [9]. This can be achievd through a multi-task fusion approach based on two types of attention modules to explore semantic correlations between action and scene in the videos. The networks contain the feature embedding networks and attention inference networks to describe the inferred action and scene feature representations [9].

2.3 Soccer Video Summarization Using Deep Learning

This paper presented a deep learning approach to summarize a long soccer videos by leveraging the spatiotemporal learning capability of three-dimensional convolutional neural network (3-D-CNN) and long short-term memory (LSTM)—Recurrent neural network (RNN) [10]. The proposed approach involved, (1) a step-by-step development of a Residual Network (ResNet)-based 3-D-CNN that detects a soccer actions, (2) Train 744 soccer clips from five soccer action manually, and (3) training an LSTM network on soccer features extracted by the proposed ResNet-based 3-D-CNN [10]. Thereafter, the 3-D-CNN and LSTM models are merged to identify soccer highlights. At last, to summarize a soccer match video, authors model the video input as a sequential concatenation of video segments based on their validated existence.

2.4 Event-Based Large Scale Surveillance Video Summarization

This paper proposed a new approach of large scale surveillance video summarization based on event detection. Initially, the vehicles and pedestrian paths are gathered using tracking-by-detection technique, and then the abnormal events are detected using the gathered paths. At last, a disjoint max-coverage algorithm is designed that generates a summarized sequence with maximum coverage of required events with minimum number of frames [11]. This approach contains a rich set of features as compared to traditional key frame-based approaches. Firstly, important information can be effectively extracted from the redundant data as the approach is event-centric and the required events contain almost all the vital information. Secondly, the abnormal events are efficiently identified by combining the Random Forest classifier and the features of object's paths and displayed [11].

3 Implementation

3.1 Methodology

The implementation methodology works in the following manner:

1. A large set of dataset available online is chosen as per the requirement.
2. Using deep learning algorithm, this dataset is trained for object detection.
3. At the same time, the video is processed and the frames are generated from the video.
4. In the next step, the trained dataset and the video frames are given to a pre-trained CNN model.
5. Using transfer learning, the CNN model learns about the trained dataset and using its own algorithm, the process of object detection is initiated.
6. The objects whose probability of detection is higer than the pre-defined threshold value are stored in the new dataset.
7. Lastly, the summary of video with time frames of each object is created automatically.

3.2 Work Flow

The complete implementation work flow can be easily described with the below mentioned steps:

Step 1: Take video as an input.
Step 2: Detect the frames with accuracy level above the threshold probability value.
Step 3: Record the first and the last frame for each detected object.
Step 4: Calculate the time in seconds for the first and the last frame and store start time, end time and object name in new data list.
Step 5: Treat this new data list as the final summary.

As discussed earlier, the pre-trained CNN model used in implementing video summarization is AlexNet. The network has deeper architecture, with more filters per layer, and with stacked convolutional layers. AlexNet is designed by the SuperVision group and its training on NvidiaGeforce GTX 580 GPU needs approximately 150 h [12].

AlexNet is deep neural network that is trained over millions of images of ImageNet dataset. AlexNet is 8 layers deep and can classify images into numerous objects like animals, birds, vehicles, pencils, pens, keyboards, etc. AlexNet contains a remarkable set of representations for large sets of images [13].

The working of the algorithm is as follows:

1. A video is chosen from the authenticated online repository.

2. The video is processed into frames and provided as an input to CNN frame by frame.
3. AlexNet labels each object in the frame with its probability and object name.
4. The predicated probabilities are used as a threshold value for object detection.
5. The details of each frame like start time, end time, object name and probability are stored in database.
6. At last, the start time and end time of the objects that are identified with more than 80% accuracy are displayed as a summary of video.

4   Results

We tested our approach on a video that contained more than 25 different birds with a dataset obtained form Kaggle. The length of the video was around 19 min. The main aim was to correctly identify the birds and generate video summary.

The level of accurate results of correct detection was measured on the probability threshold value as described in Table 1. The result with 0.7 and above threshold probability gives accuracy around 98.96% and the accuracy level decreases as the threshold probability value decreases. The decreasing accuracy level results into false detection which is discussed in the later part.

As discussed in Table 1, three birds namely, brambling, gold finch and water ouzel were selected randomly for its correct detection as described in Figs. 1, 2, 3 and 4 based on diffrent threshold probability values.

Table 1 Comparative analysis of threshold probability value versus correct detection

Probability	Total available frames	No. of frames	Correct detection	False detection	Accurate results in %
> = 0.7	27,400	9002	8900	102	98.87
0.6–0.7	27,400	11,188	10,526	662	94.08
0.5–0.6	27,400	14,782	12,701	2081	85.92
0.4–0.5	27,400	19,445	14,837	4608	76.3

Fig. 1 Bird detection with threshold probability value above 0.7

Fig. 2 Bird detection with threshold probability value between 0.6 and 0.7

Fig. 3 Bird detection with threshold probability value between 0.5 and 0.6

Fig. 4 Bird detection with threshold probability value between 0.4 and 0.5

The sample results of the exact time slots of the visibility of each bird in the video is described in Table 2.

It is possible in some cases that AlexNet may detect an incorrect bird with high threshold probability value as described in Fig. 5. This is due to some of the limitations of $11 \times 11, 5 \times 5, 3 \times 3$ matrix used by AlexNet for feature extraction and matching.

Table 2 Sample results

S. No.	Start time	End time	Bird identified
1	0.00	0.20	Brambling
2	0.46	0.52	Goldfinch
3	0.55	0.58	Brambling
4	0.57	1.08	Kite
5	1.12	1.18	Great Grey Owl
6	1.36	1.47	Goldfinch
7	1.47	2.02	Indigo Bunting
8	2.11	2.24	Water Ouzel
9	2.24	2.25	Bulbul
10	2.26	2.39	Water Ouzel
11	2.39	2.56	Kite

Fig. 5 False detection

AlexNet predicts the highlighted portion in Fig. 6. as the body structure and color simialar to bulbul and hence detects brambling as bulbul. However, such frames are very less in number and can be ignored in video summary.

Fig. 6 Reason for false detection

5 Conclusion

Creating a summary of the video saves a lot of time as the user gets exact infromation very quickly from the video. In this paper, the approach to summarize bird video with the help of pre-trained AlexNet CNN and DL, provides about 99% accurate result for identifying birds with detection probability of 0.7 and above. Moreover, 94% accuracy level of results with detection probability of 0.6 and above is achieved. Furthermore, the similar appraoch can be implemented on plant detection, vehicle detection and also summarizing special types of scenes in the movie.

References

1. C. Ujatha, U. Mudenagudi, A study on Keyframe extraction methods for video summary. in *2011 International Conference on Computational Intelligence and Communication Networks (CICN)*, pp.73 77, 7–9 Oct 2011
2. M. Baccouche, F. Mamalet, C. Wolf, C. Garcia, A. Baskurt, Sequential deep learning for human action recognition. in *Human Behavior Understanding* (Springer, 2011), pp. 29–39
3. W. Sabbar, A. Chergui, A. Bekkhoucha, Video summarization using shot segmentation and local motion estimation. in *2012 Second International Conference on Innovative Computing Technology (INTECH)*, pp. 190–193, 18–20 Sept 2012
4. A. Frome, G. Corrado, J. Shlens, S. Bengio, J. Dean, M. Ranzato, T. Mikolov, Devise: a deep visual-semantic embedding model. in *NIPS* (2013), pp. 2121–2129
5. Z. Lu, K. Grauman, Story-driven summarization for egocentric video. in *CVPR* (2013), pp. 2714–2721
6. X. Wang, L. Zhang, L. Lin, Z. Liang, W. Zuo, Deep joint task learning for generic object extraction. in *NIPS* (2014), pp. 523–531
7. J. Wan, D. Wang, S.C.H. Hoi, P. Wu, J. Zhu, Y. Zhang, J. Li, Deep learning for content-based image retrieval: a comprehensive study. in *ACM Multimedia* (2014), pp. 157–166
8. B. Zhao, M. Gong, X. Li, Audio visual video summarization. Computer Vision Open Journal (2021)
9. J. Park, J. Lee, S. Jeon, K. Sohn, Video summarization by learning relationships between action and scene. IEEE Xplore Open Access Journals (2018)
10. R. Agyeman, R. Muhammad, G.S. Choi, *Soccer Video Summarization Using Deep Learning* (IEEE Computer Society, 2019), pp. 270–274
11. X. Song, L. Sun, J. Lei, G. Yuan, M. Song, Event-based large scale surveillance video summarization. Elsevier Journal (2015)
12. C. Lampert, Deep Learning with Tensorflow, [Online]. Available: http://cvml.ist.ac.at/courses/DLWT_W17/. Accessed 12 Feb 2019
13. Alexnet-Pretrained Convolutional Neural Network, [Online]. Available: https://in.mathworks.com/help/deeplearning/ref/alexnet.html. Accessed 13 Feb 2019

Generation of Seismic Fragility Curves for RC Highways Vulnerable to Earthquake-Induced Landslides Based on ICT

Aadityan Sridharan⑩ and Sundararaman Gopalan⑩

Abstract In steep terrains, major highways are laid along steep slopes that are vulnerable to slope failures. More often than not, such mountainous regions tend to be seismically active. Structures located in steep slopes are vulnerable to earthquake-induced landslides and affect the integrity of the highways. We analyze two major landslide events that disrupted major highways in the past and generate fragility curves to estimate the hazard in such a scenario. We derive the curves based on landslides that damaged the highway by measuring the distance from the source of rockfalls. Our results indicate that the RC within 3–5 m from the crest of a landslide tend to be at a greater risk. In an event of higher magnitude, the fragility curves estimate lower probability of damage. This could be because the shaking itself causes enough damage than the triggered landslides. It was also observed for a distance of 10 m from a slope crest in a low-magnitude earthquake, the damage is the most. This shows that the momentum of the triggered rockfall increases due to increased distance and causes substantial damage on the RC highways. Damage caused by Taiwan rockfall event which occurred in April 2019 was assessed using these curves and was found to be moderate. Fragility curves generated in this study can be used to estimate the damage in a future scenario where there are highway constructions along slopes that are prone to rockfalls.

Keywords Building materials · Seismic fragility curves · Earthquake-induced landslides · RC bridges and PGA

A. Sridharan (✉)
Department of Physics, Amrita Vishwa Vidyapeetham, Amritapuri, India
e-mail: aadityans@am.amrita.edu

S. Gopalan
Department of Electronics and Communication Engineering, Amrita Vishwa Vidyapeetham, Amritapuri, India
e-mail: sundar@am.amrita.edu

1 Introduction

Highways that are constructed along major routes are an important means of transport within and among various countries. The highways that are constructed in mountainous regions are prone to seasonal road blocks due to landslides caused by heavy rainfall [1]. Certain such areas evolve into major landslide zones due to subsequent slope failures and deterioration in slope forming materials. When earthquakes occur, tens of thousands of landslides ensue and the additional effect of seismic shaking tends to further complicate the damage [2]. Reinforced concrete (RC) structures that come in the way of such major slope failures are severely damaged by the mass wastage and the incoming seismic waves. Major roads that are constructed using RC are cut off for weeks and sometimes months together [3]. The integrity of the RC highways is very important for continued access in the mountainous terrain.

Research on estimation of earthquake-induced landslide (EIL) hazard is extensive and detailed. Statistical and physical models which evaluate the spatial distribution, landslide displacement and volume, give a complete picture of hazard posed by the EILs. However, the effect of EIL on RC structures, bridges and roads have not been explored in the literature except for a few recent studies [4, 5]. The effects of seismic shaking on RC buildings is a known phenomenon. From the models that describe the damage of earthquakes on RC buildings, a similar model can be enumerated for the damage that EILs can cause to RC structures [6].

Fragility curves define the hazard resistance of a RC structure in the event of seismic shaking [7]. They denote the probability of structural damage beyond a certain state. These curves can be empirical or analytical depending on the model suitable for the problem in hand. Empirical curves are generated from past data that have recorded instances of damage to structures reported by an earthquake event. Analytical curves are based on the following factors/parameters:

1. Response spectrum of an RC structure observed through generated ground motion
2. Simulation of RC structure about to collapse
3. Through simulation of RC structures that account for uncertainty in properties of building materials [8].

Empirical curves can be used in predicting the real-time hazard for events of similar intensity in the future. Analytical curves can estimate the hazard for RC structures in steep terrain based on simulated values and response spectrum of the acceleration records [8]. Although literature point towards hazard analysis based on fragility curves for RC buildings, there have been very few attempts in obtaining a direct model that connects fragility curves to RC Highways and roads due to earthquake induced landslides [9].

In this work, we address the hazard eminent on RC highways due to EIL's by generating fragility curves that estimate the probability of damage. Various levels of hazards have been named as limit states, which will define possible damage to RC structure. The variation in strain on the RC frame for each limit state is mapped with the earthquake intensity to generate probability of hazard due to EILs. It has to be noted that the primary damage due to seismic shaking and the combined effect of shaking with landslides is ignored for this study.

2 Methodology

Various categories of slope failures are observed when landslides are caused by earthquakes. The most damaging of them is rockfall (RF). The damage levels on the highways for rockfall are explained in Table 1 (adapted from Hu et al. [6]). These are different categories of triggered landslides based on the seismic ground shaking. Each case will have a corresponding level of damage which an be defined as the limit state (LS) mentioned in Table 2.

The limit state (LS) described in Table 1 will correspond to various intensity levels of an EIL event. Each of these intensity levels will cause a variation in strain caused on RC frames in the structures. Table 2 gives a comprehensive overview of the possible strain change in RC frames for each LS. Distances of 3, 5, and 10 m from the landslide crest are considered for damage assessment of highways and roads based on literature[10].

Table 1 Description of each limit state (LS). Each limit state is increasing intensity of EILs

Limited state	Failure stage and observed RF due to the earthquake
1	A small amount of rockfall may appear near the highway road cuts
2	There are some small collapses and rockfalls along the valley and highway road cuts
3	Medium and small collapses and rockfalls are common but are rarely large
4	Collapses and rockfalls are widespread, as well as the existence of rock slides

Table 2 Structural damage levels corresponding to each limit state for RC highways

Limited state	Structural damage	Description
1	None to slight	The highway RC material shows a linear response which is elastic. This can generate hairline cracks that are flexural or shear type by nature. Crack width is < 1.00 mm
2	Moderate	There are wider cracks that are almost 1 mm in width and causes spalling in the RC
3	Extensive	Widespread cracks of more than 1 mm is observed; the cracks are flexural/shear by nature. The highway might require longitudinal reinforcement and proper buckling
4	Complete	The RC Highway is now beyond repair, and it is not practically possible. Horizontal seismic displacements shatter the vertical columns due to shear failure. Demolition will be required

Fig. 1 The Zhongbu expressway, **a** before the RF event, **b** the RF video still from dashcam video (from Sridharan et al. [11])

2.1 Limit State and Probability of Damage

Tension crack is the point of initiation of the landslide usually situated near the crest of a slope. Average distance from a tension crack or a crest to the road in mountainous terrain is 5–60 m depending on the construction of the RC Highway as shown in Fig. 1. Figure 1 is an example of a highway from Taiwan where there was a rockfall on April 25, 2019, due to an earthquake of magnitude M_W 6.1 off-coast Taiwan, (a) shows the highway unaffected before the event and (b) is snapshot of the dashcam video showing the boulder falling off [11]. To analyze the impact of RF, we consider 3 m, 5 m, and 10 m distance from highway to crest as three cases to arrive at fragility curves [10]. The probability function that defines the damage possibility at each limit state is a lognormal distribution given by Eq. 1 [10].

$$P(\text{LS} > \text{LS}_j|\text{IM}) = \phi\left(\frac{L_n(\text{IM}) - L_n(\text{IM}')}{\beta}\right) \tag{1}$$

LS$_i$ is the individual limit state where ($i = 1, 2, 3, 4$), IM is the earthquake intensity measured in terms of peak ground acceleration (PGA) in units of g, IM′ is the median value of the same in terms of PGA measured in g, ϕ is the cumulative density function (CDF) which is usually known as standard normal CDF and β is the log-standard deviation. The value of PGA for around 1200 acceleration records were recalculated for slope crests from recording stations. The details in selection of records are explained in the next section.

2.2 Strong Motion Records and Rockfall Inventory

Various records of strong ground motion were analyzed from two major earthquakes: 1. M_w 7.9 Wenchuan, China (2008) and 2. M_w 7.8 Gorkha, Nepal (2015). Landslide inventories of both these events were used to find rockfall incidents along major highways in the respective regions [12]. Over 2000 such incidents were used to

Table 3 Median PGA values for each LS with the RC Highway at a distance of 3, 5 and 10 m from the slope crest

Distance from crest (m)	LS1 (PGA'(g)) $\beta = 0.91$	LS2 (PGA'(g)) $\beta = 0.91$	LS3 (PGA'(g)) $\beta = 0.91$	LS4 (PGA'(g)) $\beta = 0.92$
3	0.26	0.51	0.59	0.7
5	0.17	0.33	0.51	0.66
10	0.0725	0.28	0.46	0.59

arrive at IM' which is chosen to be the median PGA value for the three different distances in each limit states. Table 3 summarizes the median PGA value for each LS, the β values were adopted from Fotopoulou et al. [9]. The records from the events were generated using the Seismoartif software-based input parameters such as hypocenter distance and PGA at outcrop.

Each PGA values that contributed towards damage to the RC Highway were considered in the study with a segregation based on distance from highway. All the failures falling under the three categories were mapped with the PGA map of the event from the United States Geological Survey (USGS) to compare the surface PGA with the PGA from rock outcrop. Since landslides are surficial phenomenon, we considered the surface PGA as opposed to the traditional value of PGA at outcrop.

The accelerograms are generated based on local seismic codes and available sample ground acceleration records. These accelrograms are part of a large set of accelerograms produced in this study based on PGA map to analyze the variation of the accelerations with respect to distance from crest. The variation in acceleration depends on the hypocentral distance and the seismic codes in Nepal and China.

3   Results and Discussion

The fragility curves were generated for each LS and different distance values of 3, 5, and 10 m as shown in Fig. 2. The variation in the probability of each LS with respect to the PGA that can damage the RC Highway is shown in the plot. It can be observed that with higher limit states, the probability of damage due to landslides are reducing. For all the four categories of LS, it is be observed that the 10 m curve gives higher probability of damage; this is consistent with general observation that RF initiated at a farther distance can cause immense damage on the RC Highway. Similarly, with increasing LS, the reason for reduction in probability of damage is because the seismic shaking will primarily contribute to maximum damage and the RF will reach a pre-damaged RC Highway, therefore not contributing to major damage of the structure [2, 5]. On close observation of LS1, the 10 m curve starts at a probability value higher than the 5 m and the 3 m curves. This could be because a slight tremor will retain the integrity of the RC Highway but can dislodge a weak

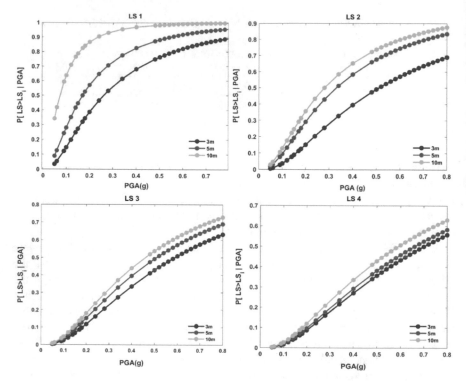

Fig. 2 Fragility curves for landslides caused by earthquakes (Wenchuan and Gorkha) for LS1, LS2, LS3, and LS4

boulder from the crest and as Sridharan et al. [11] have measured; it can cause great damage to the highway.

The LS1 and LS2 curves are showing high-probability values for low values of PGA. At a PGA value of 0.2 g in LS1 the probabilities vary from 0.4 to 0.8 over the three distances and reaches the bare minimum of 0.09 to 0.12. This signifies the effect of PGA in lower LS is profound since any slight acceleration will cause boulders/rocks to hit the highway. On close examination of LS2, we find that the 5 and 10 m curve are more similar than the 3 m curve. The 3 m curve still relates to lower probability which could be because at a short distance, the RF will not have enough momentum to cause serious damage. But in LS3 and LS4, the probability values for 3, 5, and 10 m are very similar. Albeit showing significant variation in probability values, there is not a large variation in trend among the three different distances in LS3 and LS4. As mentioned before, this could be due to damage caused by shaking being more prominent than the RF itself.

The variation in PGA from 0.05 to 0.8 g is the maximum observed range from two of the most destructive events in history [13, 14]. It has to be noted that these are not the only values of PGA recorded in these two events, but these are the values of PGA for which there was a damage caused by RF on the RC Highways. The literature points towards assessments of fragility curves from hypothetical values of PGA ranging from 0 to 2 g [2, 5]. Although such studies can give robust estimates for a wide range of acceleration that can be observed in an engineering structure, for RF originating from shallow earthquakes the fragility curves developed in this study might be more reliable. One of the important aspects of uncertainty that could occur in estimating the fragility on-site is topographic site amplification [15]. As it is outside the scope of this work, we do not show the variations in the probability with increase in amplified PGA values [15]. In such cases, it is good to analyze the stratification of the geology at site before assessing the vulnerability of the RC Highway using the fragility curves [16].

On validating these curves with the Taiwan earthquake mentioned earlier, we found that the LS is LS2 and the estimated distance from the highway to the crest is 10 m [11]. The PGA from generated records was found to be 0.2 g; hence, from the first figure in Fig. 2, there is a probability of 0.87, and the highway will be damaged moderately in reference with Table 2. The field image from the Dashcam video shows the rockfall knocks off the railing of the highway. This was one of the many RFs on the same route; there are news reports that the highway closed for more than twelve hours due to this event. It is worth noting that the results predicted by Sridharan et al. [11] based on the bounce height and the Newmark's displacement confirm that the RF was initiated by the slight tremor due to the Mw 6.1 event. This result is reflected in the fragility curves generated in this study; the low PGA value and a medium magnitude earthquake that corresponds to the lowest LS2 has a better probability of causing damage than a LS4 event with the same PGA.

On a general note, when using these curves for assessing the hazard for a future scenario, it is important to estimate the LS that the future event could possibly belong to. An estimated PGA value can be used to generate possible damage estimates along the highways. In assessing damage on highways that are along fragile slope, geotechnical slope stability models have to be used to identify regions vulnerable to RF along the highways [17]. A quick glance on the fragility curves can give us an estimate of damage possible to a RC Highway. This shows the importance of this work for quickly estimating the probability of damage when there is a possibility of RF. Since two of the largest inventories are included in the calculation, these curves will have a wide range of applications for RF events during an earthquake. We stress that the material in the highway is RC because most of the highways that are paved along steep slopes are observed to constitute the same.

4   Conclusion

Fragility curves that correspond to the various possible scenarios of earthquakes causing RFs are generated. The trend shows that for various LS described in this study, the variation in probability of damage depends on the distance. Farther the distance, more the probability of damage. Further, for small earthquake events that cause shaking just enough to dislodge loose boulders, the damage is observed to be more even though the intensity of shaking might be less. The fragility curves were validated for a RF event in Taiwan that was of LS2 and a PGA of 0.2 g. The predicted probability was as high as 0.87 for a RC Highway at a distance of 10 m from the point of initiation of RF. This was cross verified by the dashcam video shown in Fig. 1, and the damage seems to be pertinent. These fragility curves can be used on a RC Highway to analyze the probability of damage due to such rockfall events.

Acknowledgements The authors would like to thank the Chancellor of Amrita university, Sri. Mata Amritanandamayi Devi for being a constant source of inspiration behind the work.

References

1. L. Jacobs, O. Dewitte, J. Poesen, D. Delvaux, W. Thiery, M. Kervyn, The Rwenzori Mountains, a landslide-prone region? Landslides **13**, 519–536 (2016)
2. S.D. Fotopoulou, K.D. Pitilakis, Vulnerability assessment of reinforced concrete buildings subjected to seismically triggered slow-moving earth slides. Landslides **10**, 563–582 (2013)
3. S. Dhakal, P. Cui, C.P. Rijal, L. Su, Q. Zou, O. Mavrouli, C. Wu, Landslide characteristics and its impact on tourism for two roadside towns along the Kathmandu Kyirong Highway. J. Mt. Sci. **17**, 1840–1859 (2020)
4. A.M. El-Maissi, S.A. Argyroudis, F.M. Nazri, Seismic vulnerability assessment methodologies for roadway assets and networks: a state-of-the-art review. Sustainability. **13**, 61 (2020). https://doi.org/10.3390/su13010061
5. Y. Jafarian, A. Lashgari, M. Miraiei, Multivariate fragility functions for seismic landslide hazard assessment. J. Earthq. Eng. **25**, 579–596 (2021)
6. H. Hu, Y. Huang, Z. Chen, Seismic fragility functions for slope stability analysis with multiple vulnerability states. Environ. Earth Sci. **78**, 690 (2019)
7. J. Kiani, C. Camp, S. Pezeshk, On the application of machine learning techniques to derive seismic fragility curves. Comput. Struct. **218**, 108–122 (2019)
8. J.E. Padgett, R. DesRoches, Methodology for the development of analytical fragility curves for retrofitted bridges. Earthq. Eng. Struct. Dyn. **37**, 1157–1174 (2008)
9. S.D. Fotopoulou, K.D. Pitilakis, Probabilistic assessment of the vulnerability of reinforced concrete buildings subjected to earthquake induced landslides. Bull. Earthq. Eng. **15** (2017)
10. S.D. Fotopoulou, K.D. Pitilakis, Vulnerability assessment of reinforced concrete buildings at precarious slopes subjected to combined ground shaking and earthquake induced landslide. Soil Dyn. Earthq. Eng. **93**, 84–98 (2017). https://doi.org/10.1016/j.soildyn.2016.12.007
11. A. Sridharan, S. Gopalan, Predictive analysis of co-seismic rock fall hazard in Hualien County Taiwan. In: *Machine Learning and Information Processing, Advances in Intelligent Systems and Computing* (Springer Singapore, 2020), pp. 348–358
12. H. Tanyaş, C.J. van Westen, K.E. Allstadt, M. Anna Nowicki Jessee, T. Görüm, R.W. Jibson, J.W. Godt, H.P. Sato, R.G. Schmitt, O. Marc, N. Hovius, Presentation and analysis of a world-wide database of earthquake-induced landslide inventories. J. Geophys. Res. Earth Surf. **122**, 1991–2015 (2017). https://doi.org/10.1002/2017JF004236
13. K. Goda, T. Kiyota, R.M. Pokhrel, G. Chiaro, T. Katagiri, K. Sharma, S. Wilkinson, The 2015 Gorkha Nepal earthquake: insights from earthquake damage survey. Front. Built Environ. **1**, 1–15 (2015). https://doi.org/10.3389/fbuil.2015.00008
14. T. Gorum, X. Fan, C.J. van Westen, R.Q. Huang, Q. Xu, C. Tang, G. Wang, Distribution pattern of earthquake-induced landslides triggered by the 12 May 2008 Wenchuan earthquake. Geomorphology **133**, 152–167 (2011)

15. R. Ramkrishnan, K. Sreevalsa, T.G. Sitharam, Development of new ground motion prediction equation for the North and Central Himalayas using recorded strong motion data. J. Earthq. Eng. 1–24 (2019). https://doi.org/10.1080/13632469.2019.1605318
16. N. Vasudevan, K. Ramanathan, Geological factors contributing to landslides: case studies of a few landslides in different regions of India. IOP Conf. Ser. Earth Environ. Sci. (2016)
17. R. Ramkrishnan et al., Stabilization of seepage induced soil mass movements using sand drains. Geotech. Eng. **48**, 129–137 (2017)

Influence of Student–Teacher Interrelationship on Academic Achievements: A Logit Model

Monika Saini, Asha Choudhary, and Ashish Kumar

Abstract The main concentration in the present work is given to identify the effect of the student–teacher relationship on the academic achievement of university students. For this purpose, data is collected from 292 students of various universities situated in Jaipur, India. The survey was conducted from January 2018 to April 2018. A questionnaire of 36 questions was filled by all respondents enthusiastically, out of which six questions are demographic. Responses have been collected on a five-point Likert scale. In the sample, 55.5% male and 44.5% students are included with 28.1% day-boarding and 71.9% hosteller students. Maximum 52.7% of students belong to the third year, and a minimum of 2.1% students are included from fourth year from various branches like engineering, science, arts and management. The reliability of the questionnaire was checked by Cronbach's Alpha, and logistic regression analysis is performed. Two logistic regression models are developed concerning gender and locality of students, and odd ratios are identified. Finally, it is recommended that findings can be used to enhance the academic achievements of the students.

Keywords Student–teacher relationship · Academic achievements · Logistic regression · Odd ratios · Gender · Locality

1 Introduction

From the ancient era of human civilization, a teacher's role fluctuates from time to time among nations. In the beginning, teachers involved in formal teaching tasks including formulation of curriculum, teaching, preparing lectures, giving lessons and assessing the performance of students. It is always expected from a teacher that he unwraps the thinkers, touch the hearts and hold the hands of students. To achieve these objectives, the teacher must play many roles like historian, sleuth, visionary, symbol, potter, poet, actor and healer. Teaching is more about the humanity inside the teacher rather than the outside the technology. It is considered as a matter of heart instead of

M. Saini · A. Choudhary · A. Kumar (✉)
Department of Mathematics and Statistics, Manipal University Jaipur, Jaipur 303007, India
e-mail: ashish.kumar@jaipur.manipal.edu

© The Author(s), under exclusive license to Springer Nature Singapore Pte Ltd. 2023   711
Y.-D. Zhang et al. (eds.), *Smart Trends in Computing and Communications*, Lecture Notes in Networks and Systems 396, https://doi.org/10.1007/978-981-16-9967-2_67

the head. It is about the characteristics of teachers like identity, integrity, intensity and individuality. It is about commitment and connections with students. Now, teachers become facilitators, role model, assessors, planners, resource-developer and information provider more than traditional teacher. In the current competitive and globalize scenario, students and managements also expect much more from the teachers. In this situation, student–teacher interrelationship plays a key role in creating a learning environment for students. Many researchers suggested some techniques to improve student–teacher relationship like remember the name of students as soon as possible, identify some personal things of each student, discuss on some current issues, provide a positive, caring, and enthusiastic environment, avoid words like threats and punishment, remains unbiased and provide supportive atmosphere. Always try to motivate and encourage the students even though they give the wrong answer to the questions, never demoralize and humiliate them. This positive student–teacher relationship can have a long-lasting impact on the academic growth of the students. Teachers with strong relationships with students are more effective in teaching, faceless behavioural difficulties and achieve higher-level academic success. The following are the key benefits of positive student–teacher relationships:

- Improve academic success
- Prevent behaviour problems in the classroom
- Improve student attitude towards classwork
- Adding growth in and outside the classroom.

Murray [1] discovered a relationship between low-inferences classroom teaching behaviours and student ratings of college teaching effectiveness. Brophy and Good [2] studied teacher behaviour and student achievement relationship. Wang et al. [3] proposed some factors which help students to learn. Astin [4] find the factors that matter in the college life of students. Rawnsley [5] discovered the relationship between homeroom learning conditions, instructor relational conduct and understudy results in auxiliary arithmetic study halls. Evans [6] completed an investigation of understudies' social foundation and educator understudy relational conduct in optional science homerooms in Australia. Brekelmans et al. [7] set up a connection between educator experience and the instructor understudy relationship in the study hall climate. Klem and Connell [8] set up connections and connecting between instructor backing and understudy's commitment and accomplishment. Brok et al. [9] built up a numerical model between relational instructor conduct and understudy results. Bork et al. [10] considered the impact of understudy instructor relational connection on conduct and subject-explicit inspiration. This investigation welcomes insights from research on instructing, learning climate, learning in explicit subjects by associating it to educator's relational conduct with understudy's subject related demeanour. Hanushek and Rivkin [11] clarified a few speculations about utilizing esteem added proportions of instructor quality. Lewis [12] introduced pathways towards improving instructing and learning in advanced education in a global setting and foundation. Arum and Roksa [13] examined the scholastically hapless in school grounds about restricted learning. Loes et al. [14] examined the impact of educator's conduct on understudy's tendency towards long-lasting learning. It is distinguished that controlling for a battery of possible perplexing impacts, educator association was emphatically connected with gains in understudies need for discernment. Mehdipour and Balaramulu [15] researched the impact of educator's conduct on the scholarly development and accomplishments of understudies. The investigation was performed on an example of 1080 respondents taken from 13 colleges of Hyderabad, and a positive relationship has been discovered between instructor's conduct and the scholarly accomplishment of understudies. Hamre et al. [16] built up a structure for instructor adequacy through study hall communications. Hamre et al. [17] introduced some proof for general and space explicit components of instructor youngster communications that are related with preschool kids' turn of events. Chetty et al. [18] built up a model for estimating the effects of educators for assessing predisposition in instructor esteem added gauges. Hafen et al. [19] created factor structure and useful use of study hall appraisal in the scoring framework through communications. Gershenson [20] planned a model for connecting educator quality, understudy participation and understudy accomplishment. Blazar and Kraft [21] completed an investigation on the part of instructor and showing impacts on understudy's demeanour and conduct. A lot of work has been carried out in the direction of student–teacher relationship, but there is no work visualized on which influence of student–teacher relationship on academic achievements among university students. The manuscript comprises four sections. The first section is the introduction in which a detailed literature review is also given. In the second section, material and methods are discussed that are used for analysis. The third section included the results of data analysis, and the last section is the conclusion.

2 Materials and Methods

2.1 Study Design and Participants

It is a cross-sectional study accompanied by the students at various universities situated in Jaipur, the western region of India, with students studying in five faculties including management, sciences, arts, law, journalism and engineering. Respondents comprised undergraduate students at this university. All respondent's voluntarily participated in the survey. The stratified random sampling is used to select the respondents. Respondents were asked to indicate their response on a five-point Likert scale anchored to 1—"strongly agree" and 5—"strongly disagree".

2.2 Data Collection and Questionnaire Development

A questionnaire prepared by Mehdipour and Balaramulu [15] has been used to collect data from respondents. It is used due to its relevance to the objectives of the study. It also ensures that respondents understood the written language of the questionnaire. The questionnaire comprised two sections: one contains six questions related to demographic information and another one has 30 questions related to the student–teacher relationship. The 292 respondents filled the questionnaire and returned.

2.3 Statistical Analysis

SPSS Software version 21 has been used to statistically analyse the data. The categorical data were summarized as frequency and percentage from Table 1. Bartlett [22] developed a methodology for testing the sampling adequacy of questionnaire to perform factor analysis. The reliability of the questionnaire was measured by Cronbach's alpha. It is suggested that if Cronbach's alpha values are greater than 0.6, then the questionnaire is reliable. From Table 2, it is identified that Cronbach's Alpha value is 0.871 that is sufficiently large than 0.6, so data and questionnaire are reliable for statistical analysis.

Table 1 Frequency distribution of respondents

D.V	Type	Frequency	Per cent	Valid Per cent	Cumulative Per cent
Gender	Male	162	55.5	55.5	55.5
	Female	130	44.5	44.5	100
	Total	292	100	100	
Locality	Day-boarding	82	28.1	28.1	28.1
	Hosteller	210	71.9	71.9	100
	Total	292	100	100	
Class	First	46	15.8	15.8	15.8
	Second	86	29.5	29.5	45.2
	Third	154	52.7	52.7	97.9
	Fourth	6	2.1	2.1	100
	Total	292	100	100	
Stream	Engineering	114	39	39	39
	Management	80	27.4	27.4	66.4
	Science	6	2.1	2.1	68.5
	Arts	92	31.5	31.5	100
	Total	292	100	100	

Table 2 Reliability analysis of the questionnaire

Cronbach's Alpha	No. of items
0.871	30

3 Logistic Regression

A technique frequently used to build up a relapse type model for foreseeing a dichotomous ward variable. This technique uses an "s-bend" type capacity to build up a number-related model from at least one factors, which can yield the likelihood of having a place with one gathering or the other. It falls somewhere between regression and discriminant analysis. If an event has two outcomes, then it is termed as binary logit model. A maximum likelihood estimator is used to estimate the parameters in the logit model. The model fit is measured based on likelihood function, Cox and Snell R square and Nagelkerke R square. Wals's statistics are used to test the individual estimated parameters [23–25]. Mathematically,

$$\log_e\left(\frac{p}{1-p}\right) = a_0 + a_1 X_1 + a_2 X_2 + \ldots\ldots + + a_k X_k$$

Or

$$p = \frac{\exp\left(\sum_{i=0}^{k} a_i X_i\right)}{1 + \exp\left(\sum_{i=0}^{k} a_i X_i\right)}$$

where p = probability of success; X_i = Predictor variable i; a_i = unknown parameter to be estimated.

4 Data Analysis

Using statistical software SPSS 21, binary logistic regression analysis has been performed. The Omnibus test Table 3 gives the Chi-square value 142.031 and p-value = 0.000 < 0.05. By observing these values, it is analysed that the null hypothesis, i.e., intercept and all other estimated parameters are zero cannot be accepted.

Cox and Snell's R square and Nagelkerke's R square tell us about the proportion of variance explained by independent variables in the categorical dependent variable. Cox and Snell's R square is based on the log likelihood of the baseline model, and Nagelkerke's R square is the adjusted version of Cox and Snell's R square. From Table 4, it is observed that the model is good but not excellent. Here, it is interpreted

Table 3 Omnibus tests of model coefficients in respect of gender

		Chi-square	Df	Sig
Step 1	Step	142.031	30	0.000
	Block	142.031	30	0.000
	Model	142.031	30	0.000

Table 4 Model summary in respect of gender

Step	-2 Log likelihood	Cox and Snell R Square	Nagelkerke R Square
1	259.253[a]	0.385	0.516

[a]Estimation terminated at iteration number 5 because parameter estimates changed by less than 0.001

Table 5 Classification table in respect of gender

	Observed		Predicted		
			Gender		Percentage correct
			Male	Female	
Step 1	Gender	Male	132	30	81.5
		Female	32	98	75.4
	Overall Percentage				78.8

that 51.6% of students think that there are significant chances that the student–teacher relationship affects the academic achievement of students.

From Table 5, it is analysed that 78.8% correct classification of respondents according to gender-wise have been made in the logistic regression model.

By running logistic regression in SPSS, the number of variables that remains in the logit model is identified. The values for the parameters of logistic regression equation for predicting the dependent variable from the independent variable and odds ratios are estimated, and Wald values are used to test the significance of estimated parameters. By using significant values and a 0.05 level of significance, it is analysed that only seven variables and constant are statistically significant having a p-value less than the level of significance. The Omnibus test Table 6 gives the Chi-square value 96.320 and p-value $= 0.000 < 0.05$. By observing these values, it is analysed that the null hypothesis that constant and all other estimated parameters are zero cannot be accepted. So, the null hypothesis is rejected.

From Table 7, Here, it is interpreted that 40.4% of students in the locality-wise classification think that there are significant chances that student–teacher relationship affects the academic achievement of students.

From Table 8, it is analysed that 83.6% correct classification of respondents has been made in the logistic regression model according to the locality of residence.

Table 6 Omnibus tests of model coefficients in respect of locality

		Chi-square	Df	Sig
Step 1	Step	96.320	30	0.000
	Block	96.320	30	0.000
	Model	96.320	30	0.000

Table 7 Model Summary in respect of locality

Step	-2 Log likelihood	Cox and Snell R Square	Nagelkerke R Square
1	250.417[a]	0.281	0.404

[a]The cut value is 0.500

Table 8 Classification table in respect of locality

	Observed		Predicted		
			Locality		Percentage Correct
			Day-boarding	Hosteller	
Step 1	Locality	Day-boarding	48	34	58.5
		Hosteller	14	196	93.3
	Overall percentage				83.6

The number of variables that remains in the logit model is identified with respect to locality also. By using significant values and a 0.05 level of significance, it is analysed that only seven variables are statistically significant having a p-value less than the level of significance.

5 Conclusion

The assessment of the student–teacher relationship on academic achievement among university students has been made with the help of logistic regression. Two logistic regression models have been developed one for gender and the other for the locality of the students. The models are as follows:

$\log_e\left(\frac{p}{1-p}\right) = 3.398 + (1.540)^*$ your faculty have a command on their subjects – $(0.460)^*$ Your faculties provide you with relevant information/explanations to explain the points of subjects matter $-(0.524)^*$ your faculties behave more nicely to some students $-(0.656)^*$ Your faculties lay emphases on completion of course work in time $+ (0.460)^*$ Your faculties like some students more and favour them unduly $+ (0.489)^*$ Your faculties become nervous when clarifications are sort-$(0.488)^*$ Your faculties use different teaching techniques.

$\log_e\left(\frac{p}{1-p}\right) = 1.657 + (0.883)^*$ Your faculty have a command on their subjects – $(0.589)^*$ Your faculties are friendly and approachable $-(0.737)^*$ Your faculties come to class well prepared for the lecture teaching $-(0.57)$ Your faculties are efficient in their duties $+ (0.580)^*$ Your faculties become nervous when clarifications are sort-$(0.757)^*$ Your faculties use their voice effectively during teaching $+ (0.883)^*$ Your faculties relate the topic with the real-life situation through different examples.

So, this study identified seven important factors that influence the student–teacher relationship and their impact on academic achievements. From the regression analysis, it is revealed that faculty's subject command and faculty's ability to relate the topic with the real-life situation through different examples have a significant positive impact on academic growth. The developed logit models are good as Nagelkerke's R square value is 0.516. Thus, the findings of the above study help improve the student–teacher relationship that results in enhancing the academic growth of students.

References

1. H.G. Murray, Low-inference classroom teaching behaviours and student ratings of college teaching effectiveness. J. Educ. Psychol. **75**, 138–149 (1983)
2. J., Brophy, T. L. Good, Teacher behaviour and student achievement. in *Handbook of Research on Teaching*, ed. by M.C. Wittrock, 3rd edn. (MacMillan, New York, NY, 1986)
3. M.C. Wang, G.D. Haertel, H.J. Walberg, What helps students learn? Educ. Leadersh. (Alexandria, VA) **51**(4), 74–79 (1993)
4. A.W. Astin, *What matters in college? Four critical years revisited* (Jossey-Bass, San Francisco, CA, 1993)
5. D.G. Rawnsley, Associations between classroom learning environments, teacher interpersonal behaviour and student outcomes in secondary Mathematics classrooms. Unpublished Doctoral Dissertation. Curtin University, Science and Mathematics Education Centre, Perth, 1997
6. H.M. Evans, A study of students' cultural background and teacher-student interpersonal behaviour in secondary Science classrooms in Australia. Unpublished Doctoral Dissertation. Curtin University, Science and Mathematics Education Centre, Perth, 1998
7. M. Brekelmans, Th. Wubbels, P. den Brok, Teacher experience and the teacher-student relationship in the classroom environment. in Studies in Educational Learning Environments: An International Perspective, (New World Scientific, Singapore, 2002), (pp. 73–100), ed. by S.C. Goh and M.S. Khine
8. A.M. Klem, J.P. Connell, Relationships matter: Linking teacher support to student engagement and achievement. J. Sch. Health **74**(7), 262–273 (2004)
9. P. den Brok, M. Brekelmans, T. Wubbels, Interpersonal teacher behaviour and student outcomes. School Effectiveness School Improvement **15**(3/4), 407–442 (2004)
10. P. den Brok, J. Levy, M. Brekelmans, T. Wubbels, The effect of teacher interpersonal behaviour on students subject-specific motivation. J. Classroom Interact. 40(2) (2005)
11. E.A. Hanushek, S.G. Rivkin, Generalizations about using value-added measures of teacher quality. Am. Econ. Rev. **100**, 267–271 (2010)
12. K.G. Lewis, Pathways toward improving teaching and learning in higher education: international context and background. New Dir. Teach. Learn. **2010**(122), 13–23 (2010)
13. R. Arum, J. Roksa, *Academically adrift: Limited learning on college campuses* (The University of Chicago Press, Chicago, IL, 2011)
14. C.N. Loes, K. Saichaie, R.D. Padgett, E.T. Pascarella, The effect of teachers behaviour on students inclination to inquire and lifelong learning. Int. J. Scholarsh. Teach. Learn. **6**(2) (2012). Article 7
15. Y. Mehdipour, D. Balaramulu, The influence of teachers behaviour on the academic achievement. Int. J. Adv. Res. Technol. **2**, 2278–7763 (2013)
16. B.K. Hamre, R.C. Pianta, J.T. Downer, J. DeCoster, A.J. Mashburn, S.M. Jones, M.A. Brackett, Teaching through interactions: testing a developmental framework of teacher effectiveness in over 4000 classrooms. Elem. Sch. J. **113**, 461–487 (2013)

17. B. Hamre, B. Hatfield, R. Pianta, F. Jamil, Evidence for general and domain-specific elements of teacher-child interactions: associations with preschool children's development. Child Dev. **85**, 1257–1274 (2014)
18. R. Chetty, J.N. Friedman, J.E. Rockoff, Measuring the impacts of teachers I: evaluating bias in teacher value-added estimates. Am. Econ. Rev. **104**, 2593–2632 (2014)
19. C.A. Hafen, B.K. Hamre, J.P. Allen, C.A. Bell, D.H. Gitomer, R.C. Pianta, Teaching through interactions in secondary school classrooms: revisiting the factor structure and practical application of the classroom assessment scoring system-secondary. J. Early Adolesc. **35**, 651–680 (2015)
20. S. Gershenson, Linking teacher quality, student attendance, and student achievement. Educ. Finance Policy **11**, 125–149 (2016)
21. D. Blazar, M.A. Kraft, Teacher and teaching effects on student's attitudes and behaviours. Educ. Eval. Policy Anal. **39**(1), 146–170 (2017)
22. M.S. Bartlett, A note on the multiplying factors for various chi square approximations. J. Roy. Stat. Soc. B **16**, 296–298 (1954)
23. S. Chatterjee, A.S. Hadi, Regression Analysis by Examples, 5th edn. (Wiley, 2014)
24. R. Myers, *Classical and Modern Regression with Applications*, 2nd edn. (Duxbury Press, Boston, MA, 1990)
25. J.P. Stevens, *Applied Multivariate Statistics for the Social Sciences*, 4th edn. (Erlbaum, Mahwah, N.J., 2002)

Smart Water Metering Implementation

Urja Mankad⬤ and Harshal Arolkar⬤

Abstract Throughout the past years, the Indian government is putting its best efforts in developing smart cities by applying smart solutions for infrastructures and services. Smart water management is one of the key service areas where now it is a high time to work for. Many industries and researchers are contributing to incorporate smart techniques for smart water system like sensor-based monitoring, real-time data transmission and controlling, leakage management, water distribution management and many more. Nevertheless, the design and proposal of such a smart water system are still not fairly standardized. The enormous applications of smart water management still do not have systematic framework to guide real-world design and deployment of a metering system. To address this challenge, a framework is designed and tested with prototype implementation. This paper mainly focuses on minimizing smart water metering application development efforts. The framework uses open-source technologies and has mechanisms like device installation, water usage statistics, bill payments and a distributed data transmission architecture.

Keywords IoT · Smart city · Smart water metering framework · Smart water network · LoRaWAN

1 Introduction

The exponential growth in networking has given rise to Internet of things (IoT) which today touches all aspects of human life. Smart water networks (SWN) have enormous possibility to enable more efficient water usage management [1]. There are heterogeneous IoT devices, sensors and technologies available from different vendors to form smart water networks [2]. The SWN consists of mechanism devised for deployment of smart devices, control the devices, collect the data, store and analyze

U. Mankad
L.J. University, Ahmedabad, India

H. Arolkar (✉)
GLS University, Ahmedabad, India
e-mail: harshal.arolkar@glsuniversity.ac.in

© The Author(s), under exclusive license to Springer Nature Singapore Pte Ltd. 2023   721
Y.-D. Zhang et al. (eds.), *Smart Trends in Computing and Communications*, Lecture Notes in Networks and Systems 396, https://doi.org/10.1007/978-981-16-9967-2_68

the data as part of a smart water network [3]. In this context, this paper discusses different types of water meters that are available in India, how to communicate with these smart devices and collect the water usage data to generate the bill according to the individual customer's consumption of water. Different utility providers want to charge their customers for their water usage where a low-cost highly secure system design is a challenge [4]. To address this challenge, Sect. 2 discusses the related work carried out to design a smart water metering framework which gives the direction to propose a compact framework with low-cost and high-security features. Section 3 describes different components of the proposed smart water metering framework. Section 4 gives the prototype implementation and testing results. Section 5 gives future direction, and Sect. 6 concludes with the future development and scope of implementation.

2 Literature Review

Lloret et al. [5] have given an idea for an IoT-based architecture for smart metering where they have discussed different components and networking technologies which are needed to form a smart metering system for electricity, water and gas. Their review mainly focused on smart meters features, current communication protocols and big data analytics.

Pacheco et al. [1] have discussed about the security in the smart water system. Here, they have presented an IoT-based hierarchical architecture for smart water services, which mainly focuses on the connection type and communication mechanisms among smart meter devices and their end-user applications

Suresh et al. [6] have designed an application that will help to read the meter reading and generate bill for water distribution in large urban areas. Their smart meter system consists of a hardware component called EIM which communicates on radio frequency and have Ethernet stack and port for interface to TCP/IP network or Wi-Fi or GSM/3G/4G router connectivity. Proposed architecture is not for distributed network.

Shahra et al. [7] have identified the challenges of heterogeneous devices used to form a smart water network. Their main aim is to provide a SWN architecture and its components. According to them, the smart water network must have five layers which includes water meters at physical layers which have sensing, controlling layer above it which collects and communicates the water usage data which is then managed, analyzed and displayed at the top most layer.

Gonçalves et al. [8] provide an IoT framework for smart water management, having efficiency in measuring and identifying leaks. They have worked on water usage data analysis to identify the leakage problems in water supply system. They have used complex event processing (CEP) technology and declarative business processes to control the operational activities of water supply systems.

Li et al. [4] studied and classified many literatures considering smart water system framework, but none of them is complete and requires future research work. They provided a conceptual framework of smart water system with smartness and cyber wellness as their metrics. For research recommendations, they are suggesting a systematic framework, cyber security of water meter network, making water metering more robust and end-user data analysis.

Mourtzios et al. [9, 10] have designed a custom-based IoT gateway that can support multiple protocols, which helps to connect different types of water meters and water valves. The gateway communication is bidirectional with the end devices that operate on wM-Bus and LoRa as a wireless communication protocol. They are developing system infrastructure for the management of water consumption in Thessaloniki based on fix wireless network where Internet connectivity is provided via GSM and NB/IoT.

Alvisi et al. [11] have given a solution for smart water metering middleware where interoperable wireless IoT middleware interfacing with smart meters functioning with different protocols. They have used Raspbian Jessie with Raspberry Pi 3 Model B connected with wireless radio modules. M-Bus and LoRa protocols are used as wireless protocol manager. Data commands are sent over MQTT [12].

Fuentes and Mauricio [13] have proposed an architecture design for water consumption system with leak detection algorithm. Here, the data collected from the smart meter are preprocessed by the local server and sent to the cloud for leak detection. For local server, they have designed a gateway with RaspBerian OS on Raspberry Pi and have used Node-RED for preprocessing and data analysis. Mankad, Urja, and Harshal A. Arolkar has provided a detailed study of the open source framework OSGP/GXF for implementing Smart Water Metering [14].

Literature review concluded that components needed for smart water metering network [1, 5–8] are as follow.

2.1 Smart Water Meter and Gateway Devices

Smart water meter is a digital electronic device that collects water consumption at real time and provides data to the water providers. Meters are of different types, but for automatic meter reading, electromagnetic or fully electronic water meters are preferable to collect real-time data [15]. Electronic meters are best suitable for long-distance communication with better data rate as compared to others, but it requires separate power management module. As meters work on battery power, it requires power backup system and energy optimization techniques [5, 15]. Meters must be having some communication protocol implementation which can be used to send data securely on IoT platform. Data can be temporarily stored in the meter itself depending on its storage capacity and battery power. Meter indication for fault tolerance and alarm mechanism should be there. Gateways can be custom made to achieve interoperability between heterogeneous protocols and vendor-specific devices [9].

2.2 Connection and Communication Mechanism

Meter must be connected to a gateway and gateway must be connected to an IoT platform where data storage, data processing and analysis reports can be generated. Connection can be done through a GPRS, GSM, NB-IoT, UMTS, LTE network or through Wi-Fi or Bluetooth connectivity [5]. Communication can be done using RESTful and HTTPs protocols. Request and response objects are sent and receive among the metering device and gateway in a JSON format. DSS and MQTT protocols are also used which is having publish subscribed method to communicate with TLS/SSL security [7]. wM-BUS and LoRa work as IoT radio wave protocol are two noteworthy technologies for communication in smart metering applications [16, 17].

2.3 Data Storage

Water usage data need to be stored and processed at centralized location. It can be cloud-based NoSQL database, or it can be any server-based RDBMS. It will be a huge data, so enough storage capacity is required [5]. Data must be collected real-time and periodically different reports should be generated for water consumption details and billing.

3 Proposed Framework for Smart Water Metering

3.1 Architecture

The architecture of the proposed framework is divided into four layers called device layer, gateway layer, IoT platform and user application layer. Figure 1 shows the layered architecture of the said prototype. The four layers are discussed below:

Smart Water Meter—Device Layer: At this layer, smart water meters will work with flow sensors. Smart meters will have battery power. They will have wireless transceiver which will provide real-time water flow information to the gateway layer. This communication will be carried out using LoRa WAN protocol [18].

Gateway Layer: Gateway will collect water flow information from the meters on real-time basis. These data are stored temporarily in local storage where they are consolidated and then sent to the platform for further processing. It is an edge computing layer which will communicate with the IoT platform using HTTPs protocol.

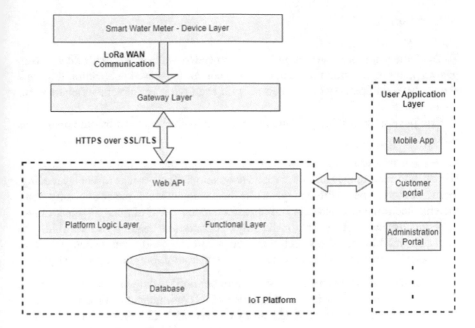

Fig. 1 Smart water metering layered architecture

IoT Platform: IoT Platform composed of further four sub-layers as follow:

Web API Layer: It will provide Web interface to internal logic. Gateways will post the usage data through Web API. User interfacing applications will access metering and other data through this layer.

Platform Logic Layer: This layer will contain logic to receive, aggregate and store usage data.

Functional Layer: It will provide all end-user functions like administration and maintenance, bill generation, reporting and customer portal.

Data Storage Layer: Database will store all the customer information along with their meter assignment details. Monthly water usage will also be recorded for each water meter and its customer. Also, the payment details will be stored against the generated and stored invoices.

User Application Layer: This layer represents different types of user applications from where end users will access the functionalities of platform. Using the mobile application, consumers can see their daily, monthly and year-wise water usage. Also, they can pay their bills at their convenience. Administration dashboard helps the admin of the utility provider to register the users and their meters. Admin can also view the water consumption and payment details of all the users.

3.2 Modules

To implement the proposed framework, we needed water meter with flow sensor, gateway, communication protocol, database and an IoT-based application. Figure 2 shows different components connected with each other to implement smart water meter system.

Components shown in Fig. 2 interact with each other in a step-by-step manner as follow:

Collection of Data: Water flow sensors will sense the amount of water flow per second, and according to that, the meter's magnetic shaft will rotate to display reading. Water meters can be of different types like displacement water meters, velocity water meters, electromagnetic meters and ultrasonic meters. Out of them all, electromagnetic meters are more suitable to display accurate flow reading as its turbidity is high compared to rest of conventional meters and can handle rapid change in flow. Magmeter's performance is not affected by temperature, pressure or viscosity.

Transmission of Data: Customer meter number and reading will be sent from the smart water meter to the gateway using LoRa WAN communication protocol. Per day, two times data are sent which is depending on network load and channel availability.

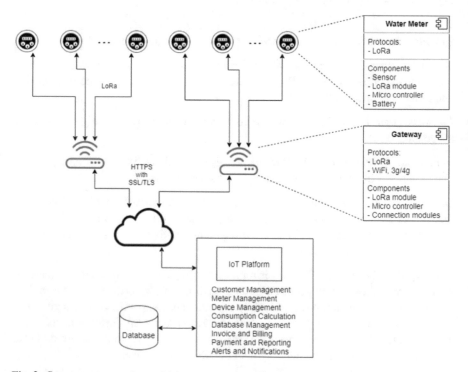

Fig. 2 Smart water metering architecture component diagram

Aggregation of Data: The data received from a meter will be sum aggregated on daily basis at the gateway. This is edge computing, which is useful for minimizing the network load and storage space on cloud server.

Processing of Data: Data will be processed at the IoT platform to generate monthly usage report, which is the bill to be paid by the customer. Processed data will be shown to the customers in the form of line chart and bar graphs.

Storing Data: Customer personal information with their location address and contact details will be stored. Meter information like meter ID, meter type and assigned to which customer will be stored. Daily water usage will be stored with customer's name and meter ID. All these data will be stored in real-time database in cloud.

3.3 Use cases

IoT-based implementation of smart water metering system will have different use cases as discussed below, and Fig. 3 represents the use cases.

Register Meter: When a meter is installed at customer site, he has to register himself in the system where personal details need to be feed. After registration, sign up process is carried out where meter need to be connected using Bluetooth connection where a random key is generated and assigned to meter as a key. That key and customer ID is stored in the database.

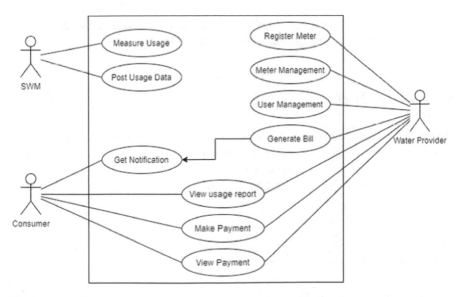

Fig. 3 Use case diagram

User Management: Admin can view each user and their history. Admin can assign meter to the registered user. If the user pays the bill or if there are pending payments, then admin can view details.

View Usage Report: Graphical representation can be used to facilitate the users to analyze their daily and monthly usage. Paid and unpaid bills can be shown in different colors to give alerts to the customers.

Payment: Online payment gateways can be used for the bill payments.

Figure 3 shows different use cases that can be implemented as a minimum requirement to test the proposed framework for smart water metering application. The system will contain variety of users who are connected to the platform for performing different types of operations using platform API. There may be variety of applications like Web portal, mobile applications and so on to facilitate the utility providers and customers.

Smart water meter will just measure and post the data. Meters will be allocated to the customers by the administrator (water provider) using Web portal or any mobile application. Using the application, admin can manage users and can view the customers' usage and payments. Customers should get notifications for their water usage and payment reminders. Different types of reports can be viewed by them for their convenience.

4 Prototype Implementation

To implement smart water metering prototype, the proposed framework is implemented with different components as discussed below.

4.1 Water Meter

Three water meters are taken to implement the prototype. The flow sensor is fitted in line with the water line and measures the flow of liquid that moves through it. Sensor detects each revolution and generates an electrical pulse for every revolution. Total flow of the water can be calculated by counting the number of pulses generated by the sensor.

4.2 Gateway

One gateway module is implemented which is connected with the three water meters and collects its meter reading and stores the data on cloud-based server. Gateway has microcontroller unit having Wi-Fi module and a LoRa module inbuilt.

4.3 Communication Protocol

The meter to gateway communication will occur using LoRaWAN protocol, and the gateway to cloud communication is using Internet.

4.4 Application

To interact with all these components of the system, mobile-based application is developed, where the customers and n domain of utility provider will have their own login to perform above mentioned use cases.

Figure 4 shows the implementation diagram for smart water metering system.

In the implemented system, the meters send the water consumption reading to the gateway periodically. The gateway will collect all three meter's data frequently and aggregate day-wise usage and send it to the database server. Implemented prototype is tested for communication delay and packet lost scenario. It is found that as the readings are collected after every 6 h and aggregated for daily consumption, if any

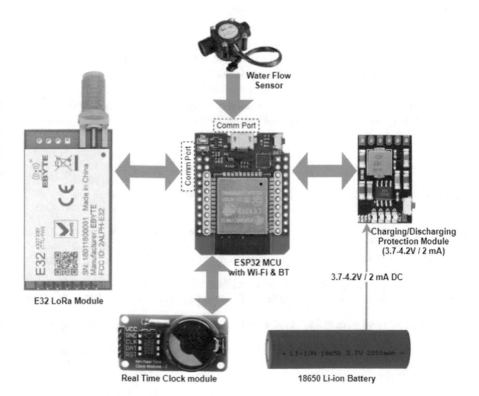

Fig. 4 Implementation diagram for smart water metering system

data packet is lost and meter readings are not taken once a while; then in the next cycle, it will get fresh reading which will not affect the overall water consumption and billing system. LoRa communication delay is found to be just 0.4 s, which is negligible.

5 Future Work

Currently, the framework is tested with a prototype model having just three water meters for metering system, but in future, the framework can be tested for meter management, leakage management, water line fault detection and many more. This prototype can be simulated to test the network load and feasibility to implement the smart water network in the cities.

6 Conclusion

An IoT-based smart water metering framework has been designed and prototype implemented to help the utility providers and water consumers to smoothen out the process of consumption tracking and billing. A mobile app is developed to easily access the data and information from both the ends. The researchers have used an open-source technology, so the prototype implementation cost is not too high and the data are stored and retrieve using a secure cloud platform. Implementing this type of system in highly populated cities of any developing countries like India, where there is much wastage and scarcity of water, we can control the usage and can save our natural resource.

References

1. J. Pacheco et al., IoT security framework for smart water system. in *2017 IEEE/ACS 14th International Conference on Computer Systems and Applications (AICCSA)* (IEEE, 2017)
2. U. Mankad, H.A. Arolkar, Survey of IoT frameworks for smart water metering. Management **24**, 26 (2018)
3. S. Sukode, S. Gite, H. Agrawal, Context aware framework in IoT: a survey. Int. J. **4**(1), (2015)
4. J. Li, X. Yang, R. Sitzenfrei, Rethinking the framework of smart water system: a review. Water **12**(2), 412 (2020)
5. J. Lloret et al., An integrated IoT architecture for smart metering. IEEE Commun. Mag. **54**(12), 50–57 (2016)
6. M. Suresh, U. Muthukumar, J. Chandapillai, A novel smart water-meter based on IoT and smartphone app for city distribution management. in *2017 IEEE Region 10 Symposium (TENSYMP)* (IEEE, 2017)

7. E.Q. Shahra, W. Wu, M. Romano, Considerations on the deployment of heterogeneous IoT devices for smart water networks. in *2019 IEEE SmartWorld, Ubiquitous Intelligence & Computing, Advanced & Trusted Computing, Scalable Computing & Communications, Cloud & Big Data Computing, Internet of People and Smart City Innovation (SmartWorld/SCALCOM/UIC/ATC/CBDCom/IOP/SCI)* (IEEE, 2019)

8. R. Gonçalves, J.J.M. Soares, R.M.F. Lima, An IoT-based framework for smart water supply systems management. Future Internet **12**(7), 114 (2020)

9. C. Mourtzios, K. Galanis, N. Papadimitriou, An experimental study of a multi-protocol IoT gateway. in *2020 Fourth World Conference on Smart Trends in Systems, Security and Sustainability (WorldS4)* (IEEE, 2020)

10. C. Mourtzios et al., Work-In-progress: SMART-WATER, a novel telemetry and remote-control system infrastructure for the management of water consumption in Thessaloniki. in *Interactive Mobile Communication, Technologies and Learning* (Springer, Cham, 2019)

11. S. Alvisi et al., Wireless middleware solutions for smart water metering. Sensors **19**(8), 1853 (2019)

12. OASIS, OASIS MQ Telemetry Transport (MQTT) Technical Committee. https://www.oasis-open.org/committees/mqtt/charter.php. Accessed 10 Nov 2020

13. H. Fuentes, D. Mauricio, Smart water consumption measurement system for houses using IoT and cloud computing. Environ. Monit. Assess. **192**(9), 1–16 (2020)

14. U. Mankad, H.A. Arolkar, A study of the opensource framework OSGP/GXF for implementing Smart Water Metering. in 2020 Fourth World Conference on Smart Trends in Systems, Security and Sustainability (WorldS4) (IEEE, 2020)

15. X.J. Li, P.H.J. Chong, Design and implementation of a self-powered smart water meter. Sensors **19**(19), 4177 (2019)

16. A. Ikpehai et al., Low-power wide area network technologies for Internet-of-things: a comparative review. IEEE Internet Things J. 1 (2018)

17. V. Slaný et al., An integrated IoT architecture for smart metering using next generation sensor for water management based on LoRaWAN technology: a pilot study. Sensors **20**(17), 4712 (2020)

18. LoRa Alliance Organisation, LoRaWAN® Specification v1.1. https://lora-alliance.org/resource-hub/lorawan-specification-v11, Accessed 12 Nov 2020

Artificial Bee Colony Optimization-Based Load Balancing in Distributed Computing Systems—A Survey

Vidya S. Handur and Santosh L. Deshpande

Abstract Distributed computing allows the interoperability of components in distributed system in which software or hardware components located at networked computers coordinate and communicate their actions by message passing. The tasks in such a system are carried out independently. In distributed computing systems, load balancing is one of the issues, which is a means to distribute the tasks such that the computational nodes are neither overloaded nor underloaded, and the performance of the system is improved. Among the various solutions proposed for load balancing, metaheuristic-based algorithms are one of them. This paper discusses the variants and recent developments of artificial bee colony optimization algorithm for solving load balancing in distributed systems. As a result of load balancing, various performance metrics measured are throughput, response time, makespan, CPU utilization, memory utilization, and network utilization. The performance of the optimization algorithms is also measured with the time taken to converge in finding the optimal solution for load balancing.

Keywords Distributed computing systems · Load balancing · Artificial bee colony optimization · Convergence rate

1 Introduction

With the development in technology and increase in the data, getting optimal solutions to data intensive applications are a challenge. To overcome this, there is need for techniques that handle the data efficiently which in turn makes it necessary that all the resources being utilized appropriately. Resource management deals with managing the resources in such a way that all the resources are utilized efficiently to improve the performance of the system. Resource management is achieved by proper resource

V. S. Handur (✉)
KLE Technological University, Hubballi, India
e-mail: vidya_handur@kletech.ac.in

S. L. Deshpande
Visveswaraya Technological University, Belagavi, India

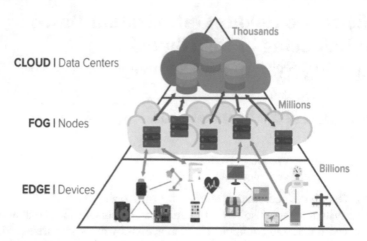

Fig. 1 Distributed system

allocation to the applications so to maximize the efficiency of the system. As the resource allocation varies with requirements of the users, there is a need to check for the availability of the resources and allocate. The resource allocation should be performed such that no resource is either underutilized or over utilized. Load balancing is a technique to discover the inefficiently used or over used resources and to balance the load among the resources proportional to their capacity [1, 2]. Load balancing is a major concern in a network that is distributed in nature. Figure 1 shows various computational nodes that communicate and coordinate in distributed manner to accomplish the task. The major concerns with distributed system are that all the computational nodes are autonomous, and there is no global clock. Other characteristics of distributed system include sharing of resources, unreliability of systems, openness, and heterogeneity of computational nodes. Due to the distributed characteristics, such systems have the advantage of better performance, availability, and scalability [3–5]. There are wide range of applications of distributed systems such as fog computing, edge computing, grid computing, cloud computing, World Wide Web, and automated banking systems.

With the increase in the technology and the data rates, there is need for techniques to handle the huge traffic to enhance the performance of the system. Also due to the differences in the computational capacities and heterogeneity of computing and network resources, the pattern of job arrival and the workloads on each node may vary significantly because of which the performance of the distributed system may deteriorate. Therefore, there is a need to distribute the workloads proportional to the capacity of the nodes. This imbalance of load can be addressed with load balancing.

The paper is systematized as follows: Sect. 2 briefs about load balancing, Sect. 3 introduces the artificial bee colony (ABC) optimization algorithm, Sect. 4 briefs the recent developments of the ABC algorithm for load balancing and followed by conclusion.

2 Load Balancing

Load balancing allows distributing the load on the system such that the computational nodes receive the load proportional to their capacity. The load balancing is either static or dynamic [6–8]. Load balancing is static when the load is assigned to the computational nodes without considering the current load of the nodes and the resources needed to carry out the tasks. Such an assignment is preferable when the load on the system is constant, and it provides good result for homogeneous environments. Due to growth in the technology usage, the traffic is not constant and is increasing at galloping speed, and static method of allocating load to the nodes is inefficient [8]. There is a need for appropriate technique to distribute the traffic among the nodes in the system. Therefore, to meet the current needs of the system, dynamic load balancing is suitable. In dynamic load balancing, the present conditions of the workload on the computational nodes are considered before executing the load. Some of the load balancing methods are round robin, min–max, game theory-based [4], throttled algorithm, etc. Nature-inspired metaheuristic techniques have also been proposed [9]. These techniques have the characteristics that they are decentralized, self-organizing, agent-based, flexible, robust, scalable, and adaptive to changes in the network. Due to ease of implementation, swarm intelligence-based techniques have been extensively employed [10].

With dynamic load balancing algorithms, there are several research questions that need to be addressed such as how frequently the load balancing should be invoked, which host initiates the decision of load balancing, how to collect the load information of the nodes, and migration of load between the hosts. Various researchers have proposed many different solutions for load imbalance. With advancement in the technology, load balancing has become more significant in fog and edge computing. The goal of load balancing is to improve the overall performance of the system by reducing the load imbalance. The various performance metrics measured are the response time, makespan, resource utilization, fault tolerance, throughput, and migration time. This paper focuses on how the artificial bee optimization algorithm and the metaheuristic-based solution can be applied for balancing the load in distributed systems.

3 Artificial Bee Colony Optimization

Artificial bee colony (ABC) optimization algorithm was originally published by Karaboga in 2005 for numerical optimization problems. ABC optimization algorithm is a nature-inspired, swarm intelligence, metaheuristic algorithm that is based on the foraging behavior of bees [11–13]. According to Ullah et al. [14], the ABC algorithm is adaptive to the heterogeneous environment and also to the varying nature of load [15]. The algorithm has few advantages over other swarm intelligence-based algorithms. The advantages are that the algorithm uses very few control parameters, and it is simple and robust, can be easily hybridized with other algorithms of optimizations, and has fast convergence rate [16]. The algorithm also has better exploration capability as compared to other swarm intelligence algorithms [17]. Some of the application areas of ABC algorithm are cluster analysis, software testing, cluster problem optimization, structural optimization, multilevel thresholding, MR brain image classification, advisory system, numerical assignment problem, bioinformatics, face pose estimation, parameter estimation in software reliability models, wireless sensors [18], big data analytics, edge and fog computing [12, 19].

3.1 Phases of Artificial Bee Colony Optimization Algorithm

Artificial bee colony optimization algorithm has three types of bees: employed bees, onlooker bees, and scout bees.

- Employed bees: The task of the employed bees is to arbitrarily search for food source which represents the potential solutions for the problem. These bees share the information of food source by dancing in the bee hive's dance area. This shared information specifies the quality of solution. The number of employed bees and the number of food sources for the bee hive are equal.
- Onlooker bees: These are the bees waiting in the dance area to assess several dances before choosing a position of food source. The selection is based on the probability proportional to the quality of that food source.
- Scout bees: These bees randomly look for possible new food source [13, 14, 20]

  Figure 2 represents the flow diagram of basic ABC optimization algorithm.

4 Recent Developments of ABC Optimization for Load Balancing

With the increase in data and variations in the data rates, there is a need to manage the varying traffic across the system. Load balancing is a technique to balance the load so that all the resources are utilized efficiently. There are several solutions proposed for load balancing. This section introduces few of the recent solutions proposed using ABC optimization algorithm.

The authors in Kruekaew et al. [6] have simulated ABC optimization algorithm for cloud computing by applying a heuristic method which uses certain rules or random methods to find an optimal solution. In the proposed approach, the heuristics is based on the priority for the process selection in the system. The priority criteria used are first come first serve, shortest job first, and largest job first. The system is tested under homogeneous and heterogeneous environments and has compared with other optimization algorithms like ant colony optimization and particle swarm

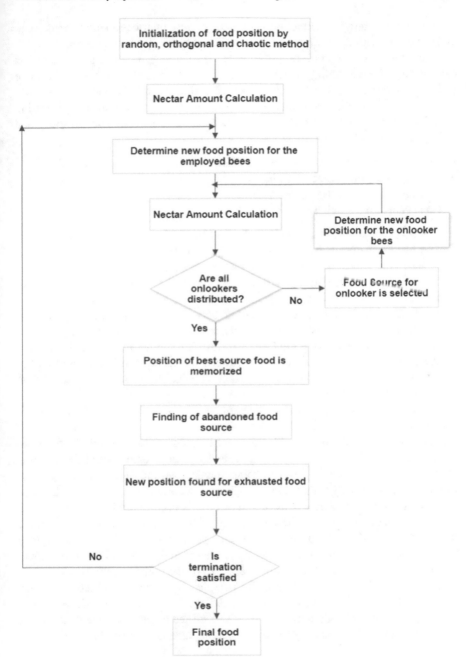

Fig. 2 Flow diagram of ABC optimization algorithm

optimization. In the proposed solution, the performance metric measured is makespan and algorithm with largest job first outperforms the other priority criteria.

According to the solution proposed in Hashem et al. [21], the task allocation is carried out by determining the processing time variation of each virtual machine (VM) with respect to the average processing time of all the virtual machines. The proposed solution determines the utilization of a virtual machine by setting a predefined threshold. The system is modeled to be non-pre-emptive. To solve load imbalance, some of the parameters determined are

- The processing time of host and each VM
- Average processing time of all hosts
- Average processing time of all VMs
- Load standard deviation.

The proposed algorithm is compared with round robin and throttled algorithms. The performance metric measured is the response time and the task execution time, and honey bee-based solution outperforms the other algorithms.

The solution proposed in Bhavya et al. [22] is that the authors assign the tasks dynamically according to the change in the user's demands. The algorithm works by grouping the virtual servers, and each virtual server maintains a queue of processes. After processing the request, the profit is determined. With high value of profit, the server stays and low value of profit causes a return to foraging. In this method, there is an overhead in computing the profit which affects the throughput. The load balancing is initiated by assigning the load to a virtual machine (VM) which is underloaded and with maximum throughput. If the load remains below the preset threshold, the load is assigned otherwise a VM with next highest value of throughput is selected. The performance metric measured is throughput.

The authors Korat and Gohe [23] proposed a solution based on multi-objective optimization Pareto dominance, and the weighted sum method is used for selecting the optimal virtual machine (VM) for balancing the load and for priority assignment to the task. All the virtual machines are grouped according to their load. The tasks considered are the pre-emptive tasks, and priority is used to migrate the tasks from one VM to another. To calculate the priority of the task, the parameters used are latency time of task, user type of task, expected priority of the task, and the length of the task. The performance metrics measured are the makespan and the throughput.

In the proposed study, Mallikarjuna et al. [24], the authors initially set various parameters for the virtual machines. The parameters set is million instructions per second (MIPS) which is a measure of the speed of CPU, storage space for each VM, RAM space, and the bandwidth. Each task is identified by the parameters task id and length of the task. The capacity of any virtual machine is calculated as in Eq. 1.

$$\text{Capacity} = \text{MIPS} \times \text{Number of CPUs} + \text{Bandwidth} \tag{1}$$

To balance the load, an iteration method is adopted to select the most appropriate virtual machine, and a newton gravity function is used. The performance metric measured is the resource utilization.

Table 1 Performance metrics measured

Publication	Performance metrics				
	Response time	Makespan	Resource utilization	Throughput	Task execution time
[22]				✓	
[21]	✓				✓
[6]		✓			
[23]		✓		✓	
[24]			✓		

Table 1 summarizes the performance metrics measured by applying ABC optimization algorithm to balance load.

5 Conclusion

With the advancement in technology such as fog, cloud, and edge computing, load balancing or the offloading plays a significant role. There are a number of solutions to reduce the load imbalance in the network, however, there is still a requirement for the load balancing. Many researchers have proposed swarm intelligence-based solutions for load balancing. This paper is an attempt to show the recent developments of artificial bee colony optimization algorithm to solve load imbalance in distributed system. Among various performance metrics measured, convergence rate of the algorithm is one of them which needs to be addressed.

References

1. S. Handur Vidya, R.M. Prakash, Response time analysis of dynamic load balancing algorithms in cloud computing. in Fourth World Conference on Smart Trends in Systems, Security and Sustainability (WorldS4), vol. 2020 (2020), pp. 371–375
2. S. Handur Vidya, S. Deshpande, P.R. Marakumbi, Particle swarm optimization for load balancing in distributed computing systems–a survey. Turk. J. Comput. Math. Educ. (TURCOMAT) **12**(1S), 257–265 (2021)
3. V.S. Handur, S. Belkar, S. Deshpande, P.R. Marakumbi, Study of load balancing algorithms for cloud computing. in *2018 Second International Conference on Green Computing and Internet of Things (ICGCIoT)* (2018), pp. 173–176
4. S. Penmatsa, A.T. Chronopoulos, Game-theoretic static load balancing for distributed systems. J. Parallel Distrib. Comput. **71**(4), 537–555 (2011)
5. S.B. Kshama, K.R. Shobha, A novel load balancing algorithm based on the capacity of the virtual machines. ICACDS CCIS **905**, 185–195 (2018)
6. B. Kruekaew, W. Kimpan, Enhancing of artificial bee colony algorithm for virtual machine scheduling and load balancing problem in cloud computing. Int. J. Comput. Intell. Syst. **13**(1), 496–510 (2020)

7. A. Hanamakkanavar, V.S. Handur, Load balancing in distributed systems—a survey. Int. J. Emerg. Technol. Comput. Sci. Electron. (IJETCSE) **14**(2) (2015). ISSN: 0976-1353

8. S.P. Belkar, V. Handur, Comparative study of static load balancing algorithms in distributed system using cloudsim. Int. J. Adv. Res. Basic Eng. Sci. Technol. (IJARBEST) **5**, 26–30, (2017)

9. S.T. Milan, L. Rajabion, H. Ranjbar, N.J. Navimipou, Nature inspired meta-heuristic algorithms for solving the load-balancing problem in cloud environments. Comput. Oper. Res. 159–187 (2019)

10. C. Bansal, A. Gopal, A.K. Nagar, Analysing convergence, consistency, and trajectory of artificial bee colony algorithm. IEEE Access **6**, 73593–73602(2018). https://doi.org/10.1109/ACCESS

11. Sudha Senthil Kumar, K. Brinda, R. Rathi, Angulakshmi, Jothi, Yash Vardhan Thirani: Honeybee foraging algorithm for load balancing in cloud computing optimization. Int. J. Eng. Sci. Comput. 7(12) (2017)

12. S.S. Ilango, S. Vimal, M. Kaliappan, Optimization using artificial bee colony based clustering approach for big data. Clust. Comput. **22**, 12169–12177 (2019)

13. D. Karaboga, An idea based on honey bee swarm for numerical optimization. *Technical Report TR06* (Erciyes University, Engineering Faculty, Computer Engineering Department, 2005)

14. A. Ullah, N.M. Nawi, J. Uddin, S. Baseer, A.H. Rashed, Artificial bee colony algorithm used for load balancing in cloud computing: review. IAES Int. J. Artif. Intell. (IJ-AI) **8**(2), 156–167 (2019)

15. A.L. Bolaji, A.T. Khader, M. Azmi Al-Betar, M.A. Awadallah, Artificial bee colony its variants and applications: a survey. J. Theor. Appl. Inf. Technol. **47**(2) (2013)

16. S. Anam, Multimodal optimization by using hybrid of artificial bee colony algorithm and BFGS algorithm. IOP Conf. Ser.: J. Phys.: Conf. Ser. **893** (2017)

17. D. Yazdani, M.R. Meybodi, A novel artificial bee colony algorithm for global optimization. in *International Conference on Computer and Knowledge Engineering* (2014)

18. S. Kaswan, S. Choudhary, K. Sharma, Applications of artificial bee colony optimization technique: survey. in *2015 2nd International Conference on Computing for Sustainable Global Development (INDIACom)* (2015), pp. 1660–1664

19. S.A. Zakaryia, S.A. Ahmed, M.K. Hussein, Evolutionary offloading in an edge environment. Egypt. Inf. J. (in press)

20. L. Shen, J. Li, Y. Wu, Z. Tang, Y. Wang, Optimization of artificial bee colony algorithm based load balancing in smart grid cloud. in IEEE Innovative Smart Grid Technologies—Asia (ISGT Asia), vol. 2019, (2019), pp. 1131–1134

21. W. Hashem, H. Nashaat, R. Rizk: Honey bee based load balancing in cloud computing. KSII Trans. Internet Inf. Syst. **11**(12) (2017)

22. V.V. Bhavya, K.P. Rejina, A.S. Mahesh: An intensification of honey bee foraging load balancing in cloud computing. Int. J. Pure Appl. Math. **114**(11), 127–136 (2017)

23. C. Korat, P. Gohe, A novel honey bee inspired algorithm for dynamic load balancing in cloud environment. Int. J. Adv. Res. Electr. Electron. Instrum. Eng. 4(8), (2015)

24. B. Mallikarjuna, P. Venkata Krishna, A nature inspired bee colony optimization model for improving load balancing in cloud computing. Int. J. Innov. Technol. Exploring Eng. (IJITEE) 8(2S2) (2018). ISSN: 2278-3075

Distributed and Scalable Healthcare Data Storage Using Blockchain and KNN Classification

Manjula K. Pawar, Prakashgoud Patil, and Amit Singh Patel

Abstract Medical data from patients is saved electronically in the healthcare industry. The paper suggests storing medical data on a distributed on-chain utilizing InterPlanetary File Storage (IPFS) and blockchain. The use of InterPlanetary File Storage can help an organization manage the storage space efficiently. The proposed method gives details about the intersection of artificial intelligence and blockchain. This proposed approach eliminates the requirement for a trusted centralized authority, intermediates, and transaction records, resulting in increased efficiency and safety while retaining high integrity, scalability, reliability, and security. It also emphasizes smart contracts to manage and control all interactions and transactions between all healthcare participants. Using artificial intelligence, a revolutionary new system can be built in which a huge amount of medical data can be processed efficiently and can be used to predict diseases. The paper utilizes the assistance of binary classification method by K-nearest neighboring algorithm for machine learning to foresee the patient's heart health. Using InterPlanetary File Storage (IPFS) improves the storage system framework by making data distributed and, subsequently, more reliable, scalable, and accessible to the clients.

Keywords Artificial intelligence · Machine learning · Blockchain · InterPlanetary File Storage · KNN classification · Scalability

1 Introduction

A lack of patient agency, slow access to the medical records, a lack of optimized system interoperability, and the quality and amount of data for medical research are all issues that blockchain and machine learning technology can help. Blockchain technology facilitates and secures data storage easier and more secure. Doctors can

M. K. Pawar (✉) · P. Patil · A. S. Patel
KLE Technological University, Hubballi, Karnataka, India
e-mail: manjulap@kletech.ac.in

P. Patil
e-mail: prakashpatil@kletech.ac.in

© The Author(s), under exclusive license to Springer Nature Singapore Pte Ltd. 2023 741
Y.-D. Zhang et al. (eds.), *Smart Trends in Computing and Communications*, Lecture Notes in Networks and Systems 396, https://doi.org/10.1007/978-981-16-9967-2_70

see a patient's whole medical history. Still, researchers can only access statistical data rather than any personal information. Machine learning may use this data to discover trends and make accurate predictions, providing more assistance to patients and research domains where accurate data is necessary to predict trustworthy outcomes. Blockchain technology offers a decentralized storage method that allows a medical record of any patient registering themselves to the network to be easily shared between peers in a healthcare system (mainly targeting hospitals having more than one center to share the data among them). This technology ensures data consistency, chain integrity, block immutability, and data privacy. Those are all crucial necessities in today's medical record storing system.

Though blockchain has better features, it faces certain challenges such as scalability [1]. There are several solutions for improving scalability for blockchain systems [1–3]. IPFS is a Merkle tree-based peer-to-peer (P2P) distributed storage format for storing huge data in blocks. IPFS preserves data with its content-addressed hash in a hash table which is distributed and using the history of version control. It removes duplicates. IPFS is a block storing system with high throughput, transaction security via hash mapping, and network peers' concurrent access to transactions. To access the data next time with guaranteed quick access, it stores local hash values of requests with the high frequency.

The convergence of AI and blockchain technology could lead to substantial breakthroughs in healthcare, commercial development, and research. The proposed method uses AI-ML and blockchain technology in the dataset. The decentralized AI is chosen because it is being utilized in many healthcare areas to protect data misuse stored in hospital records, medical research labs [4–6], etc.

Instead of using "Location-Based Addressing" (LBA) IPFS [7] uses "Content-Based Addressing" (CBA). In LBA, whenever we want to search the dataset, we tell the exact location to the computer from where to fetch the data; this is called As LBA. But if the dataset is unavailable or say that the server is down, the data won't be fetched. The dataset may be available, or some other computer must have the dataset, but it cannot be accessed if the exact location is not provided. So to resolve this issue, IPFS has moved from LBA to CBA. In "content-based addressing," it uses hash values to search the required dataset of the patients.

IPFS [8, 9] uses hash values instead of storing the entire medical data records. Files are transferred or received using the hash values. It makes the work easier than sharing the whole dataset. But there can be tampering of files. As hash is used to request files, it can be verified. The file is asked with a specific hash, and when it is found, it will be checked to see if it matches. Deduplication is another mechanism used by hashing to address stuff. When numerous people publish the same file on IPFS, it will be produced only once, resulting in a very efficient network.

In the proposed methodology, AI for blockchain is used to implement the healthcare application by applying the KNN algorithm.

2 Related Work

In Ref. [10], the work in this study is motivated by the following observation: in blockchain networks, storage is a problem due to the append-only characteristic. In this paper, they offer an IPFS-based blockchain storage system that overcomes the limitations of existing blockchain storage. The author also campaigned for transaction access that is content addressed rather than simply location-based.

In Ref. [11], the author introduces explainable artificial intelligence (XAI) that generates the explanations for auditing the responsibility by regulators in case of catastrophic failure. But since AI works in a centralized manner which has security limitations. And there may be a chance of tampering with the data for their benefit to avoid penalty. To overcome these issues author adapts the framework using blockchain as proof-of-authenticity about XAI. It stores the data in IPFS to overcome Ethereum storage limitations. Smart contracts are designed and deployed to retrieve the explanations and supervise the storage. To induce cryptographic security, it uses the hash value and is stored in blockchain. The author also performs cost and security analysis of implemented framework.

In Ref. [12], author uses a modified IPFS and is used to deploy Ethereum smart contracts. These smart contracts provide the ability to share files according to access control. To maintain the list of access controls, smart contracts are used. To enforce it, they modified the IPFS software. Its smart contracts interact with it whenever a file upload is done, a download, or a file transfer.

In Ref. [13], the author overcomes the drawbacks of POW and POS. The author implements the proof of AI with an AI algorithm for selecting the miners by training the neural network model without compromising the basic properties of blockchain technology. Furthermore, the author implements the healthcare domain dataset on top of the blockchain network to train the NN model. This implementation provides an advantage to the state-of-the-art consensus models by increasing efficiency.

In Ref. [14], the combination of blockchain, and AI is called decentralized AI. As AI is centralized and there are chances of data being tamper, there is no guarantee that source data is safe and not misused. Contrarily, blockchain is decentralized in nature and provides more authenticity to the data source, and when merged with AI, it provides unchangeable and reliable characteristics.

In Ref. [15], the author discusses that AI is decision-making on systems like human thinking, whereas blockchain is a decentralized platform in a secure way. The author discusses that integrating AI and blockchain builds decentralized AI, which is nothing but decision-making on a secured platform for sharing the ledgers without involving a third party. The author discusses issues and problems involved in the integration of AI and blockchain. It helps in identifying the malware blocks in the system.

3 Proposed Model

As in Fig. 1, the proposed model presents the use of content addressed (rather than location-based addressing) hash ID of the data as the hash of block in the blockchain. The user enters the medical data for each suggested parameter, and this data is stored in the backend. This data follows a particularly strict format of comma-separated values such that query data dimension matches training data dimension for the used machine learning model. It undergoes two stages such as data classification and data storing.

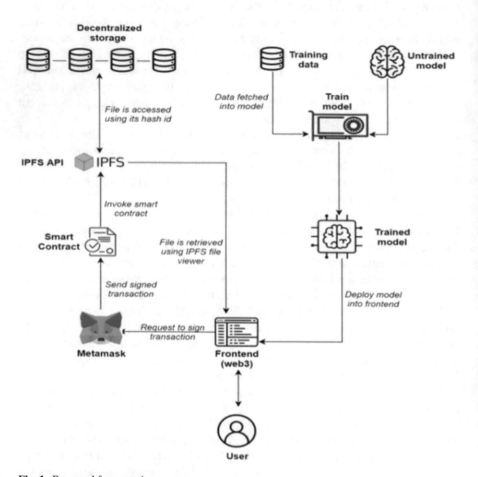

Fig. 1 Proposed framework

3.1 Data Classification

The algorithm, Algorithm 1, takes account for preparing the trained ML model. First, the raw medical data is preprocessed. Then the preprocessed medical data is divided randomly into 80% to train the model, and the remaining 20% to test the model. Using this training data, the model is KNN model is trained with k value as 3. The trained model is then saved for future use, and preprocessed test data is also saved for testing the model in the future.

Algorithm 1 for ML Model Generation
Input: Raw data
Output: Hash of the data

1. Begin
2. Split the full data into train data(80%) and test data (20%)
3. Select the preprocessed training data
4. Declare the k value equal to 3
5. Train the model with training data
6. Now save the trained model using pickle.dumps("model_ name")
7. Save the data
8. End.

3.2 Data Storing

The algorithm for data storing described below in Algorithm 2 is used to classify the medical data and show its result into the front end. First, the application's backend is connected to infura sever and used to connect to the IPFS network. The user enters the IPFS hash address of the data and stores the value to preprocess and get evaluated. Here, the saved ML model is loaded and used to classify the medical data for the patients' heart condition.

Algorithm 2 Algorithm for fetching data from IPFS
Input: IPFS hash
Output: Computed result for the patient

1. Begin
2. api=ipfsapi.Client(host='https://ipfs.infura.io/', port=5001) // connect to infura serverwith port number 5001
3. hashstring = request.POST['hashfromweb'] //Get the hash value from user from frontend
4. api.get(hashstring) //Get the data at this hash from ipfs api and download the file name hashstring
5. str = open(hashstring,'r').read() //Open and read the content of the file hashstring into str
6. preprocess(str)7 loadedmodel = pickle.load(open('modelFileName.sav','rb')) //Load saved trained model tothis django application

7.   prediction = loadedmodel.predict(str)[0] //predict the output of the model after
     fetchingtest data in it and store its result in prediction
8.   if(prediction==0): result="The Patient Is Not Having Heart Disease"
     if(prediction==1): result="The Patient Is Having Heart Disease"
9.   else
10. result = "Invalid Result"
11. End.

4 Implementation

The implementation of this model has been divided into two different modules: data
storing and data classification. The modules are designed carefully and meant to work
independently. There are two different environments for implementations, one for
React-based Web applications and the other for Django-based Web applications. The
modules are carefully planned to be separated because both models have different
functionalities. The first user has to store the data on to IPFS and attest its hash to
the smart contract. While another module contains the ML model required to run
on Python, thus Django makes a suitable choice for this module's implementation.
To implement this model a Linux-based (Ubuntu) OS was used because it is more
secure, less vulnerable, less graphic extensive, and does not collect a lot of data,
unlike Microsoft Windows.

The specification for hardware is Intel(R) i5-8300H CPU @4 GHz, OS: Ubuntu
× 64, 8 GB of primary storage, 128 GB SSD, and 1 TB HDD of secondary storage.
The details of each implemented module are given as follows.

4.1 Data Storing

This module is built on a React-based Web application [9, 16, 17], and its backend
includes the use of IPFS API, the user uploads the patient's medical data to the IPFS
network. Stores its content-addressed hash on smart on file by the name "attestor.
Sol" smart contract (written on solidity language) to be attested on blockchain for
calling and implementing various custom functionalities stated in the smart contract
to be implemented using the blockchain.

User needs to send ETH value of 0.01 ETH from faucet address to burner address.
For completing any transaction on the Ethereum chain, some amount of ETH is
required as a gas price. A burner address of zero balance cannot initiate the transaction
for creating a new block of Hash Id equals the hash value of the data uploaded to
IPFS.

4.2 Data Classification

This module is essential for classifying the medical data. This module is a Django Web application that takes the IPFS hash of the data from the frontend and uses this Hash ID to retrieve the data from the IPFS network. For testing, this data retrieved is test data and was not used to train the ML model. The trained ML model is loaded to the application's backend and used for predicting the output by fetching the test data. The output is sent to the frontend of the application and shows the result of classification done by ML.

5 Results and Discussion

The proposed model testing gets an accuracy of 88.52% when the value k value is set with 3. While uploading to the IPFS network, the size of data does not affect the upload time. Still, while uploading the same data again, the data gets uploaded in much less time because the data is already present in the network and hosted by the setup computer. Hence, uploading of the data is not needed again. Just the same hash is returned to the user.

In Fig. 2, the graph defines the comparison between data size vs. load time. Figure 2 shows that the time does not vary much as the size increases while uploading the data. There are very slight increases when re-uploading the data because it uses already uploaded hash value data, so time does not vary much on re-uploading.

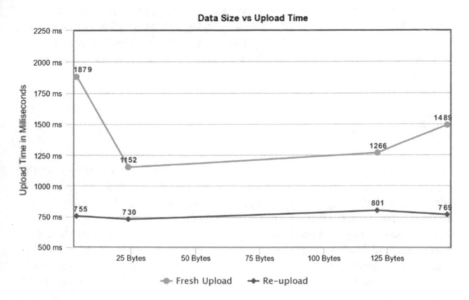

Fig. 2 Data size vs. load time

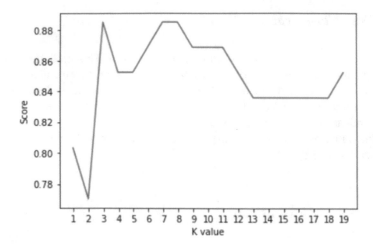

Fig. 3 KNN algorithm performance

Figure 3 shows that it will reach the maximum score when it is set with the k value equal to 3, 7, and 8. The figure shows that the maximum score is 88.52%. This plot contains the k values as the x-axis and score or accuracy as the y-axis. While training the KNN model for different values of k from 0 to 20, the model gives different accuracy for different values of k. Among that values, $k = 3, 7$, and 8 gave the highest accuracy of 88.52%. So it is feasible to train the model among any one of the three values.

The standard model with centralized data storing models, such as large data centers, is purely dependent and governed by the storage providers. Any malfunctions, hacks, and data breaches will result in losing or leaking of sensitive medical data. But the proposed decentralized model is secured from problems like server malfunction because if a node is offline, another node can process the upcoming requests.

6   Conclusion

This paper presents designing and implementing a decentralized medical data storing platform using IPFS-based blockchain-based on smart contracts and this decentralized data to classify a patient's report for their heart condition using AI. The proposed model is more efficient and scalable than the conventional data soring model because it stores only the hash of the data instead of storing data in each blockchain block. This model demonstrates the use of machine learning on blockchain data. The model demonstrates the requirement of an account ID to add the block on the chain, which showcases that the security of the data is not compromised even the data is decentralized.

6.1 Limitations of the Study

The proposed methodology uses IPFS, which is a public protocol. Using a public protocol for data storing is not secure enough as medical data is sensitive and needs data confidentiality. However, the proposed method is implemented using a P2P (peer-to-peer) network. It ensures data confidentiality, integrity, and availability for retrieving data so that accessibility is maintained to the patient's data and does not allow any other person to view medical records. Also, IPFS improves the scalability by saving the space in blocks as it stores only hash values in blocks.

The future scope of the paper is to implement a decentralized storage system of healthcare data that can be shared among a group of hospitals with the same organization or hospitals within the same city [18].

References

1. M.K. Pawar, P. Patil, P.S. Hiremath, A study on blockchain scalability. in: *ICT Systems and Sustainability*, ed. by M. Tuba, S. Akashe, A. Joshi. Advances in Intelligent Systems and Computing, vol. 1270 (Springer, Singapore, 2021)
2. J. Liu, H. Tang, R. Sun, X. Du, M. Guizani, Lightweight and privacy-preserving medical services access for healthcare cloud. IEEE Access (2019)
3. M.K. Pawar, P. Patil, P.S. Hiremath, V.S. Hegde, S. Agarwal, P.B. Naveenkumar, Scalable blockchain framework for a food supply chain. in *Advances in Computing and Network Communications*, ed. by S.M. Thampi, E. Gelenbe, M. Atiquzzaman, V. Chaudhary, K.C. Li. Lecture Notes in Electrical Engineering, vol. 735 (Springer, Singapore, 2021). https://doi.org/10.1007/978-981-33-6977-1_35
4. S. Yu, K. Lv, Z. Shao, Y. Guo, J. Zou, and B. Zhang, A high performance blockchain platform for intelligent devices. in *Proceedings of 1st IEEE International Conference on Hot Information-Centric Networking* (Shenzhen, China, 2018), pp. 260–261
5. A. Ekblaw, A. Azaria, J.D. Halamka, A. Lippman, A case study for blockchain in healthcare MedRec prototype for electronic health records and Medical research data. in *Proceedings of IEEE Open Big Data Conference* (2016), p. 13
6. T.N. Dinh, M.T. Thai, AI and blockchain: a disruptive integration. Computer **51**(9), 48–53 (2018). https://doi.org/10.1109/MC.2018.3620971
7. Q. Zheng, Y. Li, P. Chen, X. Dong, An innovative IPFS-based storage model for blockchain. in *2018 IEEE/WIC/ACM International Conference on Web Intelligence (WI)* (2018), pp. 704–708. https://doi.org/10.1109/WI.2018.000-8
8. R. Kumar, N. Marchang, R. Tripathi, Distributed off-chain storage of patient diagnostic reports in healthcare system using IPFS and blockchain. in *2020 International Conference on Communication Systems Networks (COMSNETS)* (2020), pp. 1–5. https://doi.org/10.1109/COMSNETS48256.2020.9027313
9. T.T. Huynh, T.D. Nguyen, H. Tan, A decentralized solution for web hosting. in *2019 6th NAFOSTED Conference on Information and Computer Science (NICS)* (2019), pp. 82–87. https://doi.org/10.1109/NICS48868.2019.9023837
10. R. Kumar, R. Tripathi, Implementation of distributed file storage and access framework using IPFS and Blockchain. in 2019 Fifth International Conference on Image Information Processing (ICIIP) (2019), pp. 246–251. https://doi.org/10.1109/ICIIP47207.2019.8985677

11. D. Malhotra, S. Srivastava, P. Saini, A.K. Singh,Blockchain based audit trailing of XAI deci-sions: storing on IPFS and Ethereum Blockchain. in *2021 International Conference on Commu-nication Systems Networks (COMSNETS)* (2021), pp. 1–5. https://doi.org/10.1109/COMSNE TS51098.2021.9352908

12. M. Steichen, B. Fiz, R. Norvill, W. Shbair, R. State, Blockchain-based, decentralized access control for IPFS. in *2018 IEEE International Conference on Internet of Things (iThings) and IEEE Green Computing and Communications (GreenCom) and IEEE Cyber, Physical and Social Computing (CPSCom) and IEEE Smart Data (SmartData)* (2018), pp. 1499–1506. https://doi.org/10.1109/Cybermatics2018.2018.00253

13. N. Kumar, C. Parangjothi, S. Guru, M. Kiran, *Peer Consonance in Blockchain-based Healthcare Application using AI-based Consensus Mechanism* (2020), pp. 1–7. https://doi.org/10.1109/ ICCCNT49239.2020.9225550

14. N. Nasurudeen Ahamed, P. Karthikeyan, A reinforcement learning integrated in heuristic search method for self-driving vehicle using blockchain in supply chain management. Int. J. Intell. Netw. **1**, 92–101 (2020). ISSN: 2666-6030. https://doi.org/10.1016/j.ijin.2020.09.001

15. M.K. Pawar, P. Patil, M. Sharma, M. Chalageri, Secure and scalable decentralized supply chain management using Ethereum and IPFS platform. in *2021 International Conference on Intelligent Technologies (CONIT)* (2021), pp. 1–5. https://doi.org/10.1109/CONIT51480.2021. 9498537

16. M.A. Uddin, A. Stranieri, I. Gondal, V. Balasubramanian, Continuous patient monitoring with a patient centric agent: a block architecture. IEEE Access **6**, 32700–32726 (2018)

17. R. Kumar, R. Tripathi, Traceability of counterfeit medicine supply chain through blockchain. in *2019 11th International Conference on Communication Systems Networks (COMSNETS)* (IEEE, 2019), pp.568–570

18. S.D. Smriti, M. Hooda, Possibilities at the intersection of AI and Blockchain technology. Int. J. Innov. Technol. Exploring Eng. (IJITEE) **9**(1S) (2019). ISSN: 2278-3075

A Sustainable Green Approach to the Virtualized Environment in Cloud Computing

Anjani Gupta, Prashant Singh, Dhyanendra Jain, Anupam Kumar Sharma, Prashant Vats, and Ved Prakash Sharma

Abstract Cloud technology is a dynamic industry of communication and information technology (especially the Internet) that has revolutionized current computing while also introducing new environmental concerns. The cloud is a huge improvement over the conventional computing as it has redesigned the way businesses operate. With the cloud, the information is virtualized and reduces the need of infrastructural based model. Businesses can operate through this tangible area, significantly reducing energy consumption and the need for excessive material resources. In recent years, computational technologies and ideas have transitioned to distant data centers and pay-per-use hardware and software solutions. However, as the number of information centers supporting cloud-based applications grows, vast quantities of energy are consumed, resulting to high prices and greenhouse emissions in the atmosphere. Going green emerged as a viable solution to this challenge. Green computing is a study field that intends to overcome energy and climate issues. Green sustainable cloud technology aims to accomplish not only effective processing and computing environment usage, but also to improve energy efficiency. This paper will address the possibilities of sustainable cloud technology, emerging developments, and strategies to increase data center efficiency, which also will minimize carbon emissions. This creates research hurdles when such power devices are necessary to reduce the negative environmental effect of cloud technology.

Keywords Cloud computing · Energy efficiency · Efficient data centers · CO_2 emissions

1 Introduction

The emergence of cloud-based solutions, which allows including on deployment of flexible resources via a pay-as-you-go approach, has changed the communication and information technology (especially information assurance) business. Large

A. Gupta (✉) · P. Singh · D. Jain · A. K. Sharma · P. Vats · V. P. Sharma
Dr. Akhilesh Das Gupta Institute of Technology & Management, GGSIP University, New Delhi, India
e-mail: anjaniaggarwal.06@gmail.com

© The Author(s), under exclusive license to Springer Nature Singapore Pte Ltd. 2023   751
Y.-D. Zhang et al. (eds.), *Smart Trends in Computing and Communications*, Lecture Notes in Networks and Systems 396, https://doi.org/10.1007/978-981-16-9967-2_71

companies and public agencies have transferred their information and operation infrastructure to the cloud in recent years. As we approach the latest iteration of wireless communication networks (5G), telecom operators (access network) must satisfy the growing increase in economic capacity and low-latency applications [1]. As a result, they take use of Internet computational power and operate their networking pieces as dispersed public cloud. Many businesses' use of cloud technology has led in the construction of massive data centers throughout the world, each comprising dozens of computers and communication equipment. Because of the high demand for processing resources, the number of cloud support data centers has grown dramatically in recent years. Because more and more data centers have come into existence, the energy consumption of these data centers has grown significantly [2]. The principles of green computing are similar to raw chemistry, reduce the use of hazardous substances, increase the working capacity during the life of the product, and encourage recycling or decay of non-existent products and industrial waste. Research is underway in key areas such as making the use of computers as energy efficient as possible, as well as designing algorithms and computer technology systems related to efficiency. With conventional data centers, organizations are aware of how much energy they are using in their energy bill. In the clouds, such information is not readily available. Although data centers are much more efficient than local ones, their much-needed model for rapid upgrades and their easily accessible computer resources may mean that it is easier for organizations to use additional resources indirectly. In addition to excessive energy consumption, there is an additional impact on the environment in the guise of carbon emissions (Fig. 1).

This research provides the sustainable solution to cloud computing architecture by utilizing the resources efficiently, virtualization, and enabling energy. Green cloud computing aims to incorporate energy-saving technologies, environmentally friendly production processes as well as advanced waste disposal and recycling processes.

(A)   Green Design: Designing energy-efficient servers, infrastructure, computers, software applications, and other digital devices will reduce the amount of energy used in future.

(B)   Green Production: Reducing energy and waste by recycling and recycling during
        digital device production can have a far-reaching effect on a more sustainable environment.

(C)   Green Consumption: Reducing the amount of energy a product uses while using

    it can reduce up to 27% of energy.

Sustainable public cloud gained traction after the National Institute of Standards and Technology (NIST) identified a disadvantage of cloud applications. As according NIST, "the main objective of this application would be to provide common resources, but the drawback has been its high installation cost, excessive power use, and carbon emissions, as seen in Fig. 2." As public cloud grows more popular and World Wide Web-related networking volume increases, large corporations providing Internet services use greater and greater electricity for operating data centers.

Fig. 1 Cloud computing

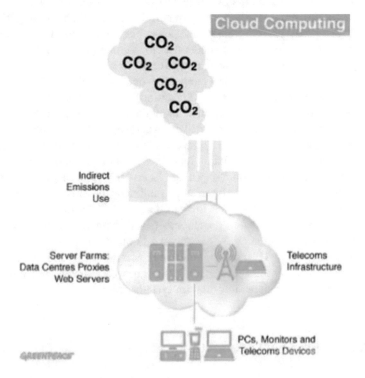

Fig. 2 Cloud Computing and its effect on climate change

To overcome this issue, data center resources must be handled in an energy-efficient way in order to support green cloud technology. Cloud storage, in instance, must be allotted not just to meet the service quality needs provided by users, but also to decrease energy consumption.

2   Key Necessity of Sustainable Cloud-Based Solutions

Computer technology must be "green," which means we must be aware of how our lives and work affect the environment. Going green, often referred to as green computing, refers to the ecologically sustainable and environmentally sensitive use of machines and their supplies. This includes the use of energy-efficient central processing unit (CPU), workstations, and accessories and also decreased resource utilization and efficient and effective disposal of wastes (e-waste). This also includes the creation of environmentally friendly materials, energy-efficient computing, and the recycle of systems and devices. Raised power usage has raised IT operating costs [3]. Modern data centers, operating under the cloud computing model, host a variety of applications ranging from those that run for a few seconds (e.g., providing Web applications such as e-commerce and temporary social networking sites) to long-term ones (e.g., simulation or large data set processing) in shared hardware forums.

The need to manage multiple applications in a data center poses a challenge to providing much-needed service and distribution in response to dynamic workloads. In general, data center resources are assigned to applications mathematically, based on heavy load factors, in order to maintain segmentation and provide performance guarantees. Green sustainable cloud computing has emerged as a result of the growing threats to the environment and the need to minimize the energy use. Sustainable computing's primary objectives are to limit use of such poisonous and hazardous substances, promote sustainable development, and recycle e-waste. This method entails the effective installation of Webserver and other components as well as the reduction of energy usage [4]. Investigation in important areas including such developing computer being used as efficient as feasible, as well as inventing computational methods for efficiency-related information systems, continued.

3   Green Cloud Services Methods

Using "virtualization" technique is vital for energy conservation in clouds since it allows for large improvements in energy efficiency while utilizing the very same architecture. Organizations that use VMware may save a lot of money in terms of space, administration, and electricity.

(A) Dynamic Cloud Provisioning

Traditionally, computer systems and commercial infrastructures were managed to meet deadliest demands. As a result, most IT firms build more capacity than is necessary. The following are the explanations behind this situation [4]:

•It is quite hard to anticipate demand in advance.

•To ensure that service continuity and to sustain a specific degree of customer satisfaction for end customers.

(B) SaaS-Based Multi-tenancy

The use of virtualized environment decreases total energy consumption and greenhouse gas emissions. Broader organization is served by SaaS application services using the same technology and equipment. This strategy is much more environmentally friendly than installing several backup copies on various infrastructure, which could also reduce the requirement for infrastructure upgrades. The lower the volatility in consumption, the greater the forecast, and the bigger the energy management system.

(C) Server Utilization

Multiple programs can indeed be maintained and operated in solitude on the very same domain controller using VMware vSphere, resulting in utilization concentrations of up to 70 percent of the overall [5]. While also increasing server utilization leads in increased power requirements, a feedforward at high utilization can perform greater workloads with the same amount of energy.

(D) Data Center Efficiency

Cloud service providers may considerably enhance the power demand performance (key performance indicator) of existing computer servers by utilizing far more energy-efficient technology. Cloud computing provides applications to be transferred across several data centers that have lower spectral efficiency levels.

4 Energy Consumed into Large Data Centers

An information center's design is complicated since it includes not just equipment but then also application that operates in the IT infrastructures. As a result, we may divide its components into 2 layers: physical and logical, as seen in Fig. 3.

Energy is usually used to operate data centers. Contemporary data centers, on the other hand, utilize green technologies such as wind and solar power, waves, and geothermal PV as part of a drive to reduce emissions and conform with proactive environmental standards. Electricity is sent from exterior power stations to inner basic infrastructures, information and technology (telecom) gear, and many other support networks. Electricity is sent to interior IT operations using uninterruptible power supply supplies (appliances and equipment) to provide uninterruptible power delivery though in the event of a power outage [6].

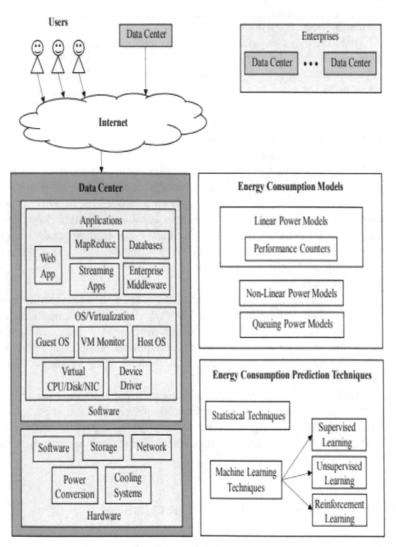

Fig. 3 Categorization of Data Centers: **Hardware and Software**

5 Initiatives Development using Sustainable Cloud Technology

There really are equipment/software technical solutions which drastically reduce power requirements and environmental effect. The following are the primary approaches for reducing energy use, CO2 emissions, and development approaching greener data centers [7]:

(A) Application Oriented

The software-as-a-service (SaaS) concept has altered how applications and software are delivered and consumed. To reduce IT costs, an increasing number of businesses are migrating to SaaS clouds. As a result, this has become necessary to resolve energy consumption somewhere at application server.

(B) Virtualization

It is just an energy-efficient way for sharing material assets across virtual server machine platforms (VMs). Variable resource distribution is a technology that allows us to use as well as allocate periodically free computational power across VMs while conserving a substantial portion of power [8]. VMware is an effective method for providing server centralization and powering off idle servers. In so many circumstances, a dormant server uses 70 percent of the overall of the electricity that a server operating at full CPU utilization requires [9].

(C) Energy-efficient Hardware

Computing system is comprised of equipment, including as processors, inbuilt displays, discs, fans, and so on. This gear must utilize as little electricity as possible. Several well-known approaches can be used to preserve computer power [10]. To begin, the program should indeed be turned off if it detects a disconnect. Secondly, we may decelerate the processor clock frequency, which would be called as clock PWM gating, and then, we can shut down the chips, which would be called as microchip inhibition. Tempo shift, energy still, cool and quiet [11], and regulation switches are some other strategies for shutting down the main CPU.

(D) Monitoring/Metering

It is claimed that what you should cannot evaluate can indeed be improved. It is critical to build electricity capabilities that help the computer to understand how much power is spent by a certain component and how it could be decreased. The sustainable grid has suggested two particular measures known as power usage effectiveness (peak power) and datacenter infrastructural effectiveness (application procedures) to assess the combined efficiency of a datacenter and increase the reaction rate per watt [12].

Power Usage Effectiveness

$$= \text{Facilities Total Available Power/Provided IT based Equipmental Power} \tag{1}$$

Datacenter Infrastructure EffeTctiveness Efficiency

$$= 1/\text{Power Usage Effectiveness}$$

$$= [\text{Provided IT based Equip mental Power/Facilities Total Available Power}] \times 100\% \tag{2}$$

(E) Network Infrastructure

Energy-saving concerns in communication are sometimes alluded to as "environmental networking," which refers to the incorporation of electricity into network infrastructure, equipment, and procedures.

6 Conclusion and Future Scope

This article provides a quick overview of sustainable green cloud computing in a public cloud scenario. It aims to accomplish not only effective processing and computing infrastructure usage, but to improve energy efficiency. Power demand control in storage systems has resulted in numerous of significant advancements in energy conservation. To summarize, no technique or approach is greener, but it is the person who employs it make it efficient by employing it properly and in accordance with its specifications. Cloud service providers must reduce their energy use and take massive moves toward embracing renewable resources instead of focusing solely on reducing costs. Sustainable computing technology, on the other hand, has to be investigated.

References

1. K. Bilal, S. Malik, S. Khan, A. Zomaya, Trends and challenges in cloud datacenters. IEEE Cloud Comput. **1**(1), 10–20 (2014)
2. P. Barham, B. Dragovic, K. Fraser, S. Hand, T. Harris, A. Ho, R. Neugebauer, Pratt and I. Warfield The art of virtualization. ACM IGOPS Operating Syst. Rev. **37**(5), 164–177
3. B. Saha, Green computing. Int. J. Comput. Trends Technol. (IJCTT) **14**(2) (2014)
4. A. Tahiliani, A. Purohit, A review on energy saving using green computing system. Int. J. Eng. Dev. Res. ISSN 2321-9939
5. A. Atrey, N. Jain, N. Iyengar, A study on green cloud computing. Int. J. Grid Distrib. Comput. **6**(6), 93–102
6. M. Aljaberi, S. Khan, S. Muammar, Green computing implementation factors: UAE case study. 1–4 (2016). https://doi.org/10.1109/ICEDSA.2016.7818528
7. B. Whitehead, D. Andrews, A. Shah, G. Maidment, Assessing the environmental impact of data centres—Part 1: background, energy use and metrics. Building Environ. **82**, 151–159 (2014)
8. H. Liu, C.-Z. Xu, H. Jin, J. Gong, X. Liao, Performance and energy modeling for live migration of virtual machines, in *Proceedings of 20th International Symposium HPDC* (2011), pp. 171–182
9. A. Beloglazov, J. Abawajy, R. Buyya, Energy-aware resource allocation heuristics for efficient management of data centers for cloud computing. Future Gener. Comput. Syst. (2012, May)
10. A. Beloglazov, R. Buyya, Energy efficient resource management in virtualized cloud data centers, in *Proceedings of 10th IEEE/ACM International CCGrid* (2010, May), pp. 826–831
11. E. Feller, C. Rohr, D. Margery, C. Morin, Energy management in IAAS clouds: a holistic approach, in *Proceedings of IEEE 5th International Conferences*.
12. P. Corcoran, A. Andrae, Emerging trends in electricity consumption for consumer ICT (National University of Ireland Galway, Ireland, Tech. Rep., 2013)
13. C. Thanmai, K.N. Narsimaha Murthy, S. Ambareesh, A survey on evolution of green computing in the cloud environment. Int. J. Innov. Res. Comput. Commun. Eng. **3**(Special Issue 5) (2015)

14. L. Krug, M. Shackleton, F. Saffre, Understanding the environmental costs of fixed line networking, in *Proceedings of 5th International Conference on Future Energy Systems* (2014), pp. 87–95

15. S. Mingay, Green IT: The New Industry Shock Wave (2007). Available online http://www.ict literacy.info/rf.pdf/Gartner_on_Green_IT.pdf. Accessed on 18 July 2017

16. C.T. Yang, K.C. Wang, H.Y. Cheng, C.T. Kuo, C.H. Hsu, Implementation of a green power management algorithm for virtual machines on cloud computing, in *Proceedings of the 8th International Conference on Ubiquitous Intelligence and Computing*, Banff, AB, Canada, 2–4 September 2011 (Springer, Berlin, Germany, 2011), pp. 280–294

17. X.L. Xu, G. Yang, L.J. Li, R.C. Wang, Dynamic data aggregation algorithm for data centers of green cloud computing. J. Syst. Eng. Electron. **34**, 1923–1929 (2012)

18. H. Lee, Y.S. Jeong, H.J. Jang, Performance analysis based resource allocation for green cloud computing. J. Supercomput. **69**, 1013–1026 (2014)

19. M. Azaiez, W. Chainbi, H. Chihi, A green model of cloud resources provisioning, in *Proceedings of the 4th International Conference on Cloud Computing and Services Science*, Barcelona, Spain, 3–5 April 2014 (SCITEPRESS—Science and Technology Publications, Lda., Setubal, Portugal, 2014), pp. 135–142

20. J. Kolodziej, S.U. Khan, L.Z. Wang, M. Kisiel-Dorohinicki, S.A. Madani, E. Niewiadomska-Szynkiewicz, C.Z. Xu, Security, energy, and performance-aware resource allocation mechanisms for computational grids. Future Gener. Comput. Syst. **31**, 77–92 (2014)

21. L. Xu, K. Wang, Z. Ouyang, X. Qi, An improved binary PSO-based task scheduling algorithm in green cloud computing, in *Proceedings of the 9th International Conference on Communications and Networking in China (CHINACOM)*, Maoming, China, 14–16 August 2014 (IEEE, New York, NY, USA, 2014), pp. 126–131

22. G. Kaur, S.A. Midha, Preemptive priority based job scheduling algorithm in green cloud computing, in *Proceedings of the 6th International Conference Cloud System and Big Data Engineering*, Noida, India, 14–15 January 2016 (IEEE, New York, NY, USA, 2016), pp. 152–156

23. Y. Liu, W. Shu, C. Zhang, A parallel task scheduling optimization algorithm based on clonal operator in green cloud computing. J. Commun. **11**, 185–191 (2016)

24. D. Zhang, Z. Chen, L.X. Cai, H. Zhou, J. Ren, X. Shen, Resource allocation for green cloud radio access networks powered by renewable energy, in *Proceedings of the Global Communications Conference*, Washington, DC, USA, 4–8 December 2016 (IEEE, New York, NY, USA, 2016), pp. 1–6

25. A. Khosravi, S.K. Garg, R. Buyya, Energy and carbon-efficient placement of virtual machines in distributed cloud data centers, in *Lecture Notes in Computer Science, Proceedings of the Euro-Par 2013 Parallel Processing*, Aachen, Germany, 26–30 August 2013, ed by F. Wolf, B. Mohr, D. Mey (Springer: Berlin, Germany, 2013), pp. 317–328.

26. F. Cao, M.M. Zhu, C.Q. Wu, Green cloud computing with efficient resource allocation approach, in *Green Services Engineering, Optimization, and Modeling in the Technological Age*, 1st edn. (IGI Global, Hershey, PA, USA, 2015), pp. 116–148. ISBN 9781466684478

27. Z. Deng, G. Zeng, Q. He, Y. Zhong, W. Wang, Using priced timed automaton to analyse the energy consumption in cloud computing environment. Clust. Comput. **17**, 1295–1307 (2014)

28. J. Huang, K. Wu, M. Moh, Dynamic virtual machine migration algorithms using enhanced energy consumption model for green cloud data centers, in *Proceedings of the International Conference on High Performance Computing & Simulation*, Bologna, Italy, 21–25 July 2014 (IEEE, New York, NY, USA, 2014), pp. 902–910

29. C.M. Wu, R.S. Chang, H.Y. Chan, A green energy-efficient scheduling algorithm using the DVFS technique for cloud datacenters. Future Gener. Comput. Syst. **37**, 141–147 (2014). (Symmetry 2017, 9, 295 16 of 20)

30. J.A. Aroca, A.F. Anta, Empirical comparison of power-efficient virtual machine assignment algorithms. Comput. Commun. **96**, 86–98 (2016)

31. F. Farahnakian, T. Pahikkala, P. Liljeberg, J. Plosila, H. Tenhunen, Utilization prediction aware VM consolidation approach for green cloud computing, in *Proceedings of the 8th International*

Conference on Cloud Computing (CLOUD), New York, NY, USA, 27 June–2 July 2015 (IEEE, New York, NY, USA, 2015), pp. 381–388

32. B. Kaur, A. Kaur, An efficient approach for green cloud computing using genetic algorithm, in Proceedings of the 1st International Conference on Next Generation Computing Technologies, Dehradun, India, 4–5 September 2015 (IEEE: New York, NY, USA, 2015), pp. 10–15

33. G. Koutsandria, E. Skevakis, A.A. Sayegh, P. Koutsakis, Can everybody be happy in the cloud? Delay, profit and energy-efficient scheduling for cloud services. J. Parallel Distrib. Comput. **96**, 202–217 (2016)

34. Z. Long, W. Ji, Power-efficient immune clonal optimization and dynamic load balancing for low energy consumption and high efficiency in green cloud computing. J. Commun. **11**, 558–563 (2016)

Threats and Challenges of Artificial Intelligence in the Healthcare Industry

Priti Ranjan Sahoo⃝, Smrutirekha⃝, and Mou Chatterjee⃝

Abstract The authors provide a crisp yet in-depth summary of the relevance of artificial intelligence (AI) in providing healthcare solutions. Artificial intelligence is a relatively new concept in the field of health care. AI aids in the prediction of disease patients for medical procedures. Patients, pharma companies, health services, insurance companies, and medical institutions benefit from AI's application in health care. Artificial intelligence supports different concepts, counting computing, computer program improvement, and information exchange. Machine learning, profound learning, normal dialect generation, discourse acknowledgment, robots, and biometric distinguishing proof are illustrations of artificial intelligence's innovation. Artificial intelligence is used in a variety of areas, including health care, manufacturing, and business. It is also used in the automotive industry. The authors have discussed the current scenario of artificial intelligence and the threats and challenges posed for AI in the healthcare industry.

Keywords Artificial intelligence · Health care · Technology · Machine learning

1 Introduction

The ability to gather, partake in, and provide data is becoming increasingly crucial as digitization eliminates all interruptions, including health care. Machine learning, big data, and artificial intelligence (AI) can all help with the issues that enormous datasets issues. Machine intellect can also assist healthcare chambers in meeting rising medical demands, improving operations, and lowering costs.

Alan Turing, a British mathematician, and philosopher questioned whether machines could think in 1950. AI was not fully conceptualized and named until the Dartmouth Summer Research Project in 1956. AI made its way into health care in the early 1970s and began to be applied to biomedical problems. Artificial intelligence in medicine, an international AI journal, was launched, and the American

P. R. Sahoo (✉) · Smrutirekha · M. Chatterjee
KIIT School of Management, KIIT University (Institution of Eminence), Bhubaneswar, India
e-mail: prsahoo@ksom.ac.in

Association for artificial intelligence was founded in 1980, focusing on medical applications. In the decades that followed, AI was introduced into clinical settings. Among the new applications of AI in healthcare are fuzzy expert systems, Bayesian networks, artificial neural networks, and crossbred intelligent systems.

Geoffrey Hinton and colleagues released AlexNet in 2012. This began the current surge in interest in significant learning. Significant neural frameworks have as of late outlined multitudinous errand execution comparable to characteristic pros. Over the decade, there has been an immediate increase in research regarding AI knowledge and it's significant capability in healthcare applications. By 2017, AI inside the treatment had finished up the preeminent common application of AI in adding up to esteem help.

At last, two sorts of fake bits of knowledge utilized in health care have risen: physical and virtual. Physical AI alludes to utilize of proclivity and robots to assist cases and suppliers in giving care. Machine, or significant capability, might incorporate virtual AI in which calculations are made through replication and inclusion [8].

AI-based innovations try to emulate human cognitive processes, and robots' ability to analyze and manage massive datasets using AI algorithms has evolved rapidly. In the era of big data, one advantage of using AI in health care is that complicated patterns, and correlations can be identified algorithmically without recruiting additional healthcare specialists for data analysis.

Artificial intelligence AI refers to machines that can perform tasks that would require intelligence if performed by humans. Machine learning refers to computer learning that is not explicitly programmed and includes a variety of techniques for achieving AI. Artificial neural networks, which simulate the human brain, are an example of machine learning technology architecture. The human brain comprises billions of nerve cells or neurons that are organized in complex interconnected networks that allow us to generate complex thought patterns and actions.

2 Role of Artificial Intelligence in Healthcare Solutions

Artificial intelligence provides numerous technology solutions for doctors, patients, and management consistently for the ease of each of the entities involved in the healthcare industry.

For example, doctors use the diagnostic imaging interpretation application for making a decision. The healthcare industry is much obliged to profound learning programs, innovation categorization, and AI-based imaging frameworks are prepared with calculations to filter pictures. MRIs and CT scans are examples of this. One of the most widely used AI-based predicated judgment checkers, buoy health, employs an algorithm that aids in treating sickness. The chatbot is aware of the case's symptoms as well as the health complications. This aids in decision-making and gives a direction for further treatment.

As the well-known saying goes, "Prevention is better than Cure." The doctors' new approach focuses on preventative care based on data collection. This includes everything from gathering in-born information to wearable devices that advance the electronic healthcare system. Fitbit, Apple, Garmin, and other companies monitor heart rate and physical activities effectively. They can send warnings to the case about exercise and medication dosage. The fashionable AI solution gathers data from mutation genetics and DNA. This facilitates the system's inspection and detection of cancer and vascular disorders in advance.

A virtual health assistant is a one-of-a-kind method for patients that can reduce frequent clinic visits and relieve the pressure on the medical staff. They also serve as a conduit between service providers and cases. Chatbots also play a significant role in serving the patients. Interactive chatbots have been created using natural language processing (NLP) and sentiment analysis [11]. They enable cases to inquire about bill payment, furniture, or pharmaceutical kitties. Cases enduring persistent ailments get it how to care for their condition. In any case, communicating with them through bots makes a difference to ease the burden on therapeutic suppliers.

Babylon, a UK-based software, uses emotional artificial intelligence to offer the best medical arguments based on a person's history of medical diagnosis. Either way, the job suggests a course of action. Another AI software, sensibly, aids healthcare workers in covering cases' circumstances and providing remedies as a result.

For the manufacturers and service providers, AI aids in maintaining cybersecurity, in the manufacturing of drugs, and detection and checking of any fraudulence in the system.

To consider the practical implications of virtual nursing assistants, in the United States, 89 percent of patient googles their symptoms before consulting a doctor, and the consequences of self-diagnosis are often frightening. Molly is an excellent example of a multi-functional virtual nurse. It tracks a patient's weight, blood pressure, and other indicators generated by monitoring equipment and provides remote support for common medical issues. Patients can also use the app's chatbot to confidentially discuss their health concerns and make an appointment with their doctor. Molly and other virtual assistants have already gained much traction, with 64 percent of patients preferring to receive instructions from them [19].

Machine learning applications can make strides in the precision of treatment conventions and well-being results through algorithmic strategies. Profound learning, for illustration, could be a sort of complex machine learning that re-enacts how the human brain capacitate [17]. Profound learning programs can identify, recognize, and assess dangerous tumors from photos by utilizing neural systems that learn from information without supervision [13].

Machine learning systems can spot irregularities in photographs that the human eye cannot because of quicker processing speeds and cloud infrastructures, assisting in disease detection and treatment.

Breakthroughs in machine learning in the medicine-related industry will continue to shape business strategies in the future. For instance, as studied by [5], machine learning applications in advancement incorporate a diabetic retinopathy symptomatic apparatus and prescient analytics to distinguish breast cancer repeat based on therapeutic information and pictures.

Artificial intelligence helps with healthcare technology record keeping, particularly electronic health records (EHRs). AI-assisted EHR management can improve the quality of health care while reducing health care and administrative costs and streamlining operations.

Normal dialect handling, for illustration, permits experts to require and record clinical notes without having to depend on manual strategies. By advertising clinical choice help, computerizing imaging procedures, and binding together innovations, machine learning methods can simplify EHR organization frameworks for doctors. Due to insufficiencies in healthcare information, machine learning calculations might create false expectations, which can critically affect clinical decision-making [16].

Because healthcare data were created for electronic health records (EHRs), it must be prepared well in advance for the machine learning algorithms to use it effectively. Health informatics experts are in charge of data integrity. Health informatics specialists are in charge of gathering, analyzing, classifying, and cleansing data. Machine learning, medical technology, and predictive analytics are all tools that can be used to improve healthcare procedures, change clinical decision support systems, and improve health satisfaction [10]. The capacity of machine learning in health care to utilize well-being informatics to anticipate well-being results through prescient analytics comes about in more exact determination and treatment and moved forward clinician bits of knowledge for custom-made and cohort medicines.

Machine learning can supplement prescient analytics by deciphering information for decision makers to distinguish handle imperfections and upgrade generally healthcare operations [1].

Machine learning has the potential to have a positive impact on patient care delivery strategies. It can, for example, assist professionals in identifying, diagnosing, and treating disease. Machine learning applications in health care can also help to streamline tasks and improve operation planning, preparation, and execution [3, 8]. By examining enormous amounts of healthcare records and other patient data, machine learning algorithms can find patterns related to diseases and health conditions in one go. This reduces the operational cost and processing time.

Machine learning technologies have the potential to increase access to health care in developing countries while also improving cancer detection and therapy [2].

Machine learning has been shown to help clinical practitioners improve their productivity and accurateness. Machine learning in medical imaging is commonly used to detect cardiovascular irregularities, diagnose musculoskeletal injuries, and screen for cancer [6].

By combining real-time information, data from earlier fruitful surgeries and therapeutic records, machine learning can move forward the accuracy of automated surgical gear. Decreased human mistakes, back amid more complex medicines, and negligibly intrusive surgeries are fair a number of the benefits. Robots, for instance, could help in spine surgery by performing precise surgeries to unblock blood vessels. Surgical robots can provide surgeons with more than just automated assistance by helping them organize surgical operations and procedures [14].

Patients' abilities can be supplemented directly by robots. Assisting paraplegic patients to regain their ability to walk and execute tasks such as blood pressure monitoring and medication reminders are just a few examples. Robot friendship can benefit even the sick and aged [9].

The quality of data input determines the output of machine learning algorithms. Erroneous data might jeopardize system reliability, raising questions about whether judgments based on the data are correct or incorrect. Flawed data can also contribute to a lack of cultural competency. In the end, issues regarding the reliability of the system and the complications due to misinformation used by machine learning algorithms could result in erroneous outputs, misinformed medical decision-making, and eventually harm people's health.

3 Threats and Challenges Faced by Artificial Intelligence in the Healthcare Industry

1. To train and optimize the performance of the algorithms, a substantial quantity of patient information is required for an AI solution to be successful. Obtaining access to these datasets in health care raises several issues. Case loneliness and data control ethics—entering idiomatic medical records is strictly prohibited. Data sharing between hospitals and AI companies have spawned competitors in recent generations, highlighting several ethical concerns.
2. Developing standards for a technology that is shadow based and constantly changing poses significant difficulties. What methods are available for securing cases? How can one ensure that a constantly learning and evolving result—rather than a unique, reading-controlled medical equipment—has adequate despotic oversight? Whether AI is a guru of treatment rather than just a device arises when AI outcomes entail direct patient relations without doctor monitoring. Will this sample expand to a desire for some medical license to operate—and would a public medical board agree to provide such a license? This raises the question of who is responsible if something goes wrong. Is the AI firm, however, responsible for the case's outcome? Will insurance firms ever invest in an AI tool, as an example?
3. Suppose this technology has control over opinion or treatment. Another barrier to utilization is the departure of a doper. With these tools, the human touch of interacting with a doctor can be lost. Are cases willing to rely on a decision made by a software algorithm rather than a human? Furthermore, in the meantime, are physicians prepared to accept these research discoveries? It may be ludicrous to expect relatively cold departing rates beyond the testimonial of sweeping generalization studies in conscientiousness that still universally uses the device.
4. Despite the revolution in incorporating AI in providing healthcare solutions, few threats as given in Fig. 1 have been found that need attention.

A risk highlighted by AI systems is that they demand enormous amounts of data from various sources, including pharmaceutical, electronic health, and insurance claims [7]. Because data are broken and cases changing insurance companies or

Fig. 1 Threats to AI in
health care

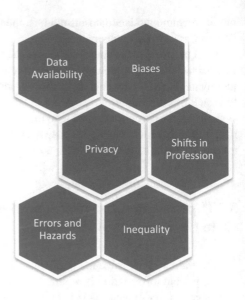

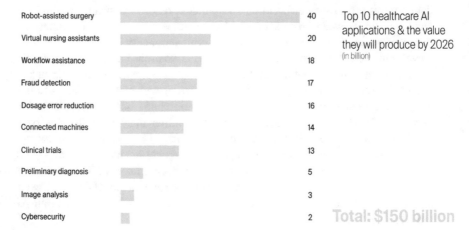

Fig. 2 Forecast for the future of the industrial implications

providers daily, the data become more convoluted and less coherent, increasing the
risk of inaccuracy and the expense of data gathering—data availability.

- Within the long run, the work of AI frameworks could lead to shifts within the
  therapeutic calling especially in ranges like radiology, where the farthest of the
  work gets computerized. This raises the concern that a tall degree of business of AI
  might lead to a drop in standard information and capacity over the dates, making
  suppliers fail in identifying AI wrongdoings and within the other advancement of
  vital information—switching of profession.

- Collecting massive datasets and exchanging information between healthcare systems and AI engineers to strengthen AI systems lead to combat cases acknowledging that this may abuse their confines, resulting in a record of claims. Another area where AI frameworks' work poses this concern is that AI can predict private data surrounding situations even when persistent has never provided this info—privacy issues.
- AI frameworks can assimilate the inclinations of the available knowledge because they assimilate and learn via the prepared material, for example, suppose the data gathered in AI is collected chiefly in scholarly therapeutic centers. In that case, the developing AI frameworks will have less information about, and as a result, will treat cases from populations that do not routinely visit scholarly therapeutic centers less successfully—biases and inequality.

For at least two reasons, AI errors are potentially distinct. While human healthcare professionals can commit mistakes at times, what makes this important is that an underlying fault, such as one in an AI system, might result in thousands of patients being injured in one single process cycle—errors and hazards [4]

4 Scope and Future Implications of Artificial Intelligence in Healthcare Solutions

Massive healthcare delivery organizations have tested and are testing AI on a broad scale, and the beauty of running an independent medical practice is that one gets to observe these giant providers test out technology before it becomes universal. Smart practices, on the other hand, are better pursuit to wait for the larger, more expensive AI applications. Keep a watch on large-scale suppliers as they increasingly include AI-powered technologies and learn from their experiences to determine which applications will be most helpful to one in the future. Figure 2 gives an overview of the forecast for the future of the industrial implications.

Digital healthcare professionals are on the cutting edge of this opportunity, assisting in integrating learning algorithms into medical and healthcare systems. In a profession that is becoming increasingly data driven, their in-depth understanding of technology and how it can improve the treatment protocol and the results are invaluable [15]. Machine learning's deep learning algorithms can reduce the time it takes to analyze a patient's medical history and the recurring disease details resulting in faster diagnosis, treatment, and recovery. The most important work of well-being informatics experts is ensuring the precision of healthcare data that becomes progressively crucial as healthcare organizations endeavor to coordinate machine learning into healthcare and restorative operations. A run of technology-driven healthcare activities will appear as a guarantee within the future in terms of making strides in care conveyance. Health informatics experts' jobs will be altered as a result of these improvements. Some of the artificial intelligence technologies that would affect and create opportunities in the healthcare industry are discussed below.

- Virtual Reality: Virtual reality (VR) alters health care by making it easier to train doctors and transform patients' lives. Surgeons wearing specific VR headsets, for example, can Webcast surgeries and provide medical trainees a unique perspective on a surgical process. In another case, virtual reality is being employed to aid in physical therapy recuperation. Physical therapy patients are frequently subjected to sensitive physical activities that might be taxing. Recovery programs can be tailored, and physical treatment activities can be made more pleasurable and engaging through VR training exercises with machine learning.
- Genome Sequencing: The physician can start creating personalized treatment plans for their patients using genomic data. Machine learning in information systems allows genetic mutations to be analyzed more quickly and aids in diagnosing conditions that can lead to disease. Deep learning applications that enable genome sequencing can improve cancer care and prognosis while minimizing the impact of infectious diseases. As genetic sequencing becomes more reasonably priced and machine learning becomes way more intelligent, healthcare technology professionals can help advance genomic medicine to treat the world's deadliest diseases.
- Augmented Reality: Augmented reality (AR) is one of the top three technologies revolutionizing health care. AR technologies in health care, like VR, can help medical students prepare better. Students can learn firsthand from surgeons doing real-life surgery using augmented reality technologies. Medical students, for example, may acquire precise, accurate renderings of human anatomy without having to study actual human beings thanks to augmented reality.
- Nanotechnology: According to the National Nanotechnology Initiative, nanotechnology is described as "the understanding and control of matter at the nanoscale, at dimensions between about 1 and 100 nm." Nano-medicine refers to the use of nanotechnology in health care. Nanotechnology can aid in executing activities involving chemicals, biological structures, and DNA, such as medicine delivery. Medications can be sent to specific sections of the human body that are not impacted by diseases.
- 3D Printing: This is an innovation that offers an array of opportunities in the healthcare sector. These techniques can benefit drug formulations, implants, prostheses, biosensor devices, and human tissues and organs. It enables the personalization of medical treatments, improves healthcare quality, lowers costs, and lowers production risks.
- Wearable Watch: Various consumer wearable devices help people become more fit, from counting steps to monitoring heart rhythms. Other wearable devices, such as heart rate, blood pressure, temperature, and heart rate, can give clinicians information regarding a patient's health. According to Pew Research Center, wearable devices, such as activity trackers and smartwatches, are used by roughly 21% of Americans. As more people adopt wearable technology, health informatics specialists can improve data communication and accuracy between these devices and the health information systems used by clinicians.

5 Conclusion

The construction of an ethical global governance structure and system and exceptional standards for frontier AI uses in medicine are needed to assure "trustworthy" AI applications in health care and medicine.

AI tools are making their way into the medical field. It is now a fact that we must face in order of their arrival. While there appears to be some hyperbole in the discussion of AI in health care, we also discovered that diverse considerations and knowledge have emerged among each stakeholder category. On the one hand, the health sector and researchers put a premium on high-quality health data; on the other side, doctors are still waiting for proof of these tools' utility, and they are concerned about being found liable in the occurrence of an injury caused by such an artificial intelligence system they do not fully comprehend [12].

Crossbred models, in which practitioners are aided in determination, treatment planning, and correlating peril variables, but retain final accountability for the case's care, are the best chances for AI in health care over the following countless areas. This will result in healthcare providers leaving faster due to the reduced perceived risk and bringing measurable advances in patient concerns and efficiency at scale.

Massive challenges continue as a result of the magnitude and complexity of health data. However, by creating new example recognition systems, adaptive computations, and unique methodology that leverage vast volumes of health data to answer broad questions, the community dwelling in the lap of artificial intelligence is well on its way to surpassing these obstacles [18].

Counterfeit insights are a science with its implementations in various fields, including the medical and healthcare system. The development of untrue mindfulness has decreased human intercessions and, as a result, has driven to fundamental and fast, the common-sense discovery of different complicated afflictions. Artificial intelligence is valuable and worth incorporating for planning one's daily life. The fundamental component of artificial intelligence in long-term care is calm finding and image examination; however, the long run holds incredible potential for applying AI to improve various aspects of long-term care management.

References

1. G.V.K.S. Abhinav, S. Naga Subrahmanyam, Artificial intelligence in healthcare. J. Drug Del. Ther. **9**(5-s), 164–66 (2019)
2. M.A. Ahmad, A. Patel, C. Eckert, V. Kumar, A. Teredesai, Fairness in machine learning for healthcare, in *Proceedings of the 26th ACM SIGKDD International Conference on Knowledge Discovery & Data Mining* (2020), pp. 3529–3530
3. R. Alugubelli, Exploratory study of artificial intelligence in healthcare. Int. J. Innov. Eng. Res. Technol. **3**(1), 1–10 (2016)
4. J. Archana, E.A. Mary Anita, A survey of big data analytics in healthcare and government. Procedia Comput. Sci. **50**, 408–413 (2015)

5. J. Bajwa, U. Munir, A. Nori, B. Williams, Artificial intelligence in healthcare: transforming the practice of medicine. Future Healthcare J **8**(2), e188 (2021)
6. A. Callahan, N.H. Shah, Machine learning in healthcare, in *Key Advances in Clinical Informatics* (Elsevier , 2017), pp. 279–91
7. P. Chowriappa, S. Dua, Y. Todorov, Introduction to machine learning in healthcare informatics, in *Machine Learning in Healthcare Informatics* (Springer , Berlin, 2014), pp. 1–23
8. S. Ellahham, N. Ellahham, Use of artificial intelligence for improving patient flow and healthcare delivery. J. Comput. Sci. Syst. Biol. **12**(3) (2019)
9. S.M. Kulkarni, G. Sundari, A review on image segmentation for brain tumor detection, in *2018 Second International Conference on Electronics, Communication and Aerospace Technology (ICECA)* (IEEE, 2018)s, pp. 552–55
10. H. Liyanage, S.-T. Liaw, J. Jonnagaddala, R. Schreiber, C. Kuziemsky, A.L. Terry, S. de Lusignan, Artificial intelligence in primary health care: perceptions, issues, and challenges. Yearb. Med. Inform. **28**(1), 41–46 (2019)
11. M. Marwan, A. Kartit, H. Ouahmane, Security enhancement in healthcare cloud using machine learning. Procedia Comput. Sci. **127**, 388–397 (2018)
12. N. Mehta, A. Pandit, S. Shukla, Transforming healthcare with big data analytics and artificial intelligence: a systematic mapping study. J. Biomed. Informatics **100**, 103311 (2019)
13. R. Miotto, F. Wang, S. Wang, X. Jiang, J.T. Dudley, Deep learning for healthcare: review, opportunities, and challenges. Brief. Bioinform. **19**(6), 1236–1246 (2018)
14. A. Panesar, *Machine Learning and AI for Healthcare.* (Springer, Berlin, 2019)
15. P. Sanjay, C. Meng, Z. Che, Y. Liu, Benchmark of Deep Learning Models on Large Healthcare Mimic Datasets (2017). ArXiv Preprint ArXiv:1710.08531
16. L. Shinners, C. Aggar, S. Grace, S. Smith, Exploring healthcare professionals' understanding and experiences of artificial intelligence technology use in the delivery of healthcare: an integrative review. Health Inf. J. **26**(2), 1225–1236 (2020)
17. S. Siddique, J.C.L. Chow, Machine learning in healthcare communication. Encyclopedia **1**(1), 220–239 (2021)
18. T.Q. Sun, R. Medaglia, Mapping the challenges of artificial intelligence in the public sector: evidence from public healthcare. Gov. Inf. Q. **36**(2), 368–383 (2019)
19. J. Waring, C. Lindvall, R. Umeton, Automated machine learning: review of the state-of-the-art and opportunities for healthcare. Artif. Intell. Med. **104**, 101822 (2020)

Exploiting Tacit Knowledge: A Review and Possible Research Directions

Pawankumar Saini and Pradnya Chitrao

Abstract In the corporate world, tacit knowledge is becoming recognised as a differentiator. The value and necessity of focusing on leveraging tacit knowledge are highlighted via a literature study. The important themes covered in this review are the knowledge generation process, information sharing, behaviours that influence tacit knowledge sharing, sharing strategies, and so on. The research examines the different factors that lead to the industrial exploitation of tacit knowledge. To address the research gaps identified, the authors attempt to propose possibilities for future research in this area of discourse. The authors feel that the provided issues for future research will aid academics in further exploring the potential of tacit knowledge exploitation in the sectors, in addition to the research work done by many scholars in the tacit knowledge domain. The authors are conscious that this article has certain limitations because it focuses on tacit knowledge in the industry, although the impact of tacit knowledge in other sectors needs to be investigated as well. This study could lead to new approaches for businesses to tap into the potential of tacit knowledge held by their staff through knowledge production and sharing.

Keywords Knowledge management · Tacit knowledge · Explicit knowledge · Knowledge sharing · Knowledge creation

1 Introduction

The markets are in a volatile state and everything, including the competition, is shifting [1]. Companies that continually develop new information, disseminate it extensively within the firm, and promptly incorporate it into new technologies and

P. Saini (✉)
Symbiosis International University, Pune, Maharashtra, India
e-mail: pawankumarsaini09@gmail.com

P. Chitrao
Symbiosis Institute of Management Studies, A Constituent of Symbiosis International University, Pune, Maharashtra, India
e-mail: pradnyac@sims.edu

© The Author(s), under exclusive license to Springer Nature Singapore Pte Ltd. 2023   771
Y.-D. Zhang et al. (eds.), *Smart Trends in Computing and Communications*, Lecture Notes in Networks and Systems 396, https://doi.org/10.1007/978-981-16-9967-2_73

products are successful. If companies want to improve the efficiency of their resource (input) utilisation and the productivity of their manufacturing processes, they must acquire knowledge [2]. The knowledge-based perspective of the firm sees a company as a knowledge-creating entity, arguing that knowledge, and the ability to develop and use it, is the most essential source of a company's long-term competitive advantage [3]. Employee participation, both willing and active, will determine the effectiveness of knowledge management methods [4]. A key element that we discuss in this review is tacit knowledge (TK) management which starts with creating TK and then sharing it for further use. Acquisition, management, and reuse of TK are more difficult than explicit knowledge (EK). While a lot of research has been done around EK areas, there is very little attention paid to TK [5]. This research will help people from the industry to apply the findings of the research.

2 Literature Review

TK, according to Michael Polanyi, is knowledge that cannot be stated or verbalised. He went on to say that one of the most important aspects of TK is that "we know more than we can tell" [6]. There is the issue of difficulty in articulating what we know, but according to Polanyi, we may not even be aware of what we know or how our TK relates to what we can demonstrate. Because TK is such a personal knowledge, it is only shared when individuals meet face to face and take action. TK is converted to EK, which is then combined with experiences to revert to tacit. To develop new knowledge, there must be a constant upward spiral of information [6]. Nonaka stated that TK is difficult to codify, making it difficult to communicate with others [7]. Because of our deeply formed mental models, beliefs, and viewpoints, we take TK for granted. As a result, articulating TK is challenging [7]. TK can also be defined as a skill acquired by actions and experience, both formally and informally. The main challenges with TK diffusion, according to Haldin-Herrgard T, are the TK's unconsciousness and difficulty in utilising it [8]. As non-codified and disembodied know-how, Howells claims that TK can be obtained by informal take-up of learned behaviour and procedures [9]. Learnings in an unstructured or semi-structured manner characterise TK acquisition and transfer. According to Jones and Leonard, we must address various factors of transforming TK to organisational knowledge in order for many more people to gain and use the knowledge [10]. According to Paolino, Lizcano, Lopez, and Lloret, TK expands with use and so is limitless [11]. According to Mohajan, TK is a dynamic process that is heavily influenced by an individual's social ties and characteristics [12].

3 Objectives of the Study

The purpose of this study is to highlight the various aspects that are significant when it comes to exploiting TK for organisational economic growth. The goal of the research is to discuss the features of TK, TK in the context of organisations, knowledge creation, sharing, and the challenges of managing TK in the workplace. The study's goal is to look at the many aspects of TK management in companies and to identify the practices that are currently being used to reap the benefits of TK. If organisations can fully utilise the potential of TK, they will benefit. We hope this article will help to open the possibilities of better TK management, its use, and to further the discourse of TK in industries.

4 Methodology of the Study

This is a review article and is prepared based on secondary data. A review of existing studies brings out the work that has so far been done and the gaps that remain in the literature on the role of TK in industries. This is followed by questions for future research.

5 Tacit Knowledge

There are two sorts of knowledge: EK and TK [3]. EK is a type of knowledge that includes facts, propositions, symbols, and so on [13]. TK refers to knowledge that is difficult to formalise, such as belief, viewpoint, mental models, concepts, and ideals [3]. While TK and EK are distinct, they are at opposite ends of a knowledge continuum and should not be viewed as antagonistic; rather, it is vital to recognise that we embrace them in various ways [14]. The shift on the spectrum happens based on the changes of the individual's ability to express and formulate knowledge tacitness [15]. Even if TK is embedded in the individual, EK, according to Polanyi, must be grasped tacitly. Polanyi concludes that there is no pure EK and that the tacit element is ingrained in all forms of knowledge [16]. When learning based on knowledge transfer takes place, both TK and EK are utilised. In this context, they are inextricably linked, and this is required to achieve specified learning outcomes [17]. There is a dynamic interaction with one another that allows for the creation of new knowledge and the development of creative individual and communal activity outcomes [18].

6 Knowledge Creation Process

Considering knowledge as a resource, it becomes important that organisations make optimum use of it. They must raise awareness of the process of knowledge generation and the importance of continual TK sharing among individuals, as well as assist in nurturing, articulating, and amplifying it [7]. Nonaka and Takeuchi proposed the SECI (socialisation, externalisation, combination, internalisation) four-step conversion model, which consists of a continuous process of knowledge creation spiral [1]. Socialisation (from TK to TK), externalisation (from TK to EK), combination (from EK to EK), and internalisation (from EK to TK) are the four modes of conversion between TK and EK [3]. At each level of the conversion, different stakeholders are involved. The organisation is responsible for developing knowledge visions, defining objectives, facilitating dialogue at the appropriate time and place, and ensuring a conducive environment for the SECI model's effective and efficient conversions [19].

7 Knowledge Sharing

Knowledge sharing is a critical link that allows new knowledge, products, and services to be created. This fact is only now beginning to be grasped by the organisation's people [14]. Other knowledge management (KM) processes and practises are brought to life by knowledge sharing [20]. When the agents share a same background and experience, mutual relationships make it easier for knowledge transfer to occur [21]. Organisations must create an environment that encourages social connections so that knowledge, experiences, and abilities may be shared across all departments [22]. Given the choice between TK and EK transfers, firms prefer TK transfers as it is difficult to replicate [23]. Knowledge is the engine that drives organisation's performance. Every organisation is trying to grow knowledge or acquire it if needed, and then work on transferring it [24, 25]. Firms that focus on knowledge transfer have a better possibility of developing more knowledge and, as a result, differentiators. These businesses are more likely to expand and become productive [26, 27].

7.1 Methods of Knowledge Sharing

TK could provide a vast array of chances and potentials for discovery and innovation. Regardless of whether it is TK or EK, firms are putting a renewed emphasis on knowledge exchange. As a result, businesses are devising new techniques to enhance knowledge sharing [14]. Some of the ways that TK can be or is being shared in organisations has been shared by researchers. Having co-working spaces can promote inter-domain learning by facilitating TK exchange, synthesising, and sharing of domain-related ideas [28]. Using metaphors and analogies can help individuals externalise their TK and convey it more easily, making it easier for the TK user [2, 29]. Having knowledge containers in place, which is the capture of knowledge in the documented form of values, rules, and procedures can help in knowledge sharing [30]. Knowledge sharing could be ingrained in everyday work routines and informal gatherings [20]. Sharing personal experiences with decisions made, difficulties addressed, and so on can help others to use the same [31, 32]. E-Mentoring (sharing TK through social media) is a great tool to pass on the tacit knowledge [33]. Putting systems in place that allow for face-to-face contacts, like as coaching and mentoring, can considerably aid TK transfer [34]. The complexity of coding and recording is eliminated by bringing individuals together to share their experiences, narratives, stories, and observations [35]. New techniques are shared through demonstrating them in pilot training courses. This aids in the conversion of TK to EK [36]. Learning by emulating the activities of professionals is the simplest, most cost-efficient, and most effective method of passing on TK to others [36–38]. Using the right questioning technique to extract tacit knowledge and then codifying it for sharing with others is of great help [34, 39–42].The method for conveying knowledge will be determined by the setting, audience, and learning domain, among other factors.

8   Tacit Knowledge in Industries

Because EK is knowledge that everyone knows and is public by nature, it is TK that distinguishes the organisation and offers it a competitive advantage [43]. Employee TK is used to drive service organisations, and TK becomes more strategic in terms of delivering performance [44]. Workers learn a lot of TK from their mentors and experienced employees by observing, mimicking, and practising, rather than through procedural manuals. This also aids in the preservation of knowledge within the organisation [1, 45, 46]. The mentor or the experienced individual plays a very important role in transferring the tacit knowledge to the newcomer and guiding him or her [47]. However, little is done to capture workers' TK, and there is a significant loss of skills when employees quit or retire [48]. It becomes vital to have a solid structure in place to capture organisational memory via KM systems, and a dedicated manager is required for this. The manager has operational experience as well as technical knowledge in managing the KM system [49]. Because TK is unevenly distributed, it is critical to set the stage and align collective action, all while looking for methods to improve the quality of TK on all levels [50].

9   Challenges in Managing Tacit Knowledge

The most difficult aspect of managing TK is keeping people who have acquired it because the TK is lost when they leave the company. The loss is further multiplied when the competitive edge is lost due to these individuals joining competing organisations. Previous study has shed light on the various problems that must be overcome. Organisations generally are not able to determine the sort of information leaving the organisation. Organisations are unable to appraise the importance and value of knowledge loss, and hence overlook the seriousness of the situation [51]. The increased use of online transactions is isolating workers who can only work with what they already know and are unable to learn new TK. It becomes challenging to share knowledge across many organisational units due to this [33]. When a large amount of TK is concentrated in the hands of a few individuals, the organisation becomes vulnerable [30]. Depending on people-based approaches alone will always pose a risk of TK loss [30]. When information is lost, it causes challenges such as duplication of effort, the need to find new qualified personnel, and the risk of not regaining all the lost knowledge [48]. When an individual departs the company, critical relationships created with external partner networks are damaged or lost [51]. When top executives leave an organisation, they take with them their working methods and knowledge of the procedures they oversaw. As a result of their lack of basic working expertise, the subordinates are susceptible [32, 48]. Because TK can only be shared through example, the number of possibilities to share it is limited [52]. Organisations are still not geared up to have a proper knowledge transfer and capture mechanism [15]. Organisations have focussed solely on EK, while TK management has gone almost unnoticed [53, 54]. There is a lack of knowledge sharing due to a lack of trust, social networking, self-awareness, and training [55].

10 Research Directions

Future study can be focussed on some of the questions raised as a result of this literature review. Is it true that organisations lose money due to a lack of TK capture? Is it tough to draw top management's attention to TK because of its difficult to express nature? How high on an organisation's strategic objective is TK capture? To what extent does an organisation's TK be captured? When new experienced people join a business, to what extent does TK flow into the organisation? When an employee quits an organisation, how much of the tacit knowledge is lost? To what extent can TK be passed on from one person to the next? To what extent can TK to EK conversion be quantified? To what extent does TK transfer differ between online and in-person working systems? To what extent does tacit knowledge transfer get hampered when units are located geographically apart? To what extent can organisations become vulnerable because of individual TK accumulation? If individuals leave the firm, how will this affect business with external partners? What impact does the departure of a senior leader have on the organisation?

11 Conclusions

One of the key duties of senior management in an organisation, according to researchers, is to create an environment in which people are motivated to share their tacit knowledge and convert it to explicit knowledge for the business's growth. Organisations are ignorant to the financial loss that results from a lack of proper TK management. There are not enough metrics for calculating the amount of TK and EK produced. For the institutionalisation of the knowledge generation process, organisations must create an organisational structure with clear duties. In organisations, there is definitely a disparity in the level of information exchange. There are a lot of strategies for knowledge sharing that have been compiled that the organisation might apply. The literature review shows that businesses are still far from obtaining an ideal level of knowledge management. To complete the TK and EK KM realm, there is still a lot of study and implementation to be done, as well as a lot of questions to be asked and addressed.

References

1. I. Nonaka, H. Takeuchi, *Knowledge-Creating Company* (Knowledge-Creating Company, 1991, 1995)
2. J.G. Woods, From decreasing to increasing returns: the role of Tacit knowledge capital in firm production and industrial growth. J. Knowl. Econ. **10**(4), 1482–1496 (2019). https://doi.org/10.1007/s13132-016-0351-2
3. I. Nonaka, R. Toyama, A. Nagata, *A Firm as a Knowledge-creating Entity: A New Perspective on the Theory of the Firm* (Japan, 2000)
4. D. Chawla, H. Joshi, Knowledge management practices in Indian industries—a comparative study. J. Knowl. Manag. **14**(5), 708–725 (2010). https://doi.org/10.1108/13673271011074854
5. J. Hao, Q. Zhao, Y. Yan, G. Wang, A review of tacit knowledge: current situation and the direction to go
6. M. Polanyi, The tacit dimension, knowledge in organizations, ed by L. Prusak (1966)
7. I. Nonaka, *The Knowledge-Creating Firm* (Harvard Business Review, 1991)
8. T. Haldin-Herrgard, Difficulties in diffusion of tacit knowledge in organizations. J. Intell. Capital **1**(4.), 357–365 (2000). https://doi.org/10.1108/14691930010359252
9. J. Howells, Tacit knowledge, innovation and technology transfer. Technol. Anal. Strate. Manag. **8**(2), 91–106 (1996). https://doi.org/10.1080/09537329608524237
10. K. Jones, L.N.K. Leonard, *From Tacit Knowledge to Organizational Knowledge for Successful KM* (2009), pp. 27–39. https://doi.org/10.1007/978-1-4419-0011-1_3
11. L. Paolino, D. Lizcano, G. López, and J. Lloret, A multiagent system prototype of a tacit knowledge management model to reduce labor incident resolution times. Appl. Sci. (Switzerland) **9**(24) (2019). https://doi.org/10.3390/app9245448
12. H.K. Mohajan, Sharing of tacit knowledge in organizations: a review (2016). [Online]. Available http://www.openscienceonline.com/journal/ajcse
13. M. Polanyi, Tacit knowing: its bearing on some problems of philosophy. Rev. Mod. Phys. **34**(4) (1962). https://doi.org/10.1103/RevModPhys.34.601
14. M.A. Fauzi, N. Paiman, A critical review of knowledge sharing in various industries and organizations. Int. J. Sci. Technol. Res. **8** (2020) [Online]. Available www.ijstr.org
15. C.H. Chuang, S.E. Jackson, Y. Jiang, Can knowledge-intensive teamwork be managed? Examining the roles of HRM systems, leadership, and tacit knowledge. J. Manag. **42**(2), 524–554 (2016). https://doi.org/10.1177/0149206313478189
16. D. Lesjak, S. Natek, Knowledge management systems and tacit knowledge. Int. J. Innov. Learn. **29**(2), 166 (2021). https://doi.org/10.1504/ijil.2021.10034239
17. J. Roberts, From know-how to show-how? Questioning the role of information and communication technologies in knowledge transfer. Technol. Anal. Strat. Manag. **12**(4), 429–443 (2000). https://doi.org/10.1080/713698499
18. I. Nonaka, G. von Krogh, Tacit knowledge and knowledge conversion: controversy and advancement in organizational knowledge creation theory. Organ. Sci. **20**(3), 635–652 (2009). https://doi.org/10.1287/orsc.1080.0412
19. I. Nonaka, R. Toyama, T. Hirata, Managing flow a process theory of the knowledge-based firm-managing flow. https://doi.org/10.1057/9780230583702preview

20. A. Abdelwhab Ali, D.D.D. Panneer Selvam, L. Paris, A. Gunasekaran, Key factors influencing knowledge sharing practices and its relationship with organizational performance within the oil and gas industry. J. Knowl Manag. 23(9), 1806–1837 (2019). https://doi.org/10.1108/JKM-06-2018-0394
21. D.J. Teece, The market for know-how and the efficient international transfer of technology. Ann. Am. Acad. Polit. Soc. Sci. 458(1), 81–96 (1981). https://doi.org/10.1177/000271628145800107
22. H.F. Lin, Knowledge sharing and firm innovation capability: an empirical study. Int. J. Manpower 28(3–4), 315–332 (2007). https://doi.org/10.1108/01437720710755272
23. D.R. Williams, Knowledge transfers in the US biopharmaceutical market during a time of transition. J. Pharma. Innov. 15(3), 445–454 (2020). https://doi.org/10.1007/s12247-019-09395-3
24. R. Teigland, M. Wasko, Knowledge transfer in MNCs: examining how intrinsic motivations and knowledge sourcing impact individual centrality and performance. J. Int. Manag. 15(1), 15–31 (2009). https://doi.org/10.1016/j.intman.2008.02.001
25. R. van Wijk, J.J.P. Jansen, M.A. Lyles, Inter-and intra-organizational knowledge transfer: a meta-analytic review and assessment of its antecedents and consequences (2008)
26. T. Kostova, Transnational transfer of strategic organizational practices: a contextual perspective (1999)
27. A.C. Inkpen, E.W.K Tsang Wayne, Knowledge transfer (2005)
28. R. Bouncken, M.M. Aslam, Understanding knowledge exchange processes among diverse users of coworking-spaces. J. Know. Manag. 23(10), 2067–2085 (2019). https://doi.org/10.1108/JKM-05-2018-0316
29. P. Busch, *Tacit Knowledge in Organizational Learning* (IGI Pub, 2008)
30. P. Wethyavivorn, W. Teerajetgul, Tacit knowledge capture in Thai design and consulting firms. J. Constr. Developing Countries 25(1), 45–62 (2020). https://doi.org/10.21315/jcdc2020.25.1.3
31. B. Saeed, A. Mahmood, and A. Saeed, Tacit knowledge sharing in technology-based firms: role of organization citizenship behavior and perceived value of knowledge. Int. J. Sci. Technol. Res. (2020) [Online]. Available www.ijstr.org
32. P. Dogra, A.E. Sparkling, Supervisors' reliance on tacit knowledge and impediments to knowledge sharing in trades
33. Z. Chen, D. Vogel, T. Yang, J. Deng, The effect of social media-enabled mentoring on online tacit knowledge acquisition within sustainable organizations: a moderated mediation model. Sustainability (Switzerland) 12(2) (2020). https://doi.org/10.3390/su12020616
34. K. Plangger, M. Montecchi, I. Danatzis, M. Etter, J. Clement, Strategic enablement investments: exploring differences in human and technological knowledge transfers to supply chain partners. Ind. Mark. Manag. 91, 187–195 (2020). https://doi.org/10.1016/j.indmarman.2020.09.001
35. A.C. Inkpen, A. Dinur, Knowledge management processes and international joint ventures. Organ. Sci. 9(4), 454–468 (1998). https://doi.org/10.1287/orsc.9.4.454
36. D. Asher, M. Popper, Tacit knowledge as a multilayer phenomenon: the 'onion' model. Learning Organ. 26(3), 264–275 (2019). https://doi.org/10.1108/TLO-06-2018-0105
37. A.S. Reber, An evolutionary context for the cognitive unconscious. Philosophical Psychol. 5(1), 33–51 (1992). https://doi.org/10.1080/09515089208573042
38. G. Csibra and G. Gergely, Natural pedagogy as evolutionary adaptation. Philosophical Trans. Roy. Soc. B: Biolog. Sci. 366(1567). 1149–1157 (2011). (Royal Society). https://doi.org/10.1098/rstb.2010.0319
39. M. Bhardwaj, J. Monin, Tacit to explicit: an interplay shaping organization knowledge. J. Knowl. Manag. 10(3), 72–85 (2006). https://doi.org/10.1108/13673270610670867
40. T. Gavrilova, T. Andreeva, Knowledge elicitation techniques in a knowledge management context. J. Knowl. Manag. 16(4), 523–537 (2012). https://doi.org/10.1108/13673271211246112
41. M. Sh, Al-Qdah, J. Salim, A conceptual framework for managing tacit knowledge through ICT perspective. Procedia Technol. 11, 1188–1194 (2013). https://doi.org/10.1016/j.protcy.2013.12.312

42. M. Supanitchaisiri, O. Natakuatoong, S. Sinthupinyo, The innovative model for extracting tacit knowledge in organisations (2020)
43. R. Seidler-de Alwis and E. Hartmann, The use of tacit knowledge within innovative companies: knowledge management in innovative enterprises. J. Knowl. Manag. **12**(1), 133–147 (2008). https://doi.org/10.1108/13673270810852449
44. J.-N. Ezingeard, S. Leigh Ernst, Y. Rebecca Chandler-Wilde, J.-N. Ezingeard, S. Leigh, R. Chandler-Wilde, Knowledge management at Ernst & Young UK: getting value through knowledge flows (2000). [Online]. Available http://aisel.aisnet.org/icis2000/93
45. I. Nonaka and A. Y. Lewin, A dynamic theory of organizational knowledge creation
46. P. Letmathe, M. Rößler, Tacit knowledge transfer and spillover learning in ramp-ups. Int. J. Oper. Prod. Manag. **39**(9–10), 1099–1121 (2019). https://doi.org/10.1108/IJOPM-08-2018-0508
47. P.M. Hildreth, C. Kimble, The duality of knowledge (2002)
48. S.M. Ferdous Azam, J. Tham, A. Albattat, Psycho-social perspectives of knowledge sharing and job performance in Malaysia: conceptual articulation. [Online]. Available www.ijstr.org
49. J.Cárcel-Carrasco, J.A. Cárcel-Carrasco, E. Peñalvo-López, Factors in the relationship between maintenance engineering and knowledge management. Appl. Sci. (Switzerland) **10**(8) (2020). https://doi.org/10.3390/APP10082810
50. I. Nonaka, R. Toyama, Strategic management as distributed practical wisdom (phronesis). Ind. Corp. Change **16**(3), 371–394 (2007). https://doi.org/10.1093/icc/dtm014
51. A. Daghfous, O. Belkhodja, Managing talent loss in the procurement function: insights from the hospitality industry. Sustainability (Switzerland) **11**(23) (2019). https://doi.org/10.3390/su11236800
52. O. Ibert, Towards a geography of knowledge creation: the ambivalences between 'knowledge as an object' and 'knowing in practice'. Reg. Stud. **41**(1), 103–114 (2007). https://doi.org/10.1080/00343400601120346
53. J.R. Gamble, Tacit vs explicit knowledge as antecedents for organizational change. J. Organ. Change Manag. **33**(6), 1123–1141 (2020). https://doi.org/10.1108/JOCM-04-2020-0121. (Emerald Group Holdings Ltd.)
54. K.E. Sveiby, Disabling the context for knowledge work: The role of managers' behaviours. Manag. Decis. **45**(10), 1636–1655 (2007). https://doi.org/10.1108/00251740710838004
55. R. Anwar, M. Rehman, K.S. Wang, M.A. Hashmani, Systematic literature review of knowledge sharing barriers and facilitators in global software development organizations using concept maps. IEEE Access **7**, 24231–24247 (2019). https://doi.org/10.1109/ACCESS.2019.2895690

An Efficient Approach to Stamp Verification

Ha Long Duy, Ha Minh Nghia, Bui Trong Vinh, and Phan Duy Hung

Abstract Stamps have become one of the most important security features in big companies where huge amounts of documents need processing daily. The stamp attached to a document is used to determine the authenticity of that document so that it is necessary to identify whether a stamp is forged or genuine. However, nowadays, it is easier for the general public to forge stamps. This paper presents a practical approach for stamp verification, based on the three stages process similar to some previous work: stamp segmentation, classification stamp or non-stamp, and stamp authenticity verification. In each stage, this work tries and tests new algorithms/methods to give a new way of solving the problem in each stage. Firstly, in our approach, an unsupervised learning machine method is implemented to detect all the objects in the input image, so all the regions including stamps and text are extracted. Next, two separate models of support vector machine classification are constructed. The first one is to distinguish between stamps and other objects in a document. The second model will determine the object which was classified as stamps in the first model whether it is genuine or not. The results show that this approach can perform the stamp verification tasks effectively.

Keywords Stamps verification · Image segmentation · Support vector machine

H. L. Duy · H. M. Nghia · P. D. Hung (✉)
FPT University, Hanoi, Vietnam
e-mail: hungpd2@fe.edu.vn

H. L. Duy
e-mail: duyhlhe141012@fpt.edu.vn

H. M. Nghia
e-mail: nghiahmhe140299@fpt.edu.vn

B. T. Vinh
Hanoi Procuratorate University, Hanoi, Vietnam
e-mail: vinhbt@tks.edu.vn

1 Introduction

Nowadays, security is an indispensable aspect in the operation of big companies, especially in banks, insurance companies, and financial companies. Every day, thousands of invoices, contracts, certificates, and documents are handled. Therefore, it is necessary to have a way to guarantee the authenticity of the content. The stamp is one of the most widely used security methods, along with handwritten signatures. However, with the popularity of the Internet these days and the availability of technology, it is easier for the general public to forge stamps. Big companies process a huge volume of documents everyday so that the risk of forged stamps is very big. Companies can lose hundreds of thousands or even millions of dollars just by errors caused by forged stamps. Moreover, forged stamps can affect the company's reputation and the customer's trust [1]. In some banks, stamp verification relies on manual handling work instead of using software [2]. Therefore, there are many cases where the forged document is accepted as genuine. In order to solve this problem, an automated stamp verification method is needed to be developed.

In this paper, the process of stamp verification inherits the process by many previous works which divide it into three stages. This work follows this three stages process closely but for each stage, the way of implementation is different with different methods and algorithms being applied. The goal is to present a new practical method for this stamp verification problem. First, the stamp must be identified and extracted from the document image. By using color space transformations and k-means clustering, the scanned color image is split according to colors and then used the XY-cut algorithm; all the components in the image are segmented into candidates which contain both stamps candidates and other elements like logos and text. The second stage is to classify the extracted candidates to identify which candidates are stamps, which are not stamp. Each candidate image is passed through a support vector machine (SVM) model to identify whether that candidate is a stamp or not. After getting all the stamps, in the final stage, another SVM model will handle the task of classifying whether the stamps are genuine or forged.

2 Related Works

Unlike handwritten signatures, stamp verification is quite limited in terms of number of published works. Many previous works only focus on the task of detecting and segmenting stamps from a document. The majority of these works applied many different color-based, [3] shaped-based features [4], or used geometric features with key point descriptors to detect and segment stamps [5]. Their ultimate goal is to separate stamps from logos, text, and other information in original document images. Chen et al. [6] developed a method to detect stamps on checks of Chinese banks with a region-growing algorithm. Micenková el at [7] presented a new method for detecting and extracting stamps of various colors, even stamps that overlap with a signature or text of a different color. There are two steps in this work, first is to separate every part of the image from text and background with color clustering; then, all the candidates obtained are classified to be whether stamp or not by using a set of features. In many works, which are dedicated to stamp detection and segmentation, the work that is worth paying attention to is the work by Younas et al. [8] with the name "D-StaR: A Generic Method for Stamp Segmentation from Document Images." They proposed a brand-new approach called D-StaR based on deep learning, capable of handling stamp segmentation in any color, shape, size, orientation, and this approach can even detect overlapping stamps. They used fully convolutional networks (FCNs) to segment stamp masks from scanned document images. Moreover, for pixel-based evaluation and reforming the original stamps, contour refinement is performed to the expected masks. D-Star is the first method to use deep learning to segment stamps and get amazing results. It is also the basis for other studies such as [7, 9, 10].

Many other previous works solved the stamp verification entirely, both stamp segmentation and stamp authentication verification. Chung et al. [11] focused on a dataset of Chinese antique seal/stamp; in the stamp segmentation step, they proposed a method based on geometric transformation and geometric transformation to find borders and align the perspective for two imprints. Then, to verify the authenticity of stamps, they calculated the similarity by PSNR and SSIM indexes as the detection metrics. Chung et al. improved their method in the next work [12]. The work by Takahiko Horiuchi [13] brings up a problem that many techniques for seal/stamp verification only solve this problem as a general pattern matching problem; therefore, the paper presented a new approach to deal with this problem. In the experiment step, an interesting way to test the method is used; the paper used both binary reference images and 3D reference to test the accuracy of the method. Works [14–17] presented some different ways in the stamp verification problem which include: using edge difference histogram, neural network, based on average relative error, and judging the percent of difference inside and difference outside to determine whether a stamp is genuine or not.

Finally, there are two works that inspired us to develop three stages of stamp verification in this paper, all by the author Micenkov et al. The first work [10] introduced three main stages of the problem. First, they segmented all every component in the original document image. Then, they used SVM to classify whether that component is stamp or not. Finally, they used features like color, shape, and print to verify stamps. With the same process, the work by Micenková el at [9] inherited two of their previous work [7, 10], and their work is improved by using k-means clustering; additionally, in the segmentation stage, they still used features to classify genuine and forgered stamps. In this study, we try and test new algorithms/methods to propose a best way for solving the problem in each stage.

3   Methodology

In this work, the hardcopy of the document is scanned in color, and the image is segmented completely. Candidates for solving the stamp identification problem are identified as rectangular segments, from which features are extracted. Candidates are classified as stamps or non-stamps using a binary classification system (logos, text etc.). Segments designated as stamps are further categorized to distinguish between legitimate and fake stamps; the process is shown in Fig. 1. Stamps are considered single-color (blue, green, red, etc.) items in this study. As a result, the primary principle behind detecting them in a picture is to group components that are the same color and are close to one another. Special examples of multiple-color stamps are detected as several stamps, which are subsequently combined.

This work identifies candidate segments by using this principle. One candidate segment is formed by each stamp in the image. If the image contains color logos or pictures, these (or single-color parts of them) are also identified as candidate segments. Then determine which parts match the genuine stamp and which are forged. The following section will give detail of this approach.

3.1   Segmentation

In the segmentation stage, the goal is to detect all the components in the original document image and crop each of those components into a new image. These new images will serve as a new material for the next classification stages. Because of the significant correlation among the channels, the RGB color model is not suitable for image segmentation. In order to work with color stamps, separate the backdrop, black text, and other roughly achromatic (white, black, and gray) components of the image

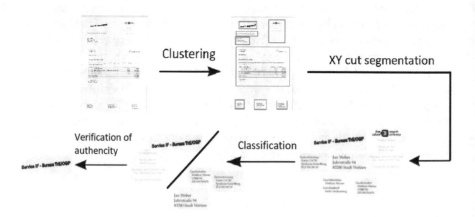

Fig. 1 Stages of the stamp verification

first, so the image is converted from an RGB image into a grayscale image. After that Otsu's binarization [18] with a threshold value of 200 is applied to split the image into background and object. The image is also blurred with Gaussian blur pixels of similar features are desired to be assigned the same label, spatially continuous pixels are desired to be assigned the same label, and the number of unique labels is desired to be large. With the image processed and ready to go, the next step is to find every cluster in the image; once again, the Otsu's binarization [18] is applied along with k-mean clustering is also applied. For each cluster, canny edge detection [19] of OpenCv is implemented to detect every edge on that cluster. Also, more OpenCv operations are applied including: cv2.dilate to make the object thicker, morphologyEx to close the cluster to the object, and then, finding the contour can be easy. Last, the largest contour is selected then crop each object detected and save it as a new image and move it to a different folders.

In order to segment every candidate properly, the XY-cut algorithm [20] is used to recursively partition the page into rectangles, resulting in candidate solutions with the smallest bounding boxes. In conclusion, the method in the first stage, we introduced above has four steps: (1) Preprocess image by resize, binarize, bur; (2) Find cluster; (3) For each cluster, detect edges, close figures, find contours; (4) Select largest contours. This method can be considered to be a good and simple way to deal with stamp segmentation problem.

3.2 SVM Classification: Stamp/non-Stamp

Now, all the candidates are segmented from the last step; those candidates contain all kinds of non-stamp candidates such as logos, parts of picture, or even forged stamps. First, the task is to classify whether that candidate is stamp or not, and finally, with all the candidates which were identified as stamp, another classification is needed to determine that stamp candidate is genuine or fraudulent. This process is clarified in Fig. 2.

This paper uses support vector machine for both classification models in each of two stages presented above. It is a supervised learning algorithm that is primarily used to categorize data. In this work, Linear SVM is used; the regularizer is applied in the final output layer, and the hinge loss is used as a loss function.

3.3 SVM Classification: Genuine Stamp/forged Stamp

Our final stage, also the most important stage, the task is to distinguish between genuine and forged stamps. In Fig. 3 is a sample of genuine and forged stamp, the difference can be recognized in hue and sharpness. Therefore, feature extraction is a suitable method in this case. This work uses several feature extractors from the scikit image library for feature extraction. These extractors help to create feature vectors,

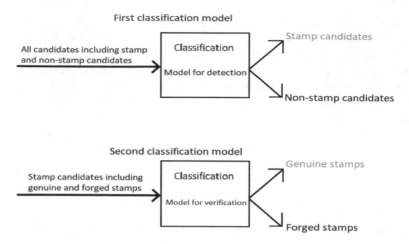

Fig. 2 Two models for detection and verification

Fig. 3 Sample genuine and
copied stamp after
segmentation

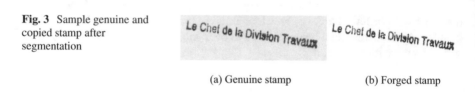

(a) Genuine stamp   (b) Forged stamp

which are then fed into the SVM classifier. The purpose is to test various scikit image feature extractor and select the ones that produce the best results. The following is the list of extractors used in this work: oriented FAST and rotated BRIEF (ORB), binary robust independent elementary features (BRIEF), center surround extremas for real-time feature detection and matching (CENSURE), histogram of the grayscale image, histogram of oriented gradients, and corner peaks.

For stamps, there are many different features such as hue, uniformity, and shape. In many cases, using a certain features extractor doesn't help to improve or even make classification performance worse. Therefore, it is necessary to determine which features extractors to combine to achieve the best result. A process needs to be developed to find out which features should be utilized. A sample of 100 images is extracted from the training set and then fed into the SVM model to correctly estimate which features reduce model accuracy. The accuracy is returned, and the features list with the highest accuracy is chosen. It was discovered that including the HOG and grayscale histogram extractors always produced more positive results, which is why they are used automatically to extract their respective features and feed them into the model. For the classifiers, the following models are chosen and implemented: logistic regression, SVM adj, SVM linear, SVC, k-nearest neighbors, decision tree, random forest, AdaBoost, Naive Bayes, gradient boosting, latent Dirichlet allocation. The feature vectors obtained from extractors are used as the input for these classifiers.

In the SVM classifier, it is necessary to choose two parameters C and gamma; these parameters have a big influence on the resulting accuracy. For this problem, a SkLearn machine learning library method called "GridSearchCV" [21] which can compute the best possible parameter is applied. With C, the value is determined from 1.0e-02 to 1.0e10. With gamma, the value is determined from 1.0e-09 to 1.0e3. This helps to increase the accuracy significantly compared to the default parameter C and gamma.

This testing process to find the best performing feature extractor is an improvement in this work. It helps to use feature extractor and classification algorithm and get the best performance in stamp verification possible.

4 Experiments and Results

4.1 Dataset and Evaluation

Evaluation is performed separately for stages: detection, classification, and verification. The public dataset of 400 document pages is used and available at http://madm. dfki.de/downloads-ds-staver. Stamped invoices with color logos and text are included in this dataset. Stamps are frequently layered with text or other objects. It was created by printing automatically generated invoices, manually stamping them, and scanning them in color at 600dpi. Lower resolutions were obtained by downscaling to reduce the effort of ground-truth labeling.

To evaluate the performance of verification, the dataset is split into training set and test set with the ratio 8:2. All of the models in this work are trained and tested on these two sets.

4.2 Result

The results obtained from two classification models are very positive. For the first SVM model, classify stamp or non-stamp, after training on a full dataset, the mean accuracy is 0.9. For the second classification, several extractors are used and tested to find the best combination of features among every possible combination of four extractors: ORB, corner peaks, BRIEF, and CENSURE. As presented in the previous section, histogram of grayscale image and HOG always produced better results so that always have to include them. After implementing with many classify algorithms, the best result of the mean accuracy is 0.96 as shown in Table 1.

The results show that this approach can be considered to be a good option when it comes to stamp verification problems. The process of determining which is the best way to utilize features extractor is a good improvement and can be applied in many other classification problems.

Table 1 Result by classifiers

Classifier name	Accuracy
Logistic regression	0.93
SVM, adj	0.82
SVM, linear	0.94
SVC	0.96
K-nearest neighbors	0.89
Decision tree	0.91
Random forest	0.97
Random forest 2	0.82
AdaBoost	0.93
Naive Bayes	0.84
Gradient boosting	0.93
Latent Dirichlet allocation	0.94

5 Conclusion and Future Works

This paper introduces stamp verification methods and algorithms. In each stage, we try, test, and recommend the best model for the process. Hence, this approach can be considered a simple and effective way to handle stamp verification problems. Other works can apply this approach, especially the works involving image segmentation or image classification.

In future work, we would like to develop this approach in more kinds of stamps. Right now, this paper mainly focuses on stamps from the public dataset that consist of entire stamps from foreign countries. The idea is to shift the focus to Vietnamese stamps, which contain Vietnamese letter patterns, symbols on them. This would make it easier to apply stamp verification methods for Vietnamese companies and organizations.

References

1. SBA Stone Forest.: Mitigating Risks Associated with Seals. Stone Forest Business Advisors. https://www.stoneforest.com.sg/business-advisors/Articles/Article/mitigating-risks-associated-with-seals. Accessed 1 Oct 2021
2. W. Gao, S. Dong, X. Chen, A system for automatic Chinese seal imprint verification, in *Proceedings of 3rd International Conference on: Document Analysis and Recognition, Montreal, Quebec* (vol. 2, 1995), pp. 660–664
3. B. Micenkov, J.V. Beusekom, Stamp detection in color document images. ICDAR 1125–1129 (2011)
4. Y. Bhalgat, M. Kulkarni, S. Karande, S. Lodha, Stamp processing with examplar features. arXiv:1609.05001 (2016)
5. P. Forczmanski, A. Markiewicz, Low-level image features for stamps detection and classification. CORES **2013**, 383–392 (2013)

6. L. Chen, T. Liu, J. Chen, J. Zhu, J. Deng, S. Ma, Location algorithm for seal imprints on Chinese bank-checks based on region growing. Optoelectron. Lett. **2**, 155–157 (2006)
7. B. Michenkova, J.V. Beusekom, Stamp detection in color document images. *Document Analysis and Recognition (ICDAR)* 1125–1129 (2011)
8. J. Younas, M.Z. Afzai, M.I. Malik, F. Shaifat, P. Lukowícs, S. Ahmed, D-StaR: a generic method for stamp segmentation from document images, in *Proceedings of 14th IAPR International Conference on Document Analysis and Recognition (ICDAR)* (2017). https://doi.org/10.1109/ICDAR.2017.49
9. B. Michenkova, J.V. Beusekom, F. Shafait, Stamp verification for automated document authentication. Comput. Forensics 117–129 (2015)
10. B. Michenkova, Verification of authenticity of stamps in documents.http://acmbulletin.fiit.stuba.sk/vol3num4/micenkovaSPY.pdf. Accessed 1 Oct 2021
11. W.H. Chung, M.E., Wu, Y.L. Ueng, Y.H. Su, Forged seal imprint identification based on regression analysis on imprint borders and metrics comparisons, in *Proceedings of IEEE Conference on Dependable and Secure Computing* (2018)
12. W.H. Chung, M.E. Wu, Y.L. Ueng, Y.H. Su, Seal imprint verification via feature analysis and classifications. Futur. Gener. Comput. Syst. **101**(12), 458–466 (2019)
13. T. Horiuchi, Automatic seal verification by evaluating positive cost, in *Proceedings of Proceedings of Sixth International Conference on Document Analysis and Recognition* (2002), pp. 572–576
14. H. Jin, Z. Hao, L. Tiegen, Seal imprint verification using edge difference histogram, in *Proceedings of Optoelectronic Imaging and Multimedia Technology II* (2012), p. 855804
15. J. Hong-yu, G. Zhen, Research of the seal imprint time verification method based on neural network, in *Proceedings of International Conference on Intelligent System Design and Engineering Application* (2010), pp. 170–173
16. J. Liang, X. Tong, Z. Yuan, The circular seal identification method based on average relative error. Appl. Mech. Mater. 513–517 (2014)
17. W. Shuang, L. Tiegen, Research on registration algorithm for check seal verification, in *Proceedings of Electronic Imaging and Multimedia Technology V* (2007), p. 68330Y
18. N. Otsu, A Threshold Selection Method from Gray-Level Histograms. IEEE Transactions on Systems, Man, and Cybernetics vol. 9, no. 1, pp. 62–66 (1979) https://doi.org/10.1109/TSMC.1979.4310076
19. J.F. Canny, A computational approach to edge detection. IEEE Trans. Pattern Anal. Mach. Intell. (PAMI) **8**(6), 679–698 (1986)
20. P. Sutheehanjard, W. Premchaiswadi, A modified recursive X-Y cut algorithm for solving block ordering problems, in *Proceedings of 2nd International Conference on Computer Engineering and Technology* (2010) , pp. V3-307–V3-311
21. G. Ranjan, A.K. Verma, R. Sudha, K-nearest neighbors and grid search CV based real time fault monitoring system for industries, in *Proceedings of IEEE 5th International Conference for Convergence in Technology* (2019), pp. 1–5

Fragmented Central Affinity Approach for Reducing Ambiguities in Dataset

Tanvi Trivedi and Mahipal Singh Deora

Abstract It is always a known fact that the role of data and its purity is very crucial in the data mining. The key role of data in the data mining is related from decision-making. It is well-known fact that if data are impure, then result will be a false picture. This crucial stage is also known as the ambiguities in datasets. Anomalous or irregular value in database is solitary of the biggest problems faced in data analysis and in data mining applications. Data preprocessing for the data mining is a key phase which is crucial place where ambiguities of database can be reduce or remove. The present study proposed an algorithm which tries to solve the problem related to an anomalous and irregular values, i.e., outliers, inliers, and missing values from a real-world imbalanced database. The study projected is based on the fragmented central affinity approach for reducing ambiguities in dataset.

Keywords Central affinity · Ambiguities · Dataset · Anomalous values · Missing values · Attribute

1 Introduction

Data in the dataset and database are denoted as the base unit. But, it is one of the most important segment of the whole data mining. Because the entire result is depends on the completeness of the data as well as the dataset. In the age of data science and analytics, the role of data is too crucial. Now, it becomes backbone of the information technology age. In the data mining or knowledge discovery in the database, the role of data is very crucial as it is base for decision taking. Data in the form of consolidated result of the analysis are more important to take decision. Likewise, the completeness of data as well as dataset is critical for any result or decision formations. Thus, the role of data preparation for data mining is the subject of key consideration.

This research paper is associated with an anomalous values which are identified and get a treatment to overcome from the characteristics of the anomalous values to

T. Trivedi (✉) · M. S. Deora
Bhupal Nobel's University, Udaipur, India
e-mail: tanvitri@gmail.com

remove ambiguities. The approach given here is an accommodation of several logics which need to reduce the ambiguities in respect of humanoid data entry approach and prone failures as well as improper evaluations. Consequently, it is valued to reconnoiter with anomalous data on or after the data sources used.

2 Methodology

The study tries to address the given problem in two consecutive section in which first section is related to identify the problem area. This section is completely associated with the process of pointing the anomalous values. The second area is associated with recovery of the missing values by the algorithm proposed for the study.

As the expansion point of view, both the stages are manage in the single elaborative algorithm which had a systematic approach to address the problem and produce conclusions to reduce the ambiguities.

The approach of reducing ambiguities covers several phage. These phage covers reading and importing dataset where missing/anomalous data are available. Afterward, the algorithm identifies the point of anomalous values. The major task which is taken in the consideration here is as follows:

- Reading and importing the dataset
- Deliberation of anomalous values
- Recover anomalous/missing values
- Test the significance.

3 Algorithm

The proposed approach in the form of algorithm is given below; this approach gives a picture to recover the missing values with the help of attribute vicinities. This approach developed the view of predicted data from the same attribute, by utilizing new concept of fragmented central affinity approach for reducing ambiguities in dataset. We can consider the details of the steps in the regular flow to reduce the ambiguities in the dataset.

Attribute/Variable/Dataset:

$V = (v_1, v_2, v_3 \ldots \ldots v_n)$.

where $V = V_{alb} + V_{aml} + V_{mis}$.

$V_{abl} = (v_1, v_2, v_3 \ldots \ldots v_k)$,

$V_{aml} = (v_{k+1}, v_{k+2}, v_{k+3} \ldots \ldots v_m)$.

$V_{mis} = (v_{m+1}, v_{m+2}, v_{m+3} \ldots \ldots v_n)$.

// after conversion of anomalous values in null values..............

Import: $V = (v_1, v_2, v_3 \ldots \ldots v_n)$.

for $(i = 1 \text{to } n)$,

If (value $(v_i) = = $ NULL), then.

v_{a2} = value (v_{i-2})// Value of second above from v_i.
v_{a1} = value (v_{i-1})// Value of first above from v_i.
v_{b1} = value (v_{i+1}) // Value of first below from v_i.
v_{b2} = value (v_{i+2})// Value of second below from v_i.
V_{temp} =0 // Initialization of temporary variable
$V_{temp}(SUM) = \sum_{i=i-2}^{i+2} vi$
$V_{temp} = V_{temp}(SUM)/n$ where n=no. of values in fragmented array
Dv1, Dv2,Dv3 =0// Initialization of temporary variable
Dv1= (val (v_{i-1}) –val (v_{i-2}))
Dv2= ((val (v_{i+1}) –val (v_{i+2})) /2)
Dv3 = (val (v_{i+2}) –val (v_{i+1})).
Dv =0 // Initialization of variable
V_{pred} =0 // Initialization of variable
Dv = ((Dv1 + Dv2 + Dv3) / 3).
V_{pred} = Dv + V_{temp}.
val (v_i) = V_{pred} // Assigning estimated value to missing value place.
$i = i + 1$
repeat until $(i > = n)$.
Stop.

4 Dataset

The dataset consists of detail about Global Carbon Dioxide Emission from fossil burning by fuel type (Million tons) from the year 1964 to 2013. Thus, it is set of 50 records in the every column. The study includes three variables/attributes, namely year, coal, and natural gas on which study is conceded.

5 Operation Procedure

The dataset with diverse variables was occupied for the study. In the strategic way, after conversion of anomalous values in the missing values, the same is executed, the ratio of the deliberated missing values is 15%.

The projected algorithm alters the anomalous values of the dataset into the null values that is finally assumed as the missing values; after that these cases are resolved through the statistical inference.

The dataset of the study has been divided into three main classes as standard datasets, second missing attribute values, and lastly dataset with predicted or recovered values.

The consequence and data consolidation acknowledged from the dataset with 15% missing values. To get the solo value for projecting whole attribute, some statistical consolidation is taken in the deliberation; these are arithmetic mean, standard deviation, and coefficient of variation. To test the significant of the result, analysis of variance ANOVA is utilized.

6 Data Analysis

Analysis with Mean: As per the Mean Values of variable of coal in different datasets are as mean received from the normal dataset about variable coal is 2351. After considering the 15% missing values in the same attribute, the mean values goes to 2049. Then, after approximation, the mean value comes on 2358.

For variable natural gas, the mean received from normal dataset is 994. Further, by including 15% of missing or null values, it goes to 845. After approximation, the consolidated mean values received are as 994. Therefore, it is observed that after approximation of missing values, the mean value of normal attribute is similar to approximated attribute for all three variables.

Standard Deviation: According result of standard deviation for variable coal, standard deviation for the normal dataset is received as 764. Similarly, the values of standard deviation for the 15% missing values of dataset are received 797 which is slightly differ from the after approximation values standard deviation is received as 766.

For variable natural gas, the values of standard deviation for the normal dataset for the variable natural gas are acknowledged as 429.07. Whereas, the standard deviation values with 15% missing values are acknowledged as 430; after approximation, value of standard deviation is observed as 429.

Analysis (Co-efficient of Variation): As per the values observed for the co-efficient of variation, it is observed that variables independently show similar nature. The likeness in the CV values for the entire datasets shows that there is homogeneousness in the variable values before and after the approximation.

ANOVA for Variable Coal

As per the hypothesis considered for the analysis of variance, it is assumed that as:

H0: $\mu 1 = \mu 2 = \mu 3$.

The alternate hypothesis assumed as at least one of the μ equivalences not kept which means one set mean value is not equal.

The test of ANOVA is applied on the dataset of coal (Table 1).

Table 1 Analysis of variance for coal

Source of variation	SS	df	MS	F
Between groups	24,232.6092	2	12,116.3	0.020196
Within groups	83,990,998.1	140	599,935.7	
Total	84,015,230.7	142		

Critical/tabulated value: $F (2, 140: 0.5\%) = 3.060$

Table 2 Analysis of variance for natural gas

Source of variation	SS	df	MS	F
Between groups	6549.09295	2	3274.546	0.017763
Within groups	25,808,554.1	140	184,346.8	
Total	25,815,103.2	142		

Critical/tabulated value: $F (2, 142: 0.5\%) = 3.0608$

7 Discussion of Result

According to the result, it is found that F (calculated) $0.020 < 3.060$ F (critical). Hence, it shows that null hypothesis (H_0) is accepted at the 0.5% level of the significance.

ANOVA for Variable Natural Gas
As per the hypothesis considered for the analysis of variance, it is assumed that as:
  H0: $\mu 1 = \mu 2 = \mu 3$
  The alternate hypothesis assumed as at least one of the μ equivalences not kept which means one set mean value is not equal.
  The test of ANOVA is applied on the dataset of natural gas (Table 2).
  As per the result, it is observes that F (calculated) $0.01776 < 3.060$ F (Tabulated). Therefore, it can be commit that null hypothesis (H_0) is accepted at the 0.5% level of the significance. It shows that there is no significant difference amid the various set of particularly variable natural gas in concerning to the mean values.

8 Conclusion

As per the result observed from the analysis the test of measurement of central tendency, this is noted that projected algorithm gives vital role to recover the missing values, and this is also observed from the result of the SD and values of CV. Thus, we can say that the result received from the algorithm supports dataset to reduce the ambiguities. According to the ANOVA result, the recommendation for the algorithm is very strong, and algorithm is very much suitable for small dataset till 300 to 500 records of time series.

References

1. A. Duraj, A. Niewiadomski, P.S. Szczepaniak, Outlier detection using linguistically quantified statements. Int. J. Intell. Syst. **33**(9), 1858–1868 (2018)
2. D.D. Pandya, S. Gaur, *Detection of Anomalous Value in Data Mining*, vol. 2 (Kalpa Publications in Engineering, 2018), pp.1–6
3. J. Lee, Y. Wonpil, Concurrent tracking of inliers and outliers. ArXiv E-prints. Retrieved 5 Oct 2015
4. B. Pratima, K. Muralidharan, Some inferential study on inliers in Lindley distribution. Int. J. Stat. Reliab. Eng. **3** (2), 108–129 (2016). ISSN (Print) 2350-0174. ISSN (online) 2456-2378
5. G. Sanjay, M.S. Dulawat, A perception of statistical inference in data mining. Int. J. Comput. Sci. Commun. **1**(II) (2010)
6. G. Sanjay, Estimation of missing value at extremes in data mining. Int. J. Adv. Found. Res. Comput. **14**(3), 13–19 (2014)
7. Y.H. Dovoedo, S. Chakraborti, Boxplot-based outlier detection for the location-scale family. Commun. Stat. Simul. Comput. **44**(6), (1492) (2015)

Correction to: Human Activity Recognition Using LSTM with Feature Extraction Through CNN

Rosepreet Kaur Bhogal and V. Devendran

Correction to:
Chapter "Human Activity Recognition Using LSTM with Feature Extraction Through CNN" in:
Y.-D. Zhang et al. (eds.), *Smart Trends in Computing and Communications*, Lecture Notes in Networks and Systems 396,
https://doi.org/10.1007/978-981-16-9967-2_24

The original version of the book was inadvertently published with an incorrect line in the Abstract in Chapter 24 (Human Activity Recognition Using LSTM with Feature Extraction Through CNN). This has now been rectified and the line has been removed.
　The chapter and book have been updated with the changes.

The updated version of this chapter can be found at
https://doi.org/10.1007/978-981-16-9967-2_24

Author Index